Advanced Structured Materials

Volume 54

Series editors

Andreas Öchsner, Southport Queensland, Australia
Lucas F. M. da Silva, Porto, Portugal
Holm Altenbach, Magdeburg, Germany

For further volumes:
http://www.springer.com/series/8611

Andreas Öchsner · Holm Altenbach
Editors

Design and Computation of Modern Engineering Materials

Springer

Editors
Andreas Öchsner
Griffith School of Engineering
Griffith University
Southport, QLD
Australia

Holm Altenbach
Fakultät für Maschinenbau, Lehrstuhl für
Technische Mechanik
Otto-von-Guericke-Universität Magdeburg
Magdeburg
Germany

ISSN 1869-8433 ISSN 1869-8441 (electronic)
ISBN 978-3-319-38377-4 ISBN 978-3-319-07383-5 (eBook)
DOI 10.1007/978-3-319-07383-5
Springer Cham Heidelberg New York Dordrecht London

Softcover reprint of the hardcover 1st edition 2014

Printed on acid-free paper

Springer is part of Springer Science+Business Media (www.springer.com)

Contents

The Effect of the Bedding Length of Lintel in Masonry Walls on Their Load Bearing Capacity

Jure Radnić, Vatroslav Čamber, Nikola Grgić and Domagoj Matešan

Abstract By using the previously developed numerical model of the authors for both, static and dynamic analysis of concrete and masonry structures, which can simulate their main nonlinear effects, the influence of the bedding length of lintel on the ultimate bearing capacity of some masonry walls with openings has been investigated. Three-storey masonry walls with door openings were analyzed. Unreinforced and confined masonry walls were considered. There were separately analyzed masonry walls under horizontal static forces at the floor levels and masonry walls under earthquake. The bedding length of the lintel reinforcement and quality of the masonry were varied. Characteristic displacements of the walls and crack states in the lintel's area are presented. Finally, main conclusions and recommendations for practical application are given.

Keywords Lintel reinforcement · Masonry wall · Numerical model · Static and dynamic loading

J. Radnić · V. Čamber (✉) · N. Grgić · D. Matešan
Faculty of Civil Engineering, Architecture and Geodesy, University of Split,
Matice Hrvatske 15, 21000 Split, Croatia
e-mail: vatroslav_camber@yahoo.com

J. Radnić
e-mail: jradnic@gradst.hr

N. Grgić
e-mail: ngrgic@gradst.hr

D. Matešan
e-mail: dmatesan@gradst.hr

A. Öchsner and H. Altenbach (eds.), *Design and Computation of Modern Engineering Materials*, Advanced Structured Materials 54,
DOI: 10.1007/978-3-319-07383-5_1, © Springer International Publishing Switzerland 2014

1 Introduction

Above the openings in masonry walls are usually low and relatively weakly reinforced beams—lintels. Lintel reinforcement is usually calculated only to take over vertical loads, without the effects of wind and earthquake.

The height from the top of the opening to the top edge of floor level is variable and depends on the building's floor. In some cases, this height is relatively large, so lintel and part of the wall above it contribute significantly to wall stiffness on the horizontal static and dynamic (seismic) activity.

The width of openings in the walls for doors and windows is variable, as well as width of supporting walls.

The bedding of rebars at the ends of the lintels is often very short (sometimes below 10 cm). In fact, lintels do not contribute much to seismic resistance of masonry walls with openings. It is well known that earthquakes may cause the hardest damage to the parts of the masonry walls above the openings. In fact, horizontal forces produce high horizontal tensile stresses and high vertical shear stresses in these wall areas. Thus, in the horizontal direction masonry walls barely have no compressive stresses from gravitational loads, and barely have no horizontal tensile bearing capacity. Therefore, the occurrence of damage in parts of the wall above openings is expected and occurs even due to small horizontal forces. Cracks typically occur at the junction of the lintel with wall supports, and are especially significant when lintel reinforcement is anchored very shortly above the opening. A significant reduction of damages in the lintel's area can be expected by increasing the bedding length of the lintel reinforcement on wall supports and by adequate increase of the amount of bottom and top lintel reinforcement. This way, greater rigidity and resistance of the wall to horizontal loads will be ensured.

Obviously, the resistance and load bearing capacity of the wall would be greatest when the horizontal lintels are extended to the neighboring vertical tie beams.

No experimental and numerical studies of the effects of the bedding length of the lintel reinforcement on the load bearing capacity of masonry walls exposed to horizontal static or dynamic load have been available to the authors of this chapter.

The influence of the bedding length of lintel reinforcement of some masonry walls on their behavior under horizontal static and dynamic (earthquake) loads has been numerically investigated in this chapter.

Three-storey masonry walls with door openings were analyzed. Unreinforced and confined masonry walls were considered. There were separately analyzed masonry walls under horizontal static forces at the floor levels and masonry walls under dynamic forces. The bedding length of the lintel reinforcement and quality of the masonry were varied. Characteristic displacements of the walls and crack states in the lintel's area are presented.

Table 1 Material nonlinearity included in numerical model

Steel	Concrete	Masonry	Soil
Nonlinear behaviour in tension and compression	Yielding in compression	Yielding in compression	Yielding in compression
	Opening of cracks in tension	Opening of cracks in tension	Opening of cracks in tension
	The mechanism of crack opening and closing under dynamic load	The mechanism of crack opening and closing under cyclic load	The mechanism of crack opening and closing under cyclic load
	Tensile and shear stiffness of cracked concrete	Tensile and shear stiffness of cracked masonry	Tensile and shear stiffness of cracked soil
	Nonlinear behaviour of the reinforcement	Transfer of shear stresses	Transfer of shear stresses
		Anisotropic properties of strength and stiffness in horizontal and vertical direction	Anisotropic properties of strength and stiffness in horizontal and vertical direction

1.1 Numerical Model

The previously developed numerical model for both static and dynamic analysis of concrete and masonry structures [1, 2], which can simulate their main nonlinear effects, was used. The main nonlinear effects are material nonlinearity (Table 1), geometrical nonlinearity of the structure (large displacements), the strain rate effects on the material properties of masonry, reinforced concrete and soil, soil yield under a foundation, soil structure dynamic interaction, construction mode—the stages of masonry walls, etc.

A macro and micro model of masonry can be used. In the macro model of the masonry, the complex behaviour of the masonry (masonry units connected by mortar) is modeled by the homogenous material of equivalent mechanical properties. In the micro model of the masonry, modelling at the level of the masonry units and mortar (joints) is possible, as well as simulation of connection of mortar and masonry units by contact elements.

Concrete behaviour is simulated with the isotropic material model. Masonry or concrete models can be used for soil simulation, with corresponding material parameters.

The basic data of the analyzed masonry walls are presented in Sect. 2, and some research results are presented in Sect. 3. Main conclusions are given in Sect. 4.

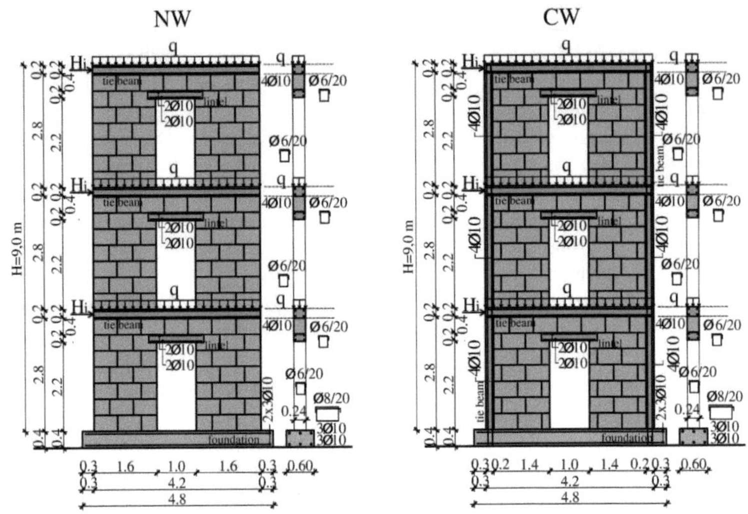

Fig. 1 Basic data about the analyzed masonry walls. **a** Unreinforced masonry wall. **b** Confined masonry wall

2 Basic Data of the Analyzed Masonry Walls

The basic data of the analyzed masonry walls are presented in Fig. 1. The walls are loaded by self weight and a uniform load $q = 35$ kN/m at floor levels. In the static analysis, the walls are additionally loaded by a horizontal force H at floor levels. The force has been applied in increments, until the collapse of the structure.

In the dynamic analysis, beside self weight and load q, the walls were exposed to a horizontal harmonic base acceleration according to Fig. 3. The period of excitation T corresponds to the first period of free oscillations of a particular wall. It was adopted that the duration of excitation is $T_p = 10T$, and the analysis was carried out for $T_a = 20T$. Implicit time integration with a time increment $\Delta t = T_1/100$ was adopted.

The walls with various lintel lengths were analyzed, i.e. with various bedding lengths of the lintel reinforcement (0.2, 0.4, 0.6, 1.6 m). Analyzed unreinforced masonry walls NW-20, NW-40, NW-60 and NW-160 are presented in Fig. 2a, and analyzed confined masonry walls CW-20, CW-40, CW-60 and CW-160 are presented in Fig. 2b. Variants of walls with rigid and soft masonry were considered. The rigid masonry has five times greater parameters of strength and stiffness compare to the soft masonry. The adopted basic material parameters for the numerical analysis are presented in Table 2.

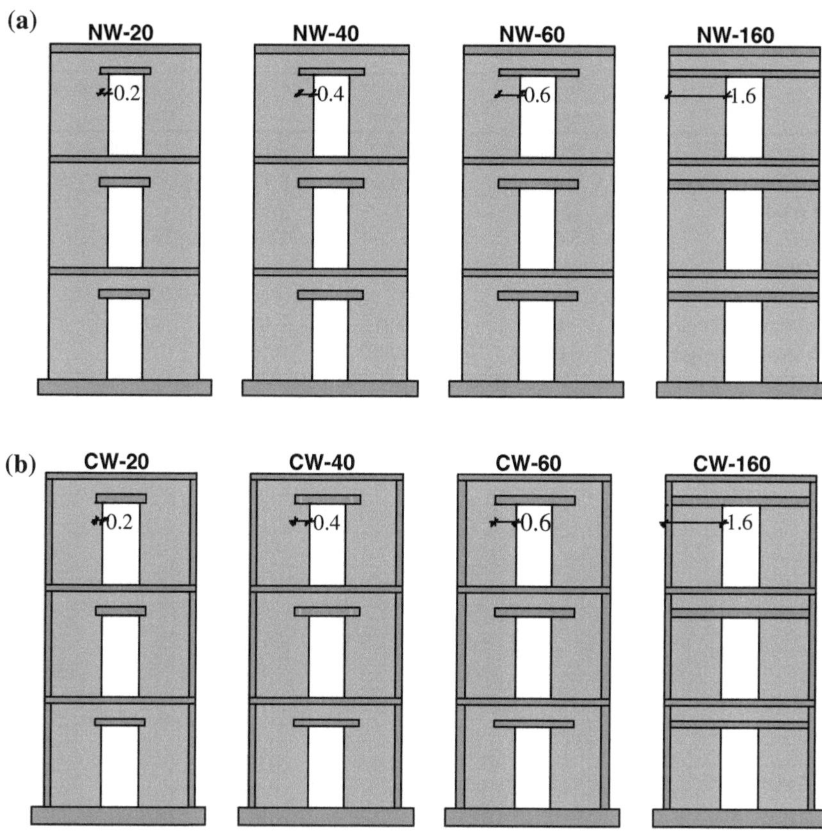

Fig. 2 Variants of analyzed masonry walls. **a** Unreinforced masonry wall. **b** Confined masonry wall

Fig. 3 Adopted base acceleration

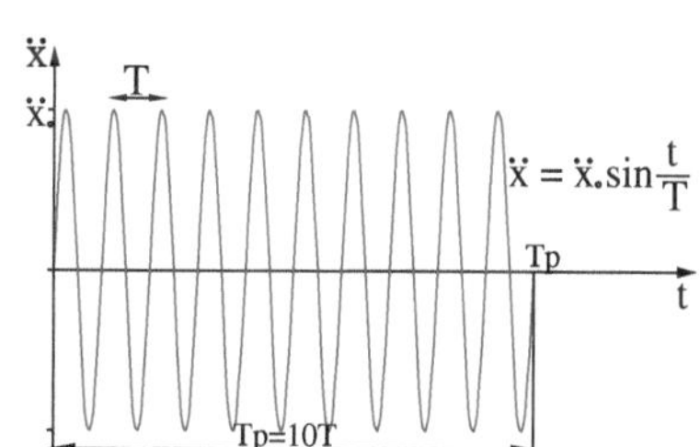

It was accepted that the wall foundation is supported by a rigid base. Possibility of the foundation lifting from the base was included. For that purpose, the thin contact elements were used between base and foundation. The relatively rough spatial discretization of the walls according to Fig. 4 was adopted, especially for the tie beams and the lintels. All longitudinal and transverse rebars were modeled.

Table 2 The adopted basic material parameters for the numerical analysis

Parameters	Unit	Material			
		Rigid masonry	Soft masonry	Concrete	Reinforcement steel
Elasticity modulus	MPa	5,000	1,000	30,500	210,000
Poisson'ratio	–	0.00	0.00	0.15	–
shear modulus	MPa	1,000	200	13,260	–
Compressive strength	MPa	5.0	1.0	25	500.0
Tensile strength	MPa	0.15	0.03	2.5	500.0
Limit comp. strength	–	−0.01	−0.01	−0.0035	−0.02
Limit tensile strength	–	0.00003	0.00003	0.0001	0.02

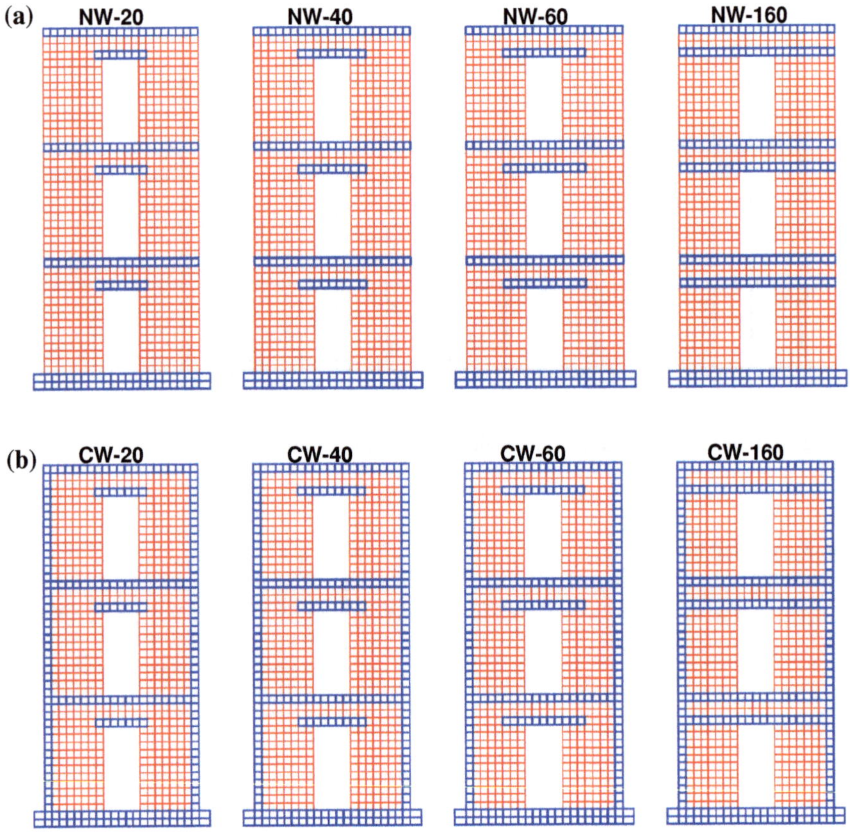

Fig. 4 Adopted spatial discretization of the walls. **a** Unreinforced masonry wall. **b** Confined masonry wall

3 Brief Comment on the Numerical Results

3.1 Static Analysis

3.1.1 Unreinforced Masonry Walls (NW)

The horizontal displacement of the top of the unreinforced masonry walls is shown in Fig. 5.

The big difference in the load bearing capacity and displacements of the wall can be noticed regarding the masonry quality, as well as the bedding length of lintel reinforcement.

For the walls with soft masonry the load bearing capacity of the wall NW-160 is more than 35 % higher than the load bearing capacity of the wall NW-20. For these walls with rigid masonry, the difference in load bearing capacity is only about 10 %.

Crack states in unreinforced masonry walls for the work levels of force H is presented in Fig. 6. As can be seen, the extension of the lintels above the openings results in reduction of the cracking zone in the lintels and supporting walls.

3.1.2 Confined Masonry Walls (CW)

The horizontal displacement of the top of the confined masonry walls is presented in Fig. 7. Analogue comments as for previously discussed unreinforced masonry can be stated. A huge difference in the load bearing capacity and displacements for these walls also depends on the masonry quality. There is also a significant difference in the load bearing capacity of the wall depending on the bedding length of lintel reinforcement. So, for rigid masonry, the difference of the load bearing capacity of the walls for CW-160 and CW-20 is about 10 %, and for soft masonry it is about 45 %. For unreinforced masonry and confined masonry, greater bedding length of lintel reinforcement contributes to the greater load bearing capacity of the walls.

Crack states in the confined masonry walls for for the work levels of force H is presented in Fig. 8. Analogue conclusions as for unreinforced masonry can be stated. Greater bedding length of lintel reinforcement has a favorable effect on the crack states in the lintel's area and at the supporting walls. Even for the work levels of loads, increase of the bedding length of lintel reinforcement results in decrease of cracking zone in lintel and supporting walls.

(a)

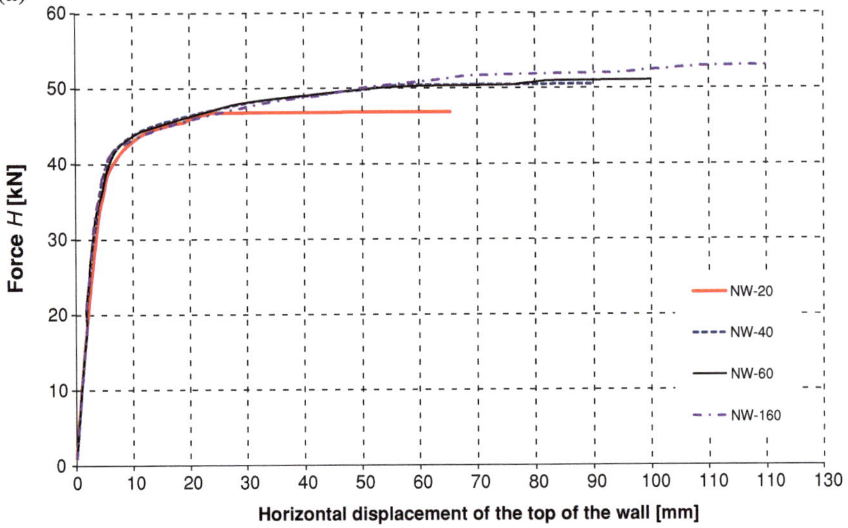

(b)

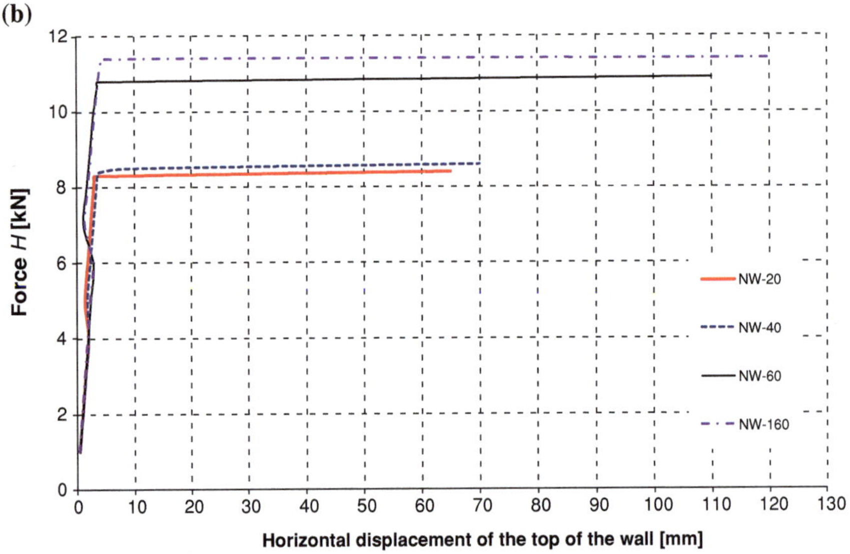

Fig. 5 Horizontal displacement of the top of the unreinforced masonry walls. **a** Rigid masonry. **b** Soft masonry

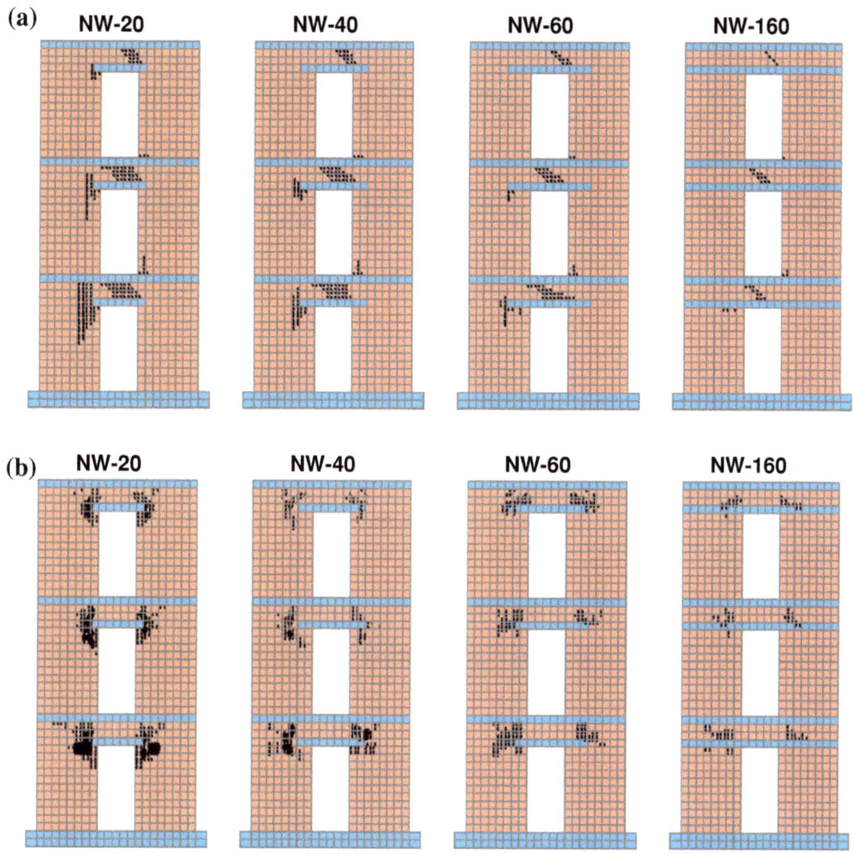

Fig. 6 Crack states in unreinforced masonry walls for work levels of force H. **a** Rigid masonry, H = 15 kN. **b** Soft masonry, H = 3 kN

3.2 Dynamic Analysis

3.2.1 Unreinforced Masonry Walls (NW)

It is obvious that the three-storey unreinforced masonry walls according to Fig. 1a have a small resistance to earthquake excitations, especially in the case of the soft masonry. At first, the dynamic analysis was performed for small values of a harmonic base acceleration $\ddot{x}_o$ according to Fig. 3. Then it was gradually increased to the walls collapse. So, it was determined which maximum base acceleration $\ddot{x}_o$ each wall could withstand (Table 3a). As it was expected, unreinforced masonry walls can withstand a low value of base acceleration, especially in the case of soft masonry.

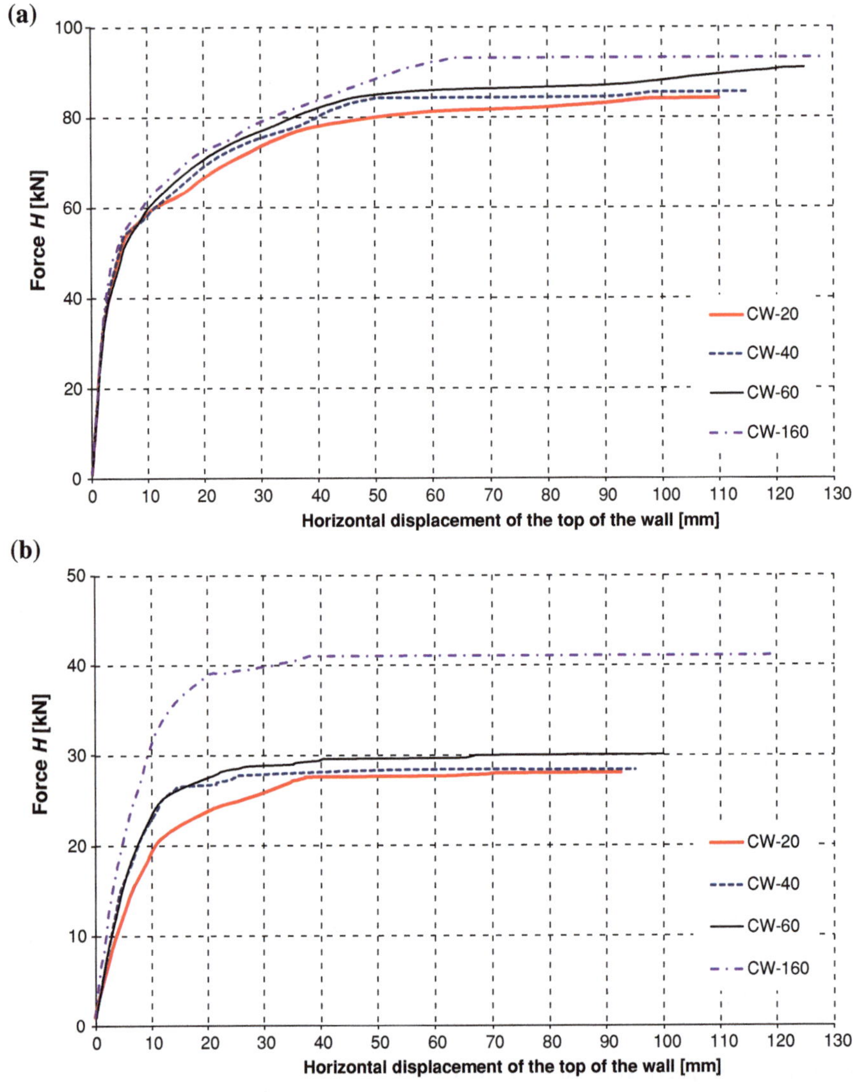

Fig. 7 Horizontal displacement of the top of the confined masonry walls. **a** Rigid masonry. **b** Soft masonry

The horizontal displacement of the top of the unreinforced masonry for the maximum base acceleration $\ddot{x}_o$ is presented in Fig. 9.

For the same dynamic excitation ($\ddot{x}_o = 0.04$ g for rigid masonry and $\ddot{x}_o = 0.001$ g for soft masonry), the walls have almost equal response (almost independent of the bedding length of lintel reinforcement).

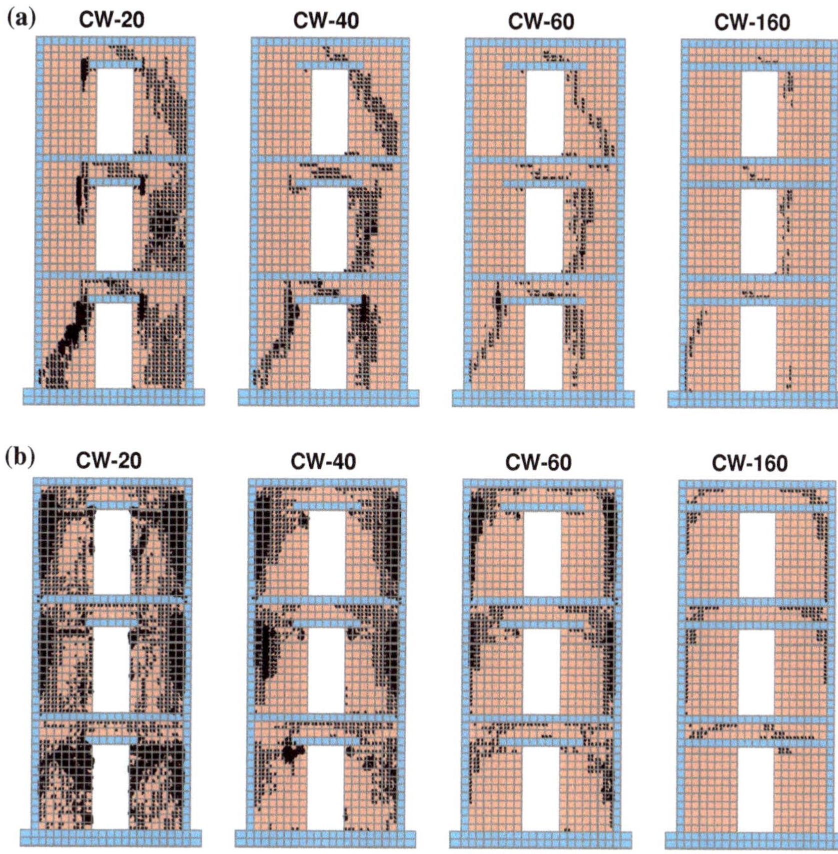

Fig. 8 Crack states in the confined masonry walls for work levels of force H. **a** Rigid masonry walls, H = 27 kN. **b** Poor masonry walls, H = 9 kN

Table 3a Maximum harmonic base acceleration $\ddot{x}_o$ which can withstand unreinforced masonry walls (NW)

	max $\ddot{x}_o$			
Masonry	NW-20	NW-40	NW-60	NW-160
Rigid	0.04 g	0.045 g	0.05 g	0.08 g
Soft	0.001 g	0.0015 g	0.002 g	0.004 g

However, for the same displacement, the walls have different crack states (Fig. 10). It is obvious that, as for the horizontal static force (Sect. 3.1), the bedding length of lintel reinforcement has a significant influence on the size of the cracking zone. Specifically, the extension of the lintel bedding length considerably narrowed the size of the cracking zone.

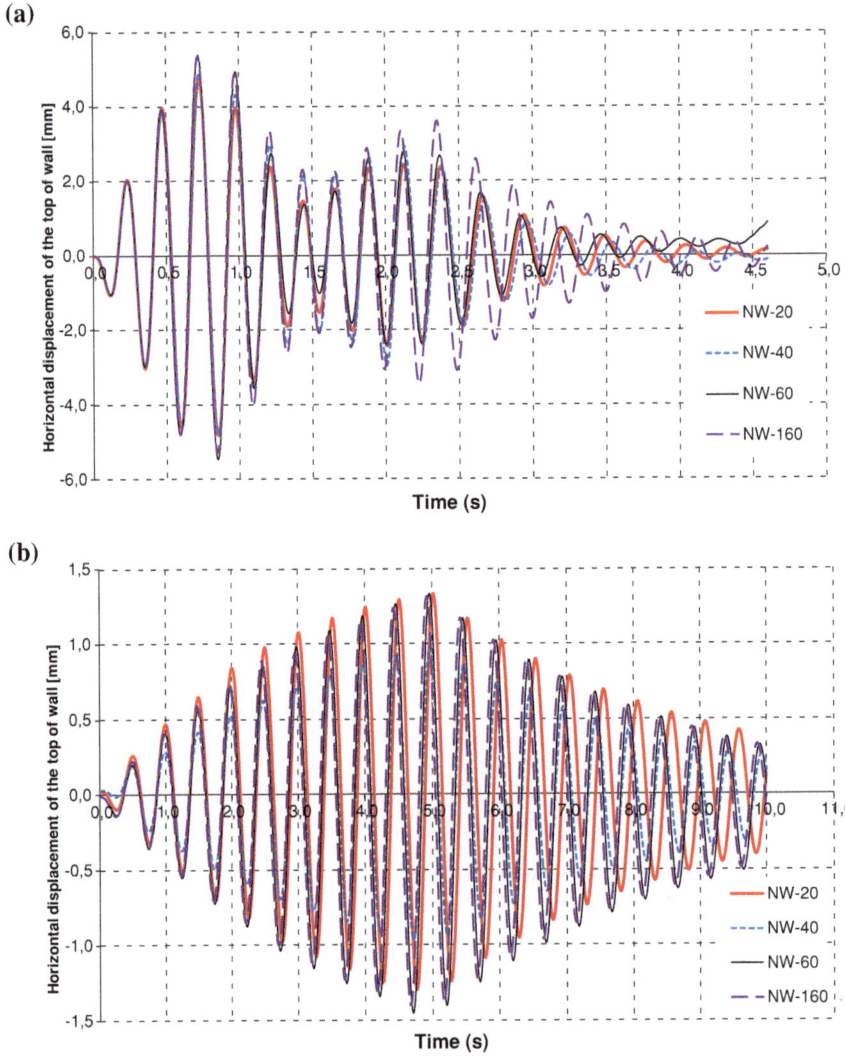

Fig. 9 Horizontal displacement of the top of the unreinforced masonry walls for the harmonic base excitation $\ddot{x}_o$. **a** Rigid masonry walls, $\ddot{x}_o = 0.04$ g. **b** Soft masonry walls, $\ddot{x}_o = 0.001$ g

3.2.2 Confined Masonry Walls (CW)

Analogously to specified in Sect. 3.2.1, it is also firstly determined which maximum base acceleration $\ddot{x}_o$ each wall could withstand (Table 3b). The horizontal displacement of the top of confined masonry wall for the maximum base

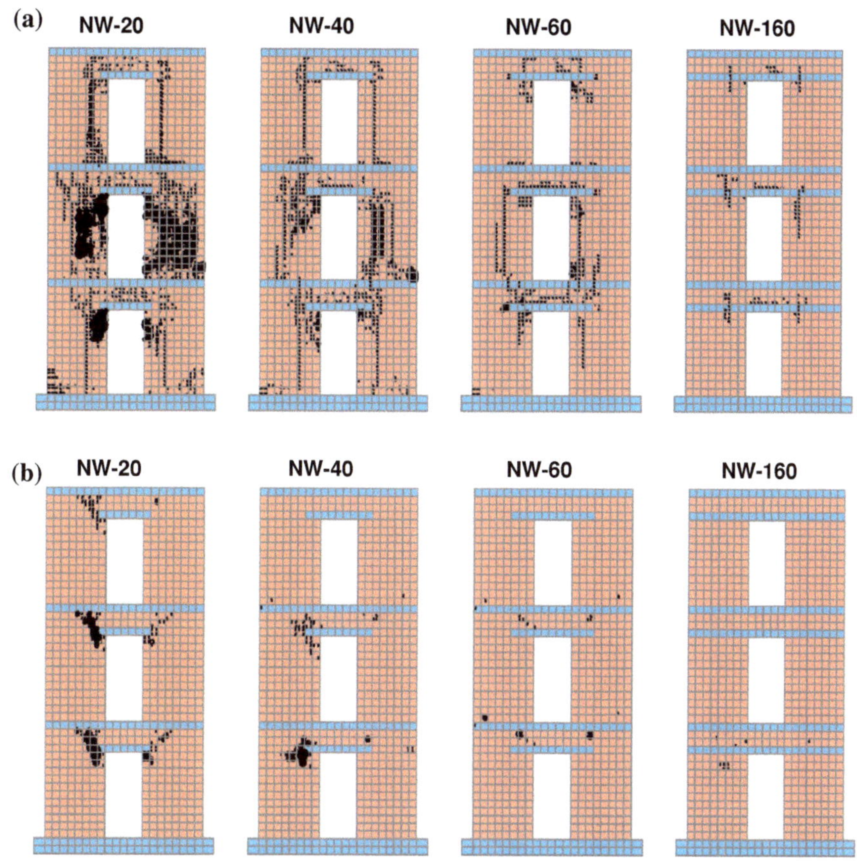

Fig. 10 Crack states in unreinforced masonry walls for the maximum base acceleration $\ddot{x}_o$. **a** Rigid masonry walls, $\ddot{x}_o = 0.04$ g, $t = 0.23$ s. **b** Soft masonry walls, $\ddot{x}_o = 0.001$ g, $t = 0.50$ s

Table 3b Maximum harmonic base acceleration $\ddot{x}_o$ which can withstand confined masonry walls (CW)

			max $\ddot{x}_o$	
Masonry	CW-20	CW-40	CW-60	CW-160
Rigid	0.24 g	0.25 g	0.27 g	0.30 g
Soft	0.02 g	0.03 g	0.04 g	0.10 g

acceleration $\ddot{x}_o$ is presented in Fig. 11. As it was expected, these walls can withstand greater maximum base acceleration $\ddot{x}_o$ compared to the unreinforced masonry walls. A huge difference in the load bearing capacity for cases of the rigid and the soft masonry is evident.

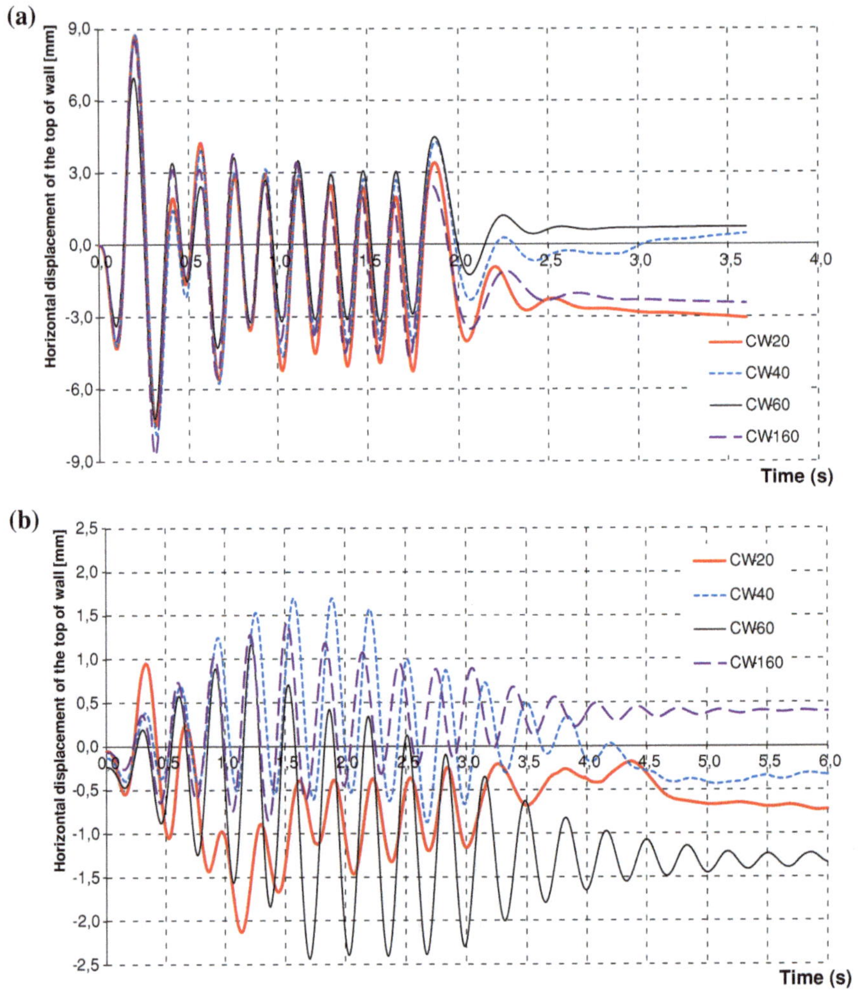

Fig. 11 Horizontal displacement of the top of the confined masonry walls for the harmonic base excitation $\ddot{x}_o$. **a** Rigid masonry walls, $\ddot{x}_o = 0.24$ g **b** Soft masonry walls, $\ddot{x}_o = 0.02$ g

Crack states in the confined masonry walls for the maximum base acceleration $\ddot{x}_o$ is presented in Fig. 12. The influence of the bedding length of lintel reinforcement on the crack states in the lintel and supporting walls is completely analogous to the previously mentioned for the unreinforced masonry walls.

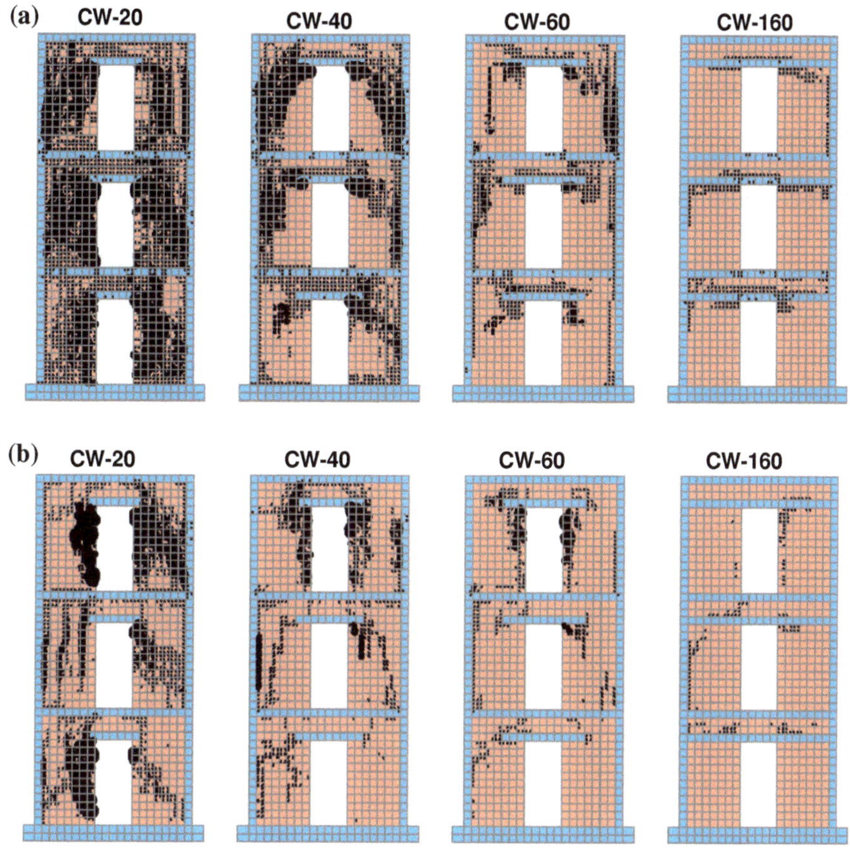

Fig. 12 Crack states in unreinforced masonry walls for the maximum base acceleration $\ddot{x}_o$. **a** Rigid masonry walls, $\ddot{x}_o = 0.04$ g, $t = 0.23$ s. **b** Soft masonry walls, $\ddot{x}_o = 0.001$ g, $t = 0.50$ s

4 Conclusions

By using the previously developed and verified numerical model for static and dynamic analysis of concrete and masonry structures, it was determined that the bedding length of lintel reinforcement may have a significant influence on the load bearing capacity and on the size of cracking zone of the unreinforced and confined masonry walls exposed to winds and earthquakes. Greater bedding length of lintel reinforcement contributes to the increase of load bearing capacity of the masonry walls, as well as to the reduction of size of cracking zones in lintel and supporting walls. That influence is greater if the quality of the masonry is softer. The quality of the masonry has a great influence on the load bearing capacity of the masonry walls, as well as on the size of cracking zones. If the masonry buildings are located in the zone with strong expected earthquakes, especially if the masonry supporting walls are relatively tight, it is recommended to extend the bedding length of lintel reinforcement to the neighboring vertical tie beams.

References

1. Radnić, J., Harapin, A., Matešan, D., Trogrlić, B., Smilović, M., Grgić, N., Baloević, G.: Numerical model for static and dynamic analysis of masonry wall. Građevinar **63**, 529–546 (2011)
2. Radnić, J., Harapin, A., Matešan, D., Smilović, M., Grgić, N.: Numerical model for static and dynamic analysis of masonry structure. In: 5th International Conference on Advanced Computational Engineering and Experimenting, Algarve (2011)

The Effect of Vertical Load on Seismic Response of Masonry Walls

Jure Radnić, Marija Smilović, Alen Harapin, Nikola Grgić and Ante Buzov

Abstract The effect of the vertical load intensity on the seismic response of masonry walls is numerically investigated. A previously developed numerical model for static and dynamic analysis of masonry structures, which can simulate the most important nonlinear effects of masonry, concrete, reinforcement and soil, was used. Two–storey unreinforced and confined masonry walls, with and without openings, with good and poor quality of masonry, are analyzed. The wall foundations are supported on a rigid base, with possibility of lifting and sliding. The walls are subjected to a real earthquake and harmonic base acceleration. The characteristic displacements of the walls top, stresses in masonry and reinforcement of the vertical ring beams, as well as the state of cracks in the walls, are presented. It was concluded that larger static vertical loads of masonry walls usually have positive effects on their behavior (reducing possible tensile stresses, as well as minimizing the possibility of lifting, sliding and overturning of the walls). However, larger vertical loads generated larger inertial forces on the structure under earthquake excitation. It is more desirable to reduce the vertical load of masonry structures in areas with expected strong earthquakes.

Keywords Masonry wall · Vertical loads · Dynamic analysis · Strength capacity

J. Radnić (✉) · M. Smilović · A. Harapin · N. Grgić · A. Buzov
Faculty of Civil Engineering, Architecture and Geodesy, University of Split,
Matice Hrvatske 15, 21000 Split, Croatia
e-mail: jradnic@gradst.hr

M. Smilović
e-mail: msmilovic@gradst.hr

A. Harapin
e-mail: aharapin@gradst.hr

N. Grgić
e-mail: ngrgic@gradst.hr

A. Buzov
e-mail: abuzov@gradst.hr

A. Öchsner and H. Altenbach (eds.), *Design and Computation of Modern Engineering Materials*, Advanced Structured Materials 54,
DOI: 10.1007/978-3-319-07383-5_2,

1 General

Behaviour of masonry structures during an earthquake significantly depends on their mass, i.e. the intensity of the vertical load. The earthquake force increase proportionally with the increase in the mass. On the other hand, the increase in vertical load reduces possible vertical tensile stresses in the masonry walls, as well as the possibility of their lifting from the base.

The effect of the vertical load intensity on the seismic response of masonry walls subjected to the earthquake Ston and the harmonic base acceleration is numerically investigated in this paper. Two–storey unreinforced and confined masonry walls, with and without openings, are analysed. Good quality and poor quality of masonry were analyzed separately. Wall foundation lifting from the rigid base is modelled.

A numerical model for static and dynamic analysis of masonry structures [1, 2], previously developed by the authors, was used. The model simulates geometric nonlinearity of the structure, as well as the main nonlinear effects of masonry, concrete, reinforcement and soil. The main conclusions of the research are given at the end of this paper. The literature that analysed the researched problem in the presented manner has not been available to the authors. Some similar researches can be found in [3–10].

2 Basis of the Used Numerical Model

Detailed description of the used numerical model can be found in [1, 2], and will be only briefly described hereinafter.

The model is intended for simulation of planar masonry structures loaded in their plane. The structure model can consist the masonry, reinforced concrete and subsoil. Simulation of all main nonlinear effects of masonry, reinforced concrete and soil is possible.

A macro and micro model of masonry can be used. In the macro model of the masonry, the complex behaviour of the masonry (masonry units connected by mortar) is modelled by the homogenous material of equivalent mechanical properties. In the micro model of the masonry, modelling at the level of the masonry units and mortar (joints) is possible, as well as simulation of connection of mortar and masonry units by contact elements. It is possible to use the isotropic model of masonry, as well as anisotropic model with different strengths (compressive, tensile, shear), modulus of elasticity, shear modulus and limit strains in horizontal and vertical directions. It can be modelled yielding and crushing of masonry in compression, opening and closing of cracks in tension, as well as tensile and shear stiffness of cracked masonry, as well as shear failure. The model of fixed orthogonal smeared cracks was used.

Concrete behaviour is simulated with the isotropic material model. It can be modelled yielding and crushing of concrete in compression, the opening and closing of cracks in tension, as well as tensile and shear stiffness of cracked concrete. A model of fixed orthogonal cracks is also used, with the crack direction corresponding to the directions of main tensile stresses. In dynamic analysis, it is possible to simulate the effect of strain rate on the mechanical properties of concrete and steel. It is also possible to simulate the nonlinear behaviour of reinforcement, with the effects of unloading.

Masonry or concrete models can be used for soil simulation, with corresponding material parameters.

The adopted numerical model also may includes geometric nonlinearity of the structure (large displacements).

3 Basic Data on Analyzed Walls

Basic data on the analysed real two-story masonry walls are shown in Fig. 1. Apart from its dead weight, the walls are loaded with variable uniform vertical load (q) at the floor levels.

Spatial discretization of the walls is shown in Fig. 2. Thin contact elements under the foundation were used to simulate wall lifting from the rigid base. The macro model of masonry [1, 2] with isotropic material properties is used. The adopted basic material parameters are presented in Table 1. In relation to poor quality of masonry, the values of material parameters of good quality of masonry are five times greater. The walls are subjected to horizontal base acceleration according to Fig. 3. The horizontal component of the earthquake Ston acceleration (Fig. 3a) and horizontal harmonic base acceleration (Fig. 3b), with a period (T_p) equal to the first period of free oscillations of corresponding wall (T_1), were used. The maximum base acceleration of 0.3 g was adopted. Implicit time integration with a time increment $\Delta t = T_1/100$ was adopted. A displacement increase of 1‰ compared to the current total displacement is adopted as the iterative procedure convergence criteria. The impact of wall buckling perpendicular to its plane has not been modelled.

4 Results of Numerical Analysis

Some results of numerical analysis, which will be only briefly commented below, are shown in Figs. 4, 5, 6, 7, 8, 9, 10, 11, 12, 13 and 14.

For all analysed masonry walls (unreinforced and confined, with and without openings, with good and poor quality of masonry) and for all analysed dynamic

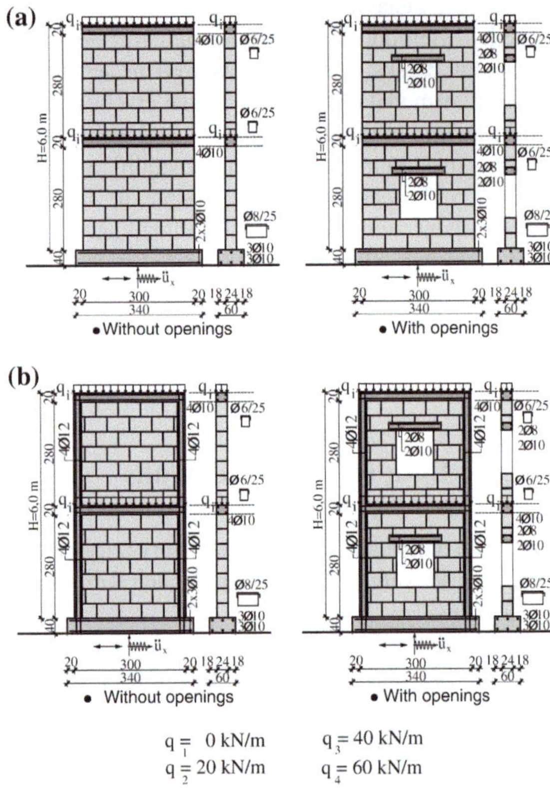

Fig. 1 Analyzed masonry walls. **a** Unreinforced masonry wall. **b** Confinedmasonry wall

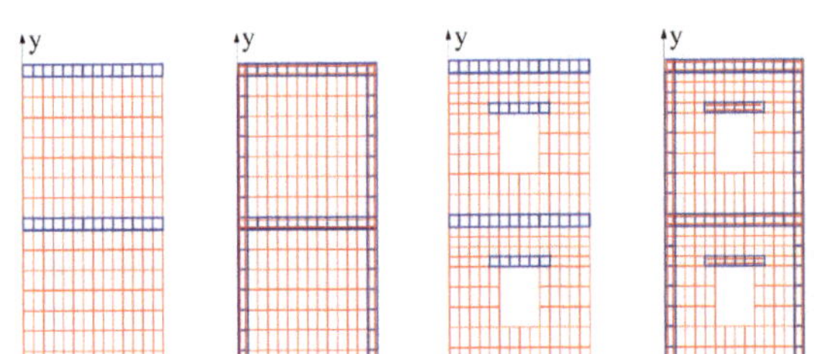

Fig. 2 Finite element discretization of analyzed masonry walls

excitations (earthquake, harmonic excitation), numerical results show that the increase of the vertical load has a negative effect on walls behaviour when they are subjected to dynamic excitation. The reason for this is the increase of inertial forces of the structure with an increase of its mass.

Table 1 The adopted basic material parameters of masonry walls

Parameters	Unit	Material				
		Good masonry	Poor masonry	Concrete	Reinforcement steel	Contact element
Elasticity modulus	MPa	5,000	1,000	30,500	210,000	30,500
Shear modulus	MPa	1,000	200	13,260	–	13,260
Compressive strength	MPa	5.0	1.0	25	560.0	25
Tensile strength	MPa	0.15	0.03	2.5	560.0	0.0

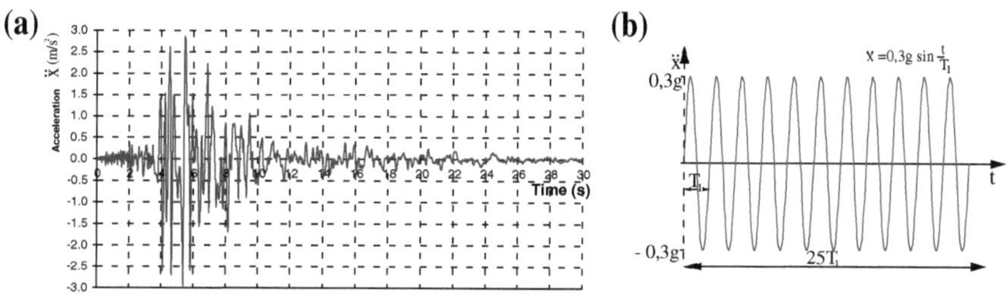

Fig. 3 The adopted horizontal base acceleration. **a** Earthquake Ston. **b** Harmonic acceleration

Although larger static vertical loads of masonry walls usually have positive effects (reducing tensile stresses in masonry and vertical reinforcement, minimizing the possibility of wall foundation lifting at bending, etc.), in earthquake condition they have adverse effect.

As was expected, the analysis indicated that the two-storey unreinforced masonry walls can not withstand the imposed dynamic excitation by maximal base acceleration of 0.3 g. Confined two-storey masonry walls can survive imposed dynamic excitations at lower vertical loads. Whereby, the overall state of the walls with openings and poor quality of masonry is worse than those without openings and with good quality of masonry (there are significant differences in the displacements, stresses and damages/cracks).

In general, regardless of the type of the walls, openings and especially the quality of masonry significantly affect the deformability, stress state, the level of damage and safety of masonry walls subjected to dynamic excitation.

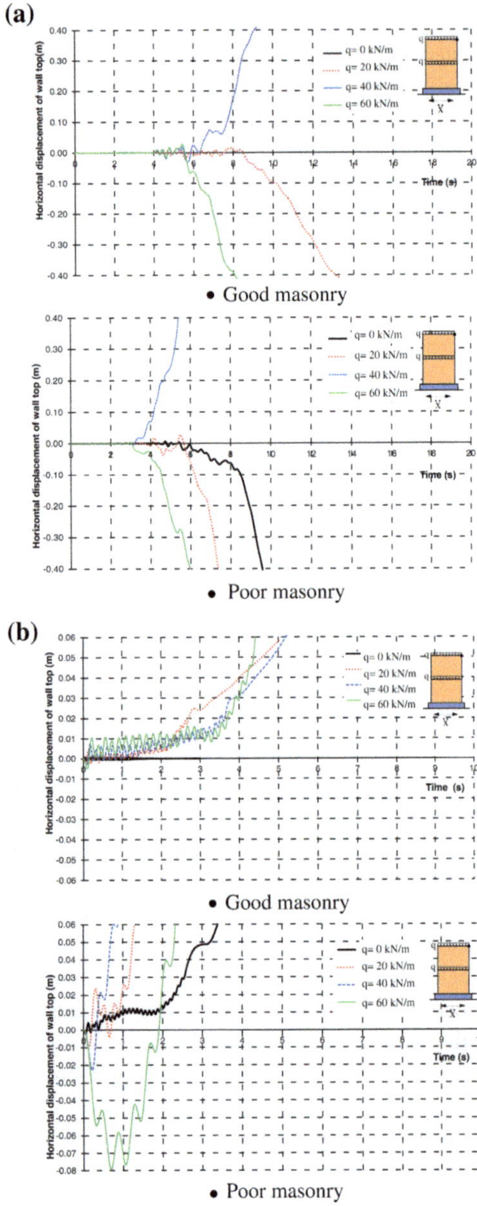

Fig. 4 Horizontal displacement of the *top* of the unreinforced masonry walls without openings.
a Earthquake Ston.
b Harmonic acceleration

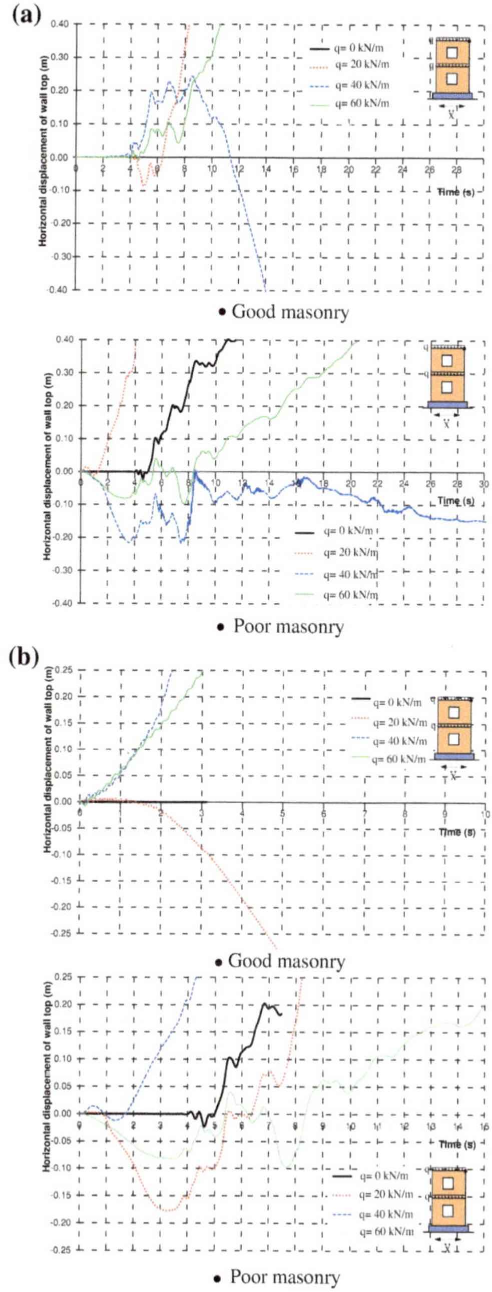

Fig. 5 Horizontal displacement of the *top* of the unreinforced masonry walls with openings. **a** Earthquake Ston. **b** Harmonic acceleration

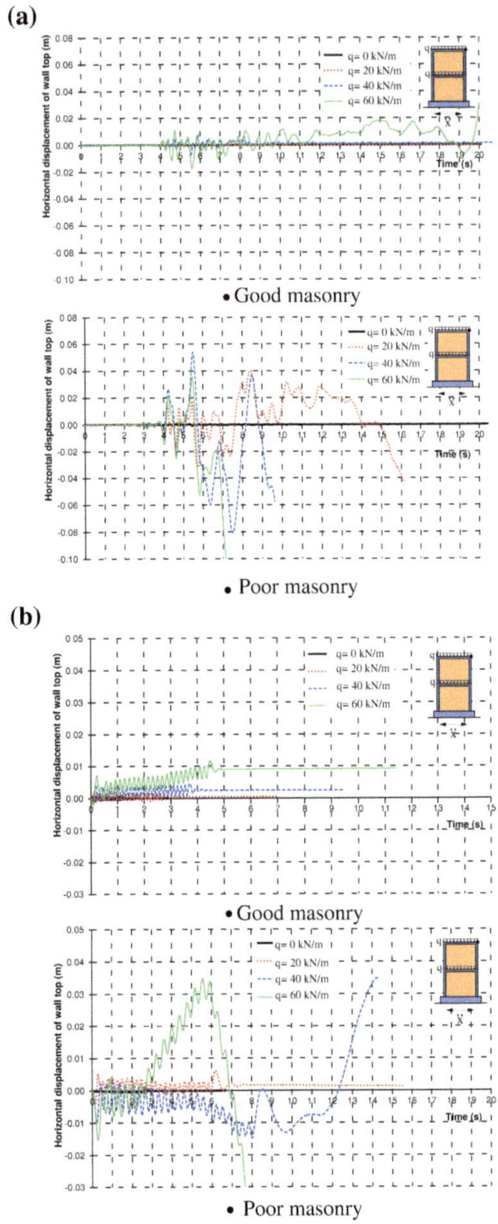

Fig. 6 Horizontal displacement of the *top* of the confined masonry walls without openings. **a** Earthquake Ston. **b** Harmonic acceleration

Fig. 7 Horizontal displacement of the *top* of the confined masonry walls with openings. **a** Earthquake Ston. **b** Harmonic acceleration

(a)

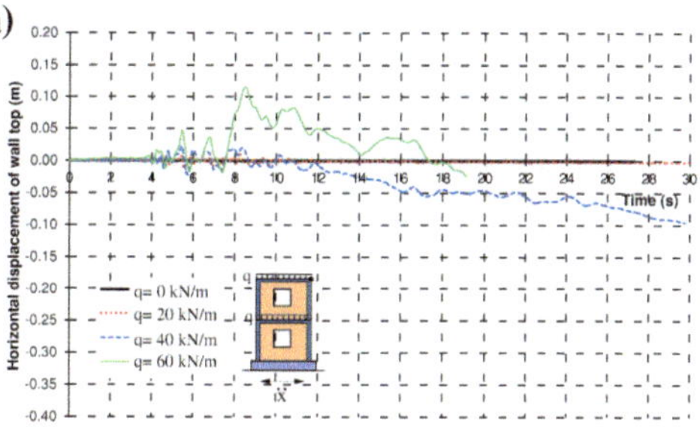

• Good masonry

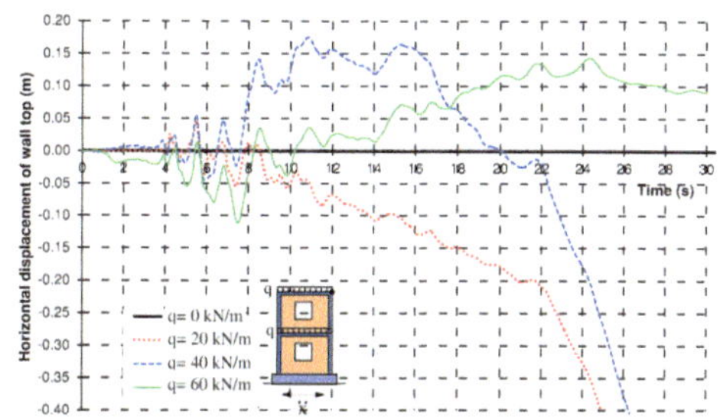

• Poor masonry

(b)

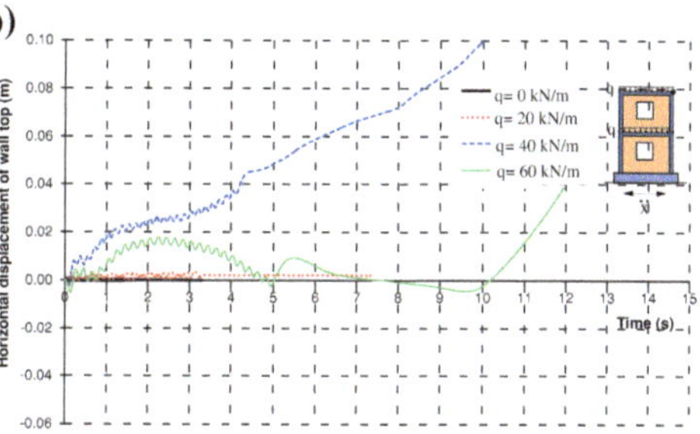

• Good masonry

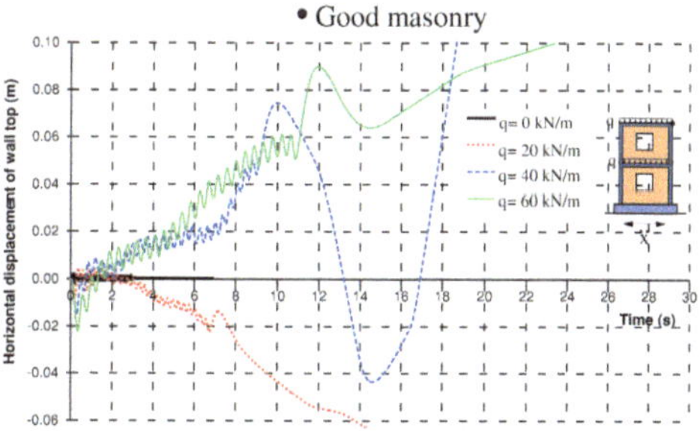

• Poor masonry

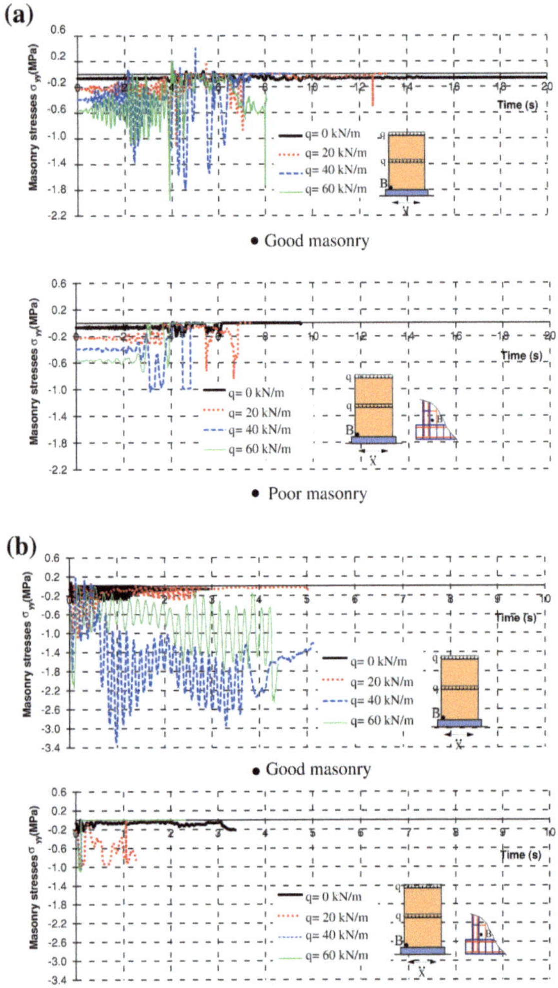

Fig. 8 Vertical stresses of masonry at the *bottom* of unreinforced masonry walls without openings.
a Earthquake Ston.
b Harmonic acceleration

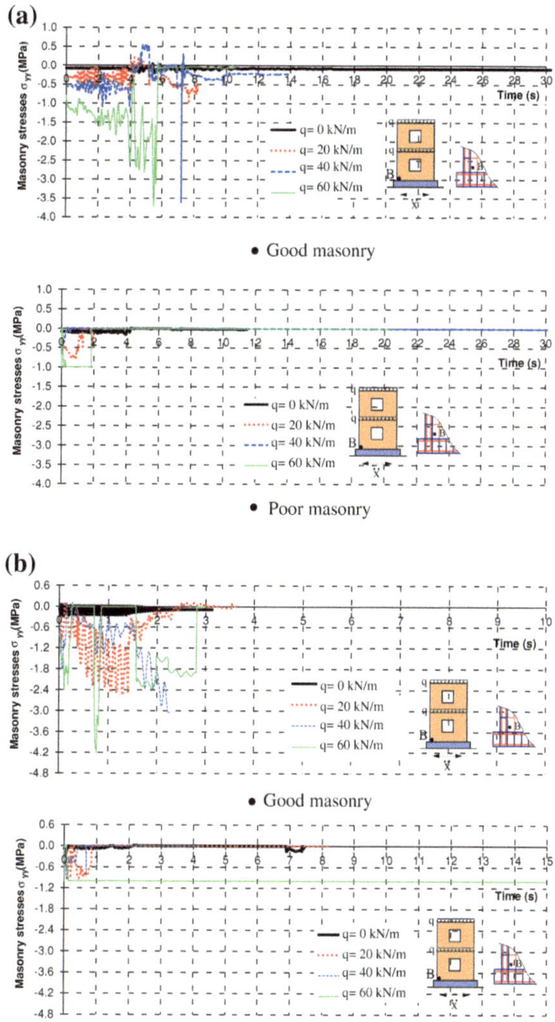

Fig. 9 Vertical stresses of masonry at the *bottom* of unreinforced masonry walls with openings. **a** Earthquake Ston. **b** Harmonic acceleration

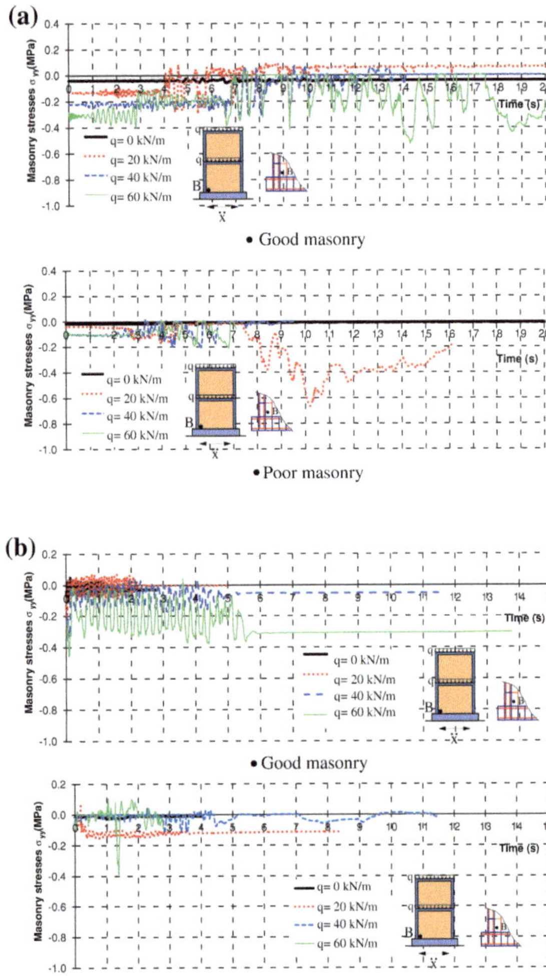

Fig. 10 Vertical stresses of masonry at the *bottom* of confined masonry walls without openings. **a** Earthquake Ston. **b** Harmonic acceleration

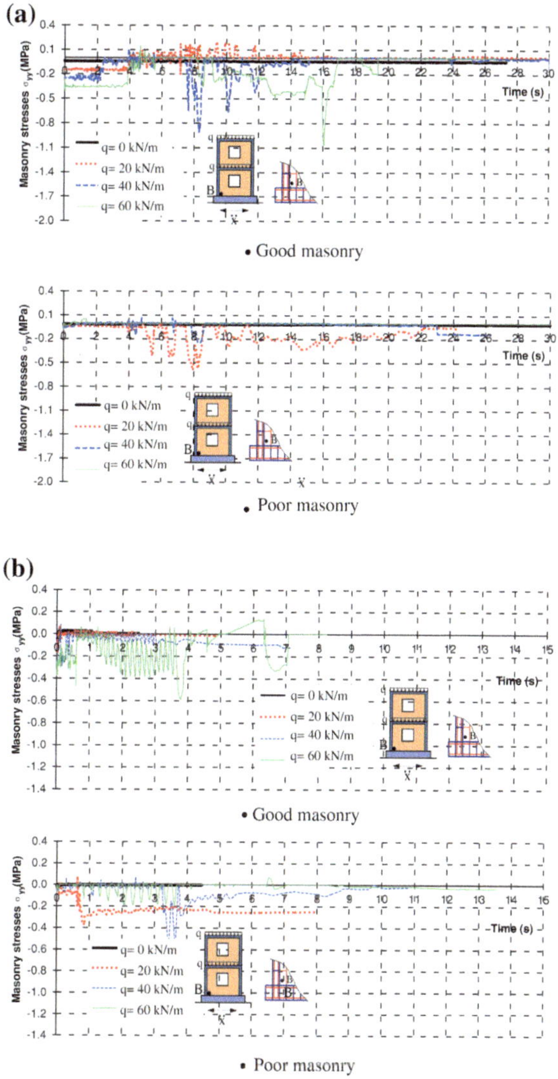

Fig. 11 Vertical stresses of masonry at the *bottom* of confined masonry walls with openings. **a** Earthquake Ston. **b** Harmonic acceleration

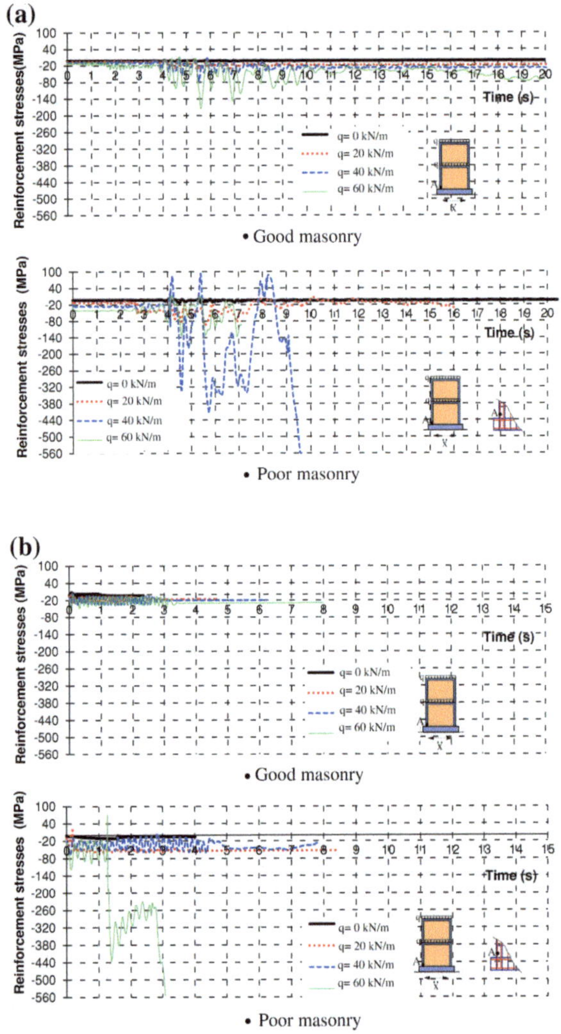

Fig. 12 Reinforcement stresses at the *bottom* of confined masonry walls without openings. **a** Earthquake Ston. **b** Harmonic acceleration

Fig. 13 Reinforcement stresses at the *bottom* of confined masonry walls with openings. **a** Earthquake Ston. **b** Harmonic acceleration

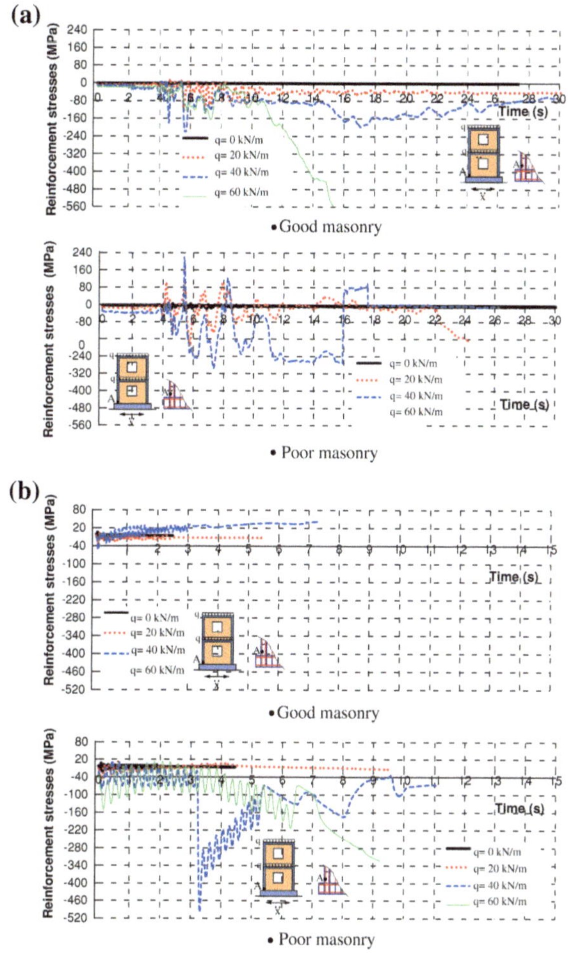

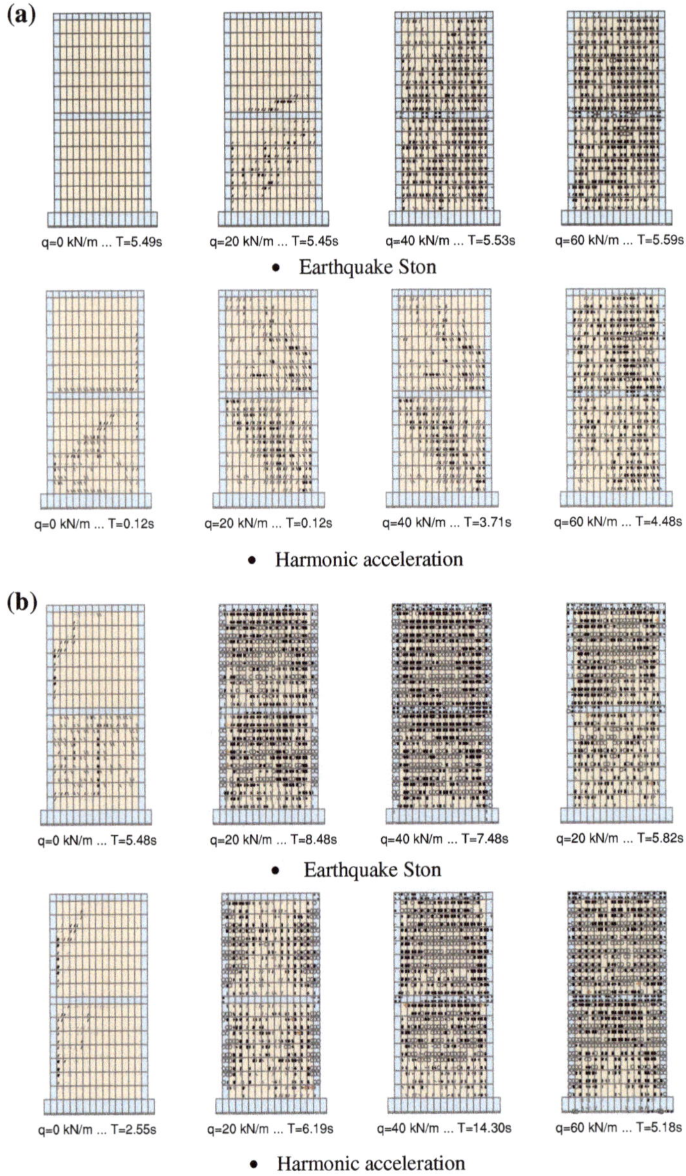

Fig. 14 The state of cracks that correspond the maximum displacements of confined masonry wall without openings. **a** Good masonry. **b** Poor masonry

5 Conclusion

Larger static vertical loads of masonry walls usually have positive effects on their behaviour (reducing possible tensile stresses in masonry, concrete and reinforcement, as well as minimizing the possibility of lifting and overturning of the walls). However, larger vertical loads have adverse effect on masonry wall behaviour in earthquake conditions since they generate larger inertial forces the structure under earthquake. Thus, weight of the buildings should be reduced as much as possible in areas with expected strong earthquakes.

A combination of high walls, great mass of the structure, large wall openings and the use of poor quality of masonry is particularly unfavourable. Unreinforced masonry should only be used for buildings of small importance, low height, in low seismicity zones and only used the masonry of good quality.

Acknowledgements This work was supported by the funds of the Ministry of Science, Education and Sport of Croatia.

References

1. Radnić, J., Harapin, A., Matešan, D., Trogrlić, B., Smilović, M., Grgić, N., Baloević, G.: Numerical model for static and dynamic analysis of masonry wall. Građevinar **63**, 529–546 (2011)
2. Radnić, J., Harapin, A., Matešan, D., Smilović, M., Grgić N.: Numerical model for static and dynamic analysis of masonry structure. In: 5th International Conference on Advanced Computational Engineering and Experimenting, Algarve (2011)
3. Yoshimura, K., Kikuchi, K., Kuroki, M., Liu, L., Kajiwara, K., Ushijima, M.: Effect of the vertical axial loads and wall reinforcements on seismic behavior of confined concrete masonry walls. In: Proceedings of the 23rd Conference on Our World in Concrete and Structures, Singapore (1998), pp. 221–228
4. Yoshimura, K., Kikuchi, K., Kuroki, M., Liu, L., Ma, L.: Effect of wall reinforcements applied lateral forces and vertical axial loads on seismic behavior of confined concrete masonry walls. In: 12th World Conference on Earthquake Engineering, New Zealand (2000)
5. Tomaževič, M.: Some aspect of experimental testing of seismic behavior of masonry walls and models of masonry buildings. IEST J. Earthq. Technol. **37**(4), 101–117 (2000) Paper No. 404
6. Drobiec, I., Kubica, J.: Influence of some types of bed joints reinforcement on mechanical properties of masonry under compression. In: Proceedings of 6th International Masonry Conference, London (2002), pp. 99–104
7. Calaveri, L., Failla, A., La Mendola, L., Papia, M.: Experimental and analytical response of masonry elements under eccentric vertical loads. Eng. Struct. **27**, 1175–1184 (2005)
8. Mura, I.: Stability of nonlinear masonry members under combined load. Comput. Struct. **86**(15–16), 1579–1593 (2008)
9. Zilch, K., Schermer, D., Scheufler, W.: Behavior of reinforced masonry walls made of hollow clay units with concrete infill under combined loading. In: Proceedings of 14th International Brick and Block Masonry Conference, Sydney, Australia (2008)
10. da Porto, F., Mosele, F., Modena, C.: Reinforced clay masonry walls under shear-compression loads: experimental behavior. In: Proceedings of 11th Canadian Masonry Symposium, Toronto (CD ROM) (2009)

Numerical Study of Flange-Web Junction of Prestressed Concrete Bridge with Corrugated Steel Webs

Xiachun Chen, Francis T. K. Au and Yu Zeng

Abstract Steel-concrete composite bridges have become popular by taking advantage of desirable properties of both materials. In particular, prestressed concrete bridges with corrugated steel webs have emerged as one of the promising bridge forms. This structural form provides excellent structural efficiency with the concrete flanges primarily taking bending and the corrugated steel webs taking shear. Most beam theories proposed for this type of bridges have assumed that the longitudinal stiffness of the corrugated web is negligible. The web is assumed to carry effectively zero longitudinal normal stress. However, longitudinal normal stresses of the web have been observed in the bonded zone of the flange-web junctions during experiments. In this chapter, the stress distribution in the bonded zone of the flange-web junctions is investigated numerically. Parametric analysis is done to identify the essential parameters. The study shows that the interaction between the web corrugations and the surrounding concrete causes longitudinal normal stresses in the bonded zones. If the bonded zone is large as compared to the whole deck section, the assumption of negligible longitudinal stiffness of web may no longer be valid. Based on the numerical study, this chapter provides some suggestions for the design of web corrugations.

Keywords Corrugated steel webs · Flange-web junction · Prestressed concrete bridge

X. Chen · F. T. K. Au (✉) · Y. Zeng
Department of Civil Engineering, The University of Hong Kong, Pokfulam Road, Hong Kong, China
e-mail: francis.au@hku.hk

X. Chen
e-mail: chenxiachun@hotmail.com

Y. Zeng
e-mail: yzeng@hku.hk

A. Öchsner and H. Altenbach (eds.), *Design and Computation of Modern Engineering Materials*, Advanced Structured Materials 54,
DOI: 10.1007/978-3-319-07383-5_3,

1 Introduction

Steel-concrete composite bridges have become popular by taking advantage of desirable properties of both materials. In particular, prestressed concrete (PC) bridges with corrugated steel webs have emerged as one of the promising bridge forms. This structural form provides excellent structural efficiency with the concrete flanges primarily taking bending and the corrugated steel webs taking shear. This type of bridge has some remarkable advantages, e.g. low weight of steel webs, efficient prestressing on concrete, high shear buckling strength of the steel webs, high ductility and easy maintenance. The configuration of one of such bridge is shown in Fig. 1. Hereafter this type of bridge is assumed unless otherwise stated.

In 1986, the first bridge of this type, Cognac Bridge, was built in France [1]. The successful application and its significant advantages over conventional PC bridges have prompted much research work and some construction companies to get involved in this new composite structure in various countries, including China, Germany, Japan and USA.

Most beam theories proposed for this type of bridges have assumed that the longitudinal stiffness of the corrugated web is negligible [2]. The web is assumed to carry effectively zero longitudinal normal stress and constant shear stress throughout the web height. However, longitudinal normal stresses of the web have indeed been observed in the bonded zone of the flange-web junctions during experiments [3, 4]. In this chapter, the stress distribution in the bonded zone of the flange-web junctions is investigated numerically. Based on the numerical study, this chapter provides some suggestions for the design of web corrugations.

Typical cross sections of the bridge are shown in Fig. 2, where t_u and t_l are the depths of upper and lower flanges respectively; b_u and b_l are the widths of upper and lower flanges respectively; h_w is the height of the web; and H is the depth of the entire beam. Figure 3 shows typical types of corrugated webs, i.e. trapezoidal and sinusoidal, where s denotes the developed length of the corrugated web; s_0 denotes the projected length of the corrugated web; s_{flat} denotes the length of the flat segment of the corrugated web; r denotes the rise of corrugations; and θ denotes the trough angle.

The analysis of these typical corrugated steel sheets is usually carried out using the theory of orthotropic shells or plates, which has been verified to adequately capture the main features of the behavior of these corrugated sheets [5]. According to [5] and [6], the shear modulus of the equivalent orthotropic web, G_e, of the above two types of corrugated web is given by

$$G_e = \frac{s_0}{s} G_w \quad (1)$$

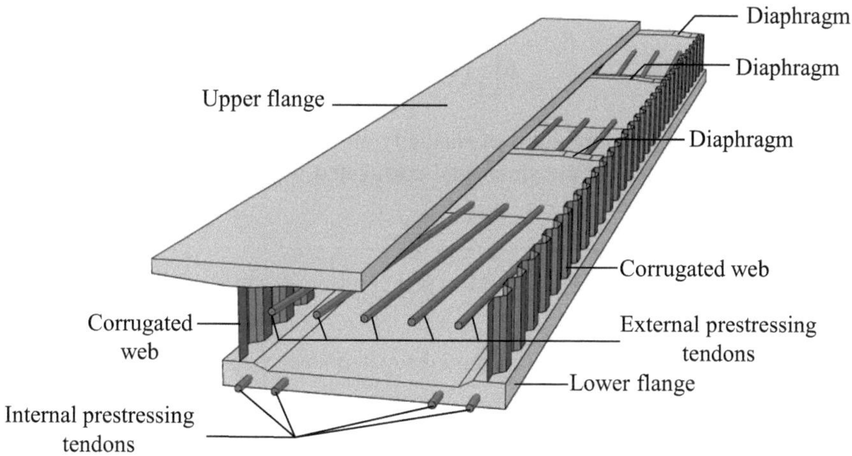

Fig. 1 Configuration of one prestressed concrete bridge with corrugated steel webs

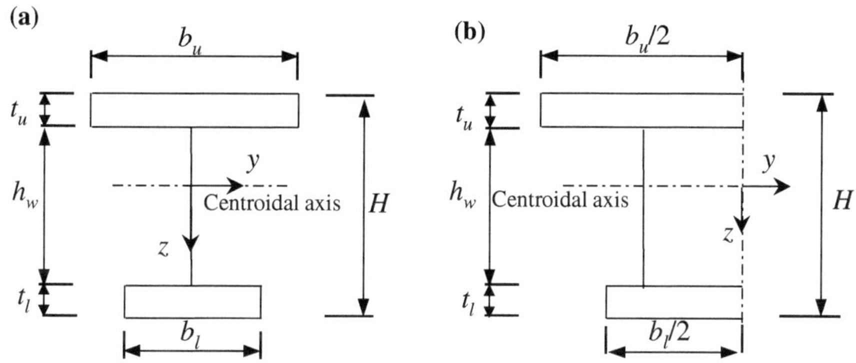

Fig. 2 Typical cross sections of the bridge: **a** with single web, and **b** with double webs

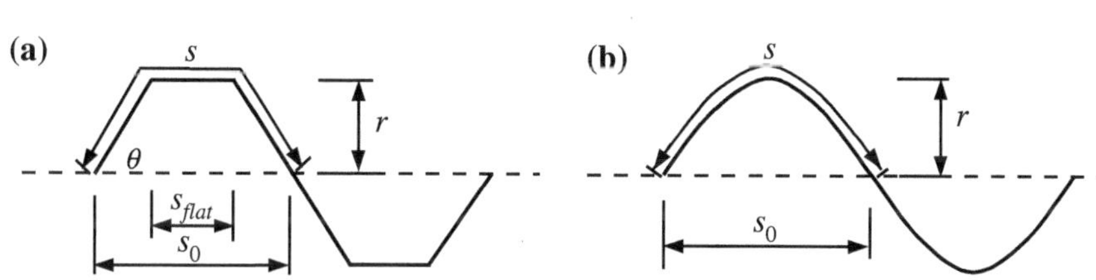

Fig. 3 One full corrugation: **a** trapezoidal, and **b** sinusoidal

where G_w is the shear modulus of the corrugated web. The modulus of elasticity of the equivalent orthotropic web, E_e, in axial direction of the trapezoidal corrugated web is given by [5]

$$E_e = \frac{E_w\left(\frac{t_w}{r}\right)^2}{6\left[\frac{2}{s_0}\left(\frac{s_0 - s\cos\theta}{1-\cos\theta}\right) + \frac{4}{3s_0}\frac{r}{\sin\theta}\right]} \tag{2}$$

where E_w is the material modulus of elasticity of the corrugated steel web and t_w is thickness of the web. For the sinusoidal corrugated web, E_e is given by [6]

$$E_e = \frac{E_w}{1 + 6\left(\frac{r}{t_w}\right)^2\left(1 - \mu_w^2\right)\left[\left(\frac{s}{s_0}\right)^2 - \frac{s}{2\pi s_0}\sin\frac{2\pi s}{s_0}\right]} \tag{3}$$

where μ_w is the Poisson's ratio of the corrugated steel web.

2 Numerical Analysis

Because of the constraints of the concrete flanges, the corrugated steel web cannot deform freely in the bonded zone of the flange-web junction. Hence, longitudinal normal stresses do exist in the corrugated steel web within the bonded zone. The values of longitudinal normal stress and depth of the bonded zone are affected by various geometric parameters of the corrugation. In this study, the parameters s_0, s_{flat}, r, θ and t_w of the trapezoidal corrugation are selected for the parametric study.

2.1 Numerical Models

A simply supported bridge subjected to four-point loading as shown in Fig. 4 is considered. By symmetry, only the left half of the bridge is taken for analysis. A finite element model of the left half of bridge, modelled by the computer programme ANSYS, is shown in Fig. 5. The details of the models are further elaborated below.

Prestressed concrete bridges with double corrugated webs as shown in Fig. 2b are considered. Three model groups are established in this study. The common dimensions and material parameters of the models are shown in Table 1. For additional comparison, a bridge model with flat webs ($t_w = 0.010$ m) is also established. In model group A as shown in Table 2, parameters r and θ vary while parameters s_0, s_{flat} and t_w remain unchanged. In other words, the lateral dimensions of corrugation are varied while the longitudinal dimensions and web thickness are kept unchanged. In model group B as shown in Table 3, the parameter t_w varies while parameters r, θ, s_0 and s_{flat} remain unchanged. In model group C as shown in Table 4, parameters s_0 and s_{flat} vary while parameters r, θ and t_w remain

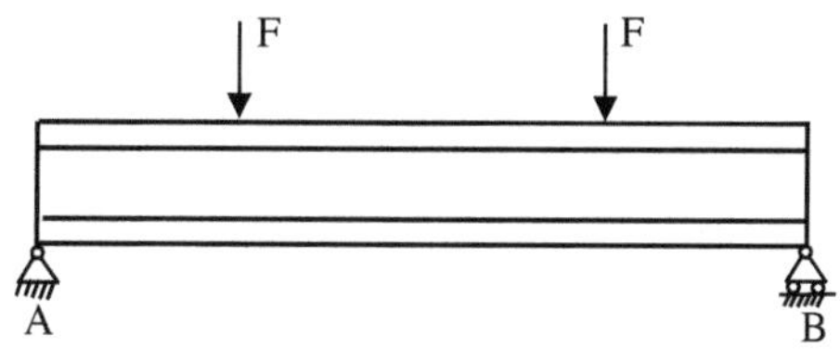

Fig. 4 A simply supported bridge with four-point loading

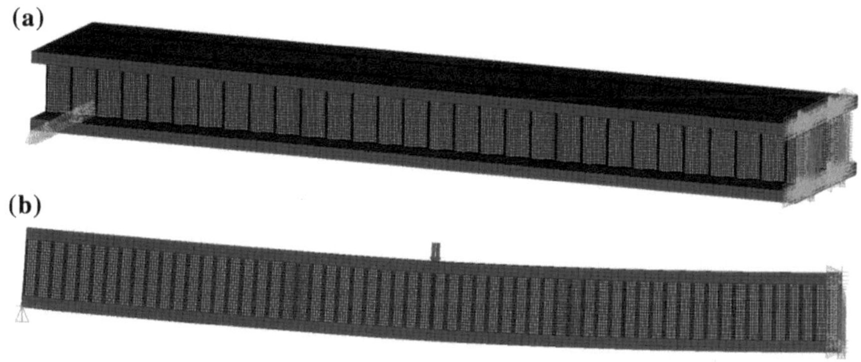

Fig. 5 Finite element model analyzed: **a** before deformation, and **b** after deformation

Table 1 Dimensions and material parameters of the bridge models

Parameter	Value	Unit	Parameter	Value	Unit
H	2.800	(m)	t_l	0.400	(m)
L	60.000	(m)	μ_w	0.3	
h_w	2.000	(m)	E_w	2.10E + 11	(Pa)
b_u	6.000	(m)	G_w	8.08E + 10	(Pa)
t_u	0.400	(m)	E_f	3.45E + 10	(Pa)
b_l	6.000	(m)			

unchanged. In other words, the longitudinal dimensions of corrugation are varied while the lateral dimensions and web thickness are kept unchanged.

The flanges of the bridge are modelled using solid elements (SOLID65 of ANSYS) and the webs are modelled using shell elements (SHELL181 of ANSYS). The element SOLID65 for 3-D modelling of solids is defined by eight nodes having three degrees of freedom at each node, i.e. translations in the nodal x, y and z directions. The element SHELL181 for analyzing thin to moderately-thick shell structures is a four-node element with six degrees of freedom at each node, i.e. translations in the x, y and z directions, and rotations about the x, y and z-axes. By symmetry, only the left half of the bridge is modelled as shown in Fig. 5. At the left end, the vertical displacements of the nodes at the bottom of the lower flange are constrained, while the lateral displacement of the node at the centre at the

Table 2 Corrugation profile and equivalent material parameters of model group A

	Model A1	Model A2	Model A3	Model A4	Model A5	Model A6
s (m)	0.510	0.517	0.525	0.566	0.660	0.618
s_0 (m)	0.500	0.500	0.500	0.500	0.500	0.340
s_{flat} (m)	0.34	0.34	0.34	0.34	0.34	0.340
r (m)	0.029	0.037	0.046	0.080	0.139	0.139
θ (degree)	20	25	30	45	60	90
t_w (m)	0.010	0.010	0.010	0.010	0.010	0.010
E_e (Pa)	1.82E + 09	1.09E + 09	7.00E + 08	2.13E + 08	5.94E + 07	3.48E + 07
G_e (Pa)	7.91E + 10	7.82E + 10	7.70E + 10	7.13E + 10	6.12E + 10	4.44E + 10
E_e/E_w (%)	0.867	0.520	0.333	0.101	0.028	0.017

Table 3 Corrugation profile and equivalent material parameters of model group B

	Model B1	Model B2	Model B3	Model B4	Model B5	Model B6
s (m)	0.566	0.566	0.566	0.566	0.566	0.566
s_0 (m)	0.500	0.500	0.500	0.500	0.500	0.500
s_{flat} (m)	0.34	0.34	0.34	0.34	0.34	0.34
r (m)	0.080	0.080	0.080	0.080	0.080	0.080
θ (degree)	45	45	45	45	45	45
t_w (m)	0.0050	0.0100	0.0150	0.0200	0.0250	0.0300
E_e (Pa)	5.33E + 07	2.13E + 08	4.79E + 08	8.52E + 08	1.33E + 09	1.92E + 09
G_e (Pa)	7.13E + 10	7.13E + 10	7.13E + 10	7.13E + 10	7.13E + 10	7.13E + 10
E_e/E_w (%)	0.025	0.101	0.228	0.406	0.634	0.913

bottom of the lower flange is constrained. At the right end, the longitudinal displacements of the nodes in the entire beam section are constrained, while the rotations of the nodes in the web section are constrained.

2.2 Results and Discussions

The longitudinal normal stresses of the web corrugation at $3L/8$ from the left support are selected for illustration. Both the longitudinal normal stresses of the flat and inclined segments of the web corrugation are discussed in this chapter. The locations, where the stresses are extracted, are shown in Fig. 6.

The longitudinal normal stresses in the flat segment of web corrugation of Model A3 in Fig. 7 show clearly two bonded zones and a non-bonded zone. In each zone, the distribution of normal stresses is approximately linear. However, those in the bonded zones are slightly curved. In the non-bonded zone, the normal stresses are mainly caused by the equivalent orthotropic modulus of elasticity of the web. In the bonded zones, the normal stresses are caused by not only the equivalent orthotropic modulus of elasticity of the web but also the interaction between the web corrugations and the surrounding concrete of the flanges, and hence the depth may vary.

Table 4 Corrugation profile and equivalent material parameters of model group C

	Model C1	Model C2	Model C3	Model C4	Model C5	Model C6	Model C7	Model C8
s (m)	0.266	0.366	0.466	0.566	0.666	0.866	1.066	1.266
s_0 (m)	0.200	0.300	0.400	0.500	0.600	0.800	1.000	1.200
s_{flat} (m)	0.040	0.140	0.240	0.340	0.440	0.640	0.840	1.040
r (m)	0.080	0.080	0.080	0.080	0.080	0.080	0.080	0.080
θ (degree)	45	45	45	45	45	45	45	45
t_w (m)	0.010	0.010	0.010	0.010	0.010	0.010	0.010	0.010
s_0/L (%)	0.333	0.500	0.667	0.833	1.000	1.333	1.667	2.000
s_0/h_w	0.10	0.15	0.20	0.25	0.30	0.40	0.50	0.60
s_{flat}/h_w	0.02	0.07	0.12	0.17	0.22	0.32	0.42	0.52
E_e (Pa)	6.89E + 07	1.28E + 08	1.77E + 08	2.13E + 08	2.37E + 08	2.64E + 08	2.76E + 08	2.81E + 08
G_e (Pa)	6.07E + 10	6.62E + 10	6.93E + 10	7.13E + 10	7.27E + 10	7.46E + 10	7.57E + 10	7.65E + 10
E_e/E_w (%)	0.033	0.061	0.084	0.101	0.113	0.126	0.131	0.134

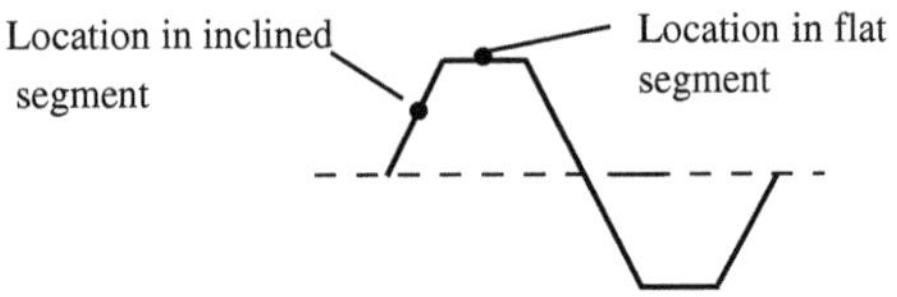

Fig. 6 Locations where the longitudinal normal stresses of the web corrugation are extracted

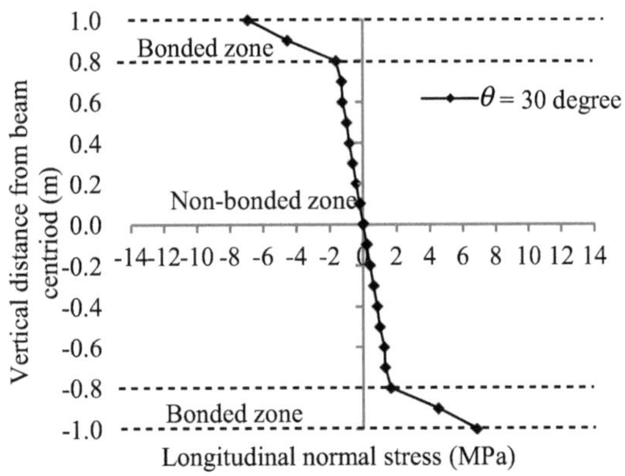

Fig. 7 Longitudinal normal stresses in the flat segment of the web corrugation at 3*L*/8 of Model A3

From the longitudinal normal stresses of the web corrugation at 3*L*/8 of model group A as shown in Fig. 8, the effects of the trough angle θ and rise r on both the normal stress and the depth of bonded zone are insignificant. The depth of bonded zone in the flat segment of web corrugation is about 10 % of the web height. In the inclined segment of web corrugation, the stresses in the bonded zone clearly decrease with increase of the parameters θ and r, but the depth remains almost the same, i.e. 10 % of the web height. Figure 8 shows that the normal stresses in the bonded zones of the corrugated web are smaller than those in the flat web. Comparing Fig. 8a, b, it is found that the longitudinal normal stresses of the bonded zones in the inclined segment are much less than those in the flat segment.

The ratio of modulus of elasticity of the equivalent orthotropic web to that of the corrugated web material of model group A is shown in Fig. 9. With the increase of modulus of elasticity of the equivalent orthotropic web, the normal stress in the non-bonded zone of the flat segment increases significantly. When the ratio of modulus of elasticity of the equivalent orthotropic web to that of the corrugated web material is less than 0.1 %, the normal stress in the non-bonded zone of the flat segment is negligible. It is also noted that the normal stress in the non-bonded zone of the inclined segment is always negligible.

As observed from the longitudinal normal stresses of the web corrugation at 3*L*/8 of model group B shown in Fig. 10, the effects of the web thickness t_w on both stress and depth of the bonded zone are insignificant. The depths of the bonded zone in both the flat and inclined segments of the web corrugation are about 10 % of the web height.

The ratio of modulus of elasticity of the equivalent orthotropic web to that of the corrugated web material of model group B is shown in Fig. 11. With the

Fig. 8 Longitudinal normal stresses of the web corrugation at 3*L*/8 of model group A: **a** flat segment, and **b** inclined segment

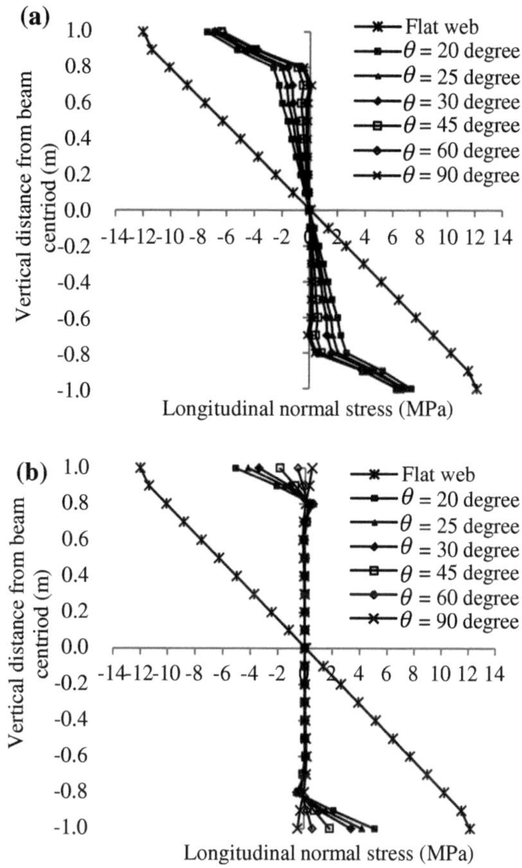

increase of modulus of elasticity of the equivalent orthotropic web, the normal stress in the non-bonded zone of the flat segment increases significantly. When the ratio of modulus of elasticity of the equivalent orthotropic web to that of the corrugated web material is less than 0.1 %, the normal stress in the non-bonded zone of the flat segment is negligible. The normal stress in the non-bonded zone of the inclined segment is always negligible.

As observed from the longitudinal normal stresses of the web corrugation at 3*L*/8 of model group C shown in Fig. 12, both the normal stresses and depth of the bonded zone increase significantly with the increase of parameters s_0 and s_{flat}. If the bonded zone is large as compared to the whole deck section, the common assumption of negligible axial stiffness of corrugated steel web may no longer be valid. When the ratio of s_0 to web height h_w is less than 0.25, the depth of the bonded zone approaches a constant of about 10 % of web height. In the inclined segment of the web corrugation, the effects of parameters s_0 and s_{flat} on both stress and depth of the bonded zone are insignificant.

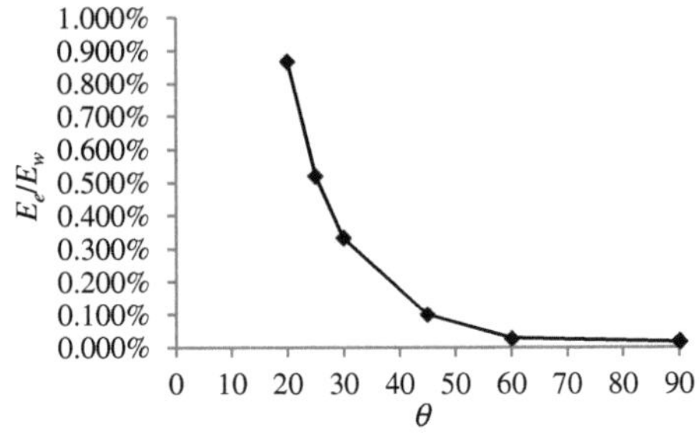

Fig. 9 Ratio of modulus of elasticity of equivalent orthotropic web to that of web material of model group A

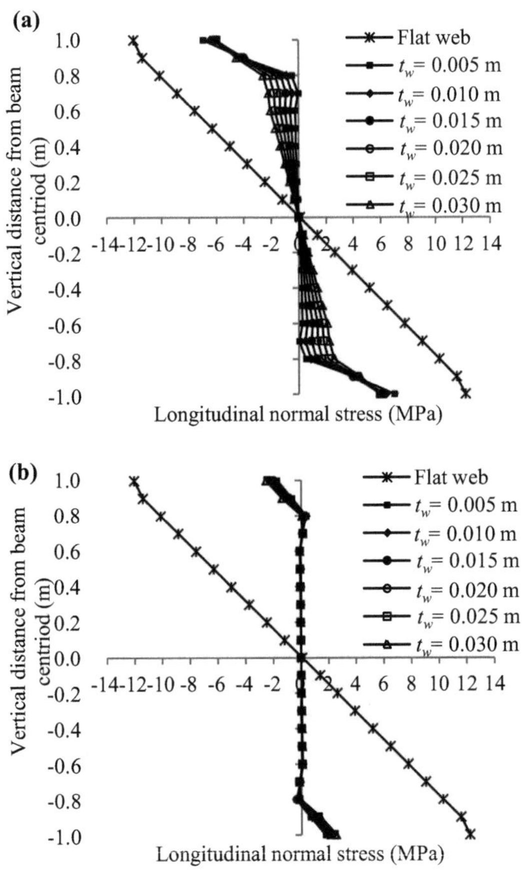

Fig. 10 Longitudinal normal stresses of the web corrugation at $3L/8$ of model group B: **a** flat segment, and **b** inclined segment

From the ratio of modulus of elasticity of the equivalent orthotropic web to that of the corrugated web material of model group C shown in Fig. 13, the change in modulus of elasticity of the equivalent orthotropic web due to change in the ratio

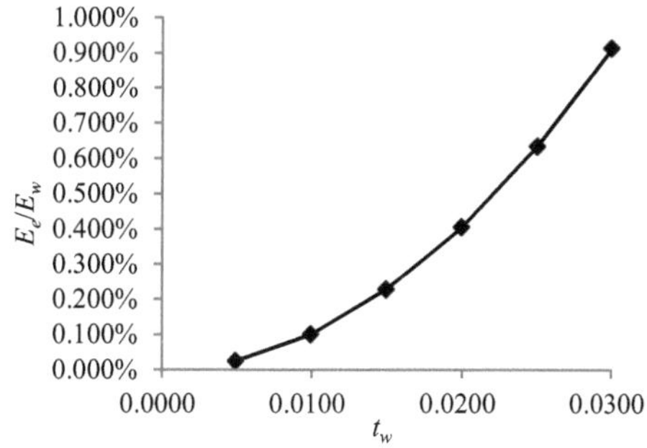

Fig. 11 Ratio of modulus of elasticity of equivalent orthotropic web to that of web material of model group B

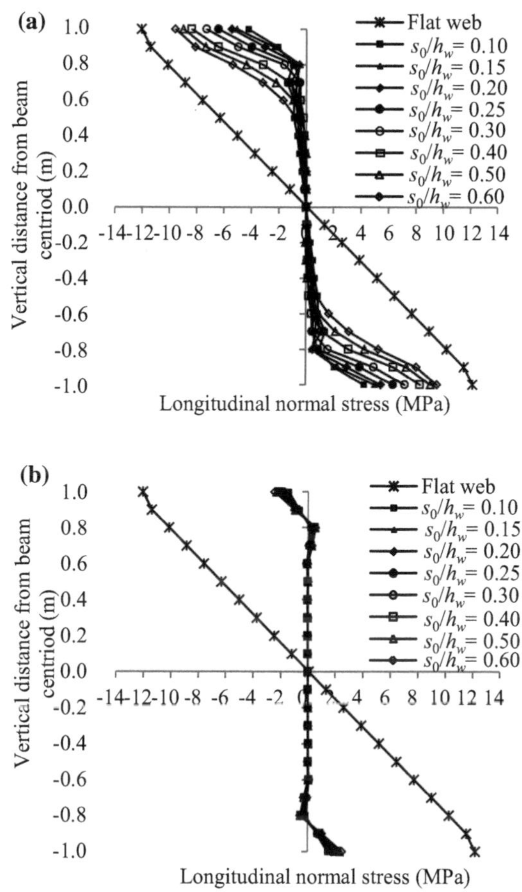

Fig. 12 Longitudinal normal stresses of the web corrugation at $3L/8$ of model group C: **a** flat segment, and **b** inclined segment

s_0/h_w is insignificant, and hence the normal stresses in the non-bonded zone of the flat segment vary a little. The normal stresses in the non-bonded zone of the inclined segment are always negligible.

Fig. 13 Ratio of modulus of elasticity of the equivalent orthotropic web to that of web material of model group C

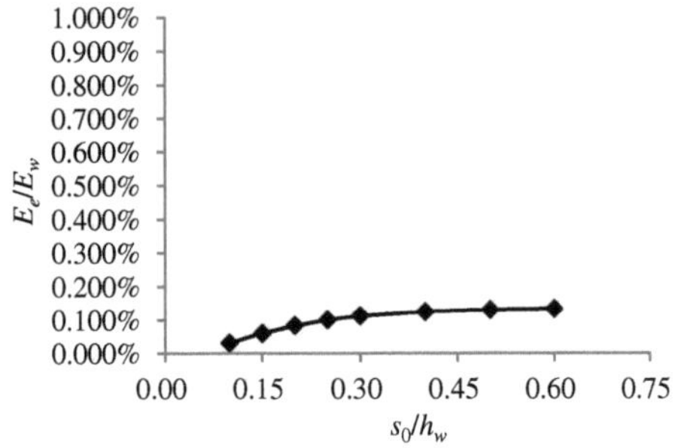

3 Conclusions

In this study, a numerical study on the flange-web junction of prestressed concrete bridge with corrugated steel webs is carried out by the finite element method. According to the parametric study, the following conclusions are drawn.

(a) The essential parameters, which affect the longitudinal normal stresses and depth of the bonded zone in the flat segment of the web corrugation, are the projected length of the corrugated web s_0 and the length of the flat segment of the corrugated web s_{flat}. With the increase of s_0 and s_{flat}, the stress and depth of the bonded zone in the flat segment of the web corrugation increase significantly. When the ratio of the projected length of the corrugated web s_0 to the web height h_w is below 0.25, the depth of the bonded zone approaches a constant of about 10 % of the web height.

(b) The essential parameters, which affect the longitudinal normal stresses of the bonded zone in the inclined segment of the web corrugation, are the trough angle θ and rise r of the web corrugation. With the increase of trough angle θ and rise r, the longitudinal normal stresses in the inclined segment of the web corrugation decrease significantly.

(c) The depth of bonded zone in the inclined segment of the web corrugation is insensitive to changes in the web corrugation profile and web thickness, and the depth is about 10 % of the web height.

(d) The longitudinal normal stresses of the bonded zone in the inclined segment are clearly below those in the flat segment. It is also noted that the longitudinal normal stresses in the bonded zones of the corrugated web are smaller than those in the flat web.

(e) With the increase of modulus of elasticity of the equivalent orthotropic web, the normal stresses in the non-bonded zone of the flat segment increase significantly. When the ratio of the modulus of elasticity of the equivalent orthotropic web to that of the corrugated web material is less than 0.1 %, the normal stresses in the non-bonded zone of the flat segment become negligible.

(f) The normal stresses in the non-bonded zone of the inclined segment are always negligible.

In accordance with the numerical study, to avoid significant longitudinal normal stresses in the corrugated web in the design of bridges, it is suggested that the ratio of the modulus of elasticity of the equivalent orthotropic web to that of the corrugated web material to be kept below 0.1 % and the ratio of the projected length of the corrugated web s_0 to the web height h_w is kept below 0.25. Under such conditions, the common assumption of negligible axial stiffness of the corrugated web can be taken as valid; otherwise more sophisticated analysis should be carried out.

Acknowledgements The work reported in this chapter is supported by the Research Grants Council (RGC) of the Hong Kong Special Administrative Region, China (RGC Project No. HKU 710111E).

References

1. Cheyrezy, M., Combault, J.: Composite bridges with corrugated steel webs—achievements and prospects. In: IABSE Symposium on Mixed Structures Including New Materials, Brussels (1990)
2. Machimdamong, C., Watanabe, E., Ustunomiya, T.: Analysis of corrugated steel web girders by an efficient beam bending theory. J. Struct. Mech. Earthq. Eng, JSCE **21**(2), 131s–142s (2004)
3. Elgaaly, M., Seshadri, A., Hamilton, R.W.: Bending strength of steel beams with corrugated webs. J. Struct. Eng, ASCE **123**(6), 772–782 (1997)
4. Yamaguchi, K., Yamaguchi, T., Ikeda, S.: Concrete Library International (C.L.I.), No.31. C.L.I. Editorial Sub-committee, JSCE, Japan (1998)
5. Briassoulis, D.: Equivalent orthotropic properties of corrugated sheets. Comput. Struct. **23**(2), 129–138 (1986)
6. Samanta, A., Mukhopadhyay, M.: Finite element static and dynamic analysis of folded plates. Eng. Struct. **21**(3), 277–287 (1999)

The Use of Ultrasonic Waves and Analytical Modeling to Estimate Elasticity Modulus of Rubber Concrete Specimen

Hadda Hadjab, Ahmed Arbia and Oussama Boulekfouf

Abstract This research work aims to evaluate the influence of the addition of rubber aggregates on elasticity modulus experimentally by using ultrasonic waves and theoretically by analytical models. Based on ultrasonic waves in concrete specimen at different percentages of rubber granules, one can evaluate the static modulus (E_{static}) from the dynamic elastic modulus (E_{dynamic}), according to British code [7]. From the obtained experimental results, one can conclude that the use of rubber granules has the potential for vibration damping capacity. In other words, the rubber granules, reduces the kinetics of ultrasonic pulses in the material. This reduction is due to the decreasing density of rubber granules (*RG*), with respect to gravel. The concrete base of its aggregates can be used such as paving of vibrating tools. Analytical modeling (Hill model) is used. The analytically obtained results converge with those from experimental procedure and give a good agreement to other researcher's works.

Keywords Concrete · Aggregates · Rubber · Modulus of elasticity · Ultrasonic

H. Hadjab (✉) · A. Arbia · O. Boulekfouf
Civil Engineering Department, University of Sciences & Technology
Houari Boumediene, Bab Ezzouar, Algiers, Algeria
e-mail: hhadjab@usthb.dz

A. Arbia
e-mail: ahmedmoon@hotmail.fr

O. Boulekfouf
e-mail: oussama-tp@hotmail.fr

A. Öchsner and H. Altenbach (eds.), *Design and Computation of Modern Engineering Materials*, Advanced Structured Materials 54,
DOI: 10.1007/978-3-319-07383-5_4,

1 Introduction

The elastic modulus is a very important property for all materials. In the field of civil engineering evaluation of the elastic modulus of concrete is to determine the scope of its use (rigid pavements, retaining walls, structural elements…). In our work we use a concrete aggregate base rubber, which has characteristics very promising in terms of reducing the kinetic ability of ultrasonic pulses through its decreased elastic modulus. The determination of the elastic modulus can be achieved by experimental methods and by analytical methods. The analytical methods are based on the use of different models, such as selected in this work as the "terminals" of Hill [1] and the bounds of Hashin-Shtrikman (BHS) [2]. These models will be used for a comparative study with the results found by the experimental method which is based on the evaluation of the modulus of elasticity from the ultrasonic test. The experimental procedure is a non-destructive technique to measure the travel time of the ultrasonic wave through the material.

2 Materials and Methods

2.1 *Experimental Aspect*

2.1.1 Principle of Ultrasonic Testing

The principle of ultrasonic testing is that an electrical signal is converted into a strain wave by a piezoelectric transducer. That wave propagates through a concrete specimen and is captured by the receiving transducer. The propagation time and the speed of the ultrasonic waves are deducted (see Fig. 1).

2.2 *Calculation Method*

Due to the heterogeneity of the material, the interpretation of ultrasonic signals is not easy. Typically, tests are often based on the measurement of velocities of longitudinal ultrasonic waves (see Fig. 2). The speed of the wave through an elastic solid is given by the following expression [3]:

$$V = \sqrt{\frac{E(1-\upsilon)}{\rho(1+\upsilon)(1-2\upsilon)}} \tag{1}$$

The Young's modulus (dynamic) E can be expressed as a function of V, and υ γ using the following formula.

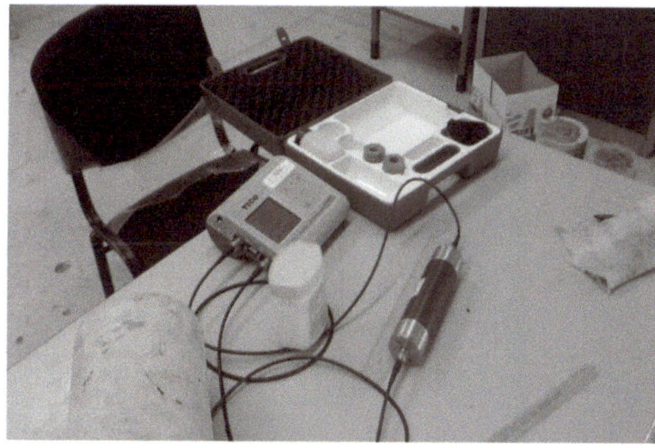

Fig. 1 Ultrasonic test

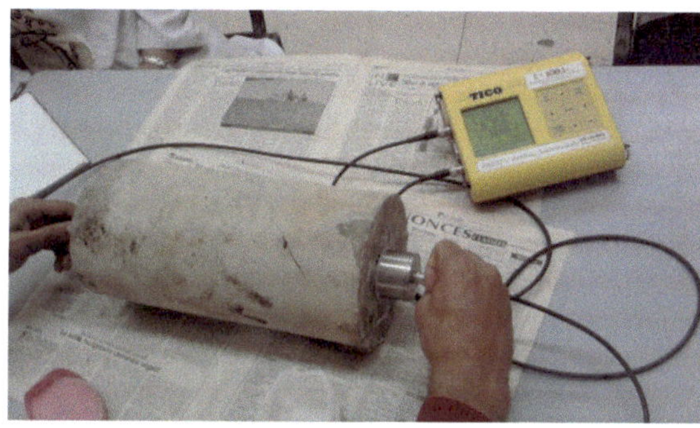

Fig. 2 Ultrasonic measurements

$$E = \frac{(1+\upsilon)(1-2\upsilon)}{1-\upsilon}\rho V^2 \tag{2}$$

V Wave velocity measured in km/s.
E Modulus of elasticity in GPa.
υ Poisson's ratio

3 Analytical Modelling

3.1 Hill Model

Hill [1] considers a materials with two phases: the inclusion and the matrix. To supervise the actual properties of these materials, it offers two models: parallel and series as shown in Fig. 3, which correspond to the "terminal" top and bottom

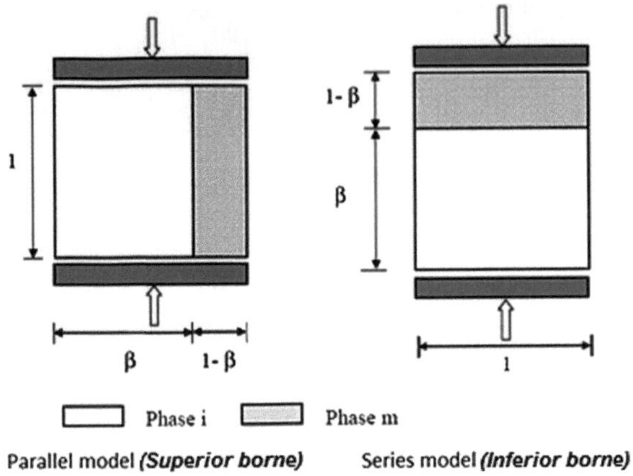

Fig. 3 Schematic representation of Hill model

respectively. Terminals Hill are best known for their simplicity and are also at the origin of models classified as parallel-series models with many applications for concrete.

Consider E_c and E_m moduls of the two phases c (rubber) and m (matrix) respectively:

$$\text{Superior borne SB}: \quad E_{sup} = \beta E_c + (1 - \beta) E_m \tag{3}$$

$$\text{Inferior borne} \quad \text{IB}: \quad \frac{1}{E_{inf}} = \frac{1}{E_C} + \frac{1-\beta}{E_m} \tag{4}$$

These models can be adapted to predict the values wrapped in the elastic modulus of composite materials.

If we consider V_c, E_c and V_m, E_m, as the volume fractions and elastic moduli of the two phases GC and cement matrix, we therefore proposed the following expressions by Hill:

$$\text{SB}: \qquad E_{sup} = E_m V_m + E_c V_c \tag{5}$$

$$\text{IB}: \qquad \frac{1}{E_{inf}} = \frac{V_m}{E_m} + \frac{V_c}{E_c} \tag{6}$$

3.2 Hashin-Shtrikman Model (BHS)

Hashin and Shtrikman [2] proposed limits established for a mixture of *n* isotropic elastic components, without any particular assumption about their form or their volume concentration. On concrete incorporating RG, if K_r, K_m, G_r, G_m, V_r, V_m are the bulk modules compressibility, shear moduli and volume fractions of each phase and rubber matrix, K_{inf} and and K_{sup} (the lower and upper bounds of the modulus) and G_{sup} and G_{inf} (terminals of the shear modulus) of the composite can be put in the form of the following equation [4]:

$$K_{\mathrm{inf}} = K_{\mathrm{r}} + \frac{Vm}{\frac{1}{Km-Kr} + \frac{3Vr}{3Kr+4Gr}} \tag{12}$$

$$K_{\mathrm{sup}} = K_{\mathrm{m}} + \frac{Vr}{\frac{1}{Kr-Km} + \frac{3Vm}{3Km+4Gm}} \tag{13}$$

$$G_{\mathrm{inf}} = G_{\mathrm{r}} + \frac{Vr}{\frac{1}{Gm-Gr} + \frac{6(\mathrm{Kr}+2\mathrm{Gr})\mathrm{Vr}}{5\mathrm{Gr}\,(3Kr+4Gr)}} \tag{14}$$

$$G_{\mathrm{sup}} = G_{\mathrm{m}} + \frac{Vm}{\frac{1}{Gr-Gm} + \frac{6(\mathrm{Km}+2\mathrm{Gm})\mathrm{Vm}}{5\mathrm{Gm}\,(3Km+4Gm)}} \tag{15}$$

We note that:

$$K_{\mathrm{r,m}} = \frac{E_{r,m}}{3(1-2\upsilon_{r,m})} \quad \text{et} \quad G_{\mathrm{r,m}} = \frac{E_{r,m}}{2(1+\upsilon_{r,m})} \tag{16}$$

where $v_{r,m}$ and Er, m are the Poisson's ratios and modulus of elasticity of each phase, i.e. rubber and matrix respectively.

So we can express the modulus of elasticity as a function of the shear modulus G and the bulk modulus K according to the relation below:

$$E = \frac{9KG}{3K+G} \tag{17}$$

And the relationship between the upper bound, lower elastic modulus and the experimental result is as follows.

$$\mathrm{BHS}_{\mathrm{inf}} = \frac{9K_{inf}G_{inf}}{3K_{inf}+G_{inf}} \leq E_{\mathrm{exp}} \leq \frac{9K_{sup}G_{sup}}{3K_{sup}+G_{sup}} = \mathrm{BHS}_{\mathrm{sup}} \tag{18}$$

4 Results and Interpretations

4.1 Experimental Results

After measurement of ultrasonic velocities, we use the formula (2) to evaluate the results of the dynamic modulus of elasticity of each concrete composition and which are illustrated in Table 1 and Fig. 4 given below:

Table 1 Variation of elasticity modulus versus rubber aggregate percentage

Concrete composition	RG	B25RG	B50RG	B100RG
Elasticity Modulus (GPa)	37.26	32.77	29.42	15.22

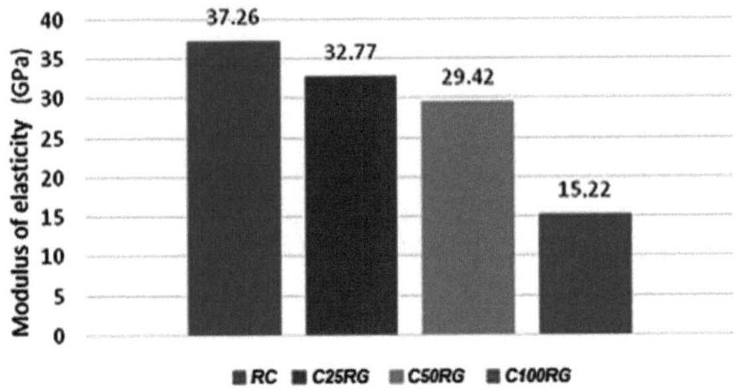

Fig. 4 Influence of rubber aggregates on elasticity modulus

It can be seen from the graphic that the addition of rubber aggregates induced a significant drop in modulus of about 59 % for concrete C100 RG, 21 % for concrete C50RG and around 12 % for concrete C25RG compared to the reference concrete.

Many studies, such as those Güneyisi [5] Ganjian [6] and Cuong [4] confirmed that the incorporation of RG induced a significant drop in modulus of elasticity.

Note that this drop is mainly due to the low stiffness of the RG, the poor quality of the transition zone between the rubber and the cement matrix and the porosity of the concrete.

In order to proceed to the analytical modeling of the dynamic modulus of elasticity and the static modulus of elasticity and basing on the British Code [7] and the work of Lydon and Balendran [8], a linear correlation between the Young's modulus static and dynamic was performed. This correlation is limited to a compressive concrete resistance less than 40 MPa concrete. The following formula defines the relationship between E_{static} and E_{dynamic}:

$$E_{static} = 0.83\ E_{dynamic} \tag{19}$$

The results are summarized and given in the following Table 2.

4.2 Results of Analytical Modeling

In this part we present only the modeling results for two models, i.e. the models according to HILL and the Hashin-Shtrikman terminals.

Table 2 Values of E_{static} for different compositions of concrete

Concrete composition	RC	C25RG	C50RG	C100RG
$E_{dynamic}$ (GPa)	37.26	32.77	29.42	15.22
E_{static} (GPa)	30.92	27.19	24.41	12.63

Table 3 Modulus of elasticity: comparison of experimental results and the model of Hill

Concrete	$V_c(\%)$	$V_m(\%)$	Modulus of elasticity (GPa)		
			IB	SB	Experimental results
RC	0	100	30.92	30.92	30.92
C25RG	9	91	12.36	28.29	27.17
C50RG	19.5	80.5	7.27	25.23	24.41
C100RG	39	61	4.13	19.53	12.63

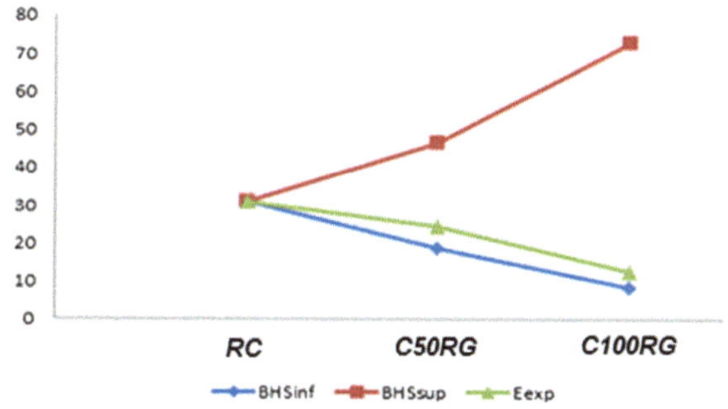

Fig. 5 Modulus of elasticity comparison of the experimental results and analytical model

4.2.1 Hill Model

Using Eqs. (5) and (6) we can calculate the lower and upper bounds and compare them with the experimental results. The modulus of elasticity of the vulcanized rubber is chosen to be 1.75 GPa.

The results are shown in Table 3 and Fig. 5 below:

The above figure shows that the experimental curve is between the two upper and lower bounds of the Hill model agrees best with the results given in relation to the upper bounds. The Hill model therefore allows to describe the variation of the elastic modulus as a function of experimental substitution rate in RG. This statement is more accurate for a percentage of the aggregate less than 50 % according to the graph in other words assays RG volume less than 20 %.

4.2.2 The Terminals Hashin-Shtrikman (BHS)

Using Eqs. (12), (13), (14), (15), (16), (17) and (18) for three concrete compositions the control concrete, concrete with a 50 % RG and concrete with 100 % RG, we find the results summarized in Table 4 below:

Table 4 Bounds of Hashin-Shtrikman model results

Concrete	V_r	V_m	K_r	K_m	G_r	G_m	K_{inf}	K_{sup}	G_{inf}	G_{sup}	BHS_{inf}	BHS_{sup}	E_{exp}
RC	0	100	14.583	18.405	0.591	12.672	396.726	18.405	1208.683	12.672	30.920	30.920	30.920
C50RG	19.5	80.5	14.583	18.405	0.591	12.672	67.189	28.064	6.504	19.032	18.903	46.568	24.410
C100RG	39	61	14.583	18.405	0.591	12.672	36.378	45.002	2.838	29.603	8.299	72.837	12.630

It can be seen that the lower Hashin-Shtrikman agreement is better with the experimental results with a small difference in contrast to the upper bound. We can say therefore that the BHS yield very significant results since they integrate at the same time the bulk modulus, shear modulus and Poisson's ratio of two constituent phases of the composite.

5 Conclusions

In this work we discussed and compared the results of two analytical methods (model BHS HILL and terminals) to calculate the elastic modulus of concrete containing rubber granules from the experimental results in order to have a reliable prediction of this important feature.

The HILL model which considers the material in two phases is one of the important tools to predict the modulus of elasticity of concretes and those incorporating rubber aggregates. This modeling approach is characterized by its simplicity and its appearance affordable based on physical characteristics and the volume fractions of each phase, but it is far from perfect.

The lower Hashin-Shtrikman bound gives results much better and more realistic because this method incorporates bulk modulus, shear and Poisson's ratio of GC phases and cement matrix and therefore we can say that it can be used as a tool for prediction of the elastic modulus.

References

1. Hill, R.: Elastic proprieties of reinforced solids: some theatrical principles. J. Mech. Phys. Solids **11**, 357–372 (1963)
2. Hashin, Z., Shtrikman, S.: A variational approach to the theory of the elastic behaviour of multiphase materials. J. Mech. Phys. Solids **11**, 127–140 (1963)
3. Zdiri, M.: Prediction and measurement of the elastic module of the RCC: case of the low cement proportioning. Int. J. Phys. Sci. **2**, 331–339 (2007)
4. Cuong, H.A.: Thèse de doctorat «Optimisation de la composition et caractérisation d'un béton incorporant des granulats issus du broyage de pneus usagés : application aux éléments de grande surface». Ph.D. Dissertation, University of Toulouse (2010)
5. Güneyisi, E., Gesoğlu, M., Özturan, T.: Properties of rubberized concretes containing silica fume. Cem. Concr. Res. **34**, 2309–2317 (2004)
6. Ganjian, E., Khorami, M., Maghsoudi, A.A.: Scrap-tyre-rubber replacement for aggregate and filler in concrete. Constr. Build. Mater. **23**, 1828–1836 (2009)
7. British Standard Institution: British code of concrete structure. London (1985)
8. Lydon, F., Balendran, R.: Some observations on elastic properties of plain concrete. Cem. Concr. Res. **16**, 314–324 (1986)

Numerical Study in Biomodels of Maxillofacial Prosthesis (Cancer and Osteonecrosis Cases)

Nayeli Camacho-Tapia, Juan Alfonso Beltrán-Fernández, Mauricio Gonzalez-Rebattú-y González, Luis Héctor Hernández-Gómez, Pablo Moreno-Garibaldi, Guillermo Urriolagoitia-Calderón and Daniel López-Liévano

Abstract This work shows a methodology to successfully generate an anatomically appropriate mandibular prosthesis for two patients with different conditions affecting the bone: cancer and necrosis. A common technique for bio-modeling was used in both cases, which includes tomographic slices of 0.5 mm for each clinical case. As a result of the processing a stereolithographic model was performed using two different techniques, were the affected area was reconstructed. In the osteonecrosis case, with the use of auxiliary guide paths and drawings in the CATIA platform a 3D model was produced, achieving a partial replacement with lateral support; while in the cancer case a left lateral replacement with the ramus was produced by reconstructing surfaces and patterns of the required area in PowerShape® and Solidworks® platforms.

N. Camacho-Tapia · J. A. Beltrán-Fernández (✉) · L. H. Hernández-Gómez · P. Moreno-Garibaldi · G. Urriolagoitia-Calderón · D. López-Liévano
Instituto Politécnico Nacional, ESIME-SEPI, Unidad Profesional "Adolfo López Mateos", Edificio 5, 3° Piso, Colonia Lindavista. Gustavo A. Madero, 07738 Mexico, Distrito Federal, Mexico
e-mail: jbeltranf@hotmail.com

N. Camacho-Tapia
e-mail: ncamacho@miners.utep.edu

L. H. Hernández-Gómez
e-mail: luishector56@hotmail.com

P. Moreno-Garibaldi
e-mail: eagle_gar@hotmail.com

G. Urriolagoitia-Calderón
e-mail: urrio332@hotmail.com

D. López-Liévano
e-mail: derive82@hotmail.com

M. Gonzalez-Rebattú-y González
Hospital Regional, Cirugía maxilofacial y cirugía plástica1°. de Octubre, ISSSTE, Av. Politécnico Nacional Colonia Lindavista #1669, 07730 Mexico, Distrito Federal, Mexico
e-mail: mauriciorebattu@yahoo.com.mx

A. Öchsner and H. Altenbach (eds.), *Design and Computation of Modern Engineering Materials*, Advanced Structured Materials 54,
DOI: 10.1007/978-3-319-07383-5_5,

The region to replace was defined by the surgical expertise of a maxillofacial physician. Both cases were printed in acrylonitrile butadiene styrene (ABS) polymer producing a three-dimensional model, which was adjusted by the maxillofacial surgeon and subsequently integrated into a compound that includes cyanoacrylate with low toxicity, calcium powder, hydroxyapatite and isophthalic resin as a fundamental basis. The models were analyzed numerically using the same conditions that will be applied mechanically and clinically in order to obtain preliminary information on stress concentration zones and material deformation.

Keywords Biomodels · Maxillofacial prosthesis · 3D prototype · Epoxic resin · Computed tomography · Cancer · Osteonecrosis · Isophthalic resin · Hydroxyapatite

1 Introduction

The mandibular bone, which is essential for speaking and chewing of food, not only provides definition to the face, but it is also connected to the lower teeth. Different conditions such as oral carcinoma of the lower gingival, of the floor of the mouth and mandible, along with osteonecrosis and odontogenic keratocysts, among others, require a segmental mandibular resection as part of the medical treatment. Damage or loss of mandibular support to the teeth, tongue, and lip produces dysfunctional mastication, swallowing, speech, airway protection, and oral competence [1]. Patients undergoing mandibular resection also suffer disfigurement because the mandible is an important aesthetic landmark [1]. The degrees to which dysfunction and disfigurement appear depend on both, the location of the mandibular segment removed and the amount of surrounding soft tissue eliminated in the surgery [1]. However, current techniques in mandibular resection only include the use of metal plates to replace the resected bone, which not only signifies the loss of osseous material, but also the loss of the teeth in that region.

The resection and replacement of the maxillofacial bones has been performed successfully for over a 100 years. Nonetheless, mandibular prosthesis design and manufacturing has proven to be a complex procedure, since it not only involves the design of the prosthesis and fixation components, but also the selection of suitable materials [2]. It has been established that the implant material must be characterized by its inertia, lightweight, ease of installation, and adaptability. Nowadays, current techniques allow the production of customized implants that can be adapted to the patients' specific needs and anatomy. One of the most notable techniques includes the use of 3D modeling and rapid 3D prototyping. Nevertheless, research efforts are focused on decreasing manufacturing time periods while providing high quality implants.

2 Background

The bones in the skeletal system provide protection, structural strength, and movement to the body. The mandible, which is the largest and strongest bone of the face, not only serves for the reception of the lower teeth, but it is also essential for the mastication and speaking processes. Multiple studies have been focused on designing mandibular prostheses in order to repair and replace segments of the jaw that have to be resected due to different clinical conditions. However, even after years of study, the most common solution for a mandibular resection is the use of metallic plates that substitute the removed bone.

The most common reason for requiring a mandibular resection includes bisphosphonate-associated osteonecrosis (see Fig. 1), oral carcinoma of different areas of the mouth (refer to Fig. 2), bone lesions (see Fig. 3 [3]) and keratocystic odontogenic tumors (see Fig. 4), among others [4]. Bisphosphonate-related osteonecrosis (BRON) is a condition found in patients who have received intravenous and oral forms of bisphosphonate therapy for various bone-related conditions or different types of cancer [5]. Bisphosphonate-related osteonecrosis reveals as exposed, nonvital bone involving the maxillofacial structures that does not heal within 8 weeks of documentation [5]. BRON is believed to be caused by trauma to dentoalveolar structures that have a limited capacity for bone healing due to the effects of bisphosphonate therapy, in other words dental extraction, dental implantation, or any other dentoalveolar surgical procedure. There is no effective therapy for this condition. However, the use of oral antibiotic treatments, hyperbaric oxygen therapy, surgery to debride dead bone, and resection of the mandibular area have proven relatively useful.

On the other hand, kerarocyst odontogenic tumors (KCOTs) are benign developmental tumors with many distinguishing clinical and histologic features such as a potential for locally destructive behavior, a relatively high recurrence rate, and designation as a consistent finding in the nevoid basal cell carcinoma syndrome, or Gorlin syndrome [6]. The KCOT is thought to arise from dental lamina and represents between 4 and 12 % off all odontogenic cyst; the lesion appears over a wide age range with a peak in the second and third decades and exhibits a fondness for white males [6]. The mandible is more commonly involved in these cases.

Mandibular reconstruction has been studied for many years. The earliest records of mandibular reconstruction date from the early 1600s when a segment of bone was replaced with a piece of the skull of a dog; the part removed from the dog skull was perfectly suited for the damaged area in shape and size and the surgery was a success [7]. Nevertheless, due to religious believes the prosthesis had to be removed [7]. Then, in 1889 Martin described the immediate restoration of a resected segment of the jaw with a prosthetic device and in 1890 a spiritual person from Massachusetts underwent the resection of half of the inferior maxillary due to a tumor located in the jaw; the prosthesis implanted was made of gold (see Fig. 5) [7].

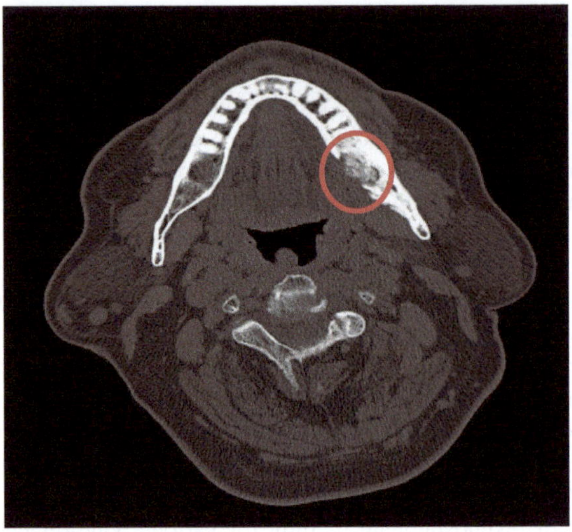

Fig. 1 Tomographic image showing bisphosphonate-induced osteonecrosis in a male patient

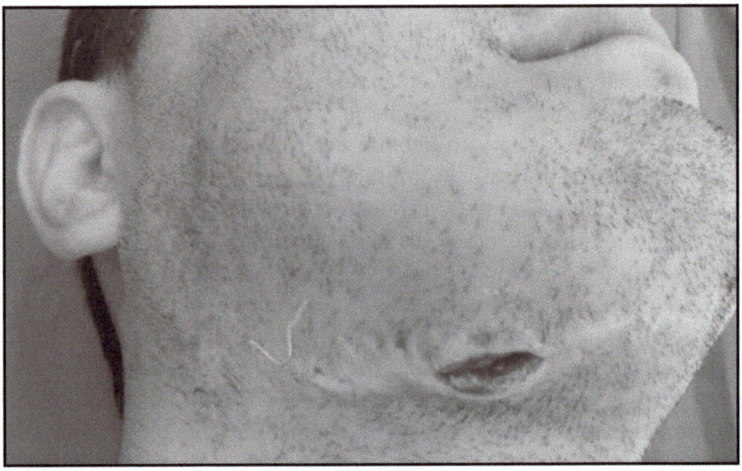

Fig. 2 Squamus carcinoma patient photograph displaying a lesion to the mandible

These plates are perceived as the basic design for current prosthesis, which only replace a small portion of the lost mandibular bone. A variety of techniques and materials used for this purpose has also been a topic addressed throughout history.

Bardenheur was the first to perform an autologous bone graft to the jaw in 1891; then, autogenous bone grafts from the rib and tibia were transferred to jaw by Bardenheuer in 1892 and Sykoff some years later. During the First World War, there was a need and great interest in the use of bone grafts to treat mandibular

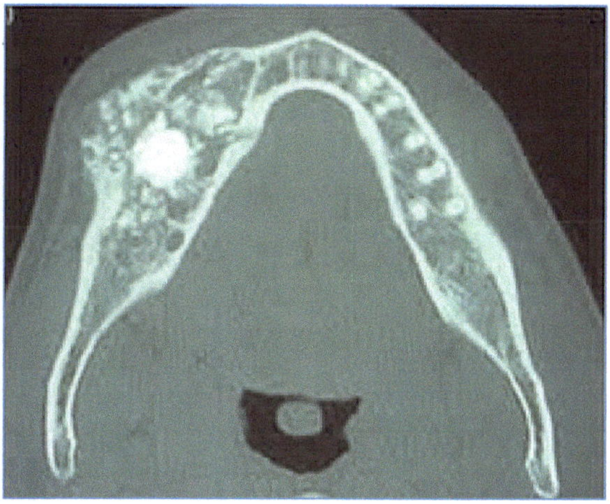

Fig. 3 Radiograph displaying a tumor causing a lesion in the mandibular bone [3]

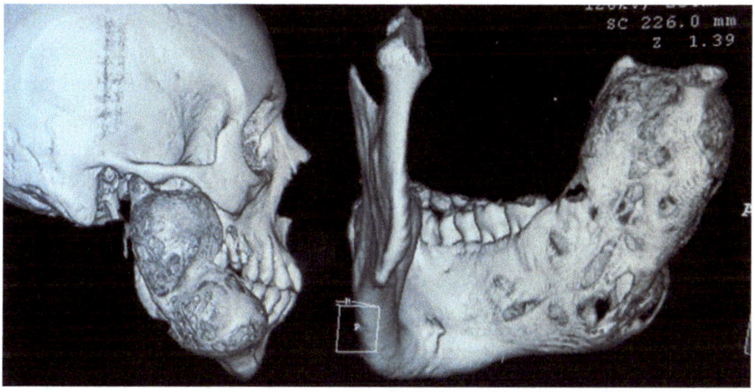

Fig. 4 Tomographic image displaying a keratocystic odontogenic tumor

defects. In 1912, Martin and Ollier replaced different jaw segments with hard rubber, while Koning used ivory to replace lost mandibular bone [7]. Gold implants were greatly used during the First World War, although gold was considered too expensive. Many other materials such as silver, aluminum, rubber, and lead were used for this purpose as well; nonetheless, these materials were soon abandoned because of the toxic effects produced when implanted inside the body. By the end of the 1920s, the use of the bone graft to replace the lost bone in mandibular resections became a routine procedure in orthopedic surgery, and it was not long before they were used in the field of plastic surgery as well [7].

Methylmethacrylate, an acrylic resin, was introduced as a bone substitute in the 1940s and still remains a popular choice for craniofacial reconstruction; this due to

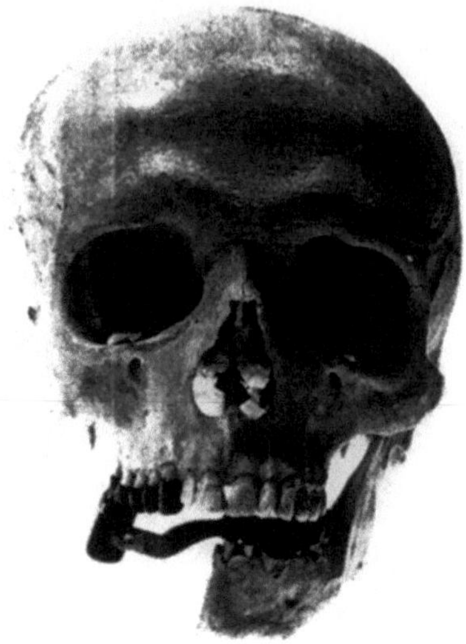

Fig. 5 Prosthesis system consisting of a gold crowns extension welded together and hooked to a joint piece [7]

its good resistance, adaptability, low cost, and relative radio luminesce. Since its introduction, the methylmethacrylate has been combined with various metal meshes in order to facilitate fixation in addition to providing additional strength to the to the implant [7, 10].

Other materials have been explored for the production and manufacturing of mandibular prostheses, for instance, metals such as vitallium were used as elements to help regenerate the mandibular bone [8]. Also, with the introduction of alloplastic materials surgeons had to look into this type of material for reconstruction of the mandible. Aubry and Pillet Edgerton used acrylic in 1950 along with Sikes and Ward used these materials [8]. Although alloplastic materials were used in mandibular reconstruction, the biocompatibility of these materials became a problem and alloplastics were only used to manufacture temporary prostheses [8]. The method of choice for the treatment of mandibular bone defects remained undoubtedly the autogenous bone replacement [7, 8]. In recent years, new techniques and materials have been developed to replace the lost bone in a mandibular resection. In 1995, Mercuri presented a series of 215 patients who had in common a degenerative disease called temporomandibular joint dysfunction (TMJ), the areas were reconstructed with prosthetic computer aided design and manufacturing (CAD /CAM) from Techmedica (Techmedica Inc., Camarillo, CA [United States]) creating customized prosthesis for each patient [7].

The use of bone substitutes in craniofacial surgery has increased due in large part to its ease of use, improvement in safety profiles, operating costs, and the advantages in terms of time and adaptability to the variety of clinical changes. A wide variety of materials used as bone substitutes are being developed. Some of the materials most commonly used as substitutes are shown in Table 1 [7].

Table 1 Common materials used in the bone replacement for mandibular regions [7]

Composite	Examples of commercial products
Hydroxyapatite	Pro-Osteon, Bio-Oss, Endobon, Calcitite
Tricalcium phosphate	Vitoss
Hydroxyapatite cement	Norian CRS, Bone source, Mimix bone void filler
Bioactive glass	NovaBone
Methylmethacrylate polymer	Hard tissue replacement
Porous polyethylene polymer	Medpor

Some advantages of patient-specific prosthesis are:

- Enables simpler, more effective treatment
- Designed to conserve similar anatomy
- Optimized use of materials
- It is well tolerated by the patient

However there are some requirements that are needed in order for the implant to be a success. The most common requirements are listed below.

- Transplantation of prosthesis made with a biocompatible material, preferably bio-active to promote bone ingrowth and ensure fixation of the device.
- Existence of an area with adequate blood supply.
- Extensive contact between adjacent bone and implant.
- Positive fixation.

3 Materials and Methods

For this chapter, three clinical cases were evaluated and solved through the use of CAD/CAM. The first case study involved a 44 year old male patient (Patient A) diagnosed with diabetes mellitus type 2 and squamous cancer of the mandible. Squamus cell carcinoma is a common form of skin cancer that develops in the thin, flat squamus cells that are present in the outer layer of the skin [9]. The patient first received treatment for a gingival lesion on the right canine and premolar area that appear to be granulomatous. Nevertheless, after an oncological evaluation, the diagnosis was established as well-differentiated epidermoid carcinoma. The patient underwent surgery in 2005 in which right supraomohyoid was dissected and there was a marginal resection in which the bone was substituted with osteosynthesis material. Subsequently, the patient was evaluated annually until 2009, when the patient was fully discharged.

In 2012 the patient started complaining about inflammation and pain in the region previously resected. After a physical and clinical evaluation, the patient underwent surgery again for a mandibular resection to extract the area affected by epidermoid carcinoma. The resected bone was replaced again with osteosynthesis material. The osteotomy was performed from the symphysis to right mandibular

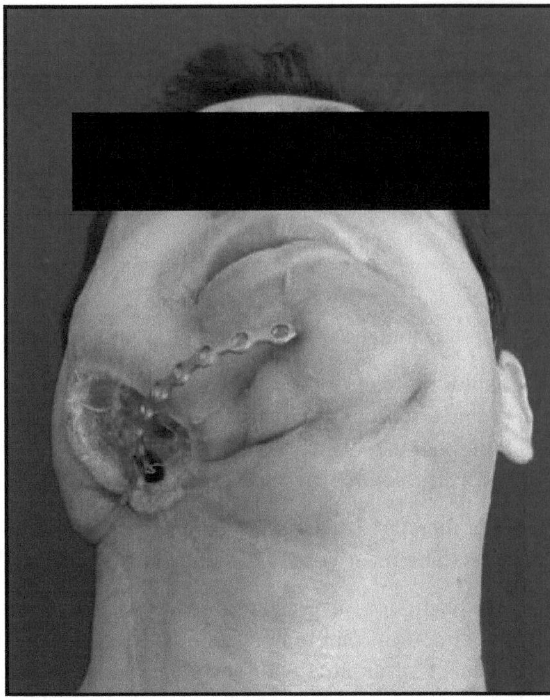

Fig. 6 Patient A's photograph displaying exposed osteosynthesis material and the lesion cause by squamous carcinoma

ramus; the removed bone block was approximately 15 cm in length. The oncology diagnosis coincided with the previous analysis and it was established that the patient suffered from squamous cell carcinoma moderately differentiated. The histopathological report of the right mandibular fragment indicated invasive poorly differentiated epidermoid carcinoma (non-keratinizing) with lymphoid-vascular permeation and focal desmoplastic reaction. The patient received radiotherapy in two phases, the first phase was 50 g in 25 fractions, while the second phase was 16 g in 8 fractions, concluding on October 2012. However, in March 2013, the osteosynthesis material became exposed (see Fig. 5) and the cancer metastasized to the right mandibular condyle. At this point, the maxillofacial surgeon decided that it was time to remove the osteosynthesis material along with the right condyle and to replace it with a customized mandibular prosthesis (Fig. 6).

The second clinical case evaluated in this chapter involve the participation of a 63 year old male patient (Patient B) diagnosed in 2009 in prostate cancer for which he received a trimestral injection of Zometa (zoledronic acid) for about 3 years and half. He was also receiving treatment for diabetes mellitus type II; he was diagnosed with this condition about 10 years ago. Patient expressed that his current condition started about 6 months ago when he started feeling pain at mandibular atrial. The pain was followed by numbness of the left lower lip and oral cavity burning. The radiographic and tomographic evaluation from the maxillofacial unit revealed a bone injury on the left side of the mandible. Because

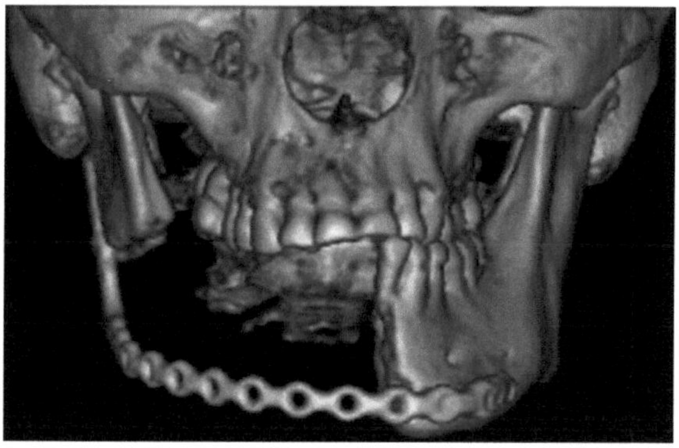

Fig. 7 Patient A tomograph image displaying the constraining points on the maxillary and jaw-bone to avoid the oral cavity

of the symptomatology, the medication for the prostate cancer and the fact that the bone exposure lasted more than 8 weeks, the patient was diagnosed with bisphosphonate-associated osteonecrosis of the jaw (BRONJ). The bone lesion was about 4 cm long and therefore and mandibular resection was the only viable option to treat this patient.

Finally, the third case involved an 18 year old male patient (Patient C) with a problem of bilateral temporomandibular hypoplasia and ankylosis. The main issues affecting the patient include null mouth opening (1–2 mm), condylar level immobility, and inclusion of foreign objects in the mandibular affected region, such as:

- Silicone tube
- Portion of titanium plate
- Screws

Because of all these problems, the treating maxillofacial surgeon decided that a customized prosthesis was the most viable option to help the patient return to a normal life style.

The first step in manufacturing customized mandibular prostheses for these three patients was to obtain a computer axial tomography (CAT) in Digital Imaging and Communication in Medicine (DICOM) format (see Fig. 7). The extent of the region to be resected was different for all three patients; therefore, there was a need to customize each implant for a better adaptability.

Processing bone reconstruction requires reproducing external geometry based on data obtained from the CAT. The image processing was performed in Scan IPTM where a 3D geometry was developed. The 3D model was created selecting the desired bone region in each CAT slice and creating a mask [7]. Then the images were rendered and a 3D model was obtained (Fig. 8a, b). The geometry was then exported as a stereolithography (STL) file into either CATIA or PowerSHAPE® software. In PowerShape®, the STL file data worked as a point cloud that was used in the cancer case as a ideal to model the missing bone region (see Fig. 8b, c). Once the model surface was completed, it was imported in a parasolid

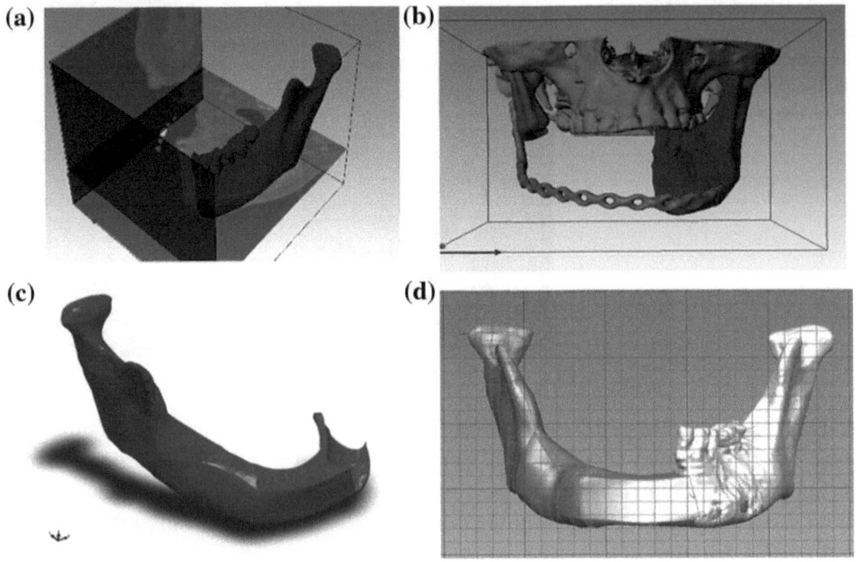

Fig. 8 Steps followed for the construction of a 3D model in different CAD software for a cancer case (Patient A)

file format (. X_t) into SolidWorks® in order to generate a solid model. This also allowed modifying the structure to ease installation (see Fig. 8c, d).

In CATIA®, auxiliary paths were created to define the shape of the mandible in different cross sectional areas. Then, planes perpendicular to the mandible were created along the bone and the auxiliary paths were used as a guide to sketch the mandible cross sectional areas [10]. Once the desired planes and sketches have been completed, the circumferences created through the auxiliary paths are joined in order to produce a surface that represents the bone region to be modeled [10]. This surface is then transformed into a solid (see Fig. 9) that can be saved as an STL file and used to print a 3D model in acrylonitrile butadiene styrene (ABS) like the one displayed in Fig. 10.

The expertise of the physicians in charge of each case was required in every step of the process, from the establishment of the area to be resected to the material selection for the prosthesis. An anionic polysaccharide mold impression was obtained using the 3D printed model, which was subsequently used to reproduce the prosthesis and the healthy regions of the mandible. The diagram presented below synthesizes the steps for each methodology and where the two methodologies converge to obtain the final prosthesis design and the required analysis.

The selected material to manufacture the prostheses was a composite that included hydroxyapatite, calcium powder, and an isophthalic resin that encapsulated the powder combination. One of the produced prosthesis by this method has already been installed and the clinical outcome has been positive (Fig. 11).

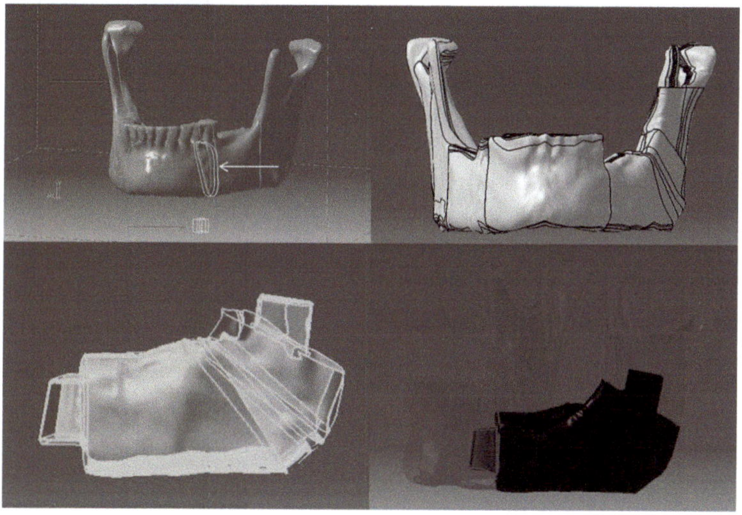

Fig. 9 3D model exported from Scan IP® displaying auxiliary paths (*green arrow*), planes (*red arrow*) and cross sectional sketches (*white arrow*) (*top left*); surface model created through a series of sketches (*top left*); final design of prosthetic device (*bottom left*) and prosthesis final design including adjacent bone regions from the mandible of patient B (*bottom right*)

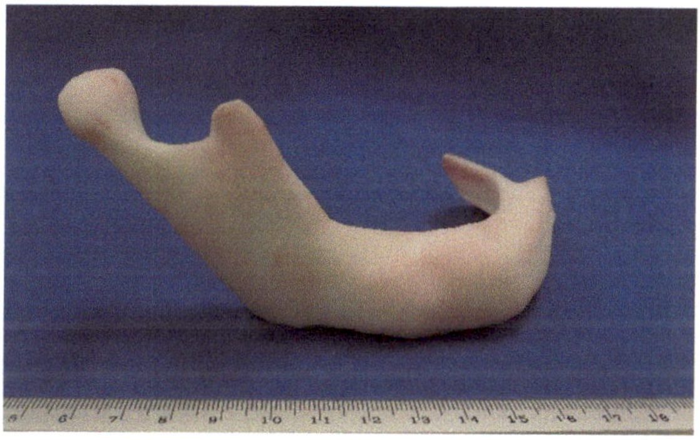

Fig. 10 Cancer case (patient A) prosthesis model printed in ABS

4 Experimental Analysis

Once the models were printed in ABS, a mold was produced with n anionic polysaccharide impression using the 3D printed model, which was subsequently used to reproduce the prosthesis and the healthy regions of the mandible in photo-polymeric resin. Validation of the design was initially accomplished experimentally by means of a descriptive photo-elastic analysis, which allowed visualizing stress concentration regions and calculating stress values. The maxillofacial surgeon used plates and screws to assemble the photo-polymeric replicas in the same

Fig. 11 Diagram displaying the methodologies used to design the prosthetic devices of patients A and B

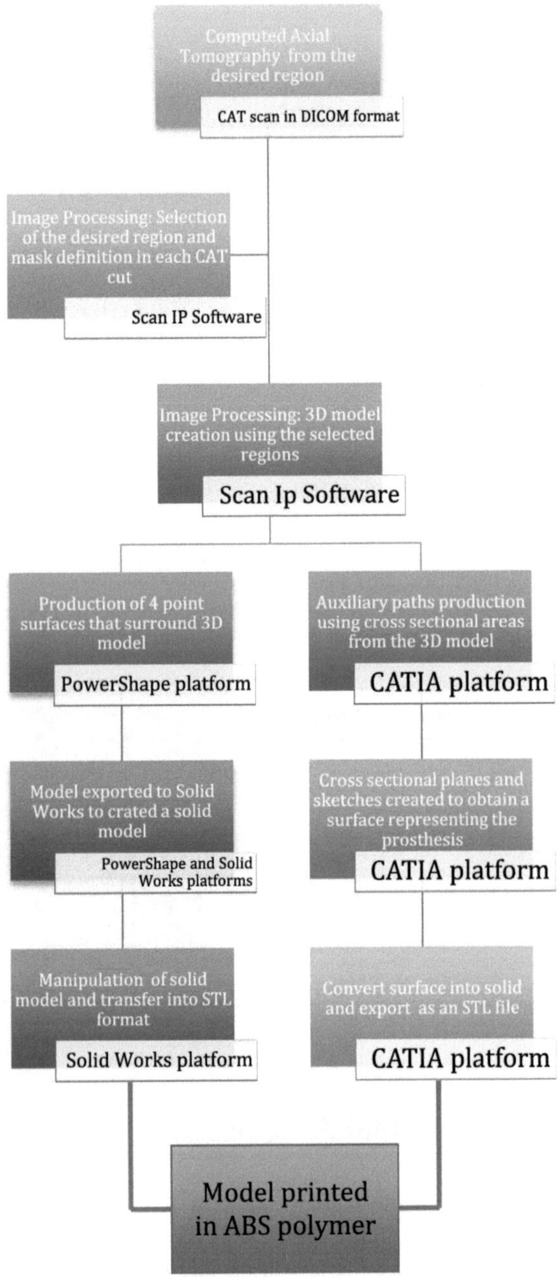

way as the prosthesis and healthy bone region would be assembled. Then, the replicas were subjected to stresses similar to those of the conventional chewing and biting processes. Furthermore, the design was also validated by means of finite

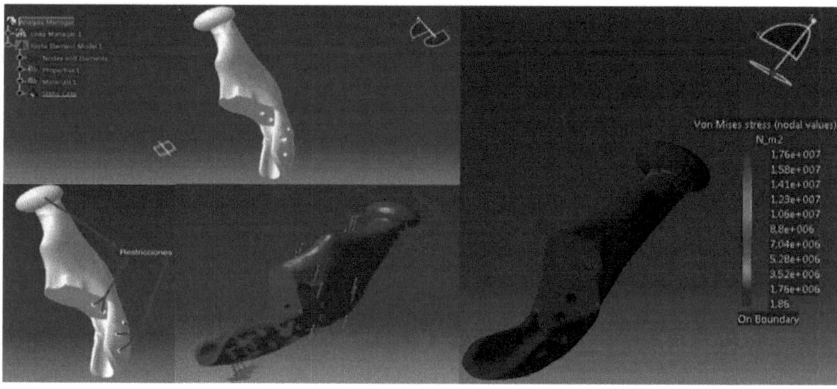

Fig. 12 Mandibular prosthesis model from patient C subjected to finite element analysis

element analysis (FEA), which allowed the observation of the deformation of the prosthesis when subjected to stresses from the chewing process; the results of the FEA are displayed in Fig. 12.

5 Discussion

Even though there are many techniques under development for the design and manufacturing of patient-specific prostheses, the methodologies employed in this chapter have proven to be not only successful but also efficient. Even though the methodologies developed in this chapter for the design of mandibular implants vary significantly, the first and final steps are exactly the same. In other words, both methodologies converge when the models are exported as STL files and printed.

On the other hand, it is important to mention that the experimental and numerical analyses displayed similar results when the mandibular prosthesis was subjected to the normal chewing and biting stresses. The mechanical response of the design showed that the stresses that the mandibular prosthesis will be subjected to are well tolerated by the material and they do not affect the integrity of the implant.

It is important to note that the material along with the prostheses needs to be evaluated clinically to guarantee the success of the design and manufacturing processes.

6 Conclusions

Computer-aided design and rapid 3D prototyping have proven useful in the design and production of patient-specific prosthesis to default that offer the best possible aesthetic and functionality [11, 12]. However, it is necessary to increase the number techniques that allow replacing the lost bone region in a rapid manner. Furthermore, testing of the material as well as mechanical analysis of the design before clinical use are essential for the prosthesis to display optimal results and in order to avoid secondary surgeries.

Acknowledgments The authors kindly acknowledge the support given to this research by the National Polytechnic Institute (IPN) and the National Council for Science and Technology (CONACYT).

References

1. Schrag, C., Chang, Y.M., Tsai, C.Y., Wei, F.C.: Complete rehabilitation of the mandible following segmental resection. J. Surg. Oncol. **94**(6), 538–545(2006)
2. Peng, L., Jia, L., Jie, L., Weidong, T.: Optimal design of an individual endoprosthsis for the reconstuction of extensive mandibular defects with finite element analysis. J. Cranio Maxill. Surg. 1–6 (2013)
3. Rubio, J., Bescós, C., Pamias, J., Sáez, M.: Pindborg tumor (Calcyfing Epithelial Odontogenic Tumor). Rev. Esp. Cir. Oral Maxilofac. **29**(5) (2007). http://dx.doi.org/10.4321/S1130-05582007000500007
4. Philip. J., Lewis, R., Wysocki, G.: Patología Oral y Maxilofacial Contemporánea, segunda edición. Elsevier, España
5. Mayo Foundation for Medical Education and Research. Bisphosphonate-associated Jaw Osteonecrosis (2013). Accessed 15 July 2013. doi:10.3928/01477447-20091020-16, http://www.mayoclinic.org/medicalprofs/bisphosphonate-associated-jaw-osteonecrosis.html
6. Grasmuck, E., Nelson, B.: Keratocyst odontogenic tumor. Head Neck Pathol. **4**(1), 94–96 (2010). doi: 10.1007/s12105-009-0146-x (Published online 2009 November 10)
7. Moreno-Garibaldi, P.: Design and Manufacturing of a Prosthesis for a young patient with articular ankylosis. Instituto Politécnico Nacional, Escuela Superior de Ingeniería Mecánica y Eléctrica (2013)
8. Komisar, A.: Mandibular Reconstruction, pp. 1–10. Thieme Medical Publishers, U.K (1997) ISBN: 0865776148
9. Mayo Foundation for Medical Education and Research. Squamus Cells Carcinoma. (2013). Accessed 13 July 2013. doi:10.3928/01477447-20091020-16, http://www.mayoclinic.com/health/squamous-cell-carcinoma/DS00924
10. Beltrán-Fernández, J.A., Hernández-Gómez, L.H., Ruiz-Muñoz, E., González-Rebattú, A., Rodríguez-Cañizo, R.G., Urriolalagoitia-Calderón, G., Urriolagoitia-Sosa, G., Hernández-Moreno, H.: Biomechanical evaluation of corporectomy in porcine lumbar specimens using flexible polymer belts. J. Phys. Conf. Ser. (2009). doi:10.1088/1742-6596/181/1/012015
11. Beltrán, J. et al.: Assessment of the structural integrity of c3–c5 cervical porcine vertebrae model based on 2-D classic cad, 3-D scanner and 3-D computed tomography. In: Öchsner, A., da Silva, L.F.M., Altenbach H. (eds.) Analysis and Design of Biological Materials and Structures, vol. 1/2010–vol 20, Springer, Berlin (2010) ISBN: 978-3-642-22130-9
12. Beltrán-Fernández, J.A., Rebattú y González, M.G., Hernández-Gómez, L.H., Rebatú y González, A.G., Calderón, G.U.: Biomechanical prosthesis design of an orbicular cranial cavity. In: Öchsner, A., da Silva, L.F.M., Altenbach, H. (eds.), Advances in Bio—Mechanical Systems and Materials, vol. 40/2013–vol 40, pp. 87–94. Springer, Berlin (2013) ISBN: 978-3-319-00478-5. doi:10.1007/978-3-319-00479-2_7

The Effect of Dynamic Loading from Routine Activities on Mechanical Behavior of the Total Hip Arthroplasty

Mohamed Mokhtar Bouziane, Smail Benbarek, Essadek Mohamed Houari Tabeti, Bel Abbes Bachir Bouiadjra, Noureddine Benseddiq, Boualem Serier and Abdelmohsen Albedah

Abstract Dynamic loads from routine activities applied to the stem create dynamic stresses varying in time and resulting in the fatigue failure of the prosthesis components. Therefore, a finite element model can be used to predict mechanical failure. The purpose of this study was to develop a three-dimensional model of the cemented hip femoral prosthesis and to carry out finite element analysis to evaluate stress distributions in the bone, the cement and the implant compounds under dynamic loads from different human activities. Linear elastic

M. M. Bouziane (✉)
Department of Mechanical Engineering, University Mustapha Stambouli of Masacra, Mascara, Algeria
e-mail: agk_bouziane@yahoo.fr

M. M. Bouziane · S. Benbarek · B. A. B. Bouiadjra · B. Serier
LMPM, Department of Mechanical Engineering, University of Sidi Bel Abbes, Ben M'hidi, BP 89, 22000 Sidi Bel Abbes, Algeria
e-mail: sma_benbarek@yahoo.fr

B. A. B. Bouiadjra
e-mail: bachirbou@yahoo.fr

B. Serier
e-mail: boualems@yahoo.fr

A. Albedah
Mechanical Engineering Department, College of Engineering, King Saud University, Riyadh, Saudi Arabia
e-mail: albedah@ksu.edu.sa

N. Benseddiq
Laboratoire de Mécanique de Lille, IUT-A, 2 rue de la Recherche, 59653 Villeneuve d'Ascq Cedex, France
e-mail: noureddine.benseddiq@univ-lille1.fr

E. M. H. Tabeti
Department of Orthopedic Surgery, University of Sidi Bel Abbes, Ben M'hidi, 22000 Sidi Bel Abbes, Algeria
e-mail: chu-sba@sante.dz

A. Öchsner and H. Altenbach (eds.), *Design and Computation of Modern Engineering Materials*, Advanced Structured Materials 54,
DOI: 10.1007/978-3-319-07383-5_6,

analysis is adapted; von Mises stress, normal stress and shear stress are the values that are of concern. Results show that the stresses distribution in the femoral arthroplasty components depends on the human activity. The analysis also showed that the stresses are high in the proximal and distal parts of the cement mantle.

Keywords Femoral prosthesis · Finite element method · Cement · Dynamic · Stress

1 Introduction

Total hip replacement (THR) is a very successful surgical technique that has become a well established procedure in current orthopedics. Patients with degenerative hip joint diseases, persistent to thigh pain and fractures of the femoral neck, can effectively be treated with an artificial hip joint reconstruction. Generally, THR leads to immediate pain relief and increased freedom of movement in the hip joint. Patients experience a substantial improvement in the quality of life, and need les support to carry out their daily activities [1]. The finite element method (FEM) is an advanced simulation technique that has been used in orthopedic biomechanics since 1972 [2]. It is an important tool used in the design and analysis of total joint replacements and other orthopedic devices [3]. Contact forces in the hip joint must be known for tests on strength, fixation, wear and friction of implants, for optimizing their design and materials by computer simulation and for giving guidelines to patients and physiotherapists as to which activities should be avoided after a replacement. The movement in the hip joint has to be known when implant wear is tested or the load directions relative to the pelvis are calculated from the forces acting at the femur [4]. The negative effects of stress singularities are also found in an FEA simulation that was derived in an earlier study [5] for the purpose of preclinical testing of cemented total hip replacement (THR) implants against the damage accumulation failure scenario. This failure scenario is often considered to be the most dominant failure scenario for the femoral component of a cemented THR reconstruction [6, 7]. Cemented hip arthroplasties are subjected to cyclic loads, which sometimes lead to the mechanical failure of components of the implant system, with the subsequent long-term failure of the whole fixation. There are usually recognized four vulnerable regions: the cement-stem interface, the bulk cement, the cement-bone interface and bone [8–10]. Higher peak stresses lead to earlier crack formation. The peak tensile stresses are usually found around sharp corners or edges in the reconstruction, and as such, crack formation is first observed at these locations [11]. The loading methods used to determine the stresses in the prosthesis design can, also,

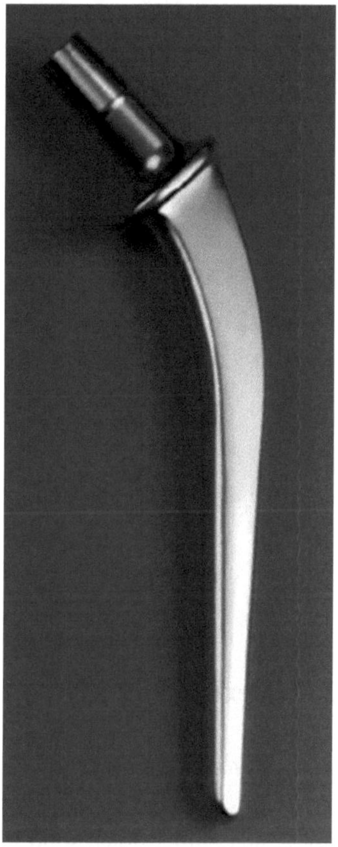

Fig. 1 Osteal femur stem

give quite different information [12]. Forces applied to the implant due to human activity generate dynamic stresses varying in time and resulting in the fatigue failure of the implant material. Therefore, it is important to ensure the hip prostheses against static, dynamic and fatigue failure [13]. Since 1979, the Ceraver-Osteal model of cemented total hip arthroplasty (Fig. 1) with a titanium femoral stem [14] has been used. This study aims to take account of the patient activity (walking, up stairs, down stairs, standing up and sitting down) when designing a total hip replacement. In this regard the stress field in the artificial hip components (prostheses, cement mantle, and bone) is analyzed dynamically. The simulations have been conducted to investigate the effect of dynamic loading from routine activities patterns on the stress-based criteria to assess implant longevity. Two quantitative measures are calculated: stress distribution and peak stress. It has been shown that each measure may lead to differing conclusions.

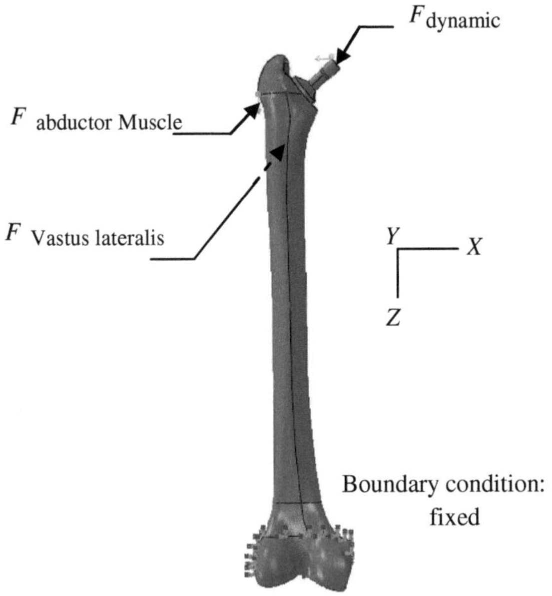

Fig. 2 Applied forces on the bone-cement–prosthesis assembly

2 Materials and Methods

2.1 Model Designs

For a three-dimensional solid model of the total hip replacement (THR), there are four major components that have to be modelled: cortical bone, cancellous bone, femoral stem and bone cement. The complete models were assembled using SolidWorks. The three-dimensional solid model assembly of femur, bone-cement and implant was transferred to Abaqus Workbench by the direct interface. Abaqus Workbench automatically recognizes the contacts existing between each part and establishes the contact conditions for corresponding contact surfaces. In this work, the Ceraver-Osteal model of the cemented total hip arthroplasty is designed (Fig. 2).

2.2 Material Properties

The material properties adopted were specified in terms of Young's modulus and Poisson's ratio for the implants and all associated components (Table 1). All materials were assumed to exhibit linear, homogeneous elastic behavior [15].

Table 1 The artificial hip components material properties [15]

Materials	Young's modulus E (MPa)	Poisson's ratio ν	Density (kg/m^3)
Cortical bone	15,500	0.28	1,990
Concelleous bone	389	0.3	500
Stem (Ti-6Al4 V)	110,000	0.3	4,430
Ciment PMMA	2,700	0.35	1,200

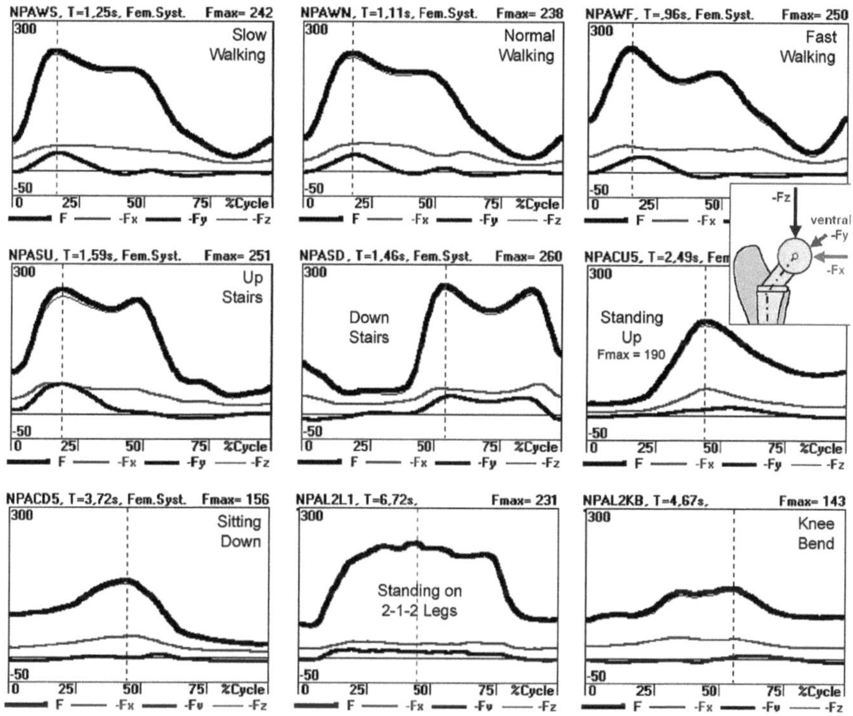

Fig. 3 Contact force F of typical patient NPA during nine activities. Contact force F and its components $-F_x$; $-F_y$; $-F_z$: F and $-F_z$ are nearly identical. The scale range is −50–300 % BW

2.3 Loading and Boundary Conditions

The contact forces F of the typical patient and their components are charted in Fig. 3 for the nine investigated activities [16]. In this study, the dynamic loads from five activities (walking, up stairs, down stairs, standing up and sitting down) were chosen from the hip contact forces, these loads for a person of 70 kg are illustrated in Fig. 4. The boundary condition was applied by fixing the distal epiphysis, which is the distal end of the femur that is connected to the knee [17]. The coordinate system used to represent the direction of the forces components is

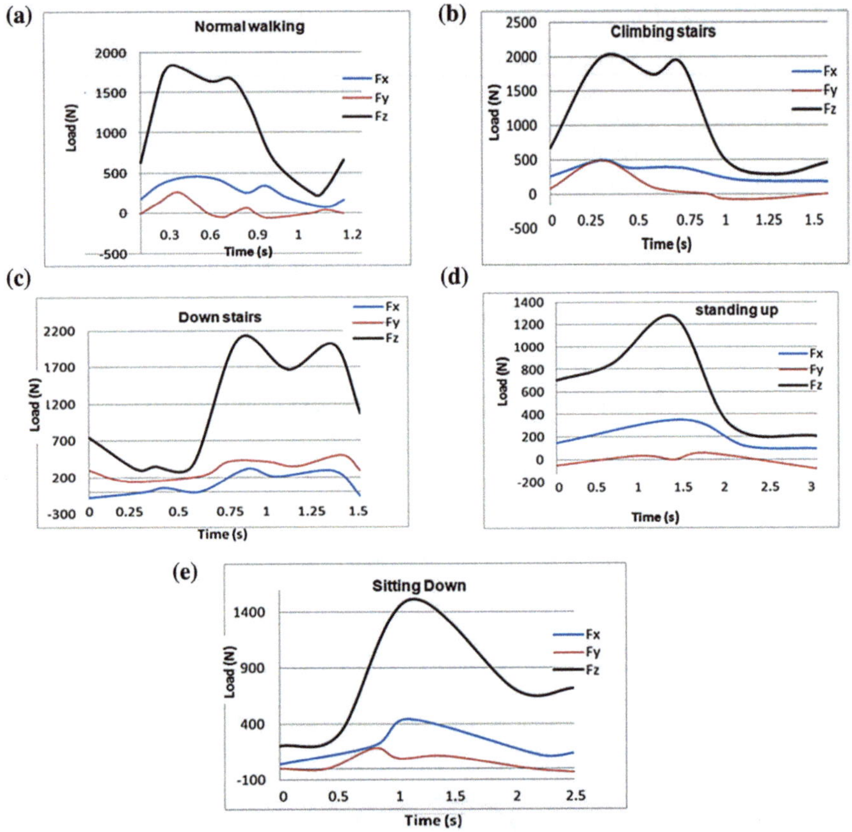

Fig. 4 The variation of forces applied on the prosthesis during five activities (**a** walking, **b** climbing up stairs, **c** down stairs, **d** standing up and **e** sitting down) for BW = 70 kg

shown Fig. 2. The femur is primarily loaded in bending [18]. The cement–bone and cement-stem interfaces were assumed rigidly fixed.

2.4 Model and Mesh

Finite element analysis (FEA) is a widely used research tool in biomechanics. A well-known problem in this type of analysis is the presence of singular points in the FEA model, causing the predicted peak stresses in particular to be dependent on the level of mesh refinement (Fig. 5). A method to reduce the mesh dependence would be of great value [11]. The model in this study is discretized by using tetrahedral elements. This is because the geometry of the femur is irregular. Tetrahedral elements are better to be suited and adjusted to curved boundaries

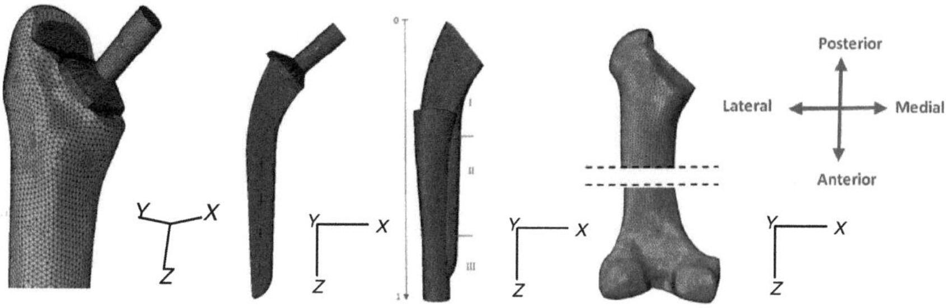

Fig. 5 Finite element meshes of hip prosthesis components: cemented hip stem, Osteal stem, cement and femur bone (from *left* to *right*) *I* Proximal part, *II* Median part and *III* Distal part

compared to others elements. Discretizing by using tetrahedral elements with four nodes makes the meshing becomes easier. The complete osteal model (stem, bone cement and femur) has in total 1,223,410 elements.

3 Result and Discussion

Hip contact forces based on gait analysis data were previously calculated using simplified muscle models and various optimization methods [19–26]. Most studies were restricted to walking or stair climbing. Typically the calculations delivered higher hip joint forces than those measured by other groups. Only Brand et al. (1994) compared calculated and measured data which were obtained, however, at different times [25]. The obtained gait data was used as an input for a musculo-skeletal model to calculate muscle forces [27]. The measured hip contact forces served to check the validity of calculated results.

For walking and stair climbing measured and calculated contact forces agreed fairly well. Their model can therefore be used to investigate clinical problems like muscle deficiencies or operative procedures. Morlock et al. (2001) measured the activity levels of 31 patients with hip implants during day-long sessions [28]. The combination of average activity numbers with the typical hip contact forces and joint movements presented here can serve to test the strength, fixation stability and wear properties of hip implants more realistically than today. Adding the muscle forces of Heller et al. (2001) will make the test conditions for hip implants, femur and pelvis even more realistic [27]. Physiological loading conditions are mandatory if bone remodelling or implant subsidence is investigated [29].

Mechanical integrity can only be maintained if the overall stress is kept below some threshold over time [30]. Another practical problem is that the influence of cement porosity may dominate the effect of the stress [31]. These stresses may occur as tensile, compressive, shear, or a stress combination known as equivalent von Mises stresses. This last one depends on the entire stress field and are widely used as an indicator of the possibility of damage occurrence [32]. During normal

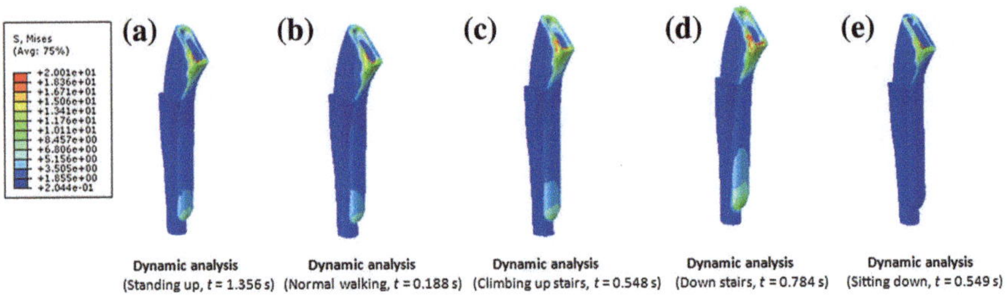

Fig. 6 von Mises stress distribution on the cement under dynamic loading from five activities: **a** Standing up (time = 1.356 s), **b** Normal walking (time = 0.188 s), **c** Climbing up stairs (time = 0.548 s), **d** Down stairs (time = 0.784 s) and **e** Sitting down (time = 1.549 s)

use, the joints experience cyclic stresses, which cause fatigue crack initiation and growth in the cement layer, leading to loss of structural integrity and eventual loosening of the implant [33].

In this study, we calculate the von Mises stresses distribution, in the components of the prosthesis (bone cement, stem and bone), for five cases of the dynamic loading (walking, up stairs, down stairs, standing up and sitting down). In addition, it is necessary to analyse the normal and shear stress distributions along the different regions of the cement mantle of the prostheses.

3.1 von Mises Stresses

Combined dynamic load was examined to determine on a phenomenological level what occurs when the hip prosthesis system is subjected to these specific loads. The stress analysis executed by Abaqus provided results that enabled the tracing of the von Mises stress field in the shape of color-coded bands. Each color band represents a particular range of stress value, which is given in Mega Pascals. Maximum stresses that occur in the cement, stem and bone under different dynamic loading conditions are shown in Figs. 6, 7, 8, 9, 10 and 11.

3.1.1 Cement Bone

By observing Figs. 6 and 7, it is found that, for all cases of the dynamic loading, the von Mises stress is still predicted to be high at proximal and distal regions, whereas the minimum stress is always found to be at the medial of the cement. Compared to the stresses, generally the stresses in the cement of the dynamic loading from the down stairs activity are higher (the maximum stress is in the order of 20 MPa for time = 0.784 s), while the stresses in the cement of the dynamic load from the sitting down activity are lower (the maximum stress is in the order of 15 MPa for time = 0.784 s). As for the results of THR with Osteal hip prosthesis, it is found

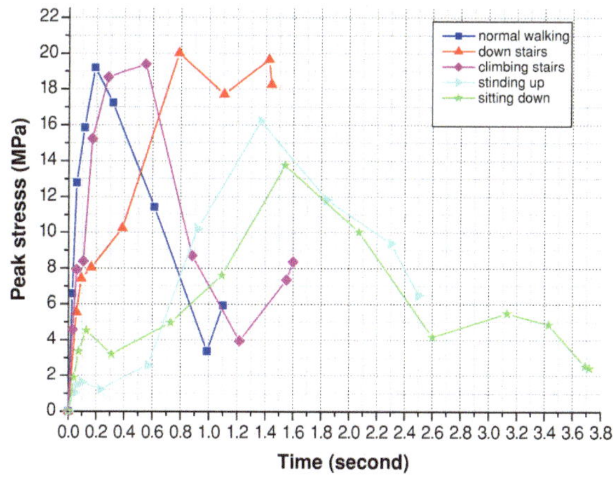

Fig. 7 Maximum von Mises stress in the cement mantle during five activities: *Standing up* (time = 1.356 s), *Normal walking* (time = 0.188 s), *Climbing up stairs* (time = 0.548 s), *Down stairs* (time = 0.784 s) and *Sitting down* (time = 1.549 s)

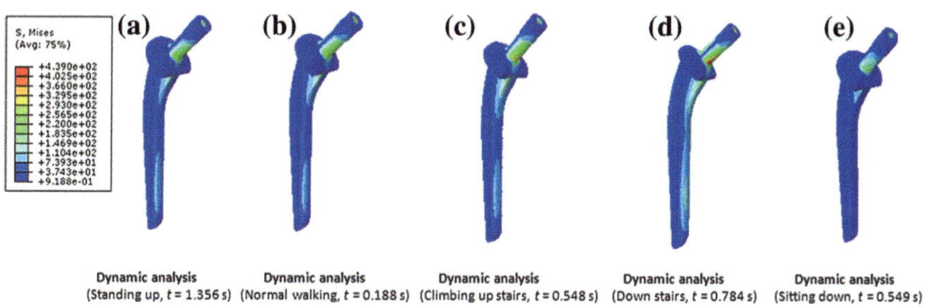

Fig. 8 von Mises stress distribution on the implant under dynamic loading from five activities: **a** Standing up (time = 1.356 s), **b** Normal walking (time = 0.188 s), **c** Climbing up stairs (time = 0.548 s), **c** Down stairs (time = 0.784 s) and **d** Sitting down (time = 1.549 s)

Fig. 9 Maximum von Mises stress in the implant during five activities: *Standing up* (time = 1.356 s), *Normal walking* (time = 0.188 s), *Climbing up stairs* (time = 0.548 s), *Down stairs* (time = 0.784 s) and *Sitting down* (time = 1.549 s)

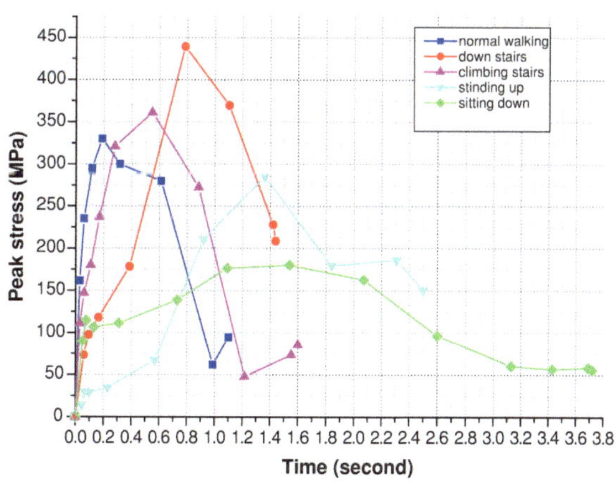

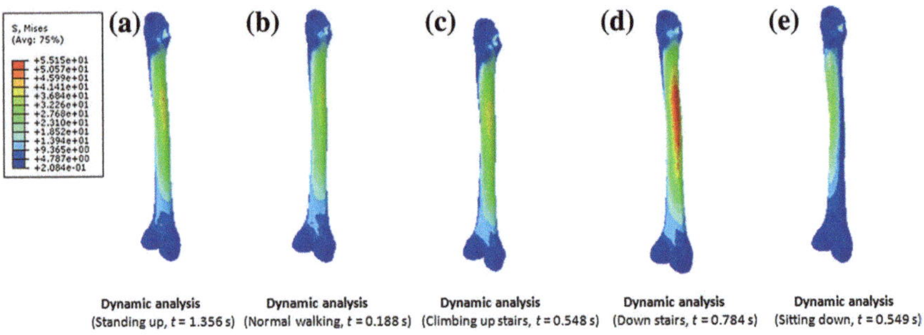

Fig. 10 von Mises stress distribution on the bone under dynamic loading from five activities: **a** Standing up (time = 1.356 s), **b** Normal walking (time = 0.188 s), **c** Climbing up stairs (time = 0.548 s), **d** Down stairs (time = 0.784 s) and **e** Sitting down (time = 1.549 s)

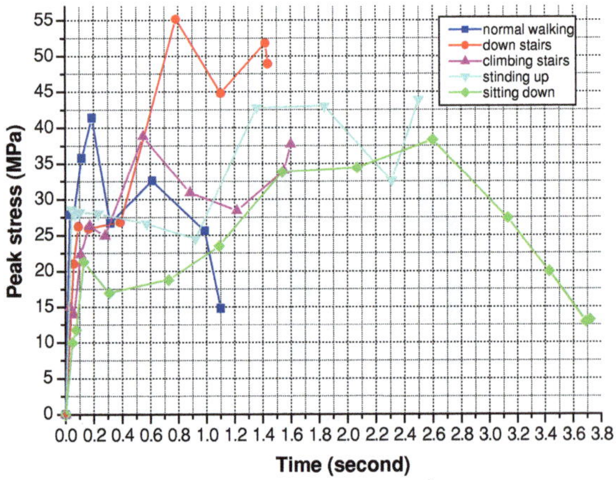

Fig. 11 Maximum von Mises stress in the bone during five activities: *Standing up* (time = 1.356 s), *Normal walking* (time = 0.188 s), *Climbing up stairs* (time = 0.548 s), *Down stairs* (time = 0.784 s) and *Sitting down* (time = 1.549 s)

that the bone cement does affect the stress distribution in the femur. Bone cement is made of polymer that has a relatively low Young's modulus, which is 2 GPa and it has a bad resistance to tensile loading (tensile strength = 25 MPa, compressive strength = 80 MPa and the shearing strength = 40 MPa) [34]. In other words, it is less stiff. Hence, when the hip prosthesis is loaded, it will transfer some of the load to the cortical through bone cement. Consequently, the stresses on the femur at that corresponding region are slightly higher. Although the difference is not much, the result is sufficient to tell us the effect of bone cement to the stresses on the femur.

3.1.2 Implant

Figures 8 and 9 show the von Mises stress distributions within the implant for five cases of the dynamic loading (walking, up stairs, down stairs, standing up and sitting down). Comparing the stress distributions on the hip prostheses, it can be observed that the stress concentration will be always at the neck area. Again, this is reasonable since there is cross section transitions at the neck are and it should always exhibit high stresses there. The higher stress is found in the prosthesis that occurred under dynamic loading from down stairs activity. The maximum stress is below 439 MPa for time = 0.784 s. If it is compared to the yield strength of Ti–6Al–4 V (880 MPa), there is still a safety factor of more than 2. Therefore, this result is still in the acceptable range. Whereas the lower stress is found in the prosthesis that occurred under dynamic loading from sitting down activity, the maximum stress is predicted to be 180 MPa for time = 0.784 s.

3.1.3 Bone

The von Mises stress distributions within the femur bone for five cases of the dynamic loading (walking, up stairs, down stairs, standing up and sitting down) is shown in Figs. 10 and 11. It is found that the stress is still predicted to be high at medial and proximal regions, whereas the minimum stress is always found to be at the distal end of the femur. Compared to the stresses, generally the stresses in the cortical bone of the dynamic loading from the down stairs activity are higher (the maximum stress is the order of 55 MPa for time = 0.784 s), while the stresses in the cement of the dynamic loading from the sitting down activity are lower (the maximum stress is the order of 33 MPa for time = 0.784 s). In biomechanical term, one say that a portion of the femur is being stress shielded. In long terms, it will cause bone sorption or bone loss. If this happens, the implant will have high possibility to loss and revision surgery is needed. The revision surgery will be more complicated than the primary surgery.

3.2 Axial Stresses

Irrespective of the method of analysis being used, maintaining the mechanical integrity is not a matter of reducing the peak stress in, e.g. the cement mantle or on the cement/bone and cement/prosthesis interfaces, although this criterion can be used to optimize a stem profile [35].

It is necessary to analyze the normal and shear stress distributions in the different regions of the cement mantle at the cement/bone and cement/prosthesis interfaces, because it is considered as the weakest component in the assembly of total hip arthoplasty. In this study we have chosen to analyze the stress distributions in the cement which was subjected to a dynamic load due to down stairs activity.

The variation of the normal stress according to *x*-direction (σ_x) along the cement/bone and cement/stem interfaces in the different regions (posterior, anterior, medial and lateral) of the cement mantle is shown in Fig. 12. Under combined dynamic load due to down stairs activity, the highest stresses of the cement were observed around the implant neck. The maximum tensile stress exists in the

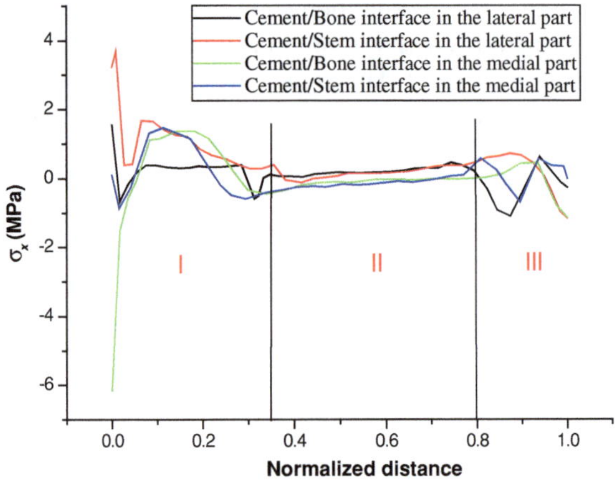

Fig. 12 Variation of the normal stress according to x-direction (σ_x) along the cement/bone and cement/stem interfaces in the different sides (lateral and medial) of the cement mantle for the static and dynamic analysis (down stairs, time = 0.784 s) *I* Proximal part, *II* Median part and *III* Distal part

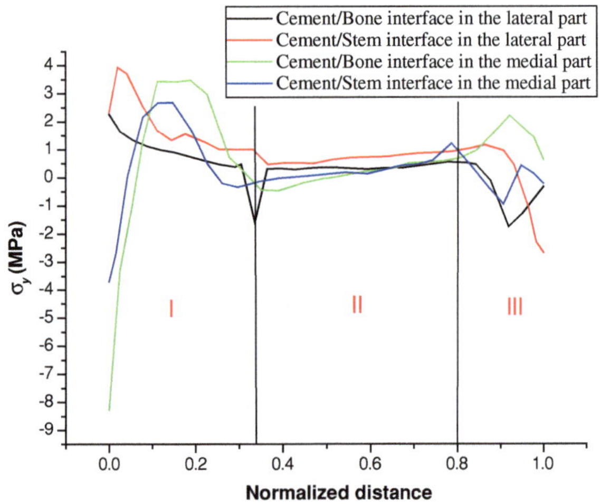

Fig. 13 Variation of the normal stress according to y-direction (σ_y) along the cement/bone and cement/stem interfaces in the different sides (lateral and medial) of the cement mantle for the static and dynamic analysis (down stairs, time = 0.784 s) *I* Proximal part, *II* Median part and *III* Distal part

posterior side at cement/stem interface with a value of 4 MPa and the maximum compressive stress exists in the anterior side at cement/stem interface with a value of 6 MPa. So, this result shows the major interaction effect between the dynamic loading and the implant neck shape.

Figure 13 shows the variation of the normal stress according to y-direction (σ_y) along the cement/bone and cement/stem interfaces in the different regions (posterior, anterior, medial and lateral). Comparing the stress distributions on the hip prostheses, it can be observed that the stress concentration will be always at the neck area. Again, this is reasonable since there is cross section transition at the neck area and it should always exhibit high stress values. The maximum tensile stress is localized in the posterior side at the cement/stem interface with a value of 4 MPa.

Fig. 14 Variation of the shear stress (τ_{xy}) along the cement/bone and cement/stem interfaces in the different sides (lateral and medial) of the cement mantle for the static and dynamic analysis (down stairs, time = 0.784 s). *I* Proximal part, *II* Median part and *III* Distal part

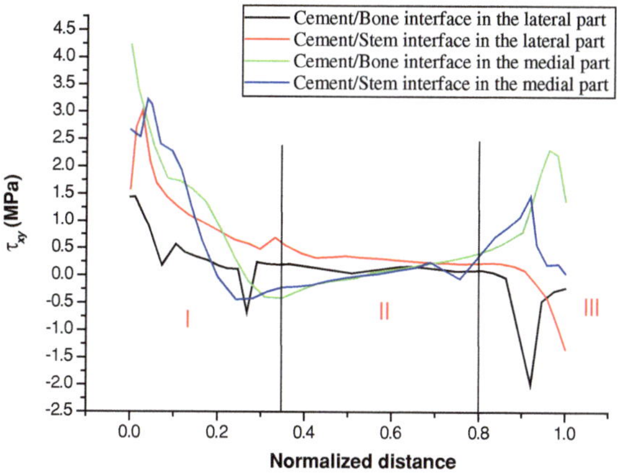

Figure 14 shows the variation of the shear stress (τ_{xy}) along the cement/bone and cement/stem interfaces in the different regions (posterior, anterior, medial and lateral).

When the prosthesis is being loaded, it will carry the entire applied load. Then, the load is transferred down along the prosthesis. When the load is transferred to the regions of the hip prosthesis, the load sharing will occur. This is due to the shear stress developed between the contact surfaces. The highest stress is observed in the anterior side at cement/stem with a value of 4 MPa.

4 Conclusion

For nearly all hip prosthesis systems, the major objective is long-term fixation of implants to bone. To achieve this goal, designers of implant systems must confront biomaterial and biomechanical problems, including in vivo forces on implants, load transmission to the interface, and interfacial tissue response. A three-dimensional finite element analysis is constructed to investigate the effect of combined dynamic load on the stress distribution in hip prosthesis. The loading methods used to determine the stresses in the prosthesis design can, also, give quite different information, and could lead to different conclusions. To be close to the reality, the dynamic load simulation is the only way to represent the effect of the patient activity on the prosthesis durability and design. This is important if finite element models are to achieve their potential as pre-clinical testing tools [36].

This study was carried out with the aim of analysing the effect of dynamic loading from routine activities on mechanical behavior of the total hip arthroplasty. The obtained results lead to the following conclusion:

- Compared to the stresses in the different components of the total hip arthoplasty, the stresses due to dynamic loading from the down stairs activity are higher, while the stresses due to dynamic loading from the sitting down activity are lower.
- In the cement mantle the critical region is still predicted to be at the neck region of the hip total arthoplasty. The critical stress is much lower than the yield strength. Hence, the design of the prosthesis is believed to be safe for use.

Acknowledgments Author Bel Abbes Bachir Bouiadjra and Abdelmohsen Albedah extends its appreciation to the Deanship of Scientific Research at King Saud University for funding the work through the research group No. RGP-VPP-035.

References

1. Stolk, J.: A Computerized Pre-Clinical Test for Cemented Hip Prostheses Based on Finite Element Techniques. Thesis University of Nijmegen, The Netherlands With summary in Dutch, p. 192 (2002)
2. Brekelmans, W., Poort, H., Sloof, T.: A new method to analyse the mechanical behavior of skeletal parts. Acta Orthop. Scand. **43**, 301–317 (1972)
3. Oguz, K., Bulent, E.: The effects of static, dynamic and fatigue behavior on three-dimensional shape optimization of hip prosthesis by finite element method. Mater. Des. **28**, 2269–2277 (2007)
4. Bergmann, G., Deuretzbacher, G., Heller, M., et al.: Hip contact forces and gait patterns from routine activities. J. Biomech. **34**, 859–871 (2002)
5. Stolk, J., Verdonschot, N., Murphy, B.P., et al.: Finite element simulation of anisotropic damage accumulation and creep in acrylic bone cement. Eng. Fract. Mech. **71**(4–6), 513–528 (2004)
6. Huiskes, R.: Failed innovation in total hip replacement diagnosis and proposals for a cure. Acta Orthop. Scand. **64**, 699–716 (1993)
7. Stauffer, R.: Ten-year follow-up study of total hip replacement. J. Bone Joint Surg. **64A**, 983–990 (1982)
8. Gruen, T., McNeice, G., Amstutz, C.: Modes of failure of cemented stem-type femoral components. Clin. Orthop. Relat. Res. **141**, 17–27 (1979)
9. Jasty, M., Maloney, W., Bragdon, C.O., Connor, D., Haire, T., Harris, H.: The initiation of failure in cemented femoral components of hip arthroplasties. J Bone Joint Surg. **73-B**, 551–8 (1991)
10. Prendergast, P.: The functional performance of orthopaedic bone cement. Key Eng. Mater. **198–199**, 291–300 (2001)
11. Stolk, J., Verdonschot, N., Mann, K., et al.: Prevention of mesh-dependent damage growth in finite element simulations of crack formation in acrylic bone cement. J. Biomech. **36**, 861–871 (2003)
12. Zafer, A., Oguz, K., Kurtaran, H.: Static, dynamic and fatigue behavior of newly designed stem shapes for hip prosthesis using finite element analysis. Mater. Des. **28**, 1577–1583 (2007)
13. El'Sheikh, H., et al.: Finite element simulation of the hip joint during stumbling: a comparison between static and dynamic loading. J. Mater. Process. Technol. **143–144**, 249–255 (2003)

14. Osorovitz, P., Goutallier, D.: Résultats cliniques et radiographiques d'une série continue de 124 prothèses totales de hanche type Céraver-Ostéal. Revue de Chirurgie Orthopédique **80**, 305–315 (1994)
15. Nuño, N., Avanzolini, G.: Residual stresses at the stem–cement interface of an idealized cemented hip stem. J. Biomech. **35**, 849–852 (2002)
16. Bergmann, G., Deuretzbacher, G., Heller, M., et al.: Hip contact forces and gait patterns from routine activities. J. Biomech. **34**, 859–871 (2001)
17. Duda, G., et al.: Influence of muscle forces on femoral strain distribution. J. Biomech. **31**, 841–846 (1998)
18. Rohlmann, A., et al.: Finite element analysis and experimental investigation in a femur with hip endoprosthesis. J. Biomech. **16**, 727–742 (1983)
19. Paul, J.:Forces transmitted by joints in the human body. In: Proceedings of the Institution of Mechanical Engineers 181, London (1967)
20. Paul, J.: The analysis of forces transmitted by joints in the human body. In: Proceedings Fifth International Conference on Stress Analysis (1974)
21. Paul, J.: Force actions transmitted by joints in the human body. In: Proceedings of the Royal Society of Medicine 163–172 (1975)
22. Crowninshield, R., Johnston, R., Andrews, J., et al.: A biomechanical investigation of the human hip. J. Biomech. **11**, 75–85 (1978)
23. Crowninshield, R., Brand, R., Johnston, R.: The effect of walking velocity and age on hip kinematics and kinetics. Clin. Orthop. Relat. Res. **132**, 140–144 (1978)
24. Röhrle, H., Scholten, R., Sigolotto, C., et al.: Joint forces in the human pelvis-legskeleton during walking. J. Biomech. **17**, 409–424 (1984)
25. Brand, R., Pedersen, D., Davy, D., et al.: Comparison of hip force calculations and measurements in the same patient. J. Arthroplasty **9**, 45–51 (1994)
26. Pedersen, D., Brand, R., Cheng, C., Arora, J.: Direct comparison of muscle force predictions using linear and nonlinear programming. J. Biomech. Eng. **109**, 192–199 (1987)
27. Heller, M., Bergmann, G., Deuretzbacher, G., Urselen, D., et al.: Musculo-skeletal loading conditions during walking and stair climbing. J. Biomech. **34**, 883–893 (2001)
28. Morlock, M., Schneider, E., Bluhm, A., Vollmer, M., et al.: Duration and frequency of everyday activities in total hip patient. J. Biomech. **34**, 873–881 (2001)
29. Duda, G., Heller, M., Albinger, J., et al.: Influence of muscle forces on femoral strain distribution. J. Biomech. **31**, 841–846 (1998)
30. Lennon, A., Prendergast, P.: Development of a physical model of a cemented hip replacement for investigation of cement damage accumulation. J. Biomech. **31**, 129 (1998)
31. Harrigan, T., Kareh, J., O'Connor, D., Burke, D., et al.: A finite element study of the initiation of failure of fixation in cemented femoral total hip components. J. Orthopaed. Res **10**, 134 (1992)
32. Sahin, S., et al.: The influence of functional forces on the biomechanics of implant-supported prostheses a review. J. Dent. **20**, 271–282 (2002)
33. Koster, et al.: Endoscopy of the femoral canal in revision arthroplasty of the hip. Arch. Orthop. Trauma Surg. **119**, 245–252 (1999)
34. Merckx, D.: Les ciments orthopédiques dans la conception des prothèses articulaires. Biomécanique et biomatériaux, Cahiers d'enseignement de la SOFCOT, Expansion scientifique française **44**, 67–76 (1993)
35. Huiskes, R., Boeklagen, R.: Mathematical shape optimization of hip-prosthesis design. J. Biomech. **22**, 793 (1989)
36. El'Sheikh, H., MacDonald, B., Hashmi, M.: Finite element simulation of the hip joint during stumbling: a comparison between static and dynamic loading. J. Mater. Process. Technol. **143–144**, 249–255 (2003)

Biomechanical Analysis of Cardiological Guidewire Geometry Forming

Witold Walke and Joanna Przondziono

Abstract The chapter presents strength analysis of selected structural forms of currently used cardiologic guidewires for introduction of endocavital electrodes. The assumptions that were made in performed biomechanical analyses considered both, the technique of treatment as well as various shapes of guidewires. Calculations were made for two alternative geometrical forms: straight and conical, made of X10CrNi18-8 steel used for medical products. Finite-element method was used for the analysis of FEM analysis enabled to obtain information about places featuring the highest effort of the material during pre-surgery premodeling. That kind of information is useful for proper selection of structure, mechanical properties of metallic biomaterial, and it is also very important for proper design of its geometry and formation of physical and chemical characteristics of the upper layer.

Keywords X10crni18-8 steel · Cardiologic guidewire · Biomechanical analysis · FEM

1 Introduction

An issue of significance in the process of functional characteristics of cardiologic guidewires formation is selection of mechanical features of metallic biomaterial as well as physical and chemical characteristics of the product. Formation of functional form is made on the ground of its properly selected biomechanical

W. Walke (✉)
Silesian University of Technology, Gen.de Gaulle'a 66, Zabrze, Poland
e-mail: witold.walke@polsl.pl

J. Przondziono
Silesian University of Technology, Krasińskiego 8, Katowice, Poland
e-mail: joanna.przondziono@polsl.pl

A. Öchsner and H. Altenbach (eds.), *Design and Computation of Modern Engineering Materials*, Advanced Structured Materials 54,
DOI: 10.1007/978-3-319-07383-5_7,

characteristics, determined with reference to treatment technique. It results from the necessity to lead the tip of guidewire to the respective place in blood and vascular system. Properly performed treatment shall result in its permanent deformation, which assures precise placement of implant or electrode [1–6]. Due to the fact that experimental determination of mutual interaction of guidewires and blood and vascular vessels in in vivo tests is difficult, researchers concentrate on virtual tests with application of computer mechanics, and in particular—finite-element method. Applicability of that method is connected with the adopted assumptions that should reflect anatomical and physiological conditions of blood and vascular system. Quality of guidewire characteristics is influenced most of all by premodeling, i.e. initial shaping of its geometry, performed by the doctor prior to treatment. Additional influence is also exerted by blood vessels in which guide wire is dislocated. Its biomechanical properties are connected among other things with the course of disease. Therefore, for proper determination of biomechanical characteristics of cardiologic guidewire, it is also necessary to prepare a numerical model of blood vessel. Only when a complex model: guidewire—blood vessel, has been prepared, taking into consideration physical and geometrical nonlinearity of guidewire and biomechanical properties of blood vessel, can proper selection of biomechanical characteristics of guidewire be made, including phenomena taking place in the course of the process [7, 8].

Many years of clinical practice enabled to determine geometrical features of various forms of guidewires that capacitate proper performance of cardiologic treatment. Not always were geometrical solutions as well as suggestions arising from location of implants supported with biomechanical analyses. Many studies also do not highlight the role of metallic biomaterial surface treatment, which is of great importance as far as corrosion resistance is concerned and which minimises blood coagulation. Explanation of those issues creates the conditions for optimum formation of functional properties, so optimum geometrical features of guidewires, spatial configuration and also their mechanical properties [9].

Against the background of unsolved problems, the chapter focused on biomechanical evaluation of the system: guidewire—blood vessel, with assumed and clinically recommended geometrical features. Properties of biomaterial were determined on the ground of biomechanical analysis.

Biomechanical analysis was made in order to determine the condition of stress, dislocation and strain of cardiologic guidewire used for electrode insertion. Obtained results are significant as far as the selection of structure and mechanical properties of metallic biomaterials used for guidewires is concerned. Due to the process of premodeling, condition of stresses initiated by plastic strain has a significant effect on the change of its geometrical features. Values of stresses that takes place in various parts of guidewire depending on deflection angle are crucial for proper design of its geometry, work hardening of biomaterial and formation of physical and chemical properties of the surface layer.

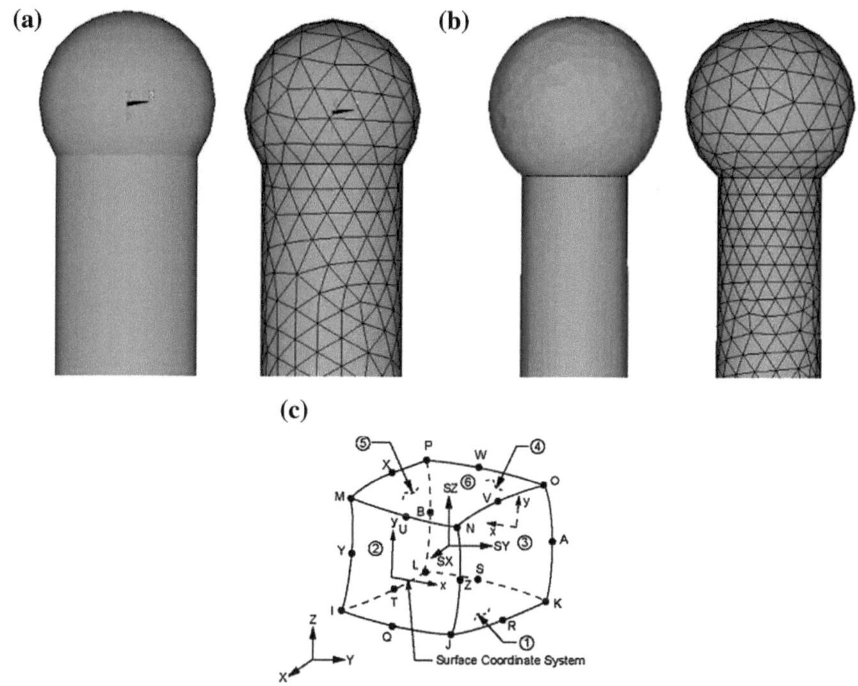

Fig. 1 Calculation model of guidewire **a** straight, **b** conical, **c** element of SOLID186 type

2 Materials and Methods

The main purpose of the chapter was strength analysis of guidewire used for implantation of atrial or ventricular endocavital electrodes. Straight guidewire with diameter of d = 0.35 mm and conical ($\varphi = 5^{\circ}$) with ball tip and operating part length of l = 400 mm were selected for the analysis.

Geometrical model of analysed structural form of guidewire was prepared with application of Inventor software. Both, its geometrical features and typical structural features which distinguish it from other forms of wire, were taken into consideration in preparation of guidewire model. Proper profile and shape of the tool tip were maintained—Fig. 1a and b.

For the purpose of performed analyses, premodeling was simulated in the form of the system of three rollers with diameter d = 5 mm. Thus, three supporting points of the tool were obtained—Fig. 2. Thus, suggested system enabled to apply variable angle and radius of bend.

On the ground of prepared geometrical models, a finite-element mesh was generated for calculations with application of FEM. Discretisation of models of both, wire and the three-roller model was made with application of a finite element of SOLID186 type—Fig. 1c.

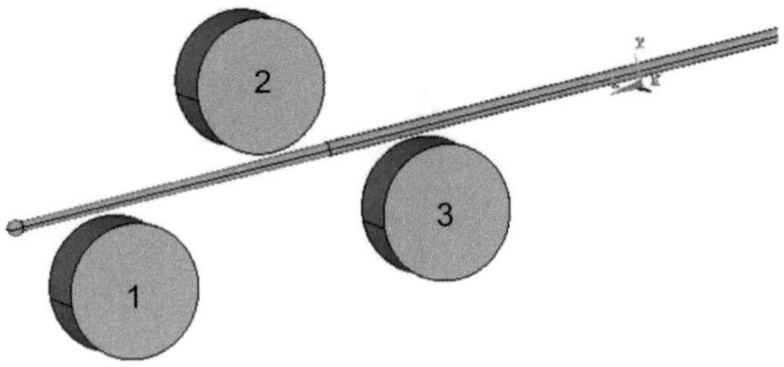

Fig. 2 Model of three rollers system which enables to simulate differentiation of bend angle

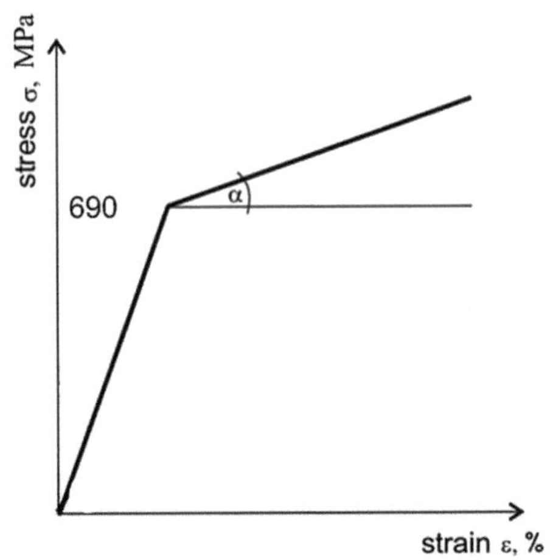

Fig. 3 Bilinear characteristics for X10CrNi18-8 steel

The scope of performed analysis covered determination of the condition of stresses generated by the applied angular displacement at the tip of wire operating part within the range $\varphi = 0 \div 120°$ for two geometrical forms of guidewire—straight and conical.

For the purpose of calculations, material data corresponding to X10CrNi18-8 steel was adopted: E = 205,000 MPa, Poisson's ratio $\nu = 0.3$, $R_m = 1{,}010$ MPa, $R_{p0,2} = 690$ MPa. For the purpose of calculations, bilinear characteristics of elastic—plastic material with isotropic hardening were made—Fig. 3.

In order to make necessary calculations, it was indispensable to determine and establish initial and boundary conditions which would reflect phenomena taking place in the real system with proper precision. The following assumptions were made for the purpose of analysis:

- all nodes belonging to roller 2 were subject to relocation along the wire axis, which enabled positioning of places where wire was bent from its tip,
- deviation of guidewire tip within angular range of $\varphi = 1 \div 120°$ was accomplished through application of relocation of roller 3 in the direction perpendicular to the axis of the tool,
- contact between surfaces and edges of elements from the system, which enabled their mutual relocation, was defined.

Table 1 Results of numerical analysis of straight and conical guidewire for differentiated values of angular displacement φ

Type of guidewire	Reduced stresses σ_{max}, MPa			
	30°	60°	90°	120°
Straight	252	504	726	859
Conical	272	545	755	907

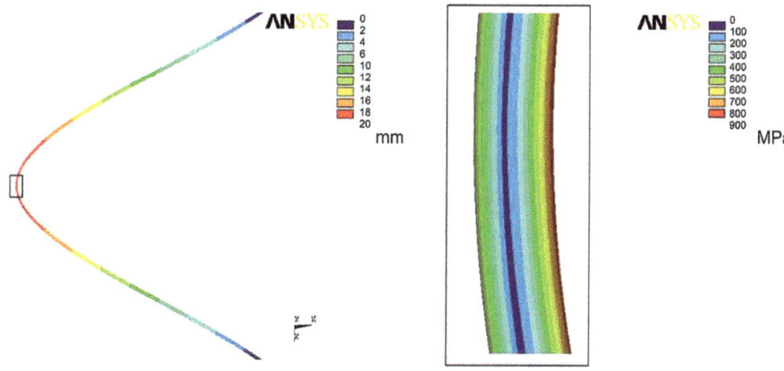

Fig. 4 Results of strength analysis for bend angle $\varphi = 120°$: **a** state of dislocations u, mm, **b** state of reduced stresses σ_{max}, MPa

All values of strain and stress are values reduced in accordance with Huber—Mises theory.

3 Results and Discussion

The first stage of performed analyses included determination of the influence of angular displacement φ on distribution of stresses generated in the tip of operating part of guidewire for straight wire—Table 1. Exemplary distribution of dislocations and stresses, obtained as the result of bending at the angle of $\varphi = 120°$, is presented in Fig. 4.

Results of performed analyses show that the biggest values of reduced stresses for the respective values of angular displacement φ were observed in the area of direct impact of roller 2 on wire operating part. Maximum values of reduced stresses were located on the internal side of bent tool area.

The second stage of the analysis was aimed at determination of the influence of guidewire geometry changes (from straight to conical) on stresses generated during its bending. Obtained results showed little increase of maximum stress values for the same bend angles in relations to straight wires—Table 1. Maximum values of reduced stresses were located on the internal side of bent guidewire area, the same as for straight wires.

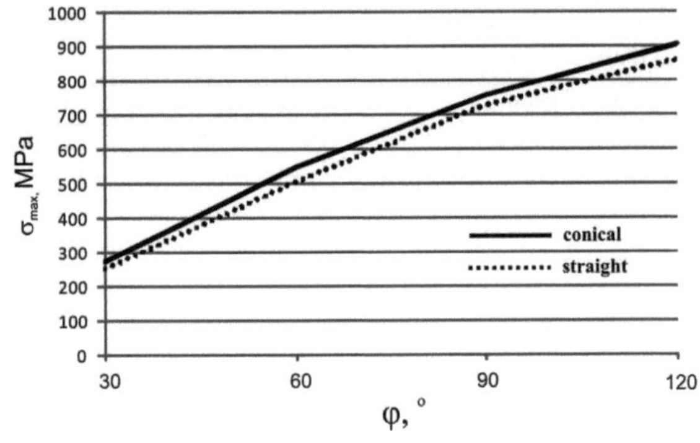

Fig. 5 Relation between maximum reduced stresses as the function of angular displacement φ for straight and conical guidewire

Obtained results additionally created the ground for determination of biomechanical characteristics of two different forms of guidewire, specifying the relation between maximum reduced stresses generated in its operating part as the function of angular displacement for differentiated values of bend angle—Fig. 5.

4 Conclusions

One of the main problems connected with application of metallic cardiologic guidewires is the possibility of their initiation of coagulation process. The main way how to limit that unfavourable phenomenon is application of atrombogeneous coatings on its surface. These coatings should, apart from featuring good adhesion to metallic substrate, be susceptible to strain conditioned by pre-surgery forming of guidewire tip. Therefore, values of stresses and strain in the respective parts of guidewire, determined in the chapter, may be useful in order to form proper mechanical properties of the surface layers and create the ground for evaluation of their susceptibility to strain.

In conclusion, numerical analysis made with application of finite-element method for cardiologic guidewire showed that its production of X10CrNi18-8 steel, with the assumed for calculations set of mechanical properties, secure proper course of pre-surgery forming, and in consequence its permanent deformation at the required angle. Such conditions guarantee proper placement of electrodes in heart wall, which has a crucial impact on surgery efficiency.

Acknowledgements This project was financed from the funds of the National Science Centre in Cracow, Poland.

References

1. Vardas, P., Auricchio, A., Merino, J.: The current status of cardiac electrophysiology in ESC member countries. EHRA White Book **2011**, 327–337 (2011)
2. Gałecka, J., Prochaczek, F., Gacek, A., Kargul, L., Wojciechowski, D.: Diagnostyczna i terapeutyczna elektrostymulacja oraz defibrylacja serca. Biocybernetyka i Inżynieria Biomedyczna 2000. Warszawa (2001)
3. Marciniak, J.: Biomateriały, Wydawnictwo Politechniki Śląskiej. Gliwice (2013)
4. Pęczalski, K. Wybrane metody diagnostyczne wykorzystywane w elektroterapii serca. Warszawa (2010)
5. Kargul, W., Młynarski, R., Piłat, E.: Implantowanie stymulatorów serca i kardiowerterów-defibrylatorów. Chir. Pol. **7**, 11–15 (2005)
6. Wieczorski, M., Stodółkiewicz, E.: Aktualne rozwiązania w zakresie defibrylacji i kardiowersji. A Bio-Optica et Inf Med **15**, 1–6 (2009)
7. Schneider, P.: Endovascular Skills. Taylor & Francis, New York (2009)
8. Hansen, D.: Metal corrosion in the human body: the ultimate bio-corrosion scenario. Electrochem. Soc. Int. **17**, 3–7 (2003)
9. Nam, N.D., Lee, S.H., Kim, J.G., Yi, J.W., Lee, K.R.: Effect of stress on the passivation of Si-DLC coating as stent materials in simulated body environment. Dia Rel Mater. **18**, 1145–1151 (2009)

Scanning Method of Temperature Distribution of Human Body by Device Registering Encircling Images

Mariusz Ciesielski, Miroslaw Dziewonski and Sebastian Freus

Abstract This chapter is related to a scanning method for measuring a skin surface temperature distribution of a patient's body with a burn wound. Results can be used for medical diagnostics in order to evaluate the skin burn depth. Within the project entitled "The diagnostic system supporting the healing process of burn and chronic wounds" funded by The Polish National Centre of Research and Development, the diagnostic device for recording thermal images was created. The chapter concerns the description of merging algorithm of these images which is based on the mathematical formulas. In the final part of this chapter, some simulation results are shown.

Keywords Thermography · Medical imaging · Wounds healing · Diagnostic system

1 Introduction

This chapter is related to a scanning method for measuring the skin surface temperature distribution of a patient's body. The temperature distribution of a skin tissue with a burn wound can be used for medical diagnostics. In emergency medicine occurs a problem of the correct evaluation of skin burn depth. Numerous

M. Ciesielski (✉) · S. Freus
Institute of Computer and Information Sciences, Czestochowa University of Technology, Dabrowskiego 69, 42-200 Czestochowa, Poland
e-mail: mariusz.ciesielski@icis.pcz.pl

S. Freus
e-mail: sebastian.freus@icis.pcz.pl

M. Dziewonski
Institute of Computational Mechanics and Engineering, Silesian University of Technology, Konarskiego 18a, 44-100 Gliwice, Poland
e-mail: miroslaw.dziewonski@polsl.pl

A. Öchsner and H. Altenbach (eds.), *Design and Computation of Modern Engineering Materials*, Advanced Structured Materials 54,
DOI: 10.1007/978-3-319-07383-5_8,

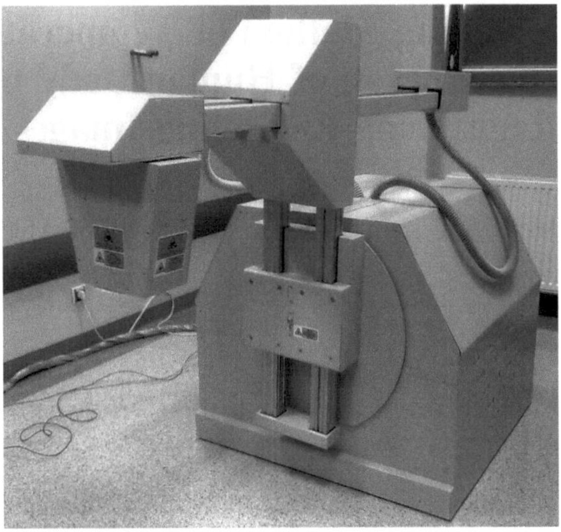

Fig. 1 Device registering encircling images

observations confirm the apparent relations between skin surface temperature and the shape of a burn wound. The tracing of temperature changes allows one to analyze the process of wound healing and a therapy choice by doctors [1, 2]. The surface temperature distribution can be obtained especially using the thermography methods, because the modern infrared thermal cameras produce images of high resolution with good precision and are easy to use. Thermal imaging provides non-invasive, correctly determine temperature differences of skin surface around burn wound.

Within the Project entitled "The diagnostic system supporting the healing process of burn and chronic wounds" (funded by The Polish National Centre of Research and Development: NR-13-0124-10/2010), the diagnostic device for recording thermal and video images has been constructed [3]—see Fig. 1. The device registering encircling images consists of, among others, a moveable outrigger with the device head (inside the thermal and video cameras, rangefinders) and image recording device. Rangefinders are coupled with the camera and allow ones to measure the distance to the patient's body in order to ensure a good diopter adjustment (preservation of the image sharpness).

The idea of recording the encircling thermograms lies in the fact that the infrared camera records the temperature of points on the body within a strip of length L, while the recording will be conducted continuously at a specified frequency, when the camera moves around at a constant angular velocity. In this way the camera registers one strip of the body. After recording the strip with a width L, the camera moves linearly by the same section with a necessary overlap and the system records another body strip in the opposite direction (see Fig. 2). During the scan, the registration of the images in certain places of the head positions takes the place—details are in Sect. 2.

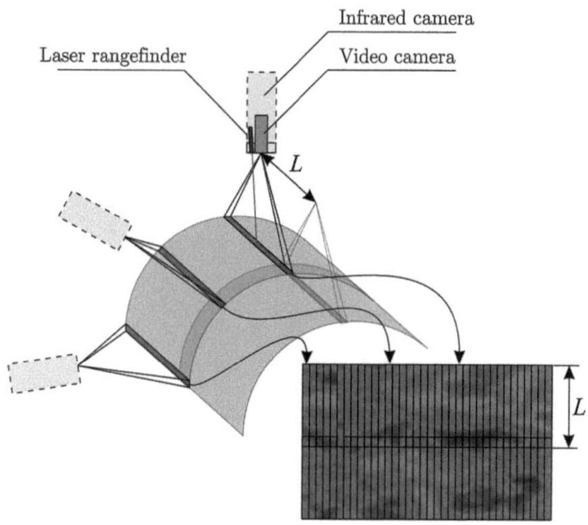

Fig. 2 Visualization of the idea of recording encircling thermograms

The data collected in this way are formed in a thermal image which corresponds to a temperature distribution field projected to a plane. Both the angle of camera movement and the total length of linear displacement are adjustable.

The first operation of the device after power on is setting the device head to the *HOME* position and reset the coordinates of all axes. For this purpose, the head is moving in three axes simultaneously. Next, the operator selects the scan-starting point on the patient's skin surface by moving the head using the buttons on the system interface. The control is done in a manual or automatic way. After the start of the scan, the head moves in a rotational motion until the end point is indicated by the operator (the scan of one strip) and then the head moves linearly with a fixed distance to scan another strip at the patient's body in the opposite direction. These steps are repeated until the end of the scan is indicated by the operator. Depending on the scanning mode, the movement of the head device can be done in reverse order, i.e. the first has linear movement, rotational then. Described motion trajectory of the head is in a schematic way shown in Fig. 3 is presented. After the scanning process, the head returns to the *HOME* position.

During the scanning process, the constant distance $\overline{dk}$ of the head to the body of the patient should be kept constant, which provides sharper images to be captured by both cameras. It is associated with the installation of three rangefinders in the head and the ability of controlling the height of the head.

To control of device, a lot of configuration parameters are used. It should also be taken into account and choose the optimal parameters of the servo, i.e. acceleration and speed of the drives in order to minimize vibration and improving the quality of registered images, while providing a reasonable scan time. In the diagnostics device control software has been used many solutions in order to improve of performance (i.e. independent threads: for servo controlling, reading data from rangefinders, and recording thermal images and video) [3].

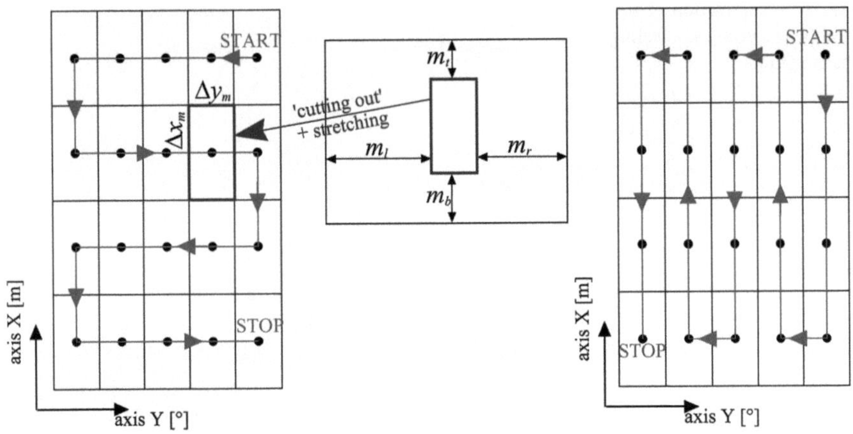

Fig. 3 Motion trajectory of the device head: rotational–linear and linear–rotational

2 Image Merging Algorithm

In order to obtain the thermographs of the surface temperature of extended burns, an algorithm of image merging is proposed. This algorithm is based on mathematical formulas that allow ones the determination of the crop size of each single image and finally the merging of all image parts together.

During scanning a selected area of the patient's body surface with burn wound, infrared and video images in the device are recorded. Images are recorded in certain discrete positions of the device head whose have been determined by the scanning algorithm. It is assumed that the step length of position head is fixed for both axes (Δx [m] for axis X, and Δy[°] for axis Y).

Before scanning, the following data should be set:

- the scan mode: rotational–linear or linear–rotational,
- the patient data, sequential number and scan date of the medical examination, the operator data, etc.,
- the geometrical position of the head of device at the selected initial scanning point: x_{init}, y_{init},
- number of scanned steps (N_x and N_y) in each direction (in order to determine of the scanned image number),
- description of the area scan (it may be supplemented after scan).

During the recording of video and thermal images in the device, for each recorded image (with indexes i and j), the following information are stored:

- the current head position (three coordinates): $x_{i,j}$, $y_{i,j}$ (or $\phi_{i,j}$), $z_{i,j}$,
- the current distance from central rangefinder to the body of the patient: $dk_{i,j}$,
- scale parameters for image,

Fig. 4 Schemas of head positions in rotational and linear motions

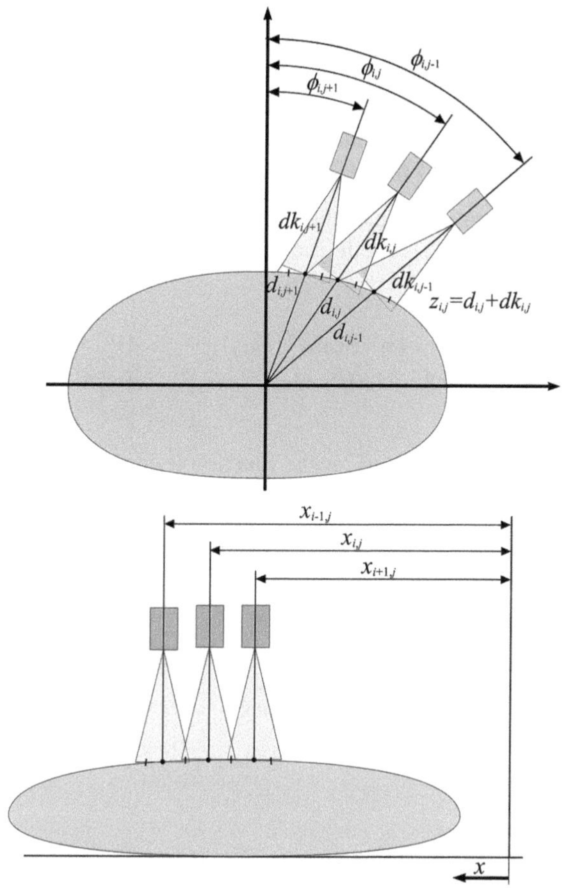

- image data in the FPF file format (FLIR System)—in the case of the infrared camera,
- image data in the standard BMP file format—in the case of the video camera.

The IR camera saves a temperature measurement array (the FPF file format). The actual temperature measurements can be brought into a PC for analysis. Software allows doing things like measure temperature points, gradients, differences, averages, assigning color palettes. After scanning, the obtained data (set of thermal and video images) are merged in the computer system.

Schemas of head positions in rotational and linear motions are presented in Fig. 4.

One can note that the distance $\overline{dk}$ of the head to the body of the patient during scanning is not always constant—because it is associated with a continuous motion of the head and insufficiently fast correction of the head position in the z–coordinate. Deviations from the intended distance were low and the images were recorded at a depth of field resulting from the optical properties of both cameras. In a similar way, this problem occurs when the images are acquired from both cameras at the determined position by the algorithm. Also, the movement of the chest during breathing causes a change in the distance between the device head and the scanned surface even by a few centimeters. For this purpose, some geometric corrections during merging images must be taken into consideration.

In order to merge images, the following algorithm is proposed:

1. determine size of the field of view of both cameras on the base of the distance rangefinder (both cameras assure single image with resolution 640 × 480 pixels)

$$\begin{aligned} \Delta x_{i,j}^{IR} &= 480/\mathrm{Scale}^{IR}\left(dk_{i,j}\right), & \Delta y_{i,j}^{IR} &= 640/\mathrm{Scale}^{IR}\left(dk_{i,j}\right),\ [\mathrm{m}] \\ \Delta x_{i,j}^{video} &= 480/\mathrm{Scale}^{video}\left(dk_{i,j}\right), & \Delta y_{i,j}^{video} &= 640/\mathrm{Scale}^{video}\left(dk_{i,j}\right),\ [\mathrm{m}] \end{aligned} \tag{1}$$

   where Scale (dk) is the function used to determine number of image pixels per 1 meter for the rangefinder distance dk.

2. determine the average size [pix] of the image fragments

$$\begin{aligned} \Delta x_{avg,pix}^{IR} &= \left\lfloor \Delta x_{avg}\ \mathrm{Scale}^{IR}\left(\overline{dk}\right) + 0.5 \right\rfloor, \\ {}_{avg,pix}^{IR} &= \left\lfloor \Delta y_{avg}\ \mathrm{Scale}^{IR}\left(\overline{dk}\right) + 0.5 \right\rfloor, \\ \Delta x_{avg,pix}^{video} &= \left\lfloor \Delta x_{avg}\ \mathrm{Scale}^{video}\left(\overline{dk}\right) + 0.5 \right\rfloor, \\ \Delta y_{avg,pix}^{video} &= \left\lfloor \Delta y_{avg}\ \mathrm{Scale}^{video}\left(\overline{dk}\right) + 0.5 \right\rfloor \end{aligned} \tag{2}$$

   where

$$\begin{aligned} &\Delta x_{avg} = \Delta x, \quad \Delta y_{avg} = 2\ d_{avg} \tan\left(\frac{\Delta y}{2}\frac{\pi}{180}\right), \\ &d_{avg} = \frac{1}{N_x N_y}\sum_{i=1}^{N_x}\sum_{j=1}^{N_y} d_{i,j}, \quad d_{i,j} = z_{i,j} - dk_{i,j}\ [\mathrm{m}] \end{aligned} \tag{3}$$

3. calculate the crop margins (m_l, m_r, m_t, m_b) (in [pix]) (see Fig. 2) of each recorded image (i, j) on the base of mathematical formulas which take into account the scale and offset correction for position of the head in the registration point $(x_{offset}, y_{offset}) = (x_{init} - i\,\Delta x - x_{i,j}, y_{init} - j\,\Delta y - y_{i,j})$ (it is related to delays resulting from the continuous motion of the head and its real position in which the image was captured)

$$
\begin{aligned}
m_l^{IR} &= \left\lfloor \left(\frac{\Delta y_{i,j}^{IR}}{2} - d_{i,j} \tan\left(\frac{\Delta y}{2} \frac{\pi}{180} \right) + d_{i,j} \tan\left(y_{offset} \frac{\pi}{180} \right) \right) \text{Scale}^{IR}(dk_{i,j}) + \frac{1}{2} \right\rfloor \\
m_r^{IR} &= \left\lfloor \left(\frac{\Delta y_{i,j}^{IR}}{2} - d_{i,j} \tan\left(\frac{\Delta y}{2} \frac{\pi}{180} \right) - d_{i,j} \tan\left(y_{offset} \frac{\pi}{180} \right) \right) \text{Scale}^{IR}(dk_{i,j}) + \frac{1}{2} \right\rfloor \\
m_t^{IR} &= \left\lfloor \left(\frac{\Delta x_{i,j}^{IR}}{2} - \frac{\Delta x}{2} - x_{offset} \right) \text{Scale}^{IR}(dk_{i,j}) + \frac{1}{2} \right\rfloor \\
m_b^{IR} &= \left\lfloor \left(\frac{\Delta x_{i,j}^{IR}}{2} - \frac{\Delta x}{2} + x_{offset} \right) \text{Scale}^{IR}(dk_{i,j}) + \frac{1}{2} \right\rfloor
\end{aligned}
\quad (4)
$$

In a similar way, the margins of the video image are determined.

4. cut out a necessary part of every image and next merge those pieces together (using stretching to the size: $\Delta x_{avg,pix}^{IR} \times \Delta y_{avg,pix}^{IR}$ or $\Delta x_{avg,pix}^{video} \times \Delta y_{avg,pix}^{video}$, respectively—if needed) into a single thermovision image and a single video image,
5. approximate the shape of the scanned area of the patient and 3D visualization of scan images (optionally).

As a result of this algorithm, the field of the temperature distribution over a large area of the body and the visual image on a two dimensional plane are obtained. In order to visualize the thermograph, a color palette is assigned. In Fig. 5 several color palettes are shown. For example, the rainbow palette has the best thermal sensitivity for displaying the differences in temperature of the human skin. Merged images will be processed and analyzed in the further project stages of diagnostic system. The mathematical models of the burn wound shape estimation on the basis of the distribution of skin surface temperature are proposed in [4, 5].

3 Example of Results

In order to verify the correctness of the proposed algorithm, a torso (chest) of a healthy male without burn wounds has been scanned. During the scan, 732 (12 × 61) images have been recorded. The field of view of both cameras (the rangefinder distance: 350 mm) are the following:

- for the IR camera: 115.94 × 86.96 mm,
- for the video camera: 147.47 × 110.60 mm.

As a result of the merging algorithm, two bitmap images were obtained. The resolution of the IR image is 2318 × 1656 pixels, and the resolution of the video image is 1830 × 1308 pixels. In Fig. 6 these images are shown. Additionally, in Fig. 7 a spatial visualizations (3D) of both images on the base of the geometric data are presented.

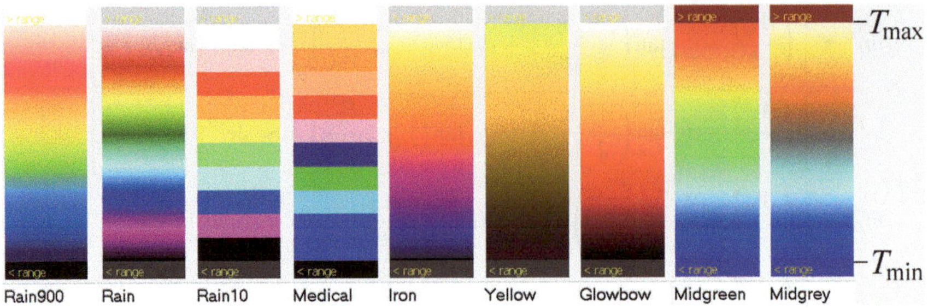

Fig. 5 Color palettes

Fig. 6 Recorded field of the temperature distribution (using color palette) and the video image of the body on a two dimensional plane

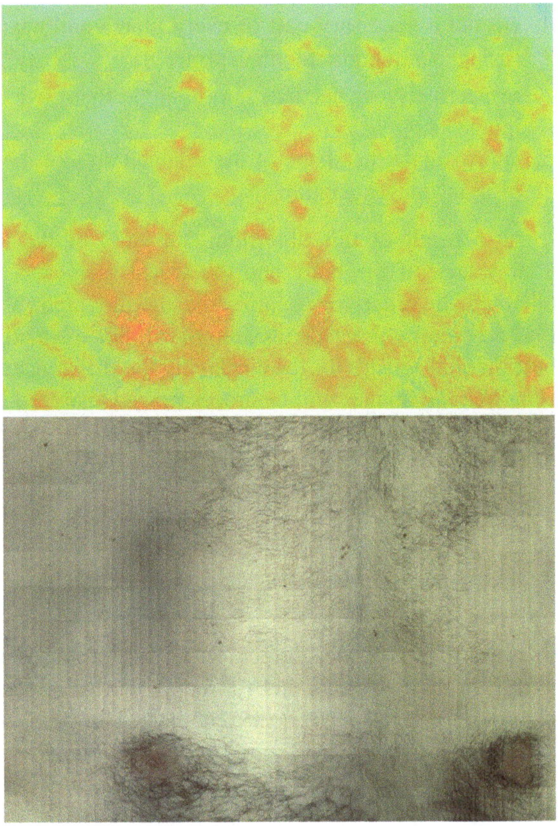

The next example presents the wound on the patient's back. The recorded field of the temperature distribution and the video image of the body on a two dimensional plane are shown in Fig. 8.

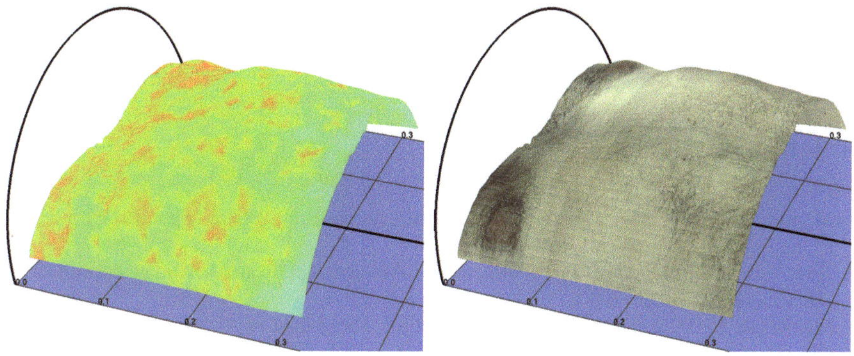

Fig. 7 Recorded field of the temperature distribution (using color palette) and the video image of the body—spatial visualization

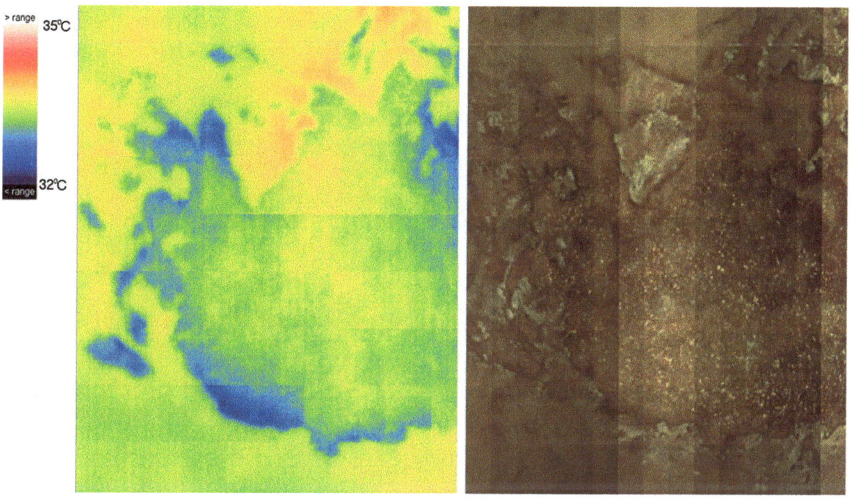

Fig. 8 Merged images (thermovision and video) with the wound on the patient's back

4 Conclusions

In the chapter, the algorithm of the encircling thermograms creation has been presented. This algorithm is part of a system supporting the treatment of the burns and wounds difficult to heal. The diagnostic system developed by the authors consists of a device for recording encircling thermal and video images and a computer system which constitutes the completely new tool intended to assist the diagnostics and the treatment process of burn and chronic wounds. A significant increase in the accuracy of measurement of the temperature field in comparison to the classic flat thermograms is observed. During further project stages, the usability of the system in clinical applications will be verified.

Acknowledgments The chapter and research were financed within the project N R13 0124 10 sponsored by Polish National Centre for Research and Development.

References

1. Monstrey, S., Hoeksema, H., Verbelen, J., et al.: Assessment of burn depth and burn wound healing potential. Burns **34**, 761–769 (2008)
2. Nowakowski, A.: Advances in Thermography—Medical Applications. Wydawnictwo Gdanskie, Gdansk (in Polish) (2001)
3. Majchrzak, E., Dziewonski, M., Nowak, M. et al.: The design of a system for assisting burn and chronic wound diagnosis. In: Pietka, E., Kawa, J. (eds) Information Technologies in Biomedicine 2012, Lect. Notes Comput. Sci. **7339**, 110–117 (2012)
4. Majchrzak, E.: Determination of burn depth on the basis of skin surface temperature—solution of inverse problem using the gradient method. In: 10th World Congress on Computational Mechanics, 8–12 July 2012, Sao Paulo, Brasil, CD ROM Proceedings:1–12 (2012)
5. Ciesielski, M., Mochnacki, B.: Numerical analysis of interrelations between skin surface temperature and burn wound shape. Sci. Res. Inst. Math. Comput. Sci. **1**(11), 15–22 (2012)

Strain Measurements Exhibited by a Steel Prosthesis Protected with Au Nanoparticles

Luis Héctor Hernández-Gómez, Joel Francisco Pava-Chipol, Martín Trejo-Valdez, Carlos Torres-Torres, Juan Alfonso Beltrán Fernández, Guillermo Urriolagotia-Sosa, Christopher René Torres San Miguel and Guillermo Urriolagotia-Calderón

Abstract The design and manufacturing of a customized femur prosthesis is presented. Besides, the application of a coating of Au nanoparticles embedded in titanium dioxide is evaluated. For this purpose, a customized femur prosthesis for a Labrador Retriever dog was manufactured. It was 7 years old and its weight was 35 kg. The main geometrical characteristics of the prosthesis are the following. Its femoral neck angle is 75.84°, the smallest and largest diameter of the shaft is 2.7 and 6 mm, respectively, while the diameter of the femur head is 20 mm. The material used for this purpose was stainless steel. Initially, a tomographic study of

L. H. Hernández-Gómez (✉) · J. F. Pava-Chipol · C. Torres-Torres · J. A. B. Fernández · G. Urriolagotia-Sosa · C. R. T. S. Miguel · G. Urriolagotia-Calderón
National Polytechnic Institute, ESIME-ZACATENCO, Section of Postgraduate Studies and Research, Unidad Profesional Adolfo López Mateos, Edificio 5, 3er Piso Colonia Lindavista, Gustavo A. Madero, 07738 Mexico, DF, Mexico
e-mail: luishector56@hotmail.com

J. F. Pava-Chipol
e-mail: joel.1286@gmail.com

C. Torres-Torres
e-mail: crstorres@yahoo.com.mx

J. A. B. Fernández
e-mail: jbeltranf@hotmail.com

G. Urriolagotia-Sosa
e-mail: guiurri@hotmail.com

C. R. T. S. Miguel
e-mail: napor@hotmail.com

G. Urriolagotia-Calderón
e-mail: urrio332@hotmail.com

M. Trejo-Valdez
National Polytechnic Institute, ESIQIE. Unidad Profesional Adolfo López Mateos, Edificio 6, Colonia Lindavista. Gustavo A. Madero, 07738 Mexico, DF, Mexico
e-mail: esiqie.fqunam@yahoo.com

A. Öchsner and H. Altenbach (eds.), *Design and Computation of Modern Engineering Materials*, Advanced Structured Materials 54,
DOI: 10.1007/978-3-319-07383-5_9,

the hip and femur was carried on. All the data was collected in .DICOM files. From this information, solid models were obtained. The new data was saved in .STL files. From such files, quick prototypes were carried out. They were made with ABS material, following a stereolithography procedure. All the dimensions were checked. Once the customized model was approved, the structural integrity was determined with the finite element method. In the next step it was manufactured. In the final part, a 250 nm coating, which was made of Au nanoparticles embedded in titanium dioxide, was applied. An indentation test was carried out. A 65° Berkovich indenter, made with diamond, was used. The modulus of elasticity of the coating was 941 MPa. The stress field during the indentation test was obtained.

Keywords Hip · Femur prosthesis · Nanostructured thin solid film coating · Canine femur · Computed tomography

1 Introduction

Prostheses required for humans must satisfy diverse requirements. In general terms, they have to be manufactured with biocompatible materials and they must have an adequate structural integrity. In this way, infections or replacements after small periods are avoided. Regarding the hip prosthesis, the femoral head is subjected to wear conditions. All these points have to be considered in its design. Also, clinical regulations have to be observed. In the case of Mexico, they are established by the medical regulatory body, COFEPRIS. They are summarized in [1].

Actually, hip prostheses have received much attention. They have been widely used in older patients, mainly. Nowadays, they are optimized continuously. Thus, diverse prototypes are developed. Several tests have to be approved before such prostheses can be used in a clinical treatment. For this reason, they are tested with some animals. Several candidates can be proposed. At the first instance, primates can be used. However, they require special cares and there are some species, which are in danger of extinction [2, 3]. In accordance with Skurla et al. [4], dogs can be used for this purpose. Goel and coworkers [5] have established the following reasons why dogs are adequate. Their femoral anatomy is similar to the human beings and their femur size is appropriate for the technique which is applied in the total reconstruction of a hip. The canine vascular anatomy is similar to the humans. Besides, dogs are sufficiently active for the use of a prosthesis and the kinematic articulations of their hip are similar to those of a human hip. However, the main disadvantage is that dogs are quadrupeds.

Accordingly, the purpose of this chapter is the design of a prototype of a customized hip prosthesis for a dog. After the design was approved, the femoral head was coated with Au nanoparticles embedded in titanium dioxide. For this reason, the mechanical behavior of the coat was evaluated with indentation tests in conjunction with the finite element method.

2 Materials and Methods

This research work was carried on with a Labrador Retriever dog. It was 7 years old and its weight was 35 kg (Fig. 1). In the case of dogs, dysplasia takes place when they are more than 6 years old. This is the reason why such a dog was selected.

Initially a model of the canine femur was developed. The procedure proposed by Beltrán-Fernández et al. [6] was followed. For this purpose, a Lightspeed computational tomograph was used. It can take 16 slices per second. For the problem at hand, 513 slices of the canine femur were taken. All the data was digitalized and saved in DICOM® files [7]. Figure 2 shows the 3D model that was obtained.

In the next step, the ScanIp® code was used. The tomographies mentioned above, were reproduced [8]. Masks were developed. They were used in the selection of the area of interest of the femur. Cortical (Fig. 3) and trabecular (Fig. 4) bone were differentiated. The first one is in blue, while the second is in purple. All the data was collected in *.STL files. The design and manufacture of a customized prosthesis require uniform and smooth surfaces, avoiding sharp edges. Thus, the surface of the model was reshaped with the Power Shape® code.

In the next step, plastic prototypes of the prosthesis of the canine femur were obtained. The clinical treatment was discussed with the orthopedic team, considering such prototypes. It was decided which part of the femur had to be cut. Therefore, it was possible to generate the prototype of the femur-prosthesis system. For this purpose, the axial axis of the diaphysis was localized. This parameter is important in order to establish the maximum and minimum diameters of the stem of the prosthesis. They are 6 and 2.7 mm, respectively. This information was used to establish the length of the stem and to determine which surfaces of the stem had to be anchored. In this way, the prosthesis does not loose easily. Also, the angle between the femoral head and the axial axis of the diaphysis and the diameter of the femoral head were established. They are 75.84° and 20 mm, respectively. Finally, the good adjustment of both elements was checked. Figure 5 illustrates how this prosthesis is coupled with the femur and Fig. 6 shows the solid model of the prosthesis.

3 Finite Element Analysis

The finite element mesh was developed from the solid model of the system femur-prosthesis (Fig. 7). For the evaluation of the structural integrity, it was considered that the prosthesis was made of stainless steel 316LVM. Its modulus of elasticity and Poissońs ratio are 210 GPa and 0.3, respectively [9]. Besides, the modulus of elasticity of the cortical bone of the femur is 10 GPa and its Poissońs ratio is 0.3. The modulus of elasticity of the trabecular bone is 1 GPa and its Poissońs ratio is 0.3 [10–12].

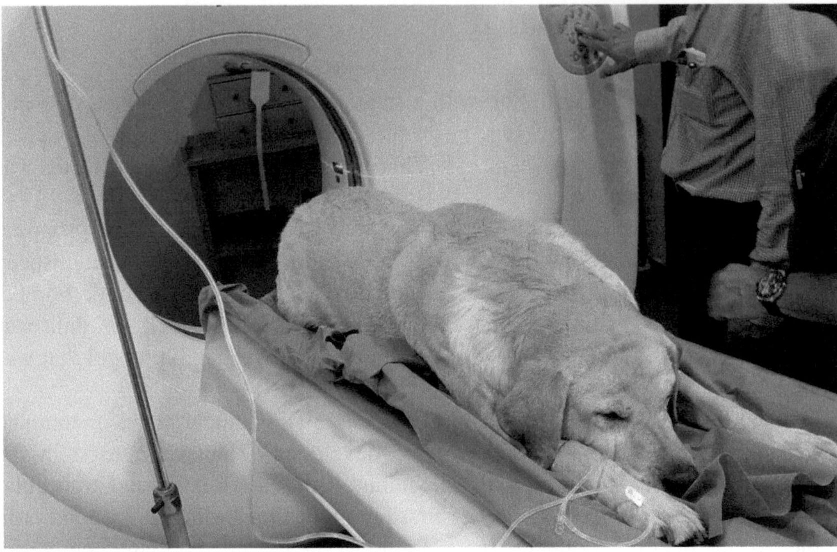

Fig. 1 Retriever dog before its femur tomography was taken

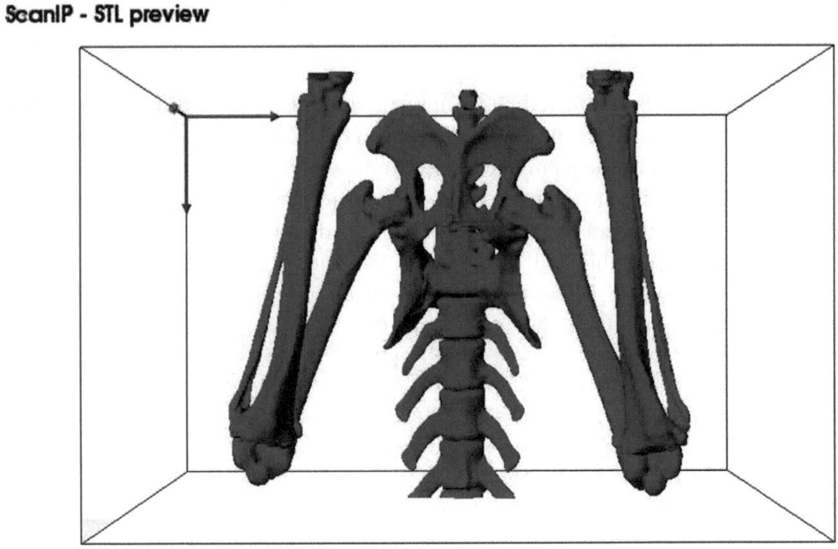

Fig. 2 3D model of the rear femurs of the dog

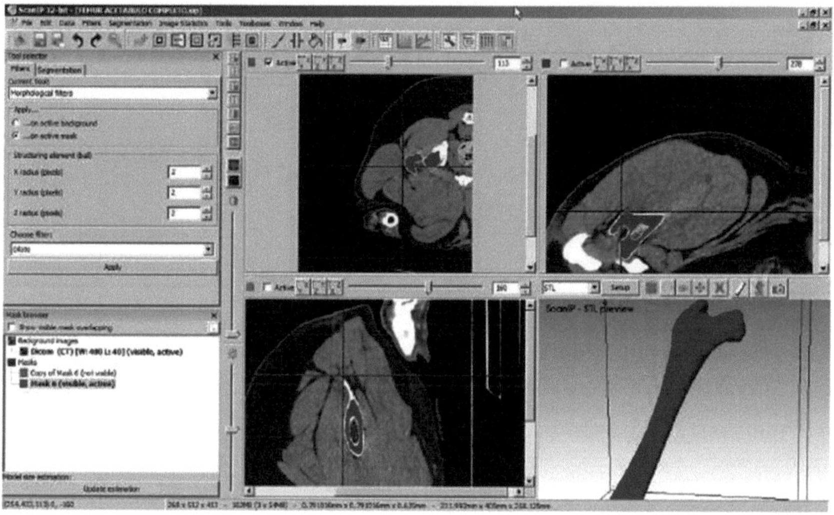

Fig. 3 3D model of the cortical bone of the canine femur

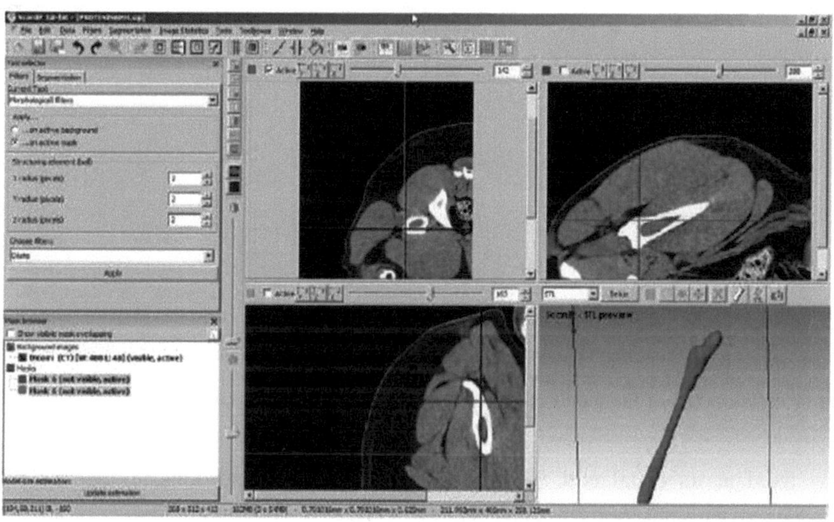

Fig. 4 3D model of the trabecular bone of the canine femur

The finite element analysis was carried out with the ANSYS code and the mesh was developed with quadratic tetrahedrons of ten nodes. Each node has three degrees of freedom. The numbers of elements and nodes are 59,827 and 75,039, respectively (Fig. 8). For the purpose of this analysis, the stem of the prosthesis

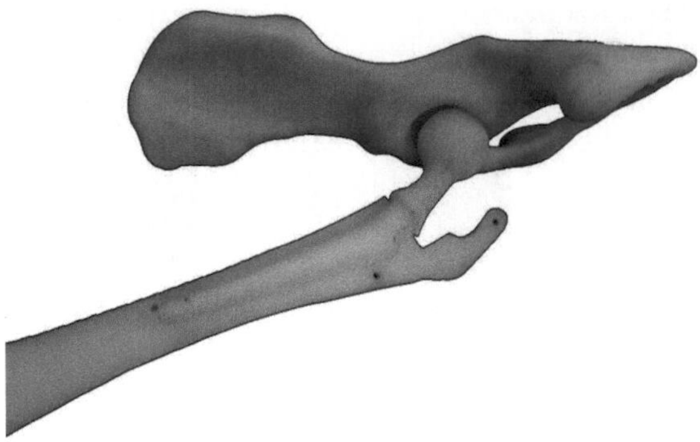

Fig. 5 System femur-prosthesis coupled with the hip

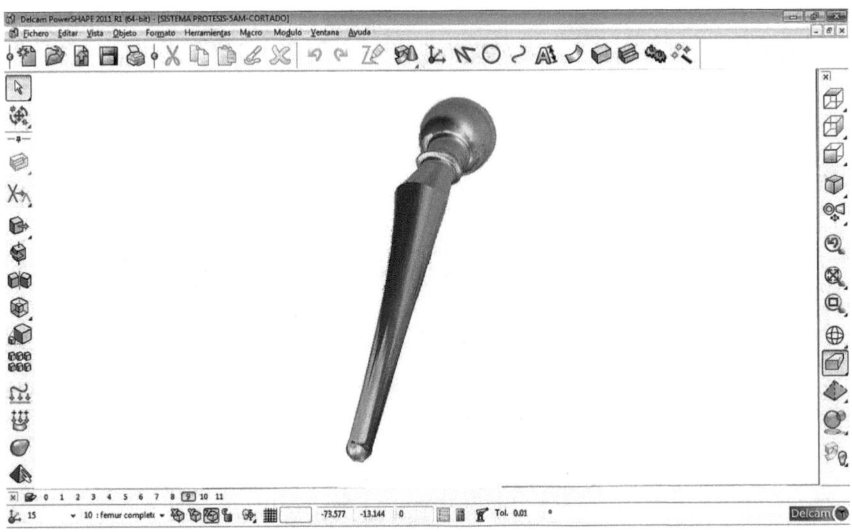

Fig. 6 Solid model of the femur prosthesis of the dog

was fixed at its distal end. Therefore, the boundary conditions applied in this analysis considered that the nodes at such end were anchored.

Regarding the loading conditions, Kim et al. [13] tested a femur prosthesis of dog under compression. The maximum load was in the range between 27 and 33 kg. In an extreme condition, a compression load of 370 N was considered. This is the weight of the dog supported only by one leg and is applied at the femur head of the prosthesis. The components of such force are $F_x = -82.75$ N, $F_y = 19.83$ N, $F_z = -360.62$ N (Fig. 9) [14].

The results of the finite element analysis show that the maximum displacement was 0.001253 mm (Fig. 10). Besides, the maximum von Mises stress was 673.214 kPa. It is located at the medial part of the diaphysis (Fig. 11). It can be

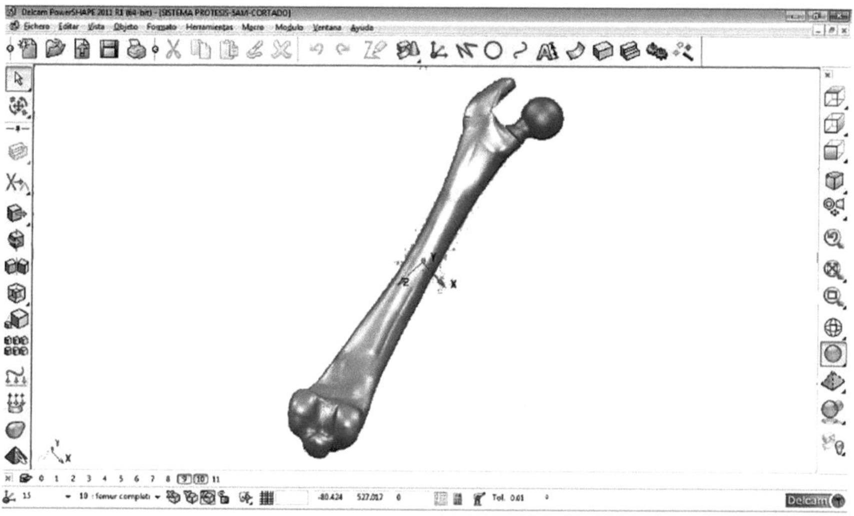

Fig. 7 Solid model of the femur-prosthesis system

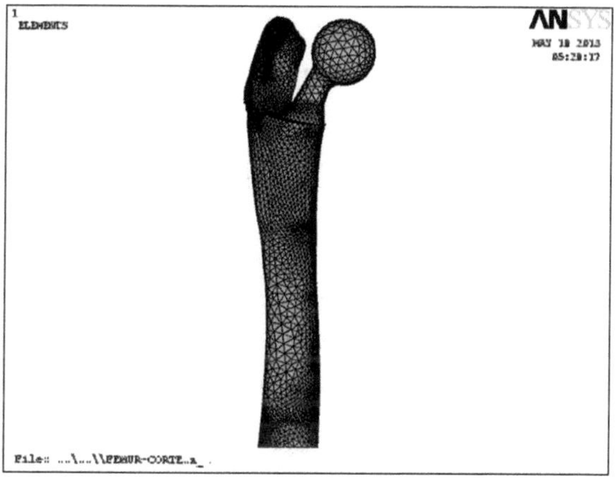

Fig. 8 Finite element mesh of the system femur-prosthesis

considered that this arrangement has an adequate structural integrity and it cannot be loosen easily.

4 Manufacture of the Prosthesis

With the information obtained in the development of the model and the prototypes, the dimensions of the customized prosthesis of the dog were obtained. It was manufactured with a conventional procedure of manufacture (Fig. 12).

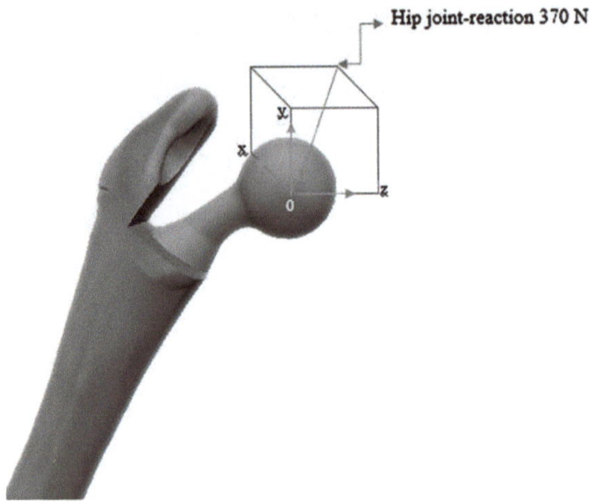

Fig. 9 Hip joint load

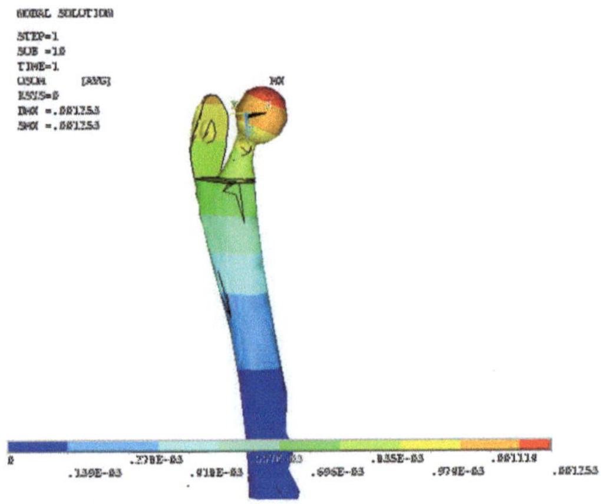

Fig. 10 Deformation in mm of the system femur-prosthesis

In the next step it was coated the femoral head with Au nanoparticles embedded in titanium dioxide (Fig. 13).

5 Nanostructured Thin Solid Film Coating

Hydroxyapatite has been widely used in diverse prostheses. However, an alternative is the use of nanometric building blocks to fabricate scaffolds. For this purpose, high quality nanoparticles with extremely high purity and crystallinity are required. Also, it is important to understand the load-bearing capacity, when these materials are used in implants. When the mechanical properties of an implant do

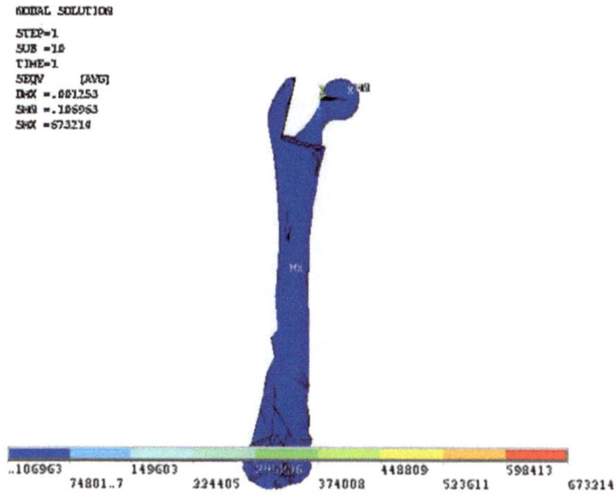

Fig. 11 Resultant von Mises stresses in Pa on the system femur prosthesis

Fig. 12 Femur prosthesis manufactured

Fig. 13 Head of the femur prosthesis coated with AuNP

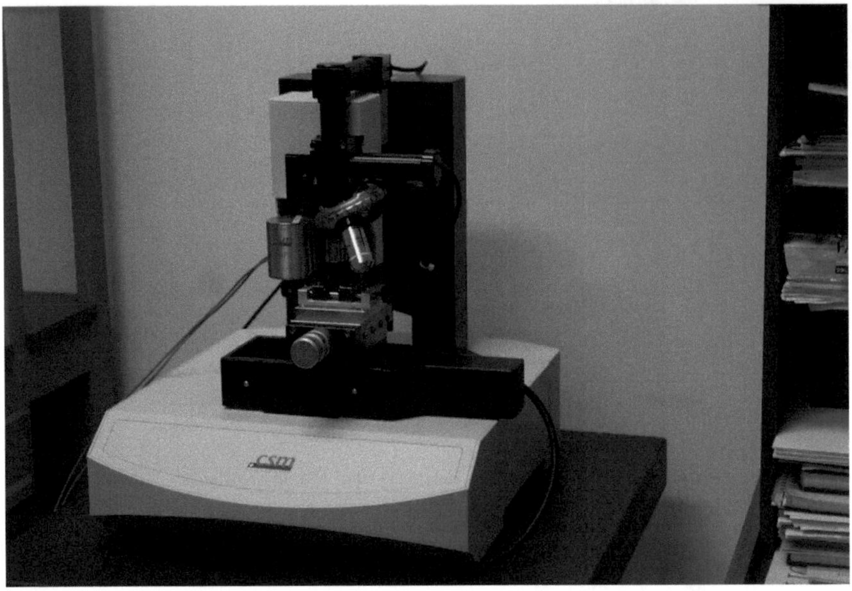

Fig. 14 Nano hardness tester

not match with bone, stress shielding can be developed. This can lead to bone resorption and loosening of the implant. In this way, biocompatible Au nanoparticles have gained considerable attention recently [15].

For the problem at hand, the femur head is under compression load and its joint is in constant wear movement. Therefore, it is expected that a film coating with Au/TiO2 can avoid the premature wear. At the same time, it is expected that the stem of prosthesis does not loose. For this purpose, a coat was applied with the sol-gel technique in the specimen. It is simple and it is cheap to obtain a thin film of metallic oxides, which can be doped with nanoparticles. This procedure can be done at room temperature.

The titanium oxide film doped with gold NPs was synthesized. Initially a SG1 solution was obtained. For this purpose, titanium i-propoxyde [Ti(OC3H7)4] solution with C = 0.05 Mol/L, pH = 1.25 and water/alkoxyde with molar ratio (rw) of 0.8 were used. The resultant solution was stored in a dark place for at least 1 week before using it in the synthesis step.

The required gold nanorods precursor solution was an Aldrich standard solution for AAS analysis with a gold nominal concentration of 1000 mg/L. This solution was used as received. It was added drop by drop into the bottle which contained the SG1 solution. Then after, it was stirred vigorously with a magnetic stirrer plate. The molar ratio of the Au/Ti(OC3H7)4 mixture was 0.76 % (mol/mol). This new solution was called SGG1.

The photocatalytic reduction of the gold ions was carried out in a homemade UV-reactor, which has twelve UV light sources. Each one has a black light blue

Table 1 Mechanical properties of the materials involved in the nano-indentation test

Material	Modulus of elasticity	Poissońs ratio
Steel substratum	200 GPa	0.33
AuNPs coating	0.941 GPa	0.27
Diamond indenter	1141 GPa	0.07

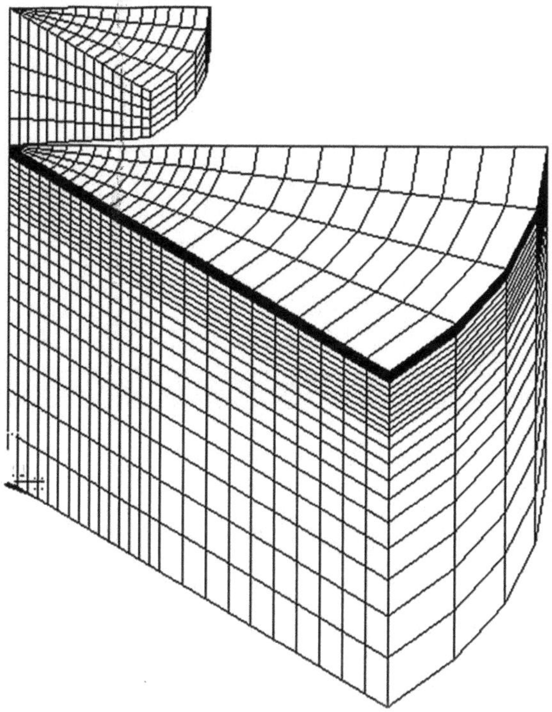

Fig. 15 Finite element mesh used in the stress analysis of the indentation process

UVA lamp (8 W, Hitachi). Under these conditions, the range of UVA light provided is between 320 to 390 nm with λ_{max} (emission) = 355 nm and a light intensity of 732 $\mu W/cm^2$.

Ten ml of the SGG1 solution was exposed to this light source. This event lasted between 15 and 20 min. Then, the light source was switched off and the irradiated sol-gel solution was used to coat the customized prosthesis. This was done with the dip coating technique. The thickness of the resulting samples was around to 250 nm. An atomic force microscope (Dimension 3,100, Nanoscope IV) was used to measure the size and density of the Au NPs [16, 17]. Figure 13 shows a spot of the AuNPs coat.

In the evaluation of the hardness of the coating, a steel substrate was coated with AuNPs. A nano hardness tester NHT of CSM instruments was used. This

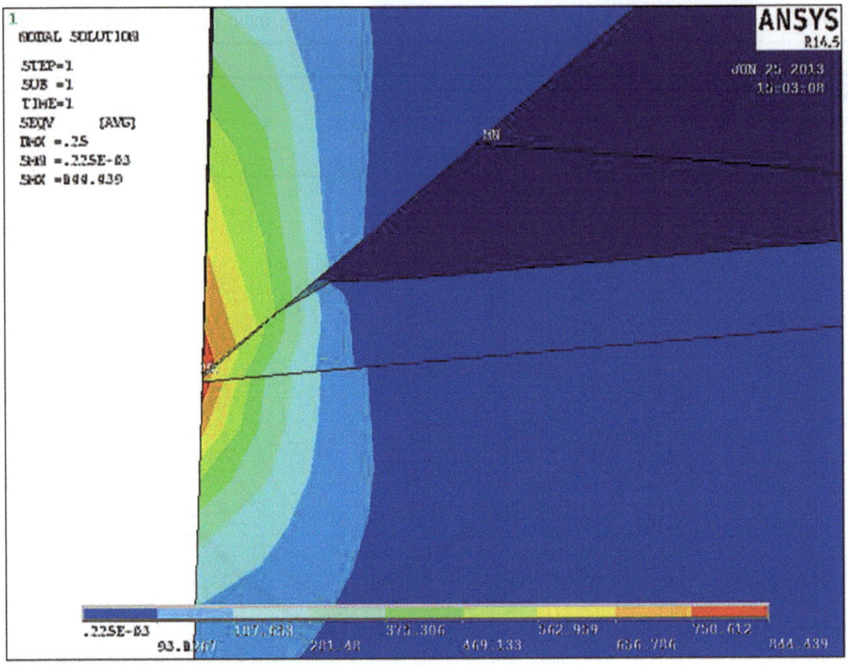

Fig. 16 von Mises stress field in Pa at the point of indentation

device has a load range of indentation between 0.1 and 500 mN and its load resolution is 0.04 μN. The maximum indentation depth is 200 μm (Fig. 14).

The mechanical behavior of the AuNPs coating was evaluated with an indentation test. 20 mN was the maximum load applied. The load and unload rates were 100 mN/mim in both cases. The thickness of the coating was 250 nm. A Berkovich indenter, made of diamond, was used. Its angle was 65°. The modulus of elasticity of the coating was obtained. It was 941 MPa.

A finite element analysis was carried out with the ANSYS code in order to obtain the resultant stress field. The mechanical properties of the involved materials are reported in Table 1. The model was considered axisymmetric. The main considerations of Bouzakis et al. [18] were followed. For this reason, only one eight of the domain of interest was modeled (Fig. 15). Symmetry conditions were considered along the symmetric planes and over the axial axis. The top surface of the indenter was anchored. The nodes on the bottom surface of the substratum were free to move along the vertical direction. In this way the nano-indentation was simulated. The top surface of the coating was pressured against the bottom surface of the indenter. The surfaces of the indenter and the coating were simulated with contact elements (CONTA174 and CONTA170). 3,745 nodes and 3,858 elements were required.

A penetration of 250 nm was simulated. In this way, all the coating thickness was penetrated. The maximum von Mises stress was 844.349 Pa (Fig. 16).

6 Conclusions

In this chapter, a manufacture process of a customized femur prosthesis has been proposed. The information technology plays an important role and optimizes the procedure in the manufacture. The main advantage is that an invasive procedure is not required. Thus, the contact with the human tissues is reduced.

The plastic prototypes of the prosthesis were useful in the discussion and the establishment of the manufacture procedure and the orthopedic treatment. Previously to the manufacture of the femur prosthesis, its correct adjustment with the femur and hip was checked. The results of the finite element method show that the system femur-prosthesis has an adequate structural integrity. It is important to observe that the deformation of the system femur prosthesis is small. Therefore, loosening will not be expected. In this case, a suitable prosthesis for the dog was obtained.

A film coating of Au nanoparticles embedded in titanium dioxide was applied to the femur head of the prosthesis at the end of the manufacture process. The mechanical behavior of this coating was evaluated with indentation tests. Encouraging results have been obtained. The technique that was followed was easy to implement and a uniform layer was obtained. Besides, it was possible to obtain the stress field at the coating during the indentation process.

However, a more detailed analysis is required in order to evaluate its biocompatibility and benefits. At the same time, the guidelines established by the clinical regulatory bodies have to be followed. More tests are demanded before a prosthesis with such coating is used in the implant of an individual.

Acknowledgments The authors kindly acknowledge the support given to the National Polytechnic Institute and the Institute of Science and Technology of the Federal District.

References

1. Farmacopea de los Estados Unidos Mexicanos Suplemento para Dispositivos Médicos. 2nd ed México:771774 (2011)
2. Bobyn, J.D., et al.: The susceptibility of smooth implant surfaces to Periimplant Fibrosis and migration of polyethylene wear debris. Clin. Orthop. Relat. **311**, 21–39 (1995)
3. Vanderby, R., et al.: Fixation stability of femoral components in a canine hip replacement model. J. Orthop. Res. **10**(2), 300–309 (1992)
4. Skurla, C.P., et al.: Assessing the dog as a model for human total hip replacement: analysis of 38 canine cemented femoral components retrieved at post-mortem. J. Bone Joint Surg. Br. **87-B**(1):120–127 (2005)
5. Goel, V.K., et al.: Selection of an animal model for implant fixation studies: anatomical aspects. Yale. J. Biol. Med. **55**(2), 113–122 (1982)
6. Beltrán-Fernández, J.A., et al.: Assessment of the structural integrity of C3-C5 cervical porcine vertebrae model based on 2D classic CAD, 3D scanner and 3D computed tomography. In: Ochsner, A., da Silva, L.F.M., Altenbach H (eds.) Analysis and design of

biological materials and structures, Advanced Structured Materials, vol. 14, pp. 3–17. Springer, Heidelberg (2012)
7. Carcedo, M.G., Brolin, K.: Generation of numerical human models based on medical imaging. Technical report—Department of Applied Mechanics, Chalmers, University of Technology, Sweden, No.1652–8549:1–50 (2012)
8. Smith, E., et al.: A generic digital imaging and communications in medicine solution for a bidirectional interface between the modality and the radiology information system. J. Digit. Imaging **12**(1), 93–95 (1999)
9. Mann, K.A., et al.: Coulomb frictional interfaces in modeling cemented total hip replacements: a more realistic model. J. Biomech. **28**(9), 1067–1078 (1995)
10. Pressel, T., et al.: Mechanical properties of femoral trabecular bone in dogs. Biomed. Eng. online **4**, 17 (2005)
11. Van Rietbergen, B., et al.: Tissue stresses and strain in trabecular of a canine proximal femur can be quantified from computer reconstructions. J. Biomech. **32**, 443–451 (1999)
12. Taylor, M.E., et al.: Stress and strain distribution within the intact femur: compression or bending? Med. Eng. Phys. **18**(2), 122–131 (1996)
13. Kim, J.Y., et al.: Biomechanical evaluation of screw-In femoral implant in cementless total hip system. Vet. Surg. **41**(1), 94–102 (2011)
14. Shahar, R., Banks-Sills, L.: Biomechanical analysis of the canine hind limb: calculation of forces during three-legged stance. Vet J **163**(3), 240–250 (2002)
15. Saji, V.S., et al.: Nanotechnology in biomedical applications: a review. Int. J. Nano. Biomaterials **3**(2), 119–139 (2010)
16. Trejo-Valdez, M., et al.: Contribution of the two-photon absorption to the third order nonlinearity of Au nanoparticles embedded in TiO2 films and in ethanol suspension. J. Phys. Chem. C **114**(22), 10108–10113 (2010)
17. Torres-Torres, D., et al.: Inhibition of the two-photon absorption response exhibited by a bilayer TiO2 film with embedded Au nanoparticles. Opt. Express **18**(16), 16406–16417 (2010)
18. Bouzakis, K., et al.: Mechanical properties, fatigue failure and milling performance of coated tools, investigated through innovative fem supported experimental procedures. Tribol. Ind. **23**(3–4), 39–50 (2001)

Design and Manufacture of a Prototype of a Testing Rig for the Evaluation of the Biomechanical Behavior of Vertebrae

Luis Héctor Hernández-Gómez, Daniel Ramón López-Liévano, Juan Alfonso Beltrán-Fernández, Guillermo Manuel Urriolagoitia-Calderón, Rafael Rodríguez Martínez and Nayeli Camacho Tapia

Abstract The objective of this work is the conceptual design of a testing rig for the biomechanical evaluation of vertebral bodies. Such a rig must test diverse vertebral specimens under compression, flexion and the combination of both loading cases. The dimensions of the porcine and human vertebrae were taken into account. Special attention was given to the design of the grips, in such a way that the range of movements of the spine is reproduced. The structural integrity of this rig was determined by means of the finite element method. In a following step, compression, bending and compression and bending tests of a C3–C7 vertebral body were simulated. This data was obtained from an individual, who was 55 years old. In all these cases, the testing rig had stresses which were within the allowable range. The resultant stresses on the vertebral bodies were in line with those reported in the open literature. In the next step, a "quick" prototype was manufactured in a 3D printer (Dimension SST 1,200). Finally, a data acquisition system has been proposed.

L. H. Hernández-Gómez (✉) · D. R. López-Liévano · J. A. Beltrán-Fernández · G. M. Urriolagoitia-Calderón · R. R. Martínez · N. C. Tapia
National Polytechnic Institute, ESIME, Unidad Profesional "Adolfo López Mateos", Section of Postgraduate Studies and Research, Edificio 5, 3er Piso, Colonia Lindavista, Gustavo A, Madero, 07738 Mexico, DF, Mexico
e-mail: luishector56@hotmail.com

D. R. López-Liévano
e-mail: derive82@hotmail.com

J. A. Beltrán-Fernández
e-mail: jbeltranf@hotmail.com

G. M. Urriolagoitia-Calderón
e-mail: urrio332@hotmail.com

R. R. Martínez
e-mail: rrodriguezm@ipn.mx

N. C. Tapia
e-mail: ncamacho@miners.utep.edu

A. Öchsner and H. Altenbach (eds.), *Design and Computation of Modern Engineering Materials*, Advanced Structured Materials 54,
DOI: 10.1007/978-3-319-07383-5_10,

Keywords Bone graft · Cervical plate · Corporectomy · Displacements · Porcine vertebrae

1 Introduction

The vertebral column has been widely analyzed. It is under different loading conditions and it has a wide range of displacements. Besides, the vertebral bodies have a peculiar geometry. For this purpose, several testing rigs have been proposed. However, most of them have been focused on specific aspects. This is the case of Panjabi et al. [1]. They analyzed the stabilization of eight spinal fixation devices. The vertebral bodies were fixed with screws over cups with polyester resin. The loads were applied with flexible wires and pneumatic actuators. Lu and coworkers [2] evaluated the stability of a vertebral column, which was fractured, after bone cement has been used. The vertebral bodies were fixed with epoxy resin over a base. Extension and bending moments were applied. McAfee et al. [3] tested vertebral bodies of goats. They evaluated the mechanical behavior of the posterior longitudinal ligament in a cervical arthroplasty. The vertebrae were fixed with jaws, which were placed over a metallic base. All these rigs were used in a universal testing machine.

Stanley et al. [4] evaluated thoracic vertebrae. They were fixed with screws over some cups. The flexion-extension loads were applied with a hydraulic system, while the compression loads were applied with wires. In a similar idea, Tomlinson and coworkers [5] tested vertebrae. They were fixed with epoxy resin to a testing rig, which was positioned eccentrically with respect to the actuator of the universal machine. In this way, the vertebral bodies were under bending loading.

Beltrán et al. [6] tested porcine specimens under axial and fatigue loading. The ends of the vertebral bodies were fixed to a pair of aluminum cups. Dental plaster and screws were used for this purpose.

In all these cases, the vertebrae have been fixed in accordance with the case of study. Special devices, such as screws, plates, wires, among others, have been used. All these experiences have shown that a previous functional analysis was required, before the testing rig was manufactured. Also, an evaluation of the structural integrity of the testing rig was needed. When these factors were determined, a good confidence about the performance of the testing devices was obtained.

Based on those experiences, a rig, which serves to test human and porcine vertebrae, has been designed. Its conceptual design is reported. Special attention has been given in the way axial and bending loads are applied. The loading conditions considered were bending, compression and the combination of both cases. Special attention was put in the design of grips.

2 Materials and Methods

2.1 Dimensions of the Testing Rig

Initially, the dimensions of the testing rig were proposed and different virtual models were developed with the Pro-Engineer code. It was considered that the biomechanical analysis of human vertebral bodies requires the fulfillment of several requirements. An alternative is to test porcine specimens, because of their geometrical similarity [7]. For this reason, the dimensions of human and porcine vertebrae were considered in the design of the jaws. Accordingly, L4 vertebra is the biggest. For this reason, its dimensions were considered. Table 1 shows the dimensional parameters that were taken in account in the design of the testing rig. The final model is shown in Fig. 1. It is illustrated with a cylindrical specimen, which was used in the initial evaluation of the structural integrity, in order to evaluate its functionality. Table 2 shows the list of its components.

2.2 Loading Conditions of the Testing Rig

Its structural integrity was evaluated with the finite element method. Two cases were taken into account. In the first instance, the compression and bending of a steel cylindrical specimen was simulated. The purpose was to evaluate the performance of the rig with a well-known geometry.

In the next step, the test of a vertebral body, which is integrated with C3–C7 vertebrae, was simulated. The loading cases considered were compression, bending and compression and bending. The results were compared with those reported in the open literature. They are related with the biomechanical evaluation of prosthesis, implants, specimens, etc. This is the case of Panjabi et al. [1], Beltrán-Fernández et al. [6] and Kunz et al. [9], among others. Table 3 classifies these parameters.

Based on those experiences, the proposed points of application of the loads and supports are illustrated schematically in Fig. 2. Table 3 suggests the loading device for the proposed points of applications of the loads. In other words, the compression loads are applied by the universal testing machine, while bending can be done with an external motor.

3 Evaluation of the Structural Integrity of the Testing Rig with a Cylindrical Specimen

The structural integrity of the testing rig was determined when higher loads are demanded. This is the case when a AISI 1045 steel specimen is tested. The transversal section of the cylindrical specimens was 1963.5 mm^2. Its radius was 25 mm and its moment of inertia was 306,796.2 mm^4.

Table 1 Dimensional parameters considered in the sizing of the positioning cavity in the testing rig for a L4 vertebra

Parameter of the vertebral body	Porcine (mm)	Human (mm)	Reference
Length of the spinous apophysis	38.4	70.1	Dath et al. [8]
Height of the spinous canal	13.5	18.6	Dath et al. [8]
Height of vertebral body	35.6	24.1	Dath et al. [8]

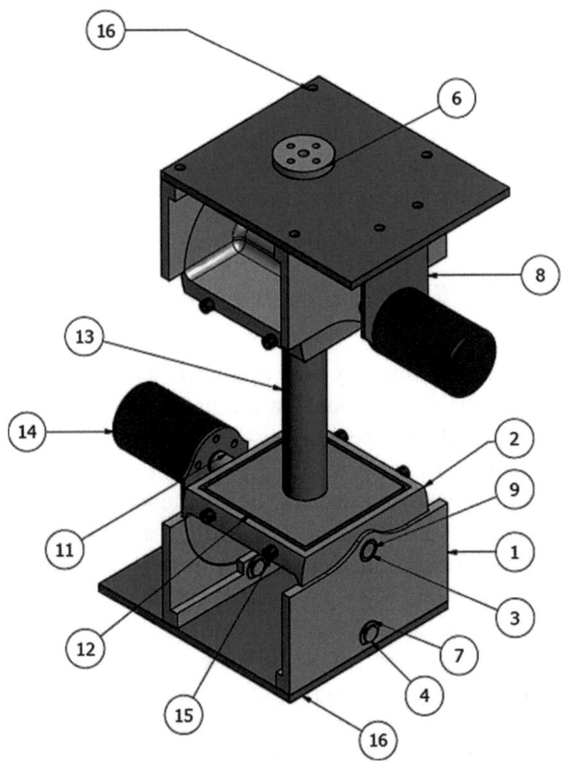

Fig. 1 Testing rig

The physical and mechanical properties of the materials of the testing rig are summarized in Table 4. Once the model of the testing rig and the specimen was obtained with a CAD code, it was exported to ANSYS WORKBENCH 12. The finite element mesh is illustrated in Fig. 3. The mesh comprised 176228 nodes and 86172 tetrahedral elements.

A compression test was simulated. Table 5 compares the resultant stresses under different loading conditions. The material of the specimen is homogeneous, continuous, isotropic and linear elastic and it has a common geometry. Therefore, it is possible to evaluate from an analytical point of view the resultant stresses and

Table 2 Components of the testing rig

Number	Quantity	Item	Material
1	4	Main frame	AISI 1045 carbon steel
2	2	Jaws	AISI 1045 carbon steel
3	12	Bushing	SAE 40 bronze
4	4	Pin	AISI 1045 carbon steel
5	2	Support plate	AISI 1045 carbon steel
6	2	Loading point	AISI 1045 carbon steel
7	8	"O" Ring	–
8	2	Support of the motor	AISI 1045 carbon steel
9	2	Motor shaft	AISI 1045 carbon steel
10	2	Key	AISI 1045 carbon steel
11	2	Coupling	AISI 1045 carbon steel
12	2	Specimen support	AISI 1045 carbon steel
13	1	Specimen	–
14	2	Motor A100FMO-T-K	–
15	8	"Screw 3/8"	–
16	1	"Screw ¼"	–

Table 3 Loading conditions and movement restrictions considered in the design of the testing rig

Point	Loading condition	Load magnitude	Loading device
A	Compression	11,318 N [10]	Actuator of the testing machine
C	Pure bending	15 Nm [1]	Motor
D	Pure bending	15 Nm [1]	Motor
B	Restriction of movement of the six degrees of freedom		
F	Restriction of movement along the horizontal plane		

determine its convergence with those obtained numerically. The maximum difference of the evaluated stresses is 2.53 %. In this case, the testing rig reproduces the resultant stress field. Figure 4 illustrates the deformation of the specimen under bending loading.

Regarding the structural behavior of the testing rig under compression loading, the maximum von Mises stress took place in one of the support plates. It was 42.26 MPa. The maximum von Mises stress under bending loading was 1.5964 MPa. It was developed at the top jaw. Under combined loading (compression and flexion), the maximum von Mises stress is close to the one obtained in compression. It was 42.2720 MPa and was located at the lower support (Fig. 5). It can be said that the dominant load condition is compression.

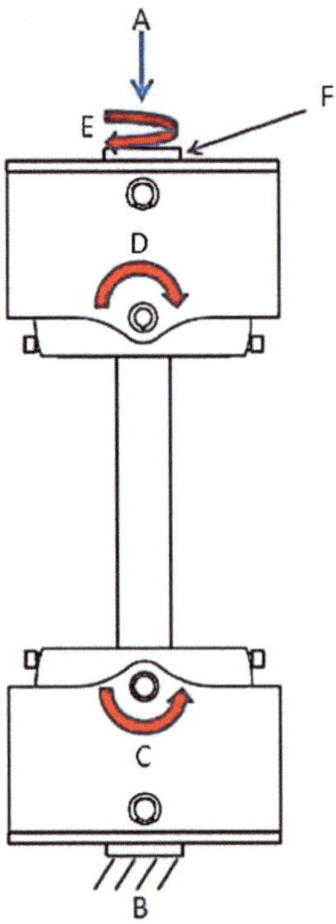

Fig. 2 Loading conditions and movement restriction of the testing rig

Table 4 Mechanical properties of the materials of the testing rig

Material	Density (kg/m^3)	Modulus of elasticity (GPa)	Yield stress (MPa)	Poisson's ratio
AISI 1045 carbon steel	7,850	200	310	0.3
SAE 40 bronze	8,820	93	206	0.29

4 Evaluation of the Structural Integrity of the Testing Rig with a Vertebral Body

The mechanical performance of the testing rig was simulated when a vertebral body was loaded in compression. For this purpose, cervical vertebrae (C1–C7) of a healthy male individual, who was 55 years old, was considered. His weight was 65 kg and he was 1.60 m tall. The density, modulus of elasticity, and Poisson's ratio of the vertebral bodies were 1,800 kg/m^3, 10 GPa, and 0.3, respectively [10].

The model was obtained with the procedure reported by Beltrán et al. [11]. The following procedure was followed: (1) The tomography of the cervical vertebral body was obtained in DICOM files. The slices of the tomography were separated by 0.5 mm; (2) These data was processed with ScanIP and .STL files were

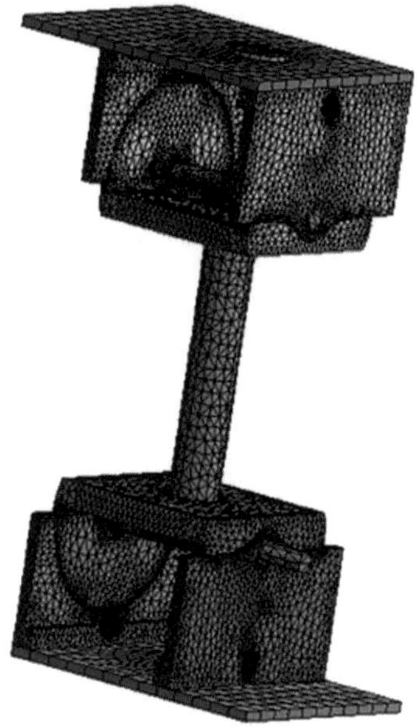

Fig. 3 Finite element mesh used in the evaluation of structural integrity of the testing rig

Table 5 Stresses of the cylindrical specimen under different loading conditions

Loading condition	Stress analytical evaluation (MPa)	Stress FEM evaluation (MPa)	Difference (%)
Compression	−5.7642	−5.7646	0.007
Bending	1.2224	1.2298	0.6
Maximum principal stress	0.0808	0.0829	2.53
Minimum principal stress	−4.6226	−4.6381	0.33

obtained; (3) ProEngineer code was used to create the .IGES files; (4) The finite element model of the system vertebral body and testing rig was obtained with ANSYS WORKBENCH; and (5) The structural integrity of the resultant model was evaluated. Figure 6 illustrates the solid model. In order to compare the results with others, the vertebral segment C3–C7 was only considered. 444273 nodes were required and 242166 tetrahedral elements were generated.

4.1 Compression Loading

The resultant stress field of the vertebral body-testing rig, when a maximum compression load of 11.318 kN was applied, is shown in Fig. 7 [10]. There was a stress concentration around the pins. The maximum stress took place on the bottom support. It was 48.358 MPa.

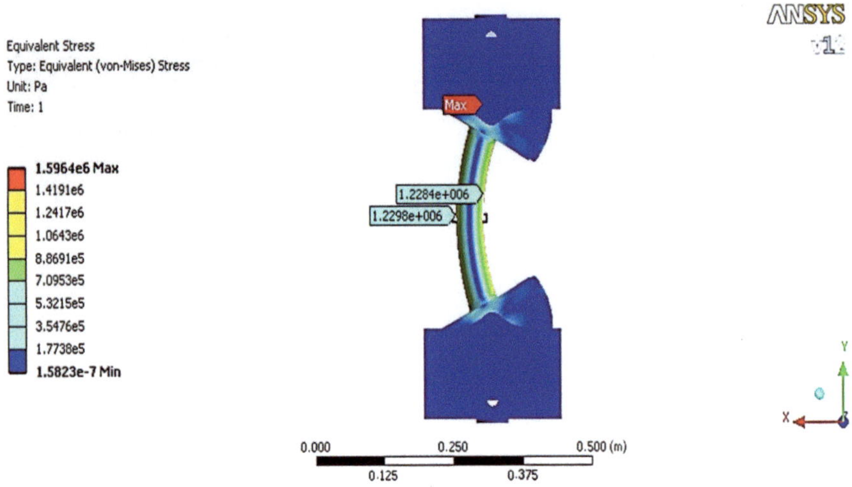

Fig. 4 Deformation of the cylindrical specimen under bending

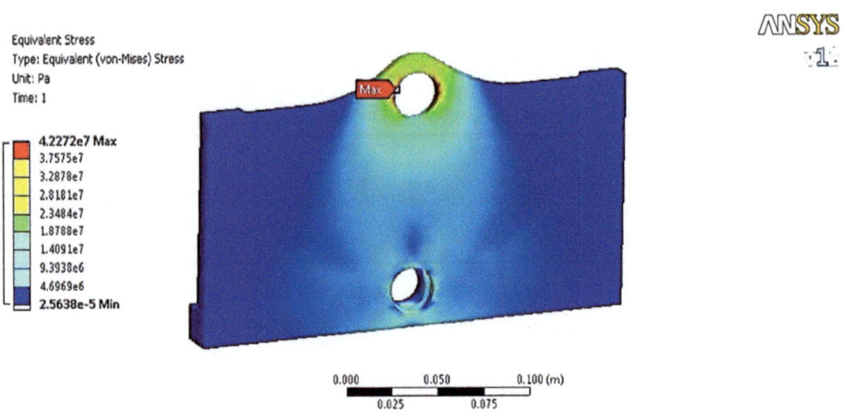

Fig. 5 Maximum von Mises stress in the testing rig

The von Mises stress field on the vertebral bodies is shown in Fig. 8. As it can be seen, a great part of the vertebral bodies is bellow 11 MPa. However, in those areas which are close to the intervertebral disks, the resultant von Mises stresses are within the range 1 and 5 MPa.

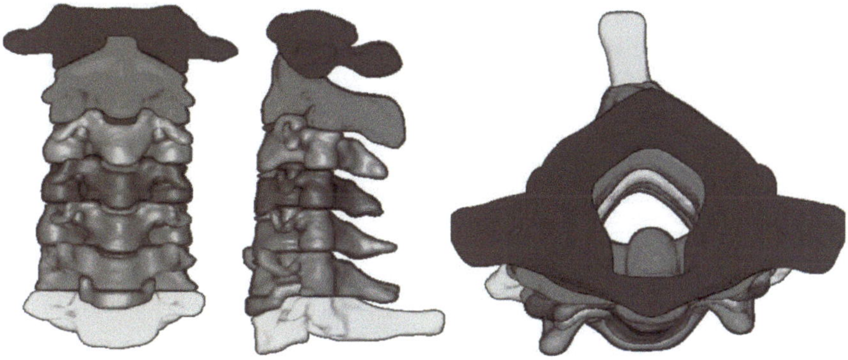

Fig. 6 Solid model of the cervical vertebrae (C1–C7) in .IGES file

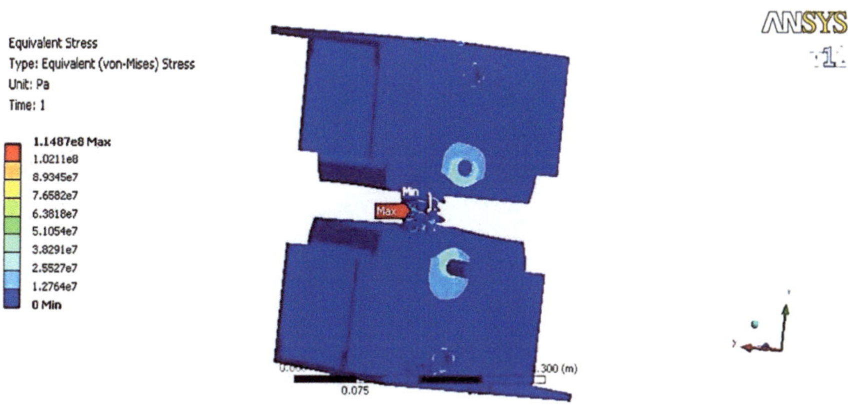

Fig. 7 Resultant von Mises stresses on the testing rig when a 11318 N compression load was applied to the C3–C7 specimen

4.2 Bending Loading

A 15 Nm bending moment was applied. The maximum von Mises stress on the testing rig was 3.57 MPa. It took place on the upper grip (Fig. 9). Under these conditions, the peak von Misses stress on the vertebral body was 27.35 MPa at C3 (Fig. 10). However, all the stresses are below 10 MPa. A similar analysis was carried out for the combine loading case (compression and bending).

5 Evaluation of the Results

Table 6 compares the maximum stresses obtained in the testing rig and vertebral specimen. In the last case, peak stresses are reported. However, all the stresses are underneath 14 MPa. As can be seen, the results are under the allowable stress

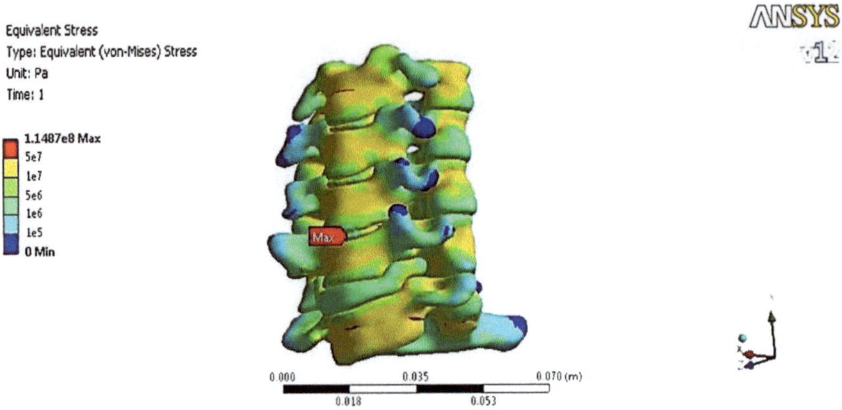

Fig. 8 von Mises stress field on the C3–C7 specimen under a 11318 N compression load

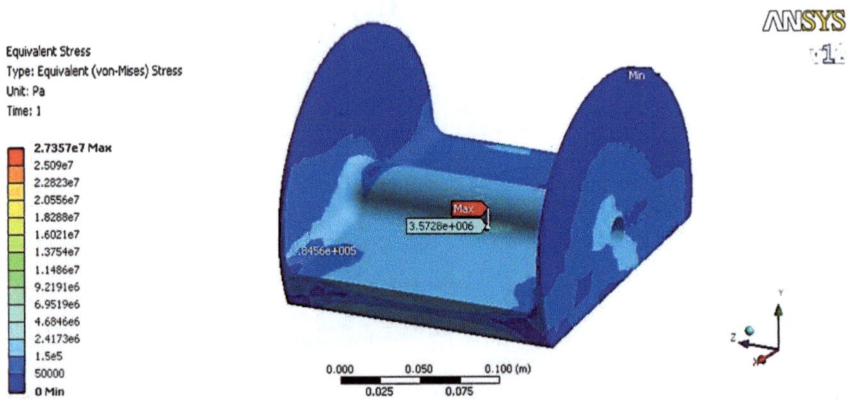

Fig. 9 Maximum von Mises stress under a 15 N-m bending moment, located on one of the jaws

range. Therefore, the structural integrity of the testing rig is guaranteed in the range of the loading cases expected.

6 Prototype

Once that the usefulness of the testing rig was established and the structural integrity was determined, a quick prototype was made with ABS plastic. A 3D Dimension SST 1,200 printer was used. The suitability of the "quick" prototype of the testing rig was evaluated in a SHIMADZU machine. Its capacity is of 10,000 kg For this demonstration, porcine specimens were used. A C3–C7 vertebral unit, which was used in a corporectomy test, was instrumented with an anterior cervical plate (Fig. 11).

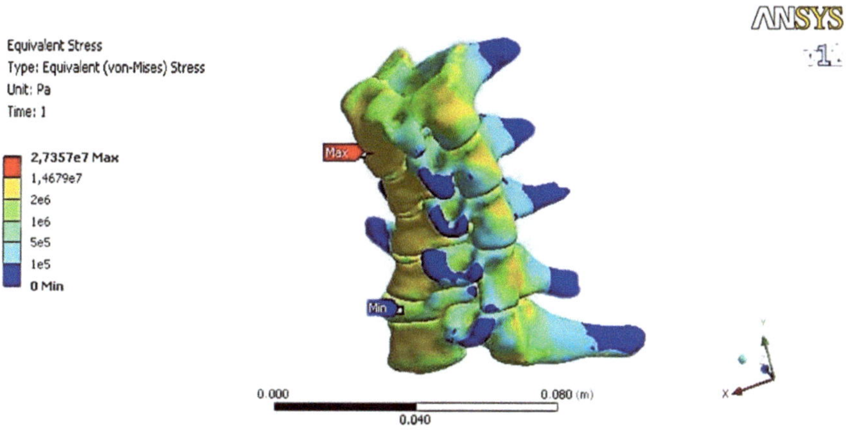

Fig. 10 Maximum von Mises stress on the C3–C7 cervical vertebra under a bending loading of 15 N-m

Table 6 Maximum von Mises stresses obtained on the testing rig and in the vertebral body under different loading conditions

Loading condition	Testing rig		Cervical vertebral body	
	Maximum von Mises stress (MPa)	Item	Maximum von Mises stress (MPa)	Location
Compression	48.3580	Lower support	114.87	Disk C5–C6
Bending	3.5728	Top grip	27.3570	C3
Combined loading conditions	48.1710	Lower support	114.8900	Disk C5–C6

7 Data Acquisition System of the Testing Rig

This system requires, at least, transducers for the evaluation of the deformations and the applied loads. A Wheatstone bridge was proposed for this purpose. The sensibility is increased when a half bridge is used. In this condition a greater change of the resistance takes place, improving the measurement. This system requires an external power supply and an amplifier, as the voltage of the transducer is low. As a general rule, higher excitation voltages generates high outputs of voltage. Nonetheless, the higher voltages can cause greater errors in the measurement. This is due to the self-heating [12]. For this reason, the gain voltage will be in the range between 3 and 10 V. On the other side, the biomechanical tests are usually carried out in quasi-static conditions. They are run with a low loading rate. In this circumstance, there is a chance for heat dissipation.

Another important aspect is the size of the strain gages. Fatigue and high resistance reduce the life of the smallest dimensions and high resistance [13]. This situation takes place when strain gages with a resistance of 350 Ω or higher are

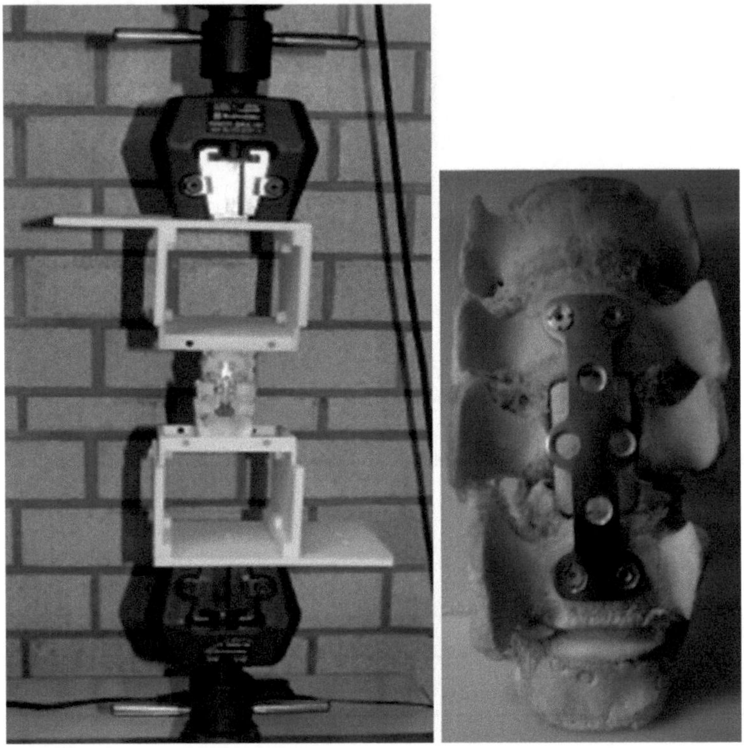

Fig. 11 Quick prototype which is used to evaluate the usefulness of the proposed testing rig in a universal testing machine

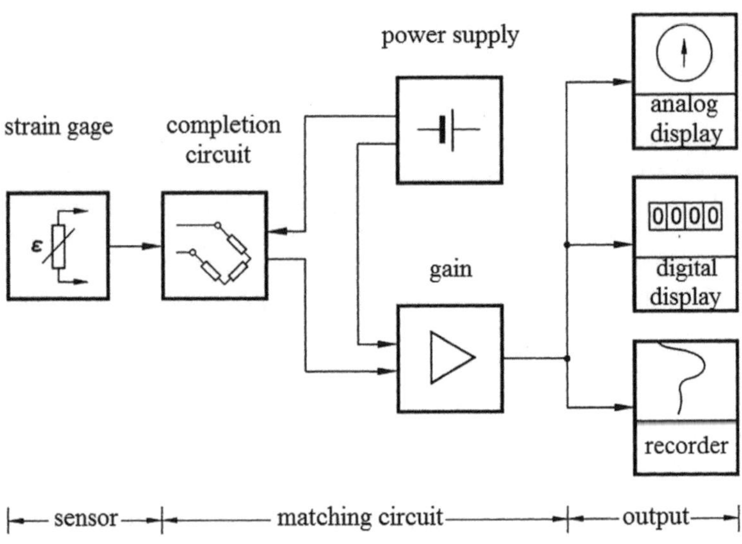

Fig. 12 General arrangement of the data acquisition system of the testing rig

used. For this reason, strain gages with a resistance of 120 Ω were selected with a gain factor (GF) equal to 2.

Regarding the dimensions of the strain gages, those, which have a length of 3 mm, were selected. They are suitable for the proposed acquisition data system. At the same time, their performance is similar to the strain gages of greater length.

A display is also required. The output of the amplifier is a voltage or the intensity of an electric current. It has been selected a digital measuring device, which shows the value of the parameter of interest. Figure 12 shows the schematic arrangement.

8 Conclusions

In accordance with the results of the conceptual design, the testing rig can be used in a universal testing machine and it can evaluate the mechanical behavior of the vertebral bodies. The results show that: (1) the convergence of the results are within the range of 2.53 % in cylindrical specimens made of steel; (2) The stress fields obtained in the vertebral bodies are similar to those reported in the open literature.

In the range of the tests which will be carried on, the structural integrity of the testing is kept. The finite element analysis showed that the maximum von Mises stress is 310 MPa in an extreme situation.

The information technology plays an important role in the design. In this case, a prototype can be obtained from the conceptual design through the development of virtual models. In the case of the vertebral bodies, the models can be obtained from the tomography of the individual.

Regarding the data acquisition system, it is independent of the universal testing machine. It will get and record the parameters directly from the specimen.

As a future work of this project, the integration of a system that loads the vertebral specimens in torsion and, also, combines it with compression and bending will be developed.

Acknowledgments The authors kindly acknowledge the support given to this research by the National Polytechnic Institute, the Institute of Science and Technology of the Federal District and CONACyT.

References

1. Panjabi, M.M., Abumi, K., Duranceu, J., Crisco, J.J., et al.: Biomechanical evaluation of spinal fixation devices II. Stability provided by eight internal fixation devices. Spine **13**(10), 1135–1140 (1988)
2. Lu, W.W., Cheung, K.M.C., Li, Y.W., et al.: Bioactive bone cement as a principal fixture for spinal burst fracture an in vitro biomechanical and morphologic study. Spine **26**(24), 2684–2691 (2001)

3. McAfee, P.C., Cunningham, B., Dmitriev, A., et al.: Cervical disc replacement—Porous coated motion prosthesis. A comparative biomechanical analysis showing the key role of the posterior longitudinal ligament. Spine **28**(20), S176–S185 (2003)
4. Stanley, S.K., Ghanayem, A.J., Voronov, L.I., Havey, R.M., et al.: Flexion–extension response of the thoracolumbar spine under compressive follower preload. Spine **29**(22), E510–E514 (2004)
5. Tomlinson, T., Chen, J., Upasani, V., Mahar, A.: Unilateral and bilateral sacropelvic fixation result in similar construct biomechanics. Spine **33**(20), 2127–2133 (2008)
6. Beltrán-Fernández, J.A., Hernández-Gómez, L.H., Ruiz-Muñoz, E., González-Rebattú, A., Rodríguez-Cañizo, R.G., Urriolalagoitia- Calderón, G., Urriolagoitia-Sosa, G., Hernández-Moreno, H.: Biomechanical evaluation of corporectomy in porcine lumbar specimens using flexible polymer belts. J. Phys. Conf. Ser. (2009). doi:10.1088/1742-6596/181/1/012015
7. Szotek, S., Szust, A., Pezowics, C., et al.: Animal models in biomechanical investigations. Bull. Vet. Inst. Pulawy **48**, 163–168 (2004)
8. Dath, R., Ebinesan, A.D., Porter, K.M., Miles, A.W.: Anatomical measurements of porcine lumbar vertebrae. J. Clin. Biomech. **22**(5), 607–613 (2007)
9. Kunz, D.R., Vanderby, Jr., Zdeblick, T. et al.: Application of pure loads in axial, torsional or flexural directions. A Multi-Axis Syst. Biomech. Test. 4–10 (1993)
10. Öchsner, A., da Silva, L.F.M., Altenbach, H. (eds.): Advanced computational engineering and experimenting. Key Eng. Mater. **478**, 103–111 (2011). doi:10.4028/www.scientific.net/KEM.478.103
11. Beltrán-Fernández, J.A., Hernández-Gómez, L.H., Urriolagoitia-Calderón, G., González-Rebatú, A., Urriolagoitia-Sosa, G., Galán- Vera, M.M., Escalante-Rodríguez, E.: Assessment of the structural integrity of C3–C5 cervical porcine vertebrae model based on 2D classic CAD, 3D scanner and 3D computed tomography. In: Ochsner, A., da Silva, L.F.M., Altenbach, H. (eds.) Analysis and design of biological materials and structures. Advanced Structured Materials, vol. 14, pp.3-17. Springer Berlin (2012)
12. Hoffmann, K.: An Introduction to Measurements Using Strain Gages. Darmstadt, Germany (1989)
13. Murray, W., Miller, W.: The Bonded Electrical Resistance Strain Gage. Oxford University Press, USA (1992)

Biomechanical Assembled Prosthesis of a Temporo Mandibular Joint Disorder Using Biocompatible Materials

Juan Alfonso Beltrán-Fernández, Eduardo Romo-Escalante, Francisco López-Saucedo, Pablo Moreno-Garibaldi, Luis Héctor Hernández-Gómez, Guillermo Urriolagoitia-Calderón and Nayeli Camacho-Tapia

Abstract In this work, the design and manufacture of a new assembled prosthesis for a jaw for a young Mexican patient with Temporo Mandibular Disorder (TMD) and articular ankilosys is reported. The required models were obtained with a computed axial tomography (CAT) in conjunction with computed aided design (CAD) and engineering codes. Several tomographic slices were taken digital format (DICOM). Each slice was processed with Scan IP, Unigraphics and Solidworks. The main objective of this study is to show the digital and a physical processing for manufacturing the desired model, taking into account the anthropometry of the young patient, who expects to receive this fixation as a maxillary implant. The parametric FEM model allows creating biomechanical testings in order to evaluate its mechanical response before the final manufacturing by using of a quick prototyping printer system (dust). This prosthesis will be created as an

J. A. Beltrán-Fernández (✉) · E. Romo-Escalante · P. Moreno-Garibaldi · L. H. Hernández-Gómez · G. Urriolagoitia-Calderón · N. Camacho-Tapia
Instituto Politécnico Nacional. ESIME-SEPI, Unidad Profesional "Adolfo López Mateos", Edificio 5, 3 Piso, Colonia Lindavista. Gustavo A. Madero, 07738 México, D.F, Mexico
e-mail: jbeltranf@hotmail.com

E. Romo-Escalante
e-mail: kleeman55@hotmail.com

P. Moreno-Garibaldi
e-mail: eagle_gar@hotmail.com

L. H. Hernández-Gómez
e-mail: luishector56@hotmail.com

G. Urriolagoitia-Calderón
e-mail: urrio332@hotmail.com

N. Camacho-Tapia
e-mail: ncamacho@miners.utep.edu

F. López-Saucedo
Departamento de Cirugia Maxilofacial, Hospital de Especialidades del Centro Médico Nacional la Raza, Calz. Vallejo y Av. Jacarandas Col. La Raza, México, D.F, Mexico
e-mail: franciscol33@hotmail.com

A. Öchsner and H. Altenbach (eds.), *Design and Computation of Modern Engineering Materials*, Advanced Structured Materials 54,
DOI: 10.1007/978-3-319-07383-5_11,

assembly, and the condylar joint will be created by a low friction metal. The expectation for this final implant requires the approval of the Hospital "La Raza" (IMSS) in order to be installed on the patient. For this purpose, a biomechanical engineering procedure and 3D manufacturing quick prototyping was used for the prototype. The clinical case discussed in this chapter is related to a 17 year-old individual. He was treated in the area of maxillofacial and plastic surgery. The severity of the deformation was based on congenital disease. Calcium hydroxyapatite, polymethylmethacrylate polymer and stainless steel, which are highly biocompatible materials, were used for this purpose. It is very important to mention that this procedure avoided the employment of bone tissue taken from the ribs and cranial regions as in other research publications referred. As a result of this research, an assembled 3D personalized jaw model could be presented in order to be fixed in the patient.

Keywords Temporomandibular joint disorder (TMJ) · Jaw · Condyllar joint · Maxillary implant · Stereolithography · Hydroxyapatite · Polymethylmethacrylate polymer · Quick prototyping

1 Introduction

The development of this research allows solving some needs of low resource people and also involves public hospitals (Centro Médico La Raza) and specifically with the surgeon Dr. Francisco López Saucedo, who is a member of the Surgery College. The main objective is to reproduce the particularized jaw prosthesis in order to be fixed to the patient, taking into account that an ideal healthy human jaw can be adjusted to the morphology of the patient (Fig. 1).

The anchylosis of the temporomandibular articulation (ATM) is a serious complication, mainly after some traumatisms and local infections pathologies [1–4]. In uncommon cases the anchylosis associates to systematic illnesses as the ankylosing spondylitis, rheumatoid arthritis and psoriasis. Accordingly to the functional restrictions and the alterations caused in the facial growth in the youth, it is desirable an early and effective therapy. There are different ways to handle the anchylosis of the ATM which go from the employment of osteochondral implants until the fixation of prosthesis [5–8]. There is not a surgical concept standardized, but all are in function of the cooperation and discipline that the patient requires in the postoperative physiotherapy [9–16].

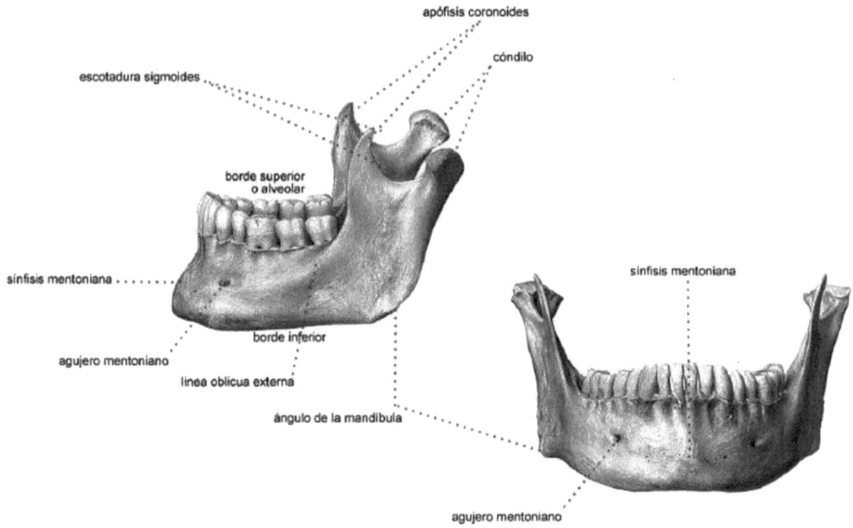

Fig. 1 Healthy human jaw

2 Materials and Methods

2.1 Biomodelling from Computed Tomography

The model was obtained taking into account medical reports and success DICOM regeneration methodologies [17–25]. The initial manipulation of information from a computerized tomography of patient was realized importing of a directory of DICOM pictures using the software ScanIP, which allows drawing several merged axial images in different planes and in a pseudovolume reconstructed. Once this information was imported, several masks were used. Each slice considered the usage of filters of grey scale and manual tools applied with a digital pad, in order to define the osseous structures as well as its axial spaces (Fig. 2). The manual brushing of the tomography was particularly laborious due to the resolution of 1 mm between each axial slice, because of the complexity to identify the boundary on each anatomic detail, for example, in the ATM.

Once the total osseous and mandibular mask were defined, it was necessary to apply a Boolean operation in order to merge all the slices and to get an initial cranium proposal.

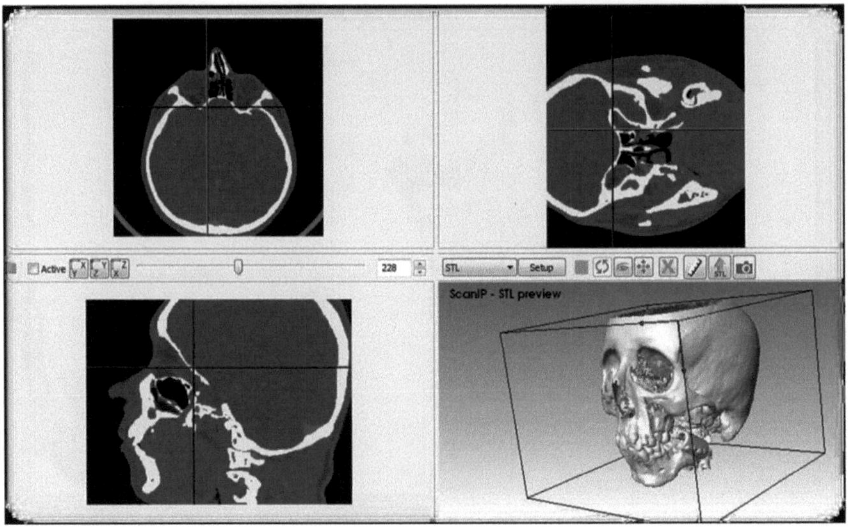

Fig. 2 Mask processing in ScanIP software

2.2 Biomodelling Exporting the Volume in SCAN-FE

Under the ScanIP software there is an attached tool called ScanFE, which is used to create specialized solids and volumes. After importing several masks files, such as: mask containing only the jaw of the patient, there was an attempt to generate a volume handled in the 3D parametric modeling software. However the computational resources required exceeded all common personal computers in which this technique was attempted (Fig. 3). Even tough, it is recognized that large memory and processing resources may be a good methodology in order to export tomographyc files into a solid format with great accuracy.

2.3 Biomodelling of the Volume Using SolidWorks and a DXF File Format

There was a similar error when trying to use a tool native SolidWorks program to generate a three-dimensional model from the point cloud file. This tool is very powerful because if the file exceeds the 10,000 polygons that make up the mesh joining the required time rises by a factor of hours or days. For the original mesh which was not tried this technique even after completion of the process of processing whole 2 days because the mesh was composed of more than 400,000 polygons.

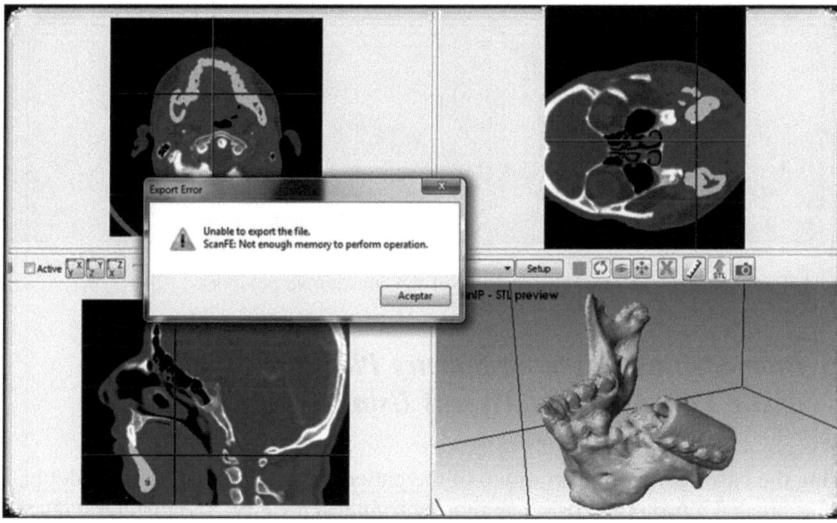

Fig. 3 Problems in the solid bioregeneration when the RAM memory is not enough in SCAN FE

This method is only recommended for scenarios where this piece consists of a simple low-resolution mesh.

2.4 Biomodelling Based on the Remeshing in Geomagic Studio and Creation of the Volume in SolidWorks Using a DXF File

Considering the problem of the previous methodology, the mesh of the models was reconstructed with Geomagic Studio. These models were created with 10 % of polygons from the original model. It represents about 40,000 polygons with satisfactory fidelity (Fig. 4). A fine remeshing was applied, however, some anatomical zones were deformed and the fidelity changed. A new rendering (meshing) considering the same methodology was repeated with models overlock 90 % and after of 8 h there was no conclusion. For this reason the process was aborted in order to consider alternative methodologies. The contribution of this methodology which prevailed throughout the development of the project was to eliminate conflict zones in the models after the remeshing.

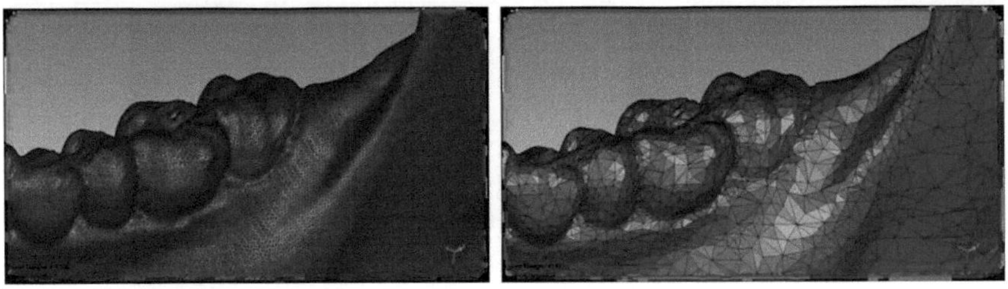

Fig. 4 Remeshing of the model using 10 % of the constitutive polygons

2.5 Biomodelling Based on Simetry Planes and Creation of the Volume in SolidWorks Using a DXF File

Using the base views of the cranium of the patient and the isolated jaw model base it was created a prototype using symmetry trying to preserve the matching respect to the ATM condyle. Otherwise using the best surgical zone (Fig. 5), and some digital tools as: Goniometers rulers and grids were used, also digital mirror planes and tools to identify and generate a symmetrical jaw which preserves the initial hypothesis of this methodology (Fig. 5).

It was generated a final prototype which guided to a best anatomy in accordance of the patient features (Fig. 6). However, there were three problems, the first one about morphology of the condyle because of a low resolution of the computed axial tomography (CAT) and the manual drawing of the jaw.

The second one is related to the lack of coincidence of the upper and lower teeth and finally the impossibility to export this file as a solid in order to fix the geometry parametrically.

2.6 Definitive Biomodelling Methodology Exporting Curves and Planes Using Geomagic Studio and Covering Surfaces Techniques with Parametrical Modelling on SolidWorks

Based on this methodology it was allowed to generate the proposed prototype. It was conserved the tomography data and its biomodelling technique in ScanIP. The final file of this technique was exported as an STL file into Geomagic Studio where it was processed using several similar models taking into account the main two sections of interest. The first one consisting of the portion of the lower jaw bone to be conserved in surgery removing the actual silicon tube (Actual state of the patient) and the opposite half part of the jaw to be conserved (Fig. 7). The second one consists of a mirror technique applied to the right condyle with a cutting detail at the last part of the first section. It means that the suggestion of the surgery about

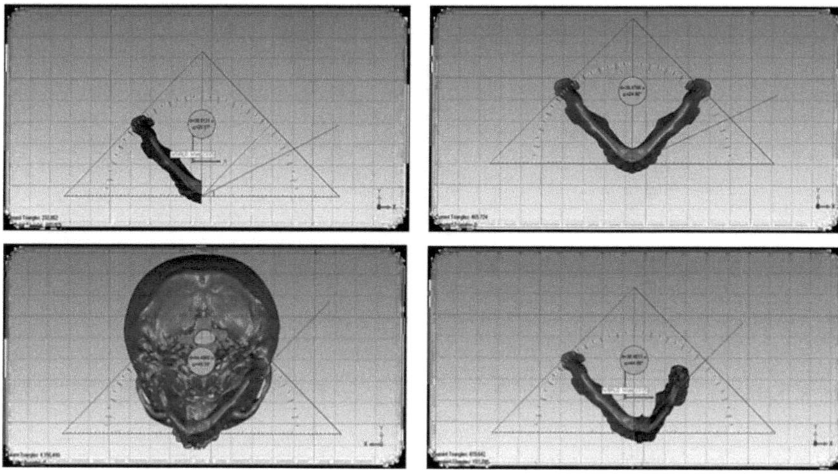

Fig. 5 Symmetric model and digital tools in order to measure angles and length over the base views on the cranium model and its isolated jaw using GeoMagics software and MB ruler

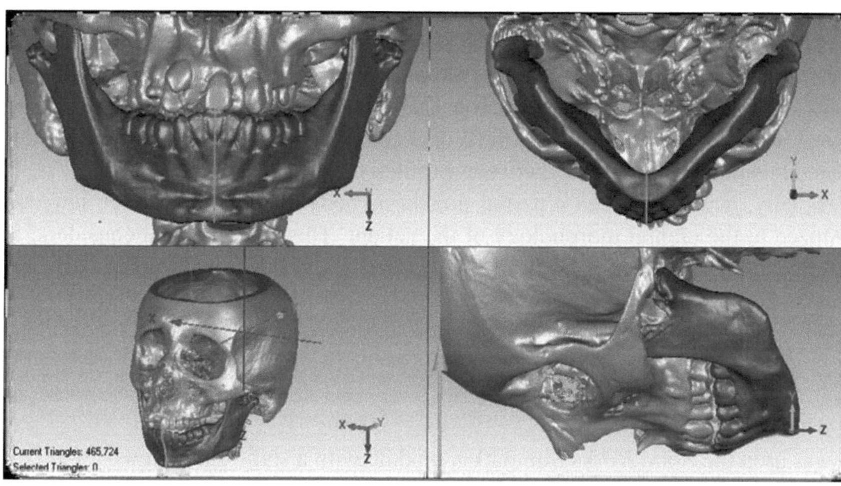

Fig. 6 3D model creation using symmetric planes and the angles from Geomagic Studio software

to cut the jaw behind the last molar of the left jaw body was conserved. Therefore this plane was the base to determine the symmetry of the right side jaw to be projected as a mirrored body.

Subsequently each one of the jaw profiles were faced in order to match transversely. Some cutting planes were created along the natural curvature of the body and its branch obtained. It was used a spacing of 0.5 mm between slices in order to accurately represent the morphology of the curves using Geomagic Studio (Fig. 8).

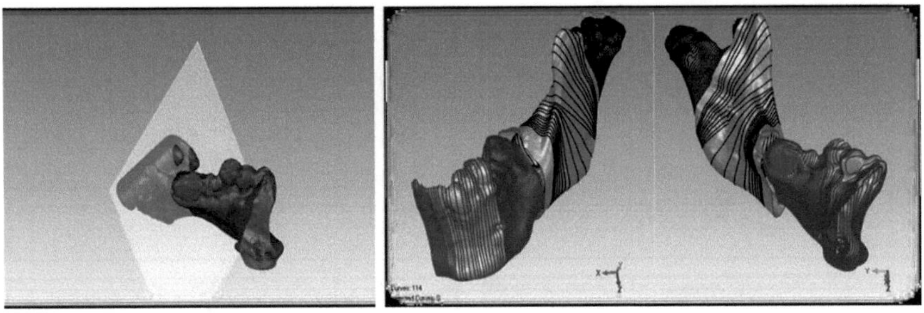

Fig. 7 Planes and slices on Geomagic Studio over the model

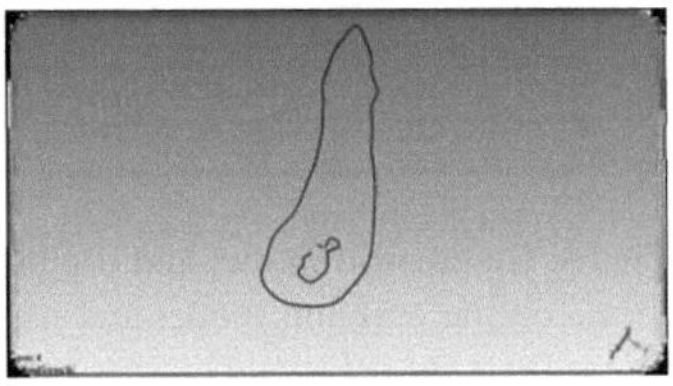

Fig. 8 Splines

These curves were exported in two blocks, the first one is the corresponding to the fraction of the jaw mandible that should be conserved and the second one the reflected branch used as construction scaffolding to build the prosthesis. Each sketch was extrapolated by the SolidWorks coating tool and manually equalized to generate suitable projections between profiles (Fig. 9).

Finally, it was obtained a partial prosthesis body and the left branch jaw customized to the cranial morphology of the patient. The final proposal considers: an intramedullar support, the body of the prosthesis and one condyle (Fig. 10).

2.7 Finite Element Analysis

It was used the SolidWorks Simulation module with a static study in which a static load of 100 kg was applied as a force over the last fixation screw (Fig. 11).

Because of there is not a definitive or final composite to manufacture the prototype, a titanium alloy called Ti-6Al 4VA was considered for the FEM simulations in all parts of the prosthesis. The FEM model considered 121047 nodes and 82120 high quality squared elements.

2.7.1 Finite Element Results

Once the simulation finished, the assembly reported a maximum displacement of 1.082 mm which was located on the fixing screw (Fig. 12).

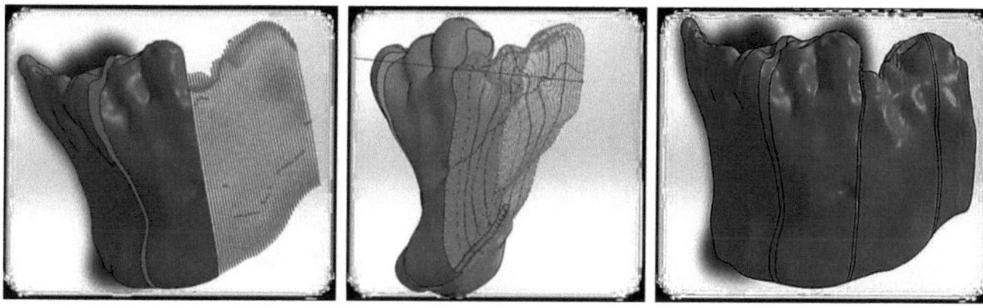

Fig. 9 Generation of the recovering in SolidWorks

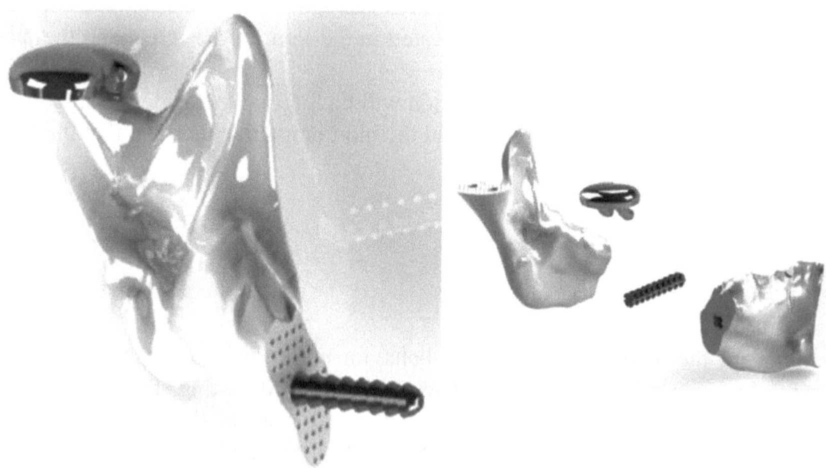

Fig. 10 Final partial assembled prosthesis of a jaw

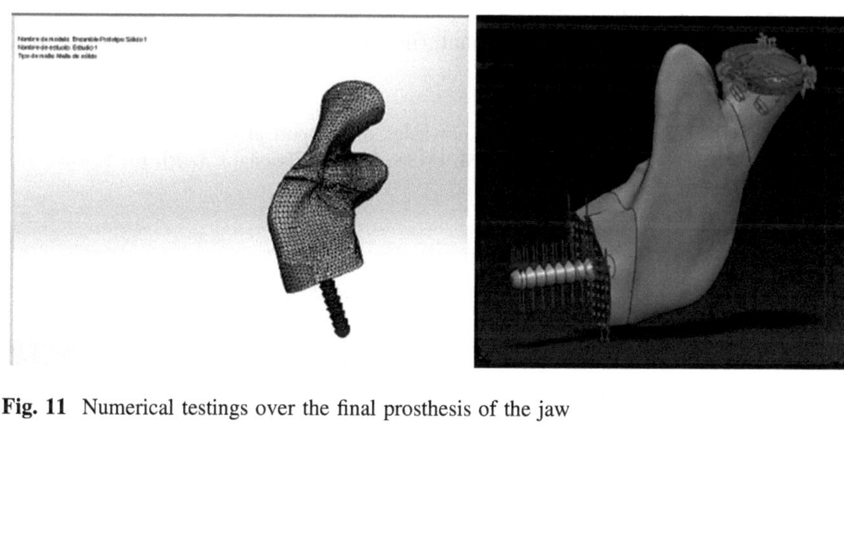

Fig. 11 Numerical testings over the final prosthesis of the jaw

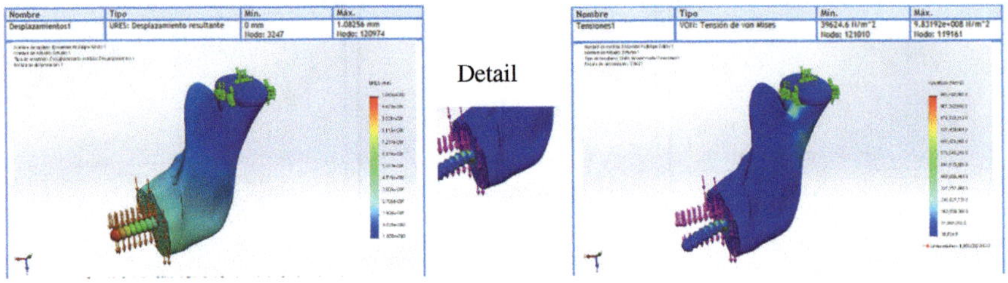

Fig. 12 Maximum displacement and von Misses Stress over the final prosthesis of the jaw

About the von Misses stresses, the assembled model reported 983.192 MPa as the maximum value. This result was located in the joint of the screw and the face of the lower edge of the jaw.

Based on this result, the final discussion will be focused to decide about the best option for the manufacturing process and the bio-compatible material that the final implant will use.

3 Conclusions

In summary, the model shows suitable behavior before to be implemented in the patient. Also considering the applied load in the FEM testing, the critical geometrical zones such as: intramedullarly support base and the condylar branch collar, which were able to be stressed. It means that the inclusion of titanium plates will increase its strength and the utility of this model will be better in order to get a stable reaction in dynamic stability.

Manufacturing processes are under review, so far depends on the final material of the body but in terms of the metal parts it can be used a mold by casting, or generated in SolidWorks or a manufacturing model using Computed Numerical Machine (CNC).

Acknowledgments The authors kindly acknowledge the support given to this research by the National Polytechnic Institute (IPN), ITESM, IMSS and the National Council for Science and Technology (CONACYT)

References

1. González, M.H.C.: Anquilosis temporomandibular. Revisión de la literatura. Protocolo de manejo e informe de un caso. Medigraphic (2011)
2. Alplay, H.K., Murat, K.O., Ortakoglut, K., Rasit, G.B., Sinan, Y.A.: Treatment of temporomandibular joint ankylosis with temporalis superficial fascia flap. Mil. Med. **170**(2), 167–170 (2005)

3. Humana, A.: Anatomía Humana. Obtained from “Maxilar Inferior y Articulacion Temporomandibular” (2011) http://www.anatomiahumana.net/maxilar-inferior-articulacion-temporomandibular.html
4. Andresen, P.R., Bookstein, F.L., Conradsen, K., Ersbøll, B.K., Marsh, J.L., Kreiborg, S.: Surface-bounded growth modeling applied to human mandibles. IEEE Trans. Med. Imag. **19**(11), 1053–1063 (2000)
5. Barrett, J.B., Fryer, M.P., Kollias, P., Ohlwiler, D.A., Templeton, J.B.: Silicone and teflon prostheses, including full jaw substitution: laboratory and clinical studies of etheron. Ann. Surg. **157**(6), 932–943 (1963)
6. Das, U.M., Keerthi, R., Ashwin, D.P., Venkata, R.S., Reddy, D., Shiggaon, N.: Ankylosis of temporomandibular joint in. J. Indian Soc. Pedod. Prev. Dent., **27**(2), 116–120 (2009)
7. Franchi, L., Baccetti, T., McNamara, J.A.: Thin-plate spline analysis of mandibular growth. Angle Orthod. **71**, 83–92 (2001)
8. Gu, Y., McNamara, J.A.: Mandibular growth changes and cervical vertebral maturation. Angle Orthod. **77**, 947–953 (2007)
9. Jain, G., Kumar, S., Rana, A.S., Bansal, V., Sharma, P., Vikram, A.: Temporomandibular joint ankylosis: a review of 44 cases. J. Oral Maxillofac. Surg. **12**, 61–66 (2008)
10. Montenegro, M.A.: Factores que regulan el crecimiento del cóndilo mandibular. Rev. Chil. Ortod. **21**, 6–16 (2004)
11. Montenegro, M.R., Rojas, M.R.: Factores que Regulan la Morfogénesis y el Crecimiento. Int. J. Odontostomat. **1**, 7–15 (2007)
12. Rabie, A.M., She, T.T., Hägg, U.: Factors regulation mandibular condylar growth. Am. J. Orthod. Dentofac. Orthop. **122**, 401–409 (2002)
13. Rabie, A.M., She, T.T., Hägg, U.: Functional appliance therapy accelerates and enhances condylar growth. Am. J. Orthod. Dentofac. Orthop. **123**, 1–10 (2003)
14. Ramfjord, S.P., Ash, M.M.: Oclusión. Nueva Editorial Interamericana, México (1972)
15. Ramirez-Yañez, G.O., Smid, J.R., Waters, M.J.: Influence of growth hormone on the craniofacial complex of transgenic mice. Euro. J. Orthod. **27**, 494–500 (2005)
16. Reynolds, M., Reynolds, M., Adeeb, S., El-Bialy, T. 3-D volumetric evaluation of human mandibular growth. Open Biomed. Eng. J. **5**, 83–89 (2011)
17. Tedaldi, J., Calderón, R., Mayora, L., Quirós, O., Farias, M., Rondón, S., Lerner, H.: Tratamiento de Maloclusiones según el estadio de maduración carpal—Revisión Bibliográfica. Revista Latinoamericana de Ortodoncia y Odontopediatría (2007)
18. Wilson, E.M., Green, J.R.: The development of jaw motion for mastication. Early Human Dev. **85**, 303–311 (2009)
19. Wolford, L.M., Mehra, P.: Custom-made total joint prostheses for temporomandibular. Baylor Univ. Med. Center Proc. **13**, 135–138 (2000)
20. World Record Academy. http://www.worldrecordacademy.com/medical/first_3D_printed_jaw_transplant_83-year-old_woman_sets_world_record_112714.html (2012)
21. Moreno-Garibaldi, P.: Diseño y Manufactura de una Protesis de Maxilar Inferior para Pacientes con Anquilosis Temporomandibular Lateral. Instituto Politécnico Nacional. Escuela Superior de Ingeniería Mecánica y Eléctrica. Maestría en Ciencias (2013)
22. Romo-Escalante, E.: Diseño y Manufactura de Prótesis Parcial de Maxilar Inferior para Paciente con Anquilosis Temporomaxilar y Malformaciones de Rama. Instituto Tecnológico y de Estudios Superiores de Monterrey, Campus Ciudad de México. Reporte de Proyecto Integrador (2012)
23. Beltrán-Fernández, J.A., Hernández-Gómez, L.H., Ruiz-Muñoz, E., González-Rebattú, A., Rodríguez-Cañizo, R.G., Urriolalagoitia-Calderón, G., Urriolagoitia-Sosa, G., Hernández-Moreno, H.: Biomechanical evaluation of corporectomy in porcine lumbar specimens using flexible polymer belts. J. Phys. Conf. Ser. (2009). doi:10.1088/1742-6596/181/1/012015

24. Beltrán, J. et al.: Assessment of the structural integrity of c3–c5 cervical porcine vertebrae model based on 2-D classic cad, 3-D scanner and 3-D computed tomography. In: Öchsner A., da Silva L.F. M., Altenbach H. (eds.) Analysis and Design of Biological Materials and Structures, vol. 1/2010– vol 20, Springer, Berlin (2010) ISBN: 978-3-642-22130-9
25. Beltrán-Fernández, J.A., Rebattú y González, M.G., Hernández-Gómez, L.H., Rebatú y González, A.G., Calderón, G.U.: Biomechanical Prosthesis Design of an Orbicular Cranial Cavity. In: Öchsner A., da Silva L.F.M., Altenbach H. (eds.), Advances in Bio—Mechanical Systems and Materials, vol. 40/2013–vol 40, pp. 87–94, Springer, Berlin (2013) ISBN: 978-3-319-00478-5. doi:10.1007/978-3-319-00479-2_7

Design Model of Damaged Steel Pipes for Oil and Gas Industry Using Composite Materials. Part II: Modeling

Zamzam A. Alsharif

Abstract This chapter presents an experimental study of a composite material for the rehabilitation of steel pipes. The damage in pipes was simulated as hole at the middle of the pipe which was made from carbon steel with internal diameter of 83 mm, thickness 12.5 mm and length of 900 mm. Cut which was chosen was (5 × 10), (7 × 14) and (9 × 18) mm for the three pipes respectively. The composite repair was fiber glass woven roving (type E) reinforced with epoxy resin. A carbon steel bolted clamp was used to clamp the pipe around the defected rectangular to minimize the delamination effect and stop the leakage of oil during the tests of the pipes. A special rig was designed to carry out pressurized tests on the repaired pipes. It was concluded that the maximum pressure obtained was 12 MPa for the pipe with a (7 × 14) mm rectangular defect. The microscopic examination showed that matrix cracking and delamination were the dominated failure mode in the most of failed pipes. Good agreement between the experimental and mode results of the radial stress-strain curves was achieved and observed at the early stages of loading [1]. However, discrepancy was observed at high strain to failure, which could be related to other relevant phenomena such as delamination and matrix macro-cracks.

Keywords Composite · Steel pipe · Delamination · Matrix cracking · Clamp · Quasi-isotropic laminate · Blister pressure · Rehabilitation · Failure mode · Stress and strain

---

Z. A. Alsharif (✉)
Faculty of Mechanical Engineering Technology, Benghazi, Libya
e-mail: zamzam_alsharif@yahoo.com

A. Öchsner and H. Altenbach (eds.), *Design and Computation of Modern Engineering Materials*, Advanced Structured Materials 54,
DOI: 10.1007/978-3-319-07383-5_12,

1 Introduction

The fast recent increase of the petroleum and natural gas production associated to a great energetic demand and primary resource provokes a significant increment in the national consumption of oil (gasoline, diesel, naphtha etc.) and gas.

As well as in other parts of the world, the discovery of relevant lid of hydrocarbonets in the Brazilian offshore coast made the Brazilian petroleum company invest more and more in the installation of petroleum extraction marine platforms. The exploration of these deposits constitutes high proportion business, as much as financial investment as return potential of the investment, regarding the production volumes involved and the load of the facilities that are necessary [2].

Metal tubular systems can be affected by internal or external corrosion or any other mechanical effects resulting in substantial damage to the systems. These lead to shutdown the plant, loss the production and increase the maintenance costs [3]. There are three options that can be chosen to solve the problem either by replacement, down rating or rehabilitation. The choice depends on the severity of the problem and the economical perspective of the option. The replacement and down rating are expensive options [4, 5].

The damages derived from corrosion processes in industrial installations produce very significant economical losses. For the gas and petroleum industry, the corrosion is responsible for 33 % of the cases [6]. The repair and reinforcement of existing structures has received a significant emphasis over the past few years due to corrosion and infrastructure aging. After some time in service, steel pipelines may be damaged, so they may be in need of repair due to the loss of carrying capacity. Alternatively, existing structures may need to have their resistance or stiffness upgraded to withstand an increased load demand or to eliminate structural design or construction deficiencies [2].

In this chapter, a novel technique for repairing metallic pipes using composite material and bolted clamp will be presented including an experimental procedure to evaluate the failure modes, stress-strain curves and microscope analysis.

Composite materials, used for repairing consist of fiber and resin. The fiber may be glass, carbon or Kevlar and the resin may be polyester, epoxy or vinyl ester [4, 5, 1].

2 Experimental Set-Up

2.1 Specimens Preparation

Three specimens of 108 mm diameter, wall thickness of 12.5 mm and length of 900 mm were cut from new carbon steel pipes(type: steel No. 10,421) as shown in Fig. 1. Each specimen was drilled at the mid length to simulate the corrosion defect. The dimensions of the rectangular defects which were chosen were 5, 7 and

Fig. 1 Photograph of pipes with rectangular defect

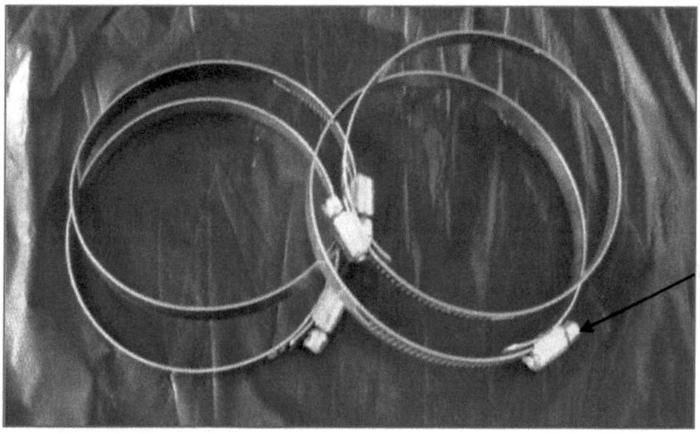

Fig. 2 Photograph of screwed clamps

9 mm for the three pipes, respectively. The specimens were flanged from their ends and provided with hydraulic fittings. The outer surface of the specimens were machined to remove the rusty skin. The warped material was fiber glass tape (type E), and the polymer was an epoxy resin. Screwed steel clamps, as shown in Fig. 2, were manufactured, in order to clamp the repaired area, in order to minimize the delamination effects and to stop the oil leakage.

2.2 Repairing Method

The scenario of repair is started by wrapping the specimens with several layers of fiber glass tape after wetting it in epoxy resin to obtain the required thickness, as shown in Fig. 3. After that an electric strain gauge is bonded, with radial direction of the pipe, on the top surface of a rectangular area. Immediately after this step, the repaired pipe is clamped by the screwed clamp as shown in Fig. 4, in order to minimize the delamination between the outer metal surface of the pipe and the inner surface of composite layer during the test stage. Finally, the pipe is lifted to cure at room temperature for 24 h and post curing at 120 °C for 2 h.

Fig. 3 Photograph of the beginning of the over-wrapping

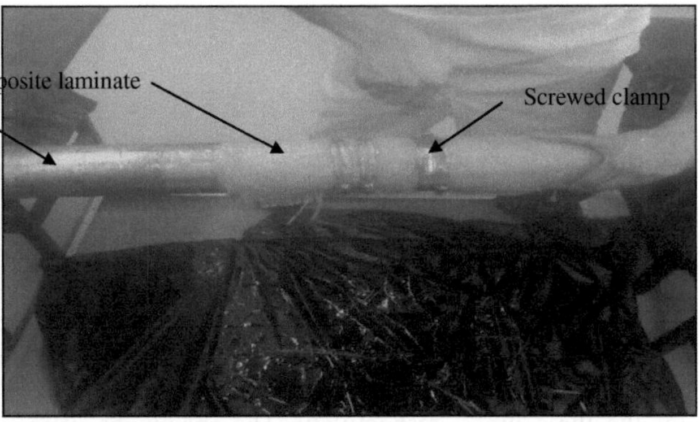

Fig. 4 Photograph of repaired pipe with screwed clamp

2.3 Testing Method

After the curing, the repaired pipe was holed on a stand and supported by two long bolts at its ends. The two ends of tube are provided by two flanges; one flange contains a hole for valve connection, and the other contains two holes for the hydraulic connections. A hydraulic oil pump with a capacity of 60 MPa is used to pressurize the repaired pipe. Figure 5 shows the assembly of the test rig with the repaired pipe. The test was started by filling the pipe with oil and switching on the pump, and then increase the pressure by small increments until failure occurs. At each pressure increment, pressure and strain readings were recorded from the pressure gauge and the strain meter, respectively.

Fig. 5 Test set-up apparatus

Table 1 Stress-Strain for rectangular defected with (5 × 10) mm

Pressure MPa	σ_x MPa	σ_y MPa	Theoretical radial strain $\varepsilon_x \mu\varepsilon$	Experimental radial strain $\varepsilon_x \mu\varepsilon$
1	0.81253	0.320362	37.29632	100
2	1.625061	0.640724	74.59264	100
3	2.437591	0.961086	111.889	100
4	3.250122	1.281448	149.1853	100
5	4.062652	1.60181	186.4816	200
6	4.875183	1.922172	223.7779	200
7	5.687713	2.242534	261.0742	300
8	6.500244	2.562896	298.3706	400
9	7.312774	2.883258	335.6669	500
10	8.125305	3.20362	372.9632	600
11	8.937835	3.523982	410.2595	700
12	9.750365	3.844344	447.5559	800

3 Results

The experimental results which were observed of the three pipes with defect lengths of 5, 7, and 9 mm and screwed clamp assembly are shown in Tables 1, 2, and 3.

4 Discussion

Figures 6, 7 and 8 show the experimental and theoretical radial stress versus radial strain curves for pipes with rectangular defect of 5, 7, and 9 mm, respectively for screwed clamp. The theoretical stress and strain were calculated using the model which was developed in the part one of this work [1]. It can be noted that all the curves exhibited the same manner. The experimental curves are linear in the early

Table 2 Stress-Strain for rectangular defected with (7 × 14) mm

Pressure MPa	σ_x MPa	σ_y MPa	Theoretical radial strain $\varepsilon_x \mu\varepsilon$	Experimental radial strain $\varepsilon_x \mu\varepsilon$
1	1.58385	0.621093	72.75582	100
2	3.167699	1.242186	145.5116	100
3	4.751549	1.863279	218.2675	200
4	6.335399	2.484372	291.0233	300
5	7.919248	3.105465	363.7791	300
6	9.503098	3.726558	436.5349	400
7	11.08695	4.34765	509.2907	500
8	12.6708	4.968743	582.0465	600
9	14.25465	5.589836	654.8024	700
10	15.8385	6.210929	727.5582	800
11	17.42235	6.832022	800.314	900
11.5	18.21427	7.142569	836.6919	1,000

Table 3 Stress-Strain for rectangular defected with (9 × 18) mm

Pressure MPa	σ_x MPa	σ_y MPa	Theoretical radial strain $\varepsilon_x \mu\varepsilon$	Experimental radial strain $\varepsilon_x \mu\varepsilon$
1	2.603991	1.022067	119.6019481	100
1.5	3.905987	1.533101	179.4029221	200
2	5.207982	2.044135	239.2038961	200
2.5	6.509978	2.555169	299.0048702	300
3	7.811974	3.066202	358.8058442	400
3.5	9.113969	3.577236	418.6068183	400
4	10.41596	4.08827	478.4077923	500
4.5	11.71796	4.599304	538.2087663	500
5	13.01996	5.110337	598.0097404	500
5.5	14.32195	5.621371	657.8107144	500
6	15.62395	6.132405	717.6116884	600
6.5	16.92594	6.643439	777.4126625	600
7	18.22794	7.154472	837.2136365	600
7.5	19.52993	7.665506	897.0146105	600
8	20.83193	8.17654	956.8155846	600
8.5	22.13392	8.687574	1016.616559	600
9	23.43592	9.198607	1076.417533	700
9.5	24.73792	9.709641	1136.218507	700
10	26.03991	10.22067	1196.019481	800
10.5	27.34191	10.73171	1255.820455	900

stages and then exhibited non linear behaviour up to failure, while the theoretical curves are linear up to failure. non-linear of the experimental curves is probably related to the matrix cracking through the thickness of the composite layers.

The non-linearity due to matrix cracks is well documented in the literature [7–9]. The immediate effect of micro cracks is to cause degradation of the stiffness due to

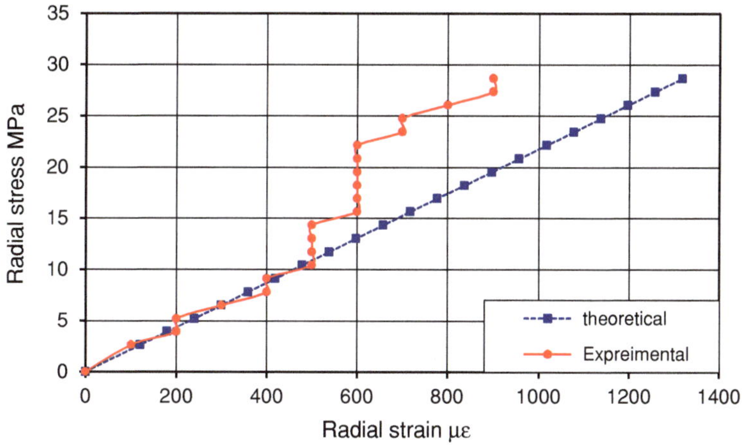

Fig. 6 Relationship between radial stress and radial strain of rectangular defect (9 × 18) mm

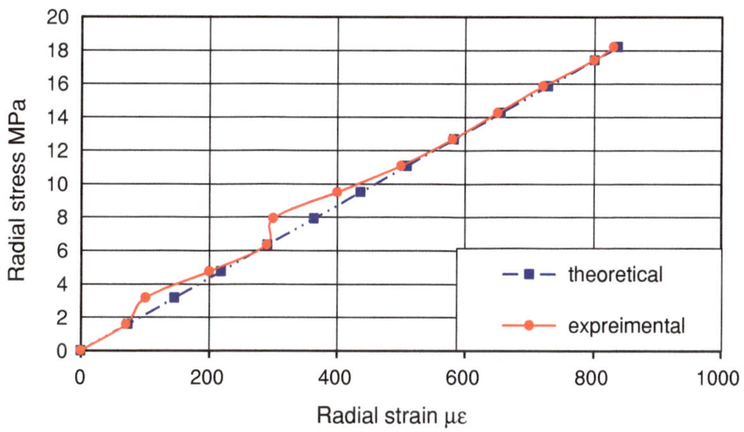

Fig. 7 Relationship between radial stress and radial strain of rectangular defect (7 × 14) mm

redistribution of stresses and variation of strain in cracked laminates [10, 11]. The matrix cracks can induce delamination which leads to fibre breakage or provides pathways for the entry of pressurised liquid between the layers and may lead to laminate failure. Good correlations between the two curves can be observed at the early stages of loading. However, discrepancies between the analytical and experimental curves can be seen at high strain to failure. This is probably due to further damage such as delamination matrix macro-cracks.

Figure 9 shows one mode of failure pipes where a rectangular defect can be observed at centre. This demonstrated that the screwed clamp works efficiently without any leakage of oil around the defected rectangular.

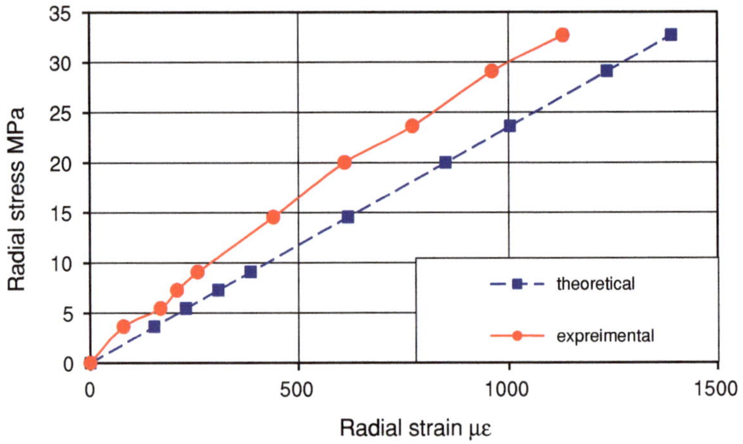

Fig. 8 Relationship between radial stress and radial strain of rectangular defect (5 × 10) mm

Fig. 9 Photograph of failure at centre of the rectangular defect (9 × 18) mm

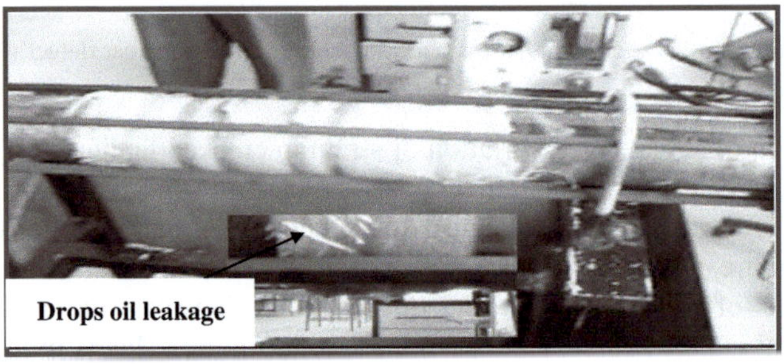

Fig. 10 Photograph of circumferential delimitation failure for rectangular defect (7 × 14) mm

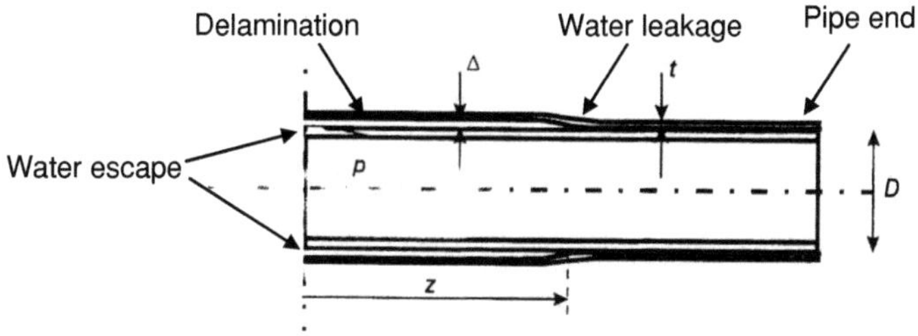

Fig. 11 Schematic diagram of the circumferential delamination mode [4]

Figure 10 shows the failure mode of the failed pipe. It can be seen that the pipe with (7 × 14 mm) exhibited a similar mode of failure which is called the circumferential delamination mode [4, 12]. This mode of failure can occur when delamination in defected regions has spread around the full circumference of the pipe as shown in Fig. 11.

5 Conclusions

- The work conclusions showed that it is possible to employ composite material for the external repair of metal pipes.
- A novel clamping technique has been used as delamination stopper and was found to work effectively.
- The maximum internal testing pressure of 12 MPa was observed by using screwed clamp for a repaired (5 × 10 mm) rectangular defect.
- A good agreement between the experimental and modelled results of the radial stress-strain curves was achieved and observed at the early stages of loading. However, discrepancy was observed at high strain to failure, which could be related to other relevant damage phenomena such as delamination and matrix macro-cracks.

References

1. Zamzam, A.A.L., Design model of damaged steel pipes for oil and gas industry using composite materials. Part I: Modeling. International Foundation for Modern Education and Scientific Research. (INFOMESR), Cairo, Egypt, WCPEE'12, 23–27 Dec 2012
2. Sampaio, R.F., Reis, J.M.L., Perrut, V.A., Costa H.S.: Rehabilitation of Corroded Steel Pipelines with Repair System, Mechanics of Solids in Brazil (2007)
3. Anwer, U.I., Hanim, M.T., Nureddin, M.A.: Steel pipe repair technique using the explosives welding. Eng. Fail. Anal. **12**(2), 181–191 (2005)
4. Mableson, A.R., Dunn, K.R., Dodds, N., Gibson, A.G.: Refurbishment of steel tubular pipes using composite materials. Plast. Rubber Compos. **29**(10), 558–565 (2000)

5. Zamzam, A.A.L., Saied, O.R., Muftah. T.A., Elarbi, M.B.: Repair of steel pipes for oil and gas industry using composite materials: part I. In: The Tenth Mediterranean Petroleum Conference and Exhibition, pp. 29–39. International Energy Foundation. Tripoli, Libya. 26–28 Feb 2008
6. Burton, M.: Applied Metallurgy for Engineers. McGraw-Hill Book Company—Inc, New York (1961)
7. Petit, P.H., Waddoups, M.E.: Method of predicting the non linear behaviour of laminated composite. J. Compos. Mater. **3**, 2–19 (1969)
8. Nahas, M.N.: Analysis of non linear stress-strain response of laminated fibre reinforced composite. Fibre Sci. Technol. **20**, 297–313 (1984)
9. Saied, R.O., Shuaeib, F.M., Modelling of the nonlinearity of stress-strain curves for composite laminates. J. Eng. Res. 7, 1–14 (2007)
10. Talreja, R.: Transverse cracking and stiffness reduction in composite laminates. J. Compos. Mater. **19**, 355–375 (1985)
11. Jones, M.L.C., Hull, D.: Microscopy of failure mechanisms in filament wound pipes. J. Mater. Sci. **14**, 165 (1979)
12. Forst, S.R.: Applications of polymer composite within the oil industry. In: Gibson, A.G. (ed.) Seventh International Conference on Fibre Reinforced Composites, pp. 84–91. University of Newcastle upon Tyne, 1998

Damage Detection in Materials Based on Computer Vision Wavelet Algorithm

Ivica Kuzmanić, Igor Vujović and Joško Šoda

Abstract Damage detection covers a wide range of different problems in a variety of technological, scientific and engineering fields. Solutions and reliability of results vary depending on the application and the field. This chapter explores the possibilities of application of computer vision in damage detection in materials. An example of damage detection in electronic components is presented, in which damage detection is based on the proposed computer vision wavelet algorithm. Furthermore, it proposes the incorporation of the algorithm in cloud service for industrial quality control. Results from the application of different wavelets are presented in the chapter.

Keywords Image processing · Damage detection · Wavelet transform · Quality control · Cloud applications

1 Introduction

The problem of damage detection is within the scope of research in many areas, such as quality control in factories, structural health monitoring (SHM) in civil engineering and maintenance of installed products (in marine, traffic, industry, electronics, machine engineering, etc.), to name a few.

I. Kuzmanić (✉) · I. Vujović · J. Šoda
Faculty of Maritime Studies, University of Split, Zrinsko-Frankopanska 38,
21000 Split, Croatia
e-mail: ikuzman@pfst.hr

I. Vujović
e-mail: ivujovic@pfst.hr

J. Šoda
e-mail: jsoda@pfst.hr

A. Öchsner and H. Altenbach (eds.), *Design and Computation of Modern Engineering Materials*, Advanced Structured Materials 54,
DOI: 10.1007/978-3-319-07383-5_13,

Generally speaking, the most non-destructive damage detection techniques are based on:

- The recording and studying of vibration signals or
- Use of computer vision methods.

Since vibration techniques rest on the assumption that damage changes the frequency response or harmonics, all kinds of spectral analysis techniques are applied.

Computer vision techniques are based on the observational difference between the damaged and undamaged sample. These algorithms can include not only visual sensors, but infrared as well. Furthermore, ultrasound visualization techniques can also be used for computer vision-based detection.

In this chapter, we deal with damage detection in electronic components/devices with the goal of detection of the damaged part. Such parts are expected to be found during maintenance, but similar approaches could likewise be used in the factory quality control service.

The motivation to investigate quality control is based on many factors. For example, surface finish is important for:

- Product's fatigue considerations,
- Strength,
- Creep life and/or
- Corrosion resistance.

The chapter is organized as follows.

Literature overview is given in Sect. 2.

Section 3 deals with the mathematical foundations of the proposed method, which is elaborated in Sect. 4.

Section 5 presents the results and the final section the conclusions.

2 Literature Overview

This section deals with trends in damage detection and the latest achievements in this area. We use the term "area" in the sense of principles and mathematical outline and not in the sense of applications. Applications are not as relevant for overview techniques and mathematical outline.

The section is divided into:

- Techniques using wavelets as kernel, and
- Techniques using computer vision.

The third part of the section deals with quality control.

2.1 Wavelet Techniques

There are a lot of methods for damage detection. For a real-time on-line health monitoring, piezoelectric, optical or electrical resistance methods are widely used.

Vibration-based damage detection is investigated in [1], but without wavelets in the results section.

Damage detection based on electrical resistance measurement is a popular method.

In [2] it was found that discrete wavelet transform (DWT) is more appropriate and that wavelet approximation coefficients are better than the original signal. Also, DWT offers better competitive information. It is argued that the advantage of the electrical resistance method in comparison to others is in fact that other techniques, such as piezoelectric sensor or optical fiber, require attachment of an external sensor.

Damage assessment of bridge based on the energy contained in wavelet coefficients is proposed in [3].

Damage sensitivity index is defined as the total energy of wavelet coefficients [3], which is close to the research presented in this chapter. Continuous wavelet transform (CWT) is performed on the time-series response data. The authors used Matlab and symlet wavelets (Matlab designation: sym8) as a tool in order to validate their research and technique.

On the other hand, DWT is performed in [4]. They used coiflets,[1] Daubechies and biorthogonal wavelets for damage detection and presented the comparison of the results of these wavelets.

The first method proposed in [5] implements the relative energy of wavelet coefficients in displacements, acceleration and velocities signals. This method is used to identify the physical parameters of an analytical model in SHM. The second proposed method in [5] improved the empirical mode decomposition (EMD) with DWT in the identification of damage in the structure.

In [6], authors correlated gearbox damage (detected in the vibration signal) in time with wavelet scale plots.

Bajaba and Alnefaie [7] used wavelet transform because their proposed method does not require the knowledge of the undamaged state. Furthermore, wavelets enabled clear visualization of signal discontinuities, they made it possible to pinpoint the damaged area.

Wavelet transform was used to detect and characterize defects in structures [8].

An approach to structural health monitoring (SHM) as a pattern recognition problem is derived in [9]. Guided waves are used for SHM and the formalism is based on distributed piezoelectric sensor network, which detects structural dynamic responses.

---

[1] Built by Daubechies at the request of Coifman. The wavelet function has 2N moments equal to zero. The scaling function has 2N − 1 moments equal to zero.

Reddy and Swarnamani [10] showed that local perturbations in mode shapes and strain energy data can be found in the finer scale of wavelet decomposition. The authors used a complex Gaussian wavelet with four vanishing moments. They successfully identified the location of damage.

Silva and others [11] used wavelets to detect damage in a cantilever beam subjected to static and modal analysis. Wavelets were used for numerical computations in the finite element method. Matlab's Wavelet Toolbox was used to compute the wavelet coefficients for two different mother wavelets—biorthogonal 6.8 (Matlab designation bior6.8) and Daubechies of order 2 (Matlab designation db2).

The method, which can be used in wireless sensor networks, was presented in [12]. The purpose is to improve SHM by the application of the "signature" of wavelet coefficients. It was shown that under 5 % of the data volume of an original signal should be enough to detect damage.

Yu and Giurgiutiu [13] used several advanced signal processing methods to improve damage detection with a piezoelectric wafer active sensor. Embedded ultrasonic structural radar was used as the signal source, which was further processed by:

- Short-time Fourier transform (STFT),
- Wavelet transform (WT),
- Hilbert transform.

The experimental algorithm [13] was programmed in LabVIEW.

2.2 Computer Vision Techniques

The use of computer vision is increasing every day. In this subsection, we review computer vision techniques of interest for this research with and without wavelets.

The application of satellite imagery to damage estimation is reported and analyzed in [14]. Damage estimation was performed on data-link satellite images. Rapid damage detection is obtained by calculating the differences between aerial images before and after disaster. The content of the chapter does not include wavelets.

Song and others [15] used a wavelet damage detection technique to locate damage in cantilevered composite beams by a non-destructive evaluation. The mentioned chapters investigated two types of damage: defects occurring during the removal of material from a metallic beam and manufactured delaminations in composite laminate.

Contactless acquisition of modal shapes is performed by a computer vision algorithm [15], which was a significant breakthrough. The authors used a conventional slow motion camera, which saved money, because there was no need for expensive equipment (e.g. laser doppler vibrometers).

A vision method for in-plane deflection curves is presented in [16]. The proposed method does not use wavelets, but a digital image correlation coefficient. Most of transforms can be expressed as a correlation between the signal and the transform's window.

Damage detection based on optical measurement was presented in [17]. They used wavelets in their research.

Structural damage assessment using image processing is presented in [18] by using the Lucas-Kanade optical flow algorithm and OpenCV environment. The main advantages of the proposed were:

- Short setup time,
- No elaborate fixture of transducers,
- No crest factor problems,
- No destruction of structure,
- No ambient or unmeasured excitation required, etc.

Drawback was the un-scaled modal model, advantage if we had some a priori knowledge, large time histories, which means higher computational costs, etc. The relative displacement is measured from a real-time video.

Kohut and others [19] monitored structures in civil engineering by computer vision.

Continuous wavelet transform (CWT) was used for damage detection and localization. Detection and localization of damage were performed by finding the maxima of the CWT coefficients and the corresponding scale and shift values. A Gauss wavelet of the 4th order (Matlab designation *gaus*4) was used in the paper [19]. The Gauss wavelet is the wavelet family, which is built from the *n*th derivate of the Gaussian function:

$$f(x) = C_p \cdot e^{-x^2} \tag{1}$$

where:

x is independent variable,
$f(x)$ function of the variable x,
p is the family parameter, which is an integer number and it is the order of derivate,
C_p is such that:

$$\left\| f^{(p)} \right\|^2 = 1 \tag{2}$$

Lima and others [20] also used peak values, but only in details coefficients of wavelet decomposition, to locate the damage.

Measuring of structural vibration is explained in [21]. They used a high resolution camera and demonstrated the technique of sub-pixel edge detection. Mode shape difference functions are subjected to CWT to determine damage locations.

2.3 Computer Vision Quality Control

Computer vision applications, using wavelet compressed Zernike moments, are presented in [22]. They referred to successful classification and object characterization in quality control systems based on visual inspection.

The normalized wavelet reconstruction error was computed by using the following expression [23]:

$$\bar{e} = \sqrt{\frac{\sum_i \sum_j [f(i,j) - \hat{f}(i,j)]^2}{\sum_i \sum_j [f(i,j)]^2}} \tag{3}$$

where:

$f(i, j)$ is the intensity function of the original signal, and

$\hat{f}(i,j)$ is the intensity function of the reconstructed image.

They suggested that the most prominent examples of the application include defect detection in visual inspection quality control systems. The application of wavelets shortened the time necessary to perform the task.

Surface quality of a machine part is monitored by surface roughness in [24]. They stated that the surface metrology with image processing is a very challenging task, with wide application in industry.

In relation, wavelet transforms was used to force images in [25]. The reason to use computer vision in such tasks is surface finishing [24]. It significantly improves:

- Fatigue strength,
- Corrosion resistance,
- Creep life, etc.

Furthermore, surface roughness affects:

- Friction,
- Wear and tear,
- Light reflection,
- Heat transmission,
- Ability of distributing and holding a lubricant, coating, etc.

The desired surface finish is required to maintain the quality.

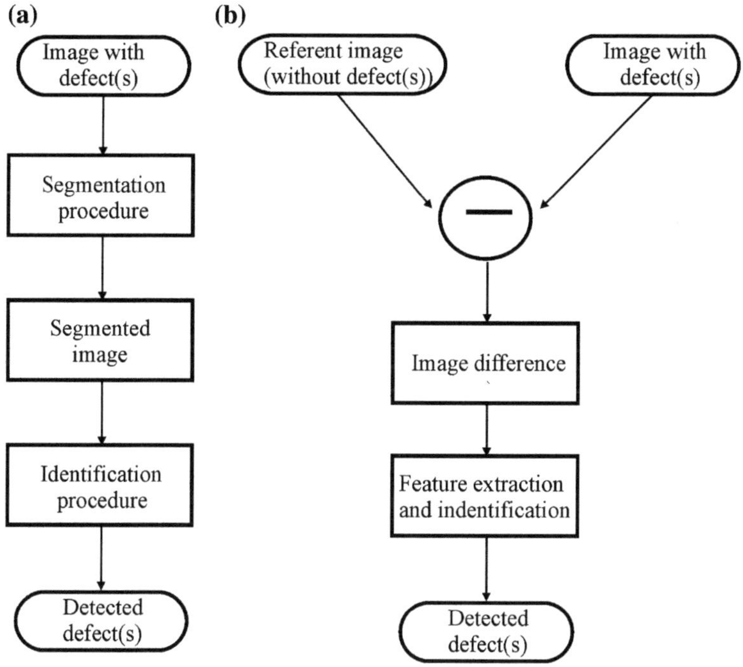

Fig. 1 Two basic defect detection procedures: **a** from the observed image (reconstructed according to [26]), **b** from the difference between referent and observed image

Optical measuring methods are sensitive to lighting conditions and noise. The measure used for quality assessment was surface roughness, defined in gray-level images as [24]:

$$G_a = \frac{\sum (|g_1 - g_m| + |g_2 - g_m| + \cdots + |g_n - g_m|)}{n} \tag{4}$$

where:

$g_1, g_2, \ldots, g_n$ are gray level values of a surface image along one line and
g_m is the mean gray level.

A rough surface can be interpreted as follows. The higher peak is brighter than the corresponding area in the image. This leads to texture analysis, which is still a subject of many researches.

An experience from industry is described in [26]. The described system is based on computer vision technique that recognizes different defect types in ceramic plates.

Figure 1 shows two basic defect detection procedures.

In Fig. 1a, the procedure is performed on the observed image. The algorithm is reconstructed from [26]. Firstly, an image with defects is processed by segmentation

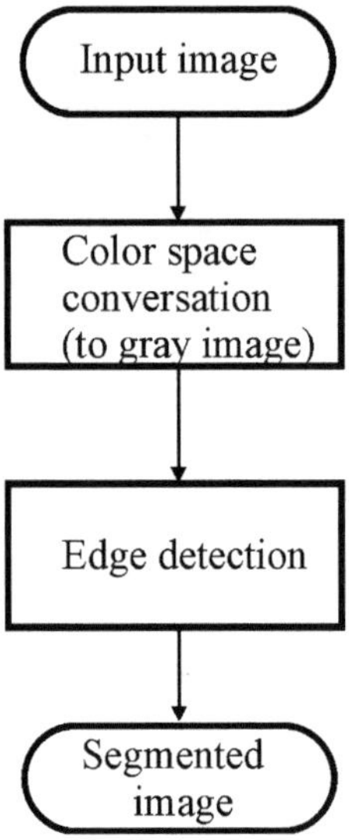

Fig. 2 Segmentation algorithm

procedure. The result is a segmented image. The obtained segmented image is used for identification procedure and, finally, defects are detected.

The second defect detection procedure (Fig. 1b) is based on image differencing. In this case, the reference image (image without defects) and the image with defect(s) must both be known. Computer vision techniques are used in image differencing to extract and identify features of interest. This scheme was conceived by the authors of the paper you are currently reading.

Figure 2 shows the segmentation algorithm similar to the one proposed in [26], but in a more general sense.

Edge detection is performed from the gray image (Fig. 2). However, the algorithm has a problem in application when the color camera serves as the input. Conversion from color space to gray image can be questionable. Due to different thresholding methods, results can be bad images or even blanks.

Figure 3 shows a classic example of identification algorithm. After image segmentation process, the segmented image is passed to a morphological filter. Usually, morphological dilatation operation is performed.

Vision quality control can be performed by a high resolution camera in the visual spectrum. In this case packaging is controlled.

Instead of the visual spectrum, the infrared spectrum and infrared cameras can be used. In this case, outside and inside of the product can be controlled with non-invasive method.

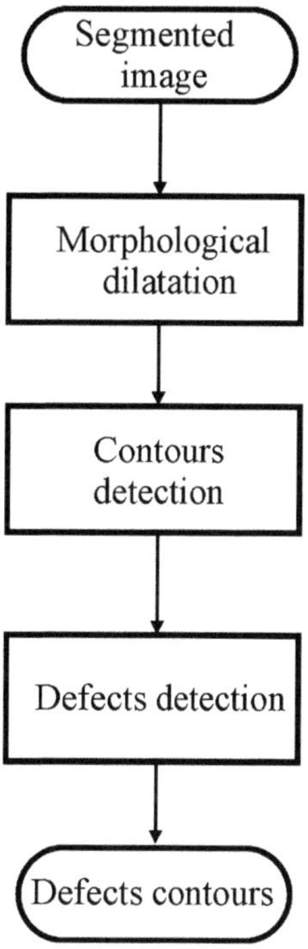

Fig. 3 An example of the identification algorithm

3 Mathematical Foundations

Wavelet theory is well known [27–35] and there is no need to repeat the basics. In this chapter, we will present only issues essential for understanding the basics of the proposed method.

Several conditions have to be satisfied in order to obtain and define functional wavelet family and transform.

The first condition is [35]:

$$\int \psi(t)dt = \Psi(\omega = 0) = 0 \tag{5}$$

where:

$\psi(t)$ is the wavelet in time domain, and

$\Psi(\omega)$ the same wavelet described in frequency domain.

The wavelet $\psi(t)$ is called mother wavelet.

The set of functions, which are obtained from the mother wavelet by shifting and translating, are:

$$\psi_{a,b}(t) = \frac{1}{\sqrt{|a|}}\psi\left(\frac{t-b}{a}\right) \tag{6}$$

where:
$\psi(t) \in L_2(\Re)$ and
$a, b \in \Re$, with $a \neq 0$.

Further condition is:

$$\|\psi(t)\|^2 = \int_{-\infty}^{+\infty} |\psi(t)|^2 dt = \frac{1}{2\pi}\int_{-\infty}^{+\infty} |\Psi(\omega)|^2 d\omega = 1 \tag{7}$$

If conditions (5), (6) and (7) are satisfied, it is possible to define wavelet transform. Continuous Wavelet Transform (CWT) is defined by [34]:

$$CWT(a,b) = \frac{1}{\sqrt{a}}\int_{-\infty}^{\infty} f(t)\cdot\psi*\left(\frac{t-b}{a}\right)dt = \langle\psi_{a,b}(t), f(t)\rangle \tag{8}$$

where:
$a \in \Re^{+}$ designates scaling factor (parameter which directly influence resolution in frequency domain),
$b \in \Re$ designates the position (influences time resolution of the signal),
$f(t)$ the analyzed signal,
$\psi_{a,b}(t)$ basic functions.

Discrete Wavelet Transform (DWT) was developed for calculations in discrete-time and computer applications. DWT is obtained by discretization of scale parameter, a, and shift parameter, b. In some references, a is replaced with s and b with τ. Designations s and τ are usually used for DWT.

The usual approach is to choose parameters a (or s) and b (or τ) as positive numbers with [35]:

$$s = s_0^m \tag{9}$$

where $s_0 \in \Re$, $m \in Z$ and:

$$\tau = n\tau_0 s_0^m \tag{10}$$

where $n \in Z$ and $\tau_0 \in \Re$.

During parameter discretization, scaling parameter is discretized firstly. Parameters choice is dependent. The wavelet basis function becomes:

$$\psi_{a,b}(t) = |a_0|^{-\frac{m}{2}} \cdot \psi\left(\frac{t - b_0 \cdot a_0^m \cdot n}{a_0^m}\right) \tag{11}$$

and finally, we have:

$$\psi_{a,b}(t) = |a_0|^{-\frac{m}{2}} \cdot \psi\left(a_0^{-m} \cdot t - b_0 \cdot n\right) \tag{12}$$

So, DWT can be represented as:

$$W(s,\tau) \cong \int x(t) \cdot \psi_{j,k}^*(t)dt = d_{j,k} \tag{13}$$

where:

s is the scale parameter,
τ is the shift parameter
$x(t)$ is the signal which is decomposed,
$\psi_{j,k}$ is the basic wavelet function, and $\psi_{j,k}^*$ its conjugate,
$d_{j,k}$ is the wavelet transform coefficient.

The inverse DWT can be expressed as:

$$x(t) = \frac{1}{c} \sum_{j=-\infty}^{\infty} \sum_{k=-\infty}^{\infty} d_{j,k} \cdot \psi_{j,k}(t) \tag{14}$$

where the constant, c, equals to:

$$c = \int_{-\infty}^{\infty} \frac{|\psi(\omega)|}{|\omega|} d\omega \tag{15}$$

under condition that:

$$\int_{-\infty}^{\infty} \psi(t)dt = 0 \tag{16}$$

To understand energy methods one must understand Definitions 3.1 and 3.2.

Definition 3.1 If f and h are in $L^1(R) \cap L^2(R)$, the following definition applies [34]:

$$\int_{-\infty}^{+\infty} f(t)h^*(t)dt = \frac{1}{2\pi} \int_{-\infty}^{+\infty} \hat{f}(\omega)\hat{h}^*(\omega)d\omega \tag{17}$$

where $\hat{f}$ and $\hat{h}$ are Fourier's transformants of f(t) and h(t), respectively.

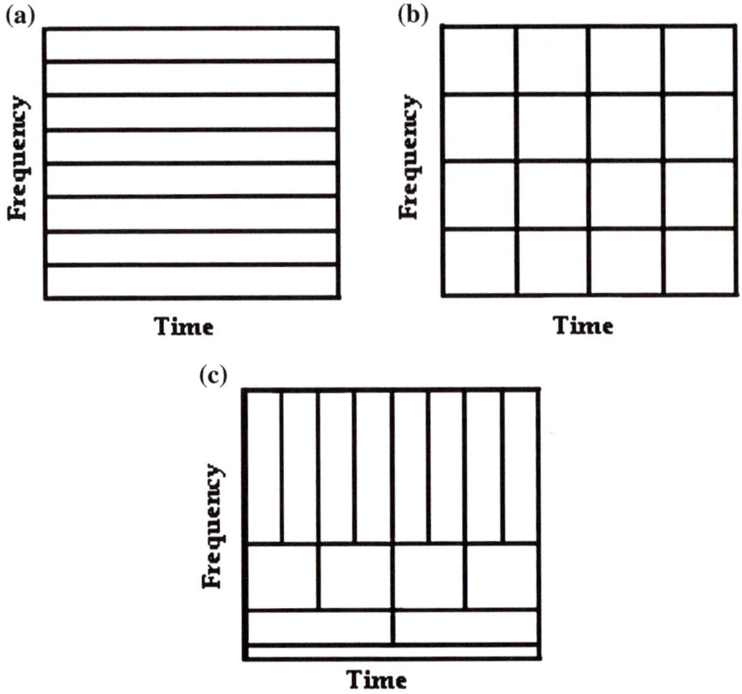

Fig. 4 Tilling of the time-frequency plane for: **a** FT, **b** STFT, **c** wavelets [31]

Definition 3.2 The energy of a signal can be calculated from the time-domain as well as from the frequency-domain as an integral in time interval or as sum of energies of discrete spectrum components, if discrete Fourier transform (DFT) is known [34]:

$$\frac{1}{2\pi}\int_{-\pi}^{\pi}\left[\sum_{n=-\infty}^{+\infty} f(n)e^{-j\omega n}\right]F^{*}\left(e^{j\omega n}\right)d\omega = \sum_{n=-\infty}^{+\infty} |f(n)|^{2} \tag{18}$$

Definitions 3.1 and 3.2 are general. They can be applied to filter realizations of the wavelet transform.

It is important to notice that the most commonly used transform is still the Fourier transform.

If it is not sufficiently good for a specific application, other transforms are used. Therefore, the tilling of the Fourier transform (FT) is usually compared with the wavelet's tilling.

In Fig. 4, we also added short time Fourier transform (STFT), which is the common approach for solving non-stationary problems.

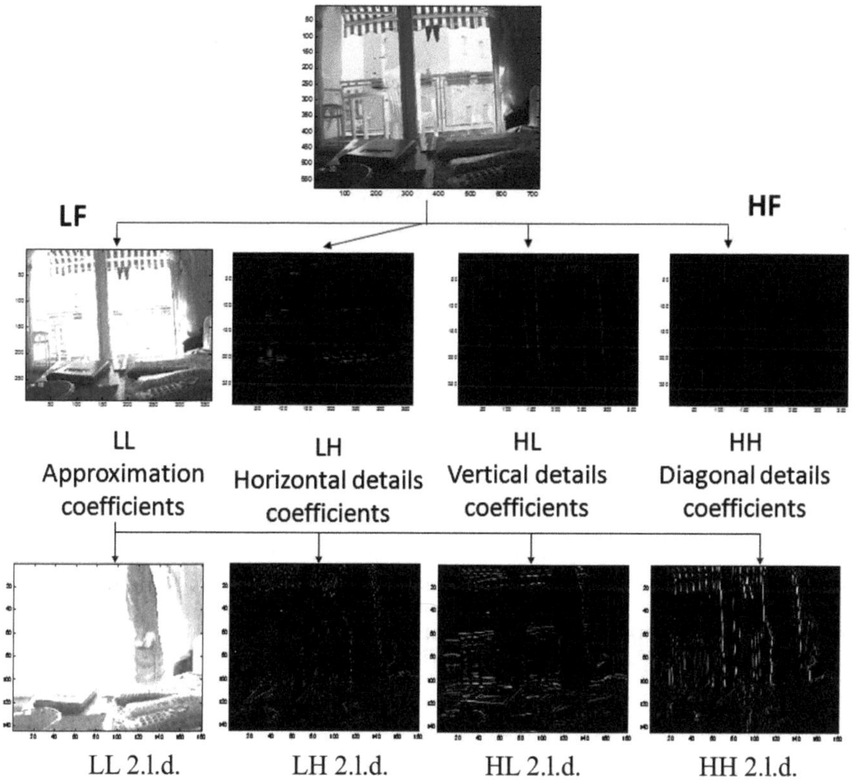

Fig. 5 Wavelet decomposition of image

The wavelet transform splits the signal into the approximation coefficients and details coefficients. When, only one value is of interest in time, then we talk about a wavelet one-dimensional problem, which is incorrect, because there are two axes. Observing the cutting depth of a turning machine would be an example of such a signal.

When dealing with computer vision, we are dealing with the so called two-dimensional problem, because we have two spatial components plus the time component. It is actually a three-dimensional problem. When dealing with a one-dimensional problem, there are only approximation and details. However, in case of images, details can be obtained from several planes.

In case of wavelets, the details coefficients are usually calculated in the horizontal, vertical and diagonal plane. Nowadays, there are bleeding-edge transforms, which calculate details in several different angles. It could be said that they are modified wavelet transforms, regardless of the name difference.

Figure 5 shows the wavelet decomposition of an image. At the first level, LL, LH, HL and HH coefficients are obtained.

If we decompose LL further, we obtain new LL, LH, HL and HH coefficients with less elements (smaller dimensions). This is the advantage of the wavelet transform, because almost all data are described by a smaller number of coefficients (less data for the computer to process).

As shown above, most of the energy remains at the first level of decomposition in approximation details. It should be noted that the second level details are obtained by first level approximation. Decomposition can be performed in the details of the first level as well.

It is common to choose coefficients which will be decomposed further. It is called a decomposition tree. In this case, we choose to decompose the approximation coefficients to the second level of decomposition.

The details coefficients are mostly full of noise. Therefore, it is usual to discard them from calculations. Taking into account all coefficients slows down execution time by four times. Furthermore, more than 90 % of the total energy is present in the approximation coefficients [33].

Figure 6 shows the chosen wavelet tree for two levels of decomposition applied in the results section. The third level is continued work from previous levels.

As can be seen from Fig. 6, the input image is decomposed at the 1st level of decomposition into four data sets (matrixes):

- A1 stands for approximation coefficients at the first level,
- Dh1 stands for details coefficients in the horizontal direction,
- Dv1 means details coefficients in the vertical direction,
- Dd1 means details in the diagonal direction.

Our tree includes only A1 coefficients at the first level of decomposition. Further decomposition of the products of first level decomposition (A1, Dh1, Dv1 and Dd1) can be decomposed further, which is called 2nd level of decomposition.

Coefficients A1 are decomposed in:

- Aa2 (approximation of the 1st level approximation),
- Ha2 (horizontal details of the 1st level approximation),
- Va2 (vertical details of the 1st level approximation) and
- Da2 (diagonal details of the first level approximation).

Coefficients Dh1 are decomposed to:

- Ah2 (approximation of the horizontal details),
- Hh2 (horizontal details of the horizontal details at the first level),
- Vh2 (vertical details of the horizontal details) and
- Dh2 (diagonal details of the horizontal details).

Coefficients Dv1 are decomposed to:

- Av2 (approximation of the vertical details),
- Hv2 (horizontal details of the vertical details),
- Vv2 (vertical details of the vertical details) and
- Dv2 (diagonal details of the vertical details).

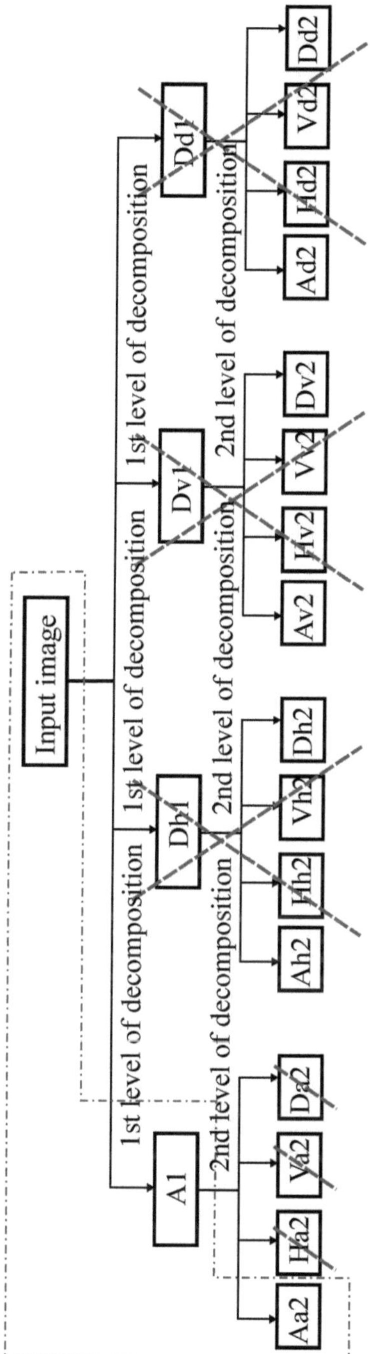

Fig. 6 Chosen wavelet tree

Coefficients Dd1 are decomposed to:

- Ad2 (approximation of the diagonal details),
- Hd2 (horizontal details of the diagonal details),
- Vd2 (vertical details of the diagonal details) and
- Dd2 (diagonal details of the diagonal details at the first level).

As stated previously, tree is chosen by the application.

4 Proposed Algorithm

In this chapter, we propose an overall algorithm for cloud visual quality control as well as the computer vision algorithm for damage detection. The overall algorithm is shown in Fig. 7. The computer vision algorithm is shown in Fig. 8. The computer vision algorithm can be implemented in the cloud as proposed, but also by on-site installation.

The overall algorithm assumes the computer network and relays the images from the web camera to the computer vision algorithm installed at a distant facility. Feedback from the computer vision algorithm is the action that should be taken by the actuators.

Possible problems in the implementation of the overall algorithm include: communication safety, data transmission rate on the internet and the algorithm's rate of data acquisition.

The damage detection computer vision algorithm consists of several steps:

- First, the image acquisition process should be implemented.
- The second step is the conversion of the usual color space to a gray image, because edge detection algorithms works only in gray images.
- The third step is edge detection. This feature is necessary in the automatic procedure. However, this step can be skipped if human operator manually chooses the region of interest (ROI). After, ROI is chosen and identified, the energy is calculated for the ROI in the corresponding area of the color image.
- The next step is to compare the calculated energy with the reference energy. The reference energy can be obtained by the knowledge base and/or image without defects by the same procedure.
- The formation of the knowledge base leads to a rudimentary expert system. In this case, we have a knowledge base which can be improved by the reference image and vice versa. Therefore, learning rules can be defined and the system becomes an expert system for visual quality control. The knowledge base can be improved by expert knowledge about a particular plant, materials, devices and products.
- The final step is to suggest action to be taken. The suggestion is then sent over the internet connection to the plant, where the suggestion is executed.

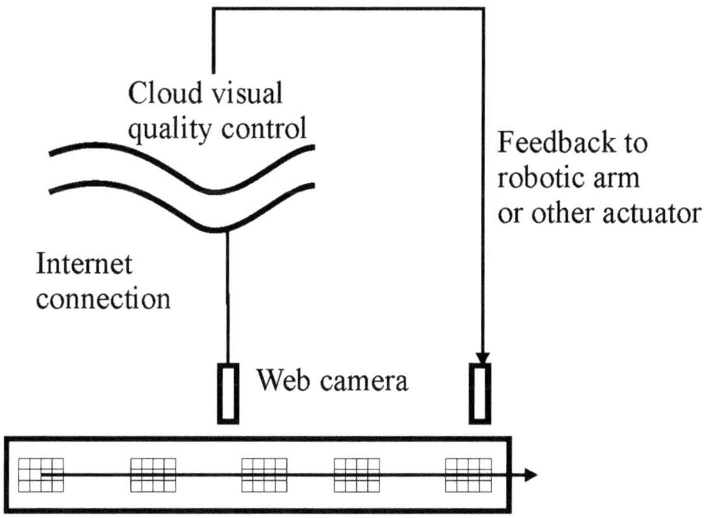

Fig. 7 Cloud visual quality control

One can notice that the algorithm does not explicitly show the point of wavelet usage. The reason is in the fact that the algorithm can be implemented with or without wavelets.

A wavelet transform can be implemented in the beginning, before color conversation. In such a case, only approximation coefficients are used in further steps.

Another place to use wavelets is in the energy calculation from the color image, when original color space is replaced by approximation coefficients. The energy is then calculated in the wavelet domain.

The algorithm leaves many issues for optimization. Firstly, what kind of edge detection should be used, i.e. Prewitt or Canny. Furthermore, conversation to gray image can lead to difficulties in correct threshold selection.

It should be pointed out that the energy is calculated in Hilbert space using Parseval relation [32]. The calculated energy is divided by the number of coefficients in ROI and the energy measure is obtained. The energy measure (*EM*) can be expressed as:

$$EM = \frac{1}{N}\sum_{i}\sum_{j}\sum_{k} a(i,j,k)^2 \tag{19}$$

where:

EM	denotes the energy measure,
N	is the number of pixels in the ROI for the considered level of decomposition,
i and *j*	are the spatial coordinates of the pixel for which *a* is the energy,
k	is the color designation (red 1, green 2, and blue 3),

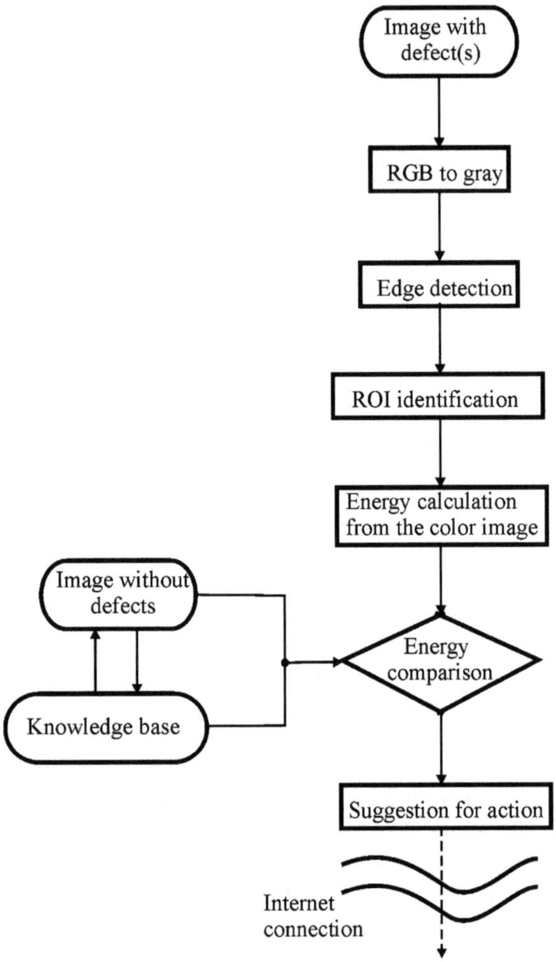

Fig. 8 Proposed algorithm

a (i, j, k) is the energy of the pixel contained in *k*-th color calculated in Hilbert space (wavelet domain).

The energy measure is used as a number for comparison of energies for damaged and undamaged ROIs.

Possible problems in the computer vision algorithm are in:

- edge detection,
- colour conversation for edge detection purposes,
- ROI shape modelling,
- ROI recognition,
- ROI analysis with conclusion about quality, etc.

Fig. 9 Example of processed images

5 Results

Energy measure (*EM*) was calculated in the experiment firstly. *EM* for the ROI without damage in case of Fig. 9 is equal to 1.6036×10^4. On the other hand, *EM* for the ROI with damage in the same case is equal to 2.2430×10^4. However, this is obtained when normalized energies are not used. If we use normalized energies, *EM* for ROI without damage is greater than with damage.

In any case, it is possible to distinguish between damaged and undamaged ROIs.

Figures 10, 11, 12, 13 and 14 show results for normalized energies.

Figures 10, 11, 14 show different trends obtained by different wavelets and different trends between even and odd order wavelets of the same family.

Figures 16 and 17 show that an increase of energy exponentially tends to make the resulting image darker. After several powers, only damage remains different than dark.

Figure 10 shows the average pixel energy in a ROI. The energy is normalized and therefore there are no units.

As it can be seen in Fig. 10a, the average energy per pixel is higher if ROI is not damaged (red line). Energy is plotted versus wavelet orders for biorthogonal family.

The results lead to the conclusion that the energy of odd moments decreases in the middle of the order's range, and that the energy of even moments increases in the middle of the order's range (Fig. 10b).

In this experiment, the energy is calculated for the first level of decomposition and takes into account only approximation coefficients.

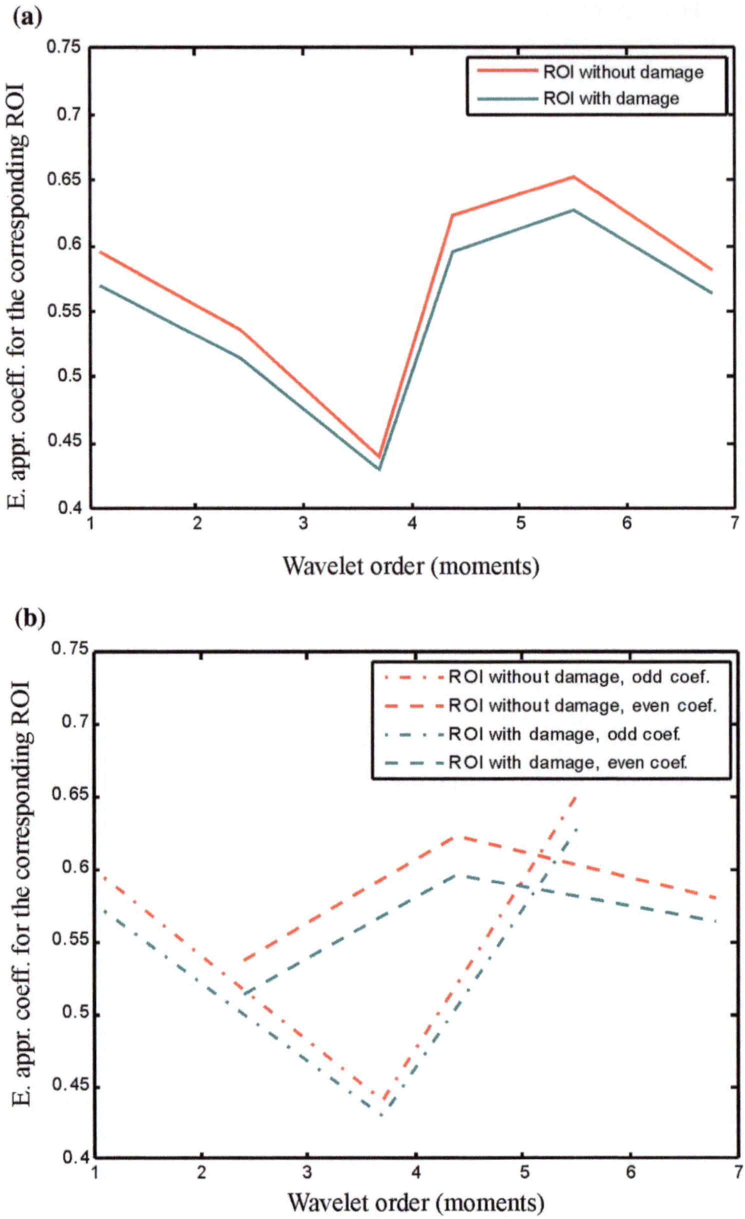

Fig. 10 Results of the proposed for biorthogonal wavelets: **a** average energy contained in the ROI for damaged and undamaged component, **b** comparison of even and odd moments

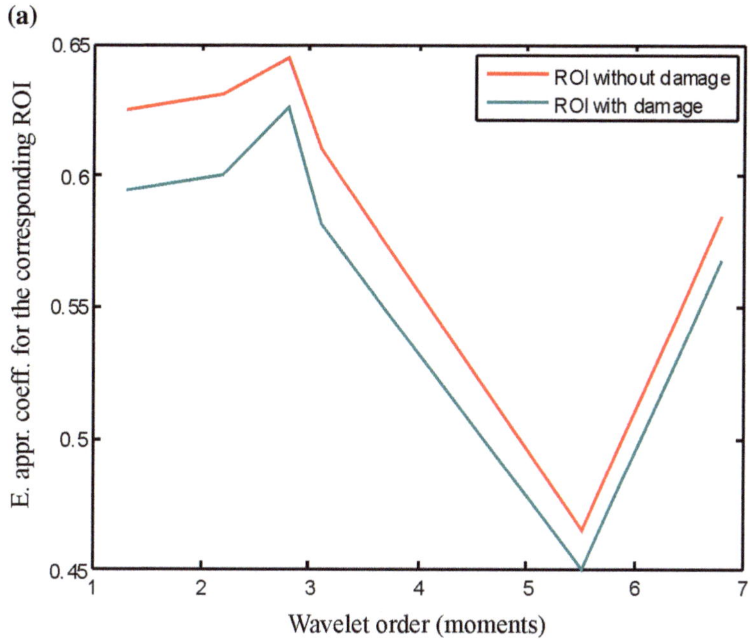

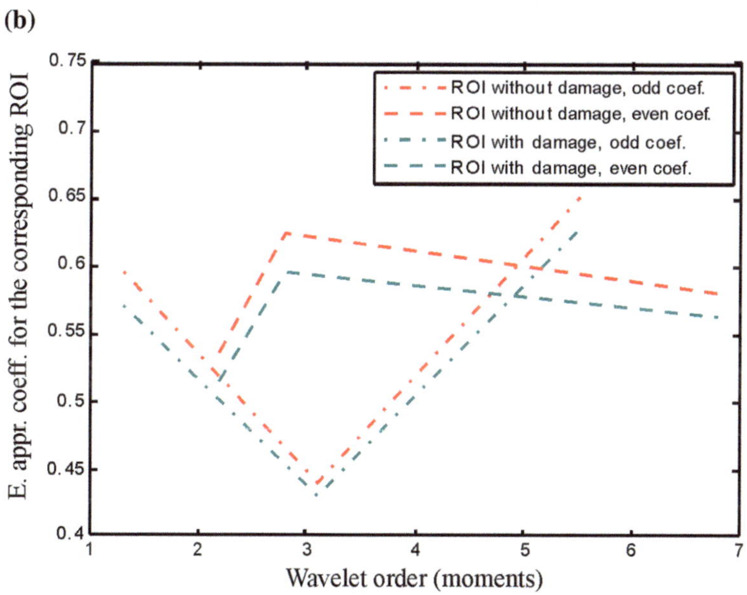

Fig. 11 Results of the proposed for reverse biorthogonal wavelets: **a** average energy contained in the ROI for damaged and undamaged component, **b** comparison of even and odd moments

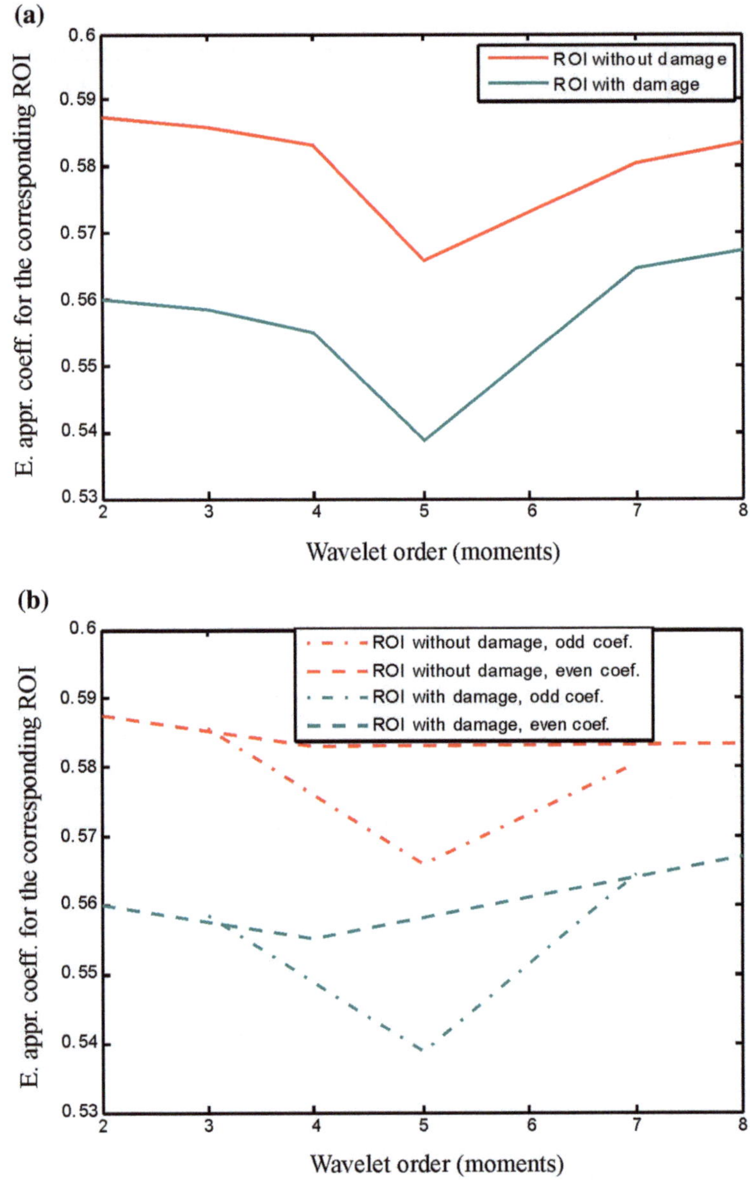

Fig. 12 Results of the proposed for symlet wavelets: **a** average energy contained in the ROI for damaged and un-damaged component, **b** comparison of even and odd moments

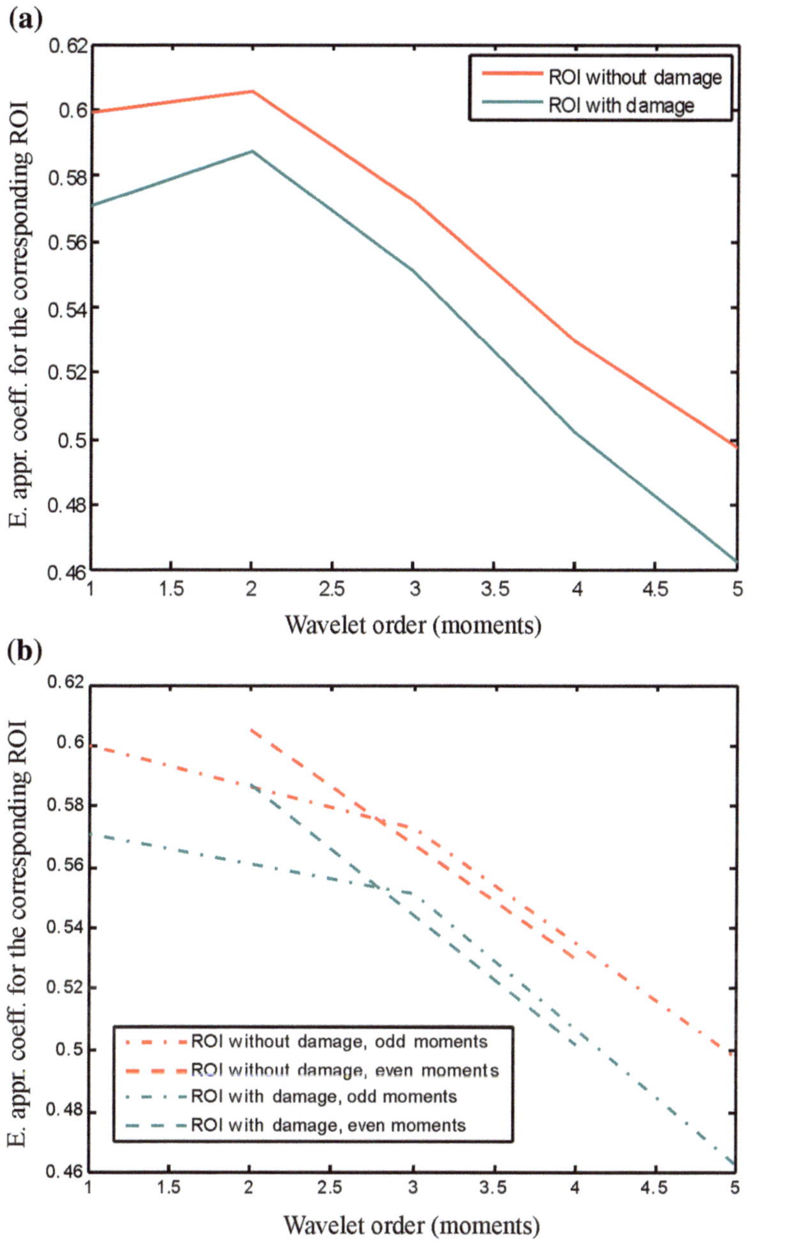

Fig. 13 Results of the proposed strategy for coiflet wavelets: **a** average energy contained in the ROI for damaged and undamaged component, **b** comparison of even and odd moments

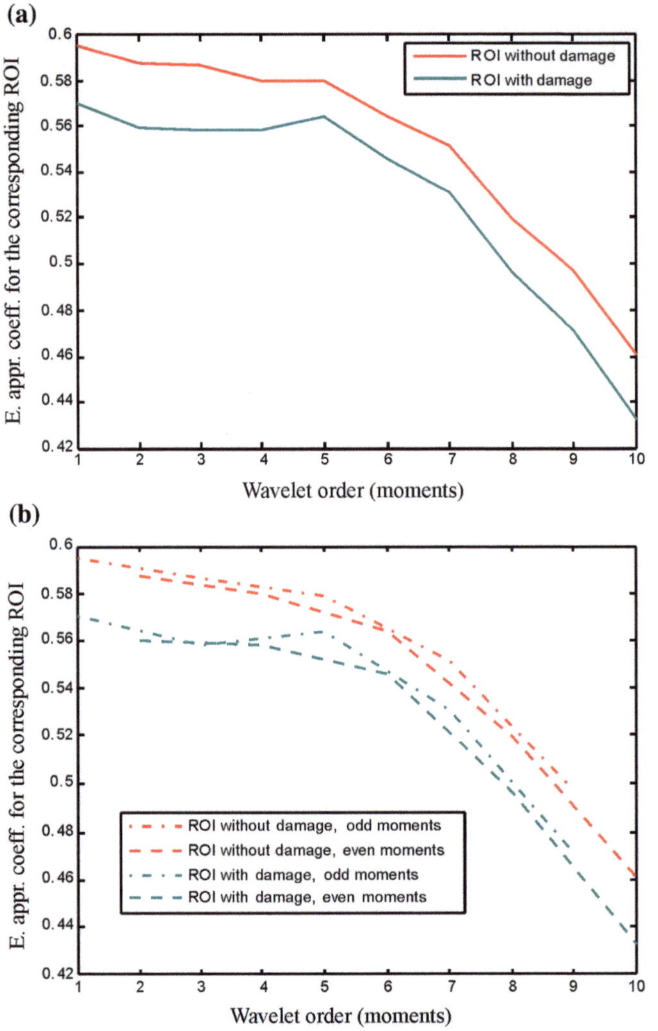

Fig. 14 Results of the proposed strategy for Daubechies wavelets: **a** average energy contained in the ROI for damaged and undamaged component, **b** comparison of even and odd moments (Matlab designations: db 1–10)

Figure 11 shows the average normalized pixel energy in the ROI calculated with reverse biorthogonal wavelets at the first level of decomposition. Only approximation coefficients were taken into account.

It can be seen (Fig. 11a) that the average energy per pixel is higher if ROI is not damaged (red line).

The results lead to the conclusion that the energy of odd moments decreases in the middle of the order's range, and that the energy of even moments increases in the middle of the order's range (Fig. 11b).

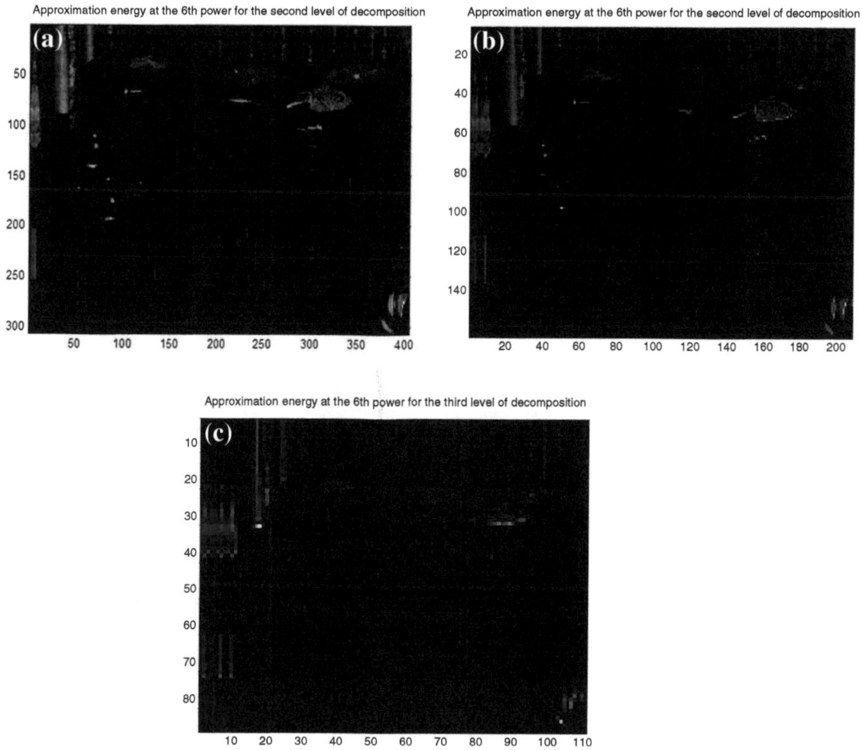

Fig. 15 Energy contained in the approximation coefficients at the sixth power for: **a** the first level of approximation, **b** the second level of decomposition, **c** the third level of decomposition

Figure 12 the shows average normalized pixel energy in the ROI calculated with symlet wavelets at the first level of decomposition. Only approximation coefficients were taken into account.

It can be seen (Fig. 12a) that the average energy per pixel is higher if ROI is not damaged (red line).

The results lead to the conclusion that the energy of odd moments decreases in the middle of the order's range, and that the energy of even moments increases in the middle of the order's range (Fig. 12b).

Figure 13 shows the average normalized pixel energy in the ROI calculated with coiflet wavelets at the first level of decomposition. Only approximation coefficients were taken into account.

It can be seen (Fig. 13a) that the average energy per pixel is again higher if ROI is not damaged (red line).

The results lead to the conclusion that the energy of both odd and even moments decreases for higher orders (Fig. 13b).

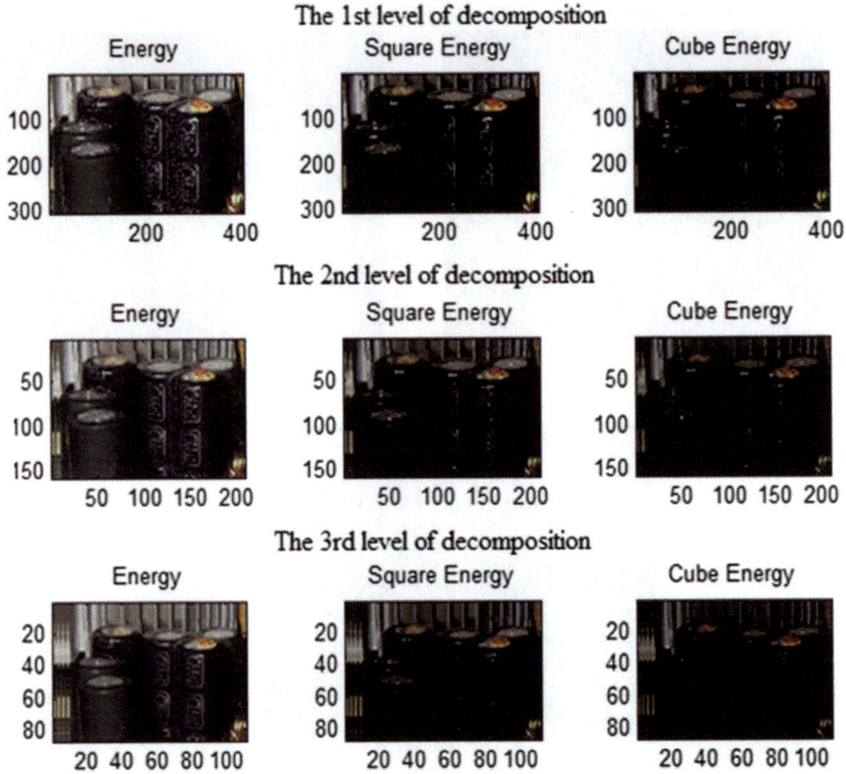

Fig. 16 Results of energy and powers of energy calculation in wavelet domain for the first three levels of decomposition (approximation coefficients)

Figure 14 shows the average normalized pixel energy in the ROI calculated with Daubechies wavelets at the first level of decomposition. Only approximation coefficients were taken into account.

It can be seen (Fig. 14a) that the average energy per pixel is again higher if ROI is not damaged (red line).

The results lead to the conclusion that the energy of both odd and even moments decreases for higher orders (Fig. 14b).

Figure 15 shows the energy contained in approximation coefficients at the sixth power for the first, second, and third level of decomposition for Daubechies wavelet of 7th order.

Figure 16 shows the energy, square energy and cube energy contained in approximation coefficients for the first, second and third level of decomposition for Daubechies wavelet of the 7th order.

Figure 17 shows energy at the 4th, 5th and 6th power for the first, second, and third level of decomposition.

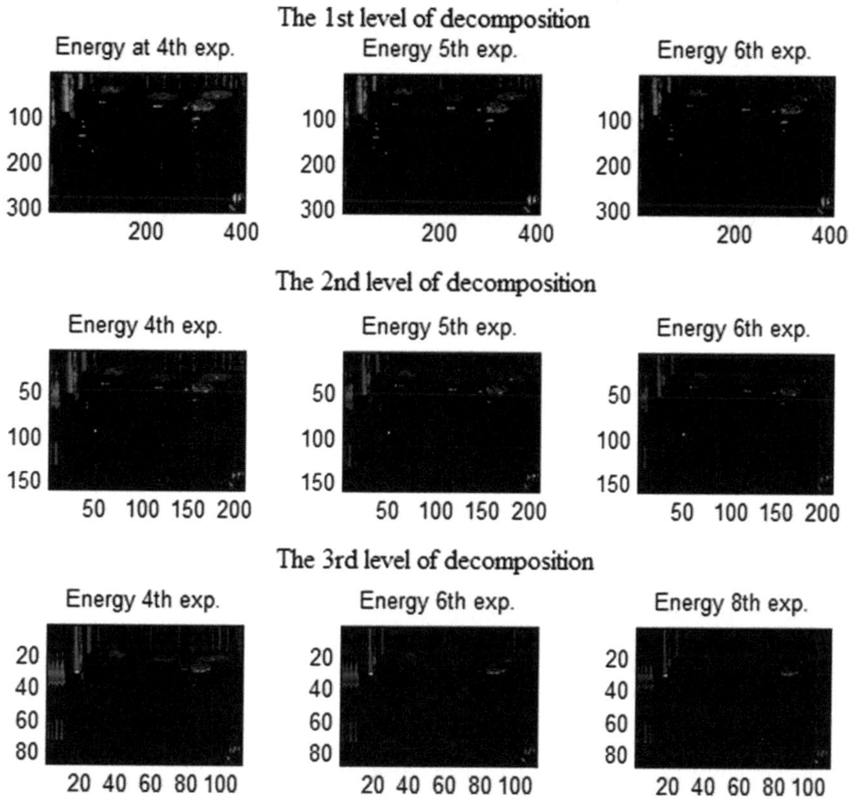

Fig. 17 Results of powers of energy calculation in wavelet domain for the first three levels of decomposition (approximation coefficients)

It can be seen that almost all phenomena in the image vanish in the dark, except for the damaged ROI. Other background anomalies can be suppressed by a background removal technique. Finally, we would obtain a clear image of the damaged ROI. A damaged ROI can be interpreted, depending on application, as material damage in the manufacturing process or damage during operation in maintenance or other.

6 Conclusions

In this chapter, a DWT method for damage detection is proposed. It is based on energy contained in the wavelet coefficients.

Energy differences can be used for damage detection, but the threshold should be carefully chosen. The obtained results are interesting and promising.

The proposed method is still in the early stages and many things can be taken as disadvantages. For example, there is no extensive research and huge experimental support.

There are several ways to further the research and there is no guarantee for success. It is shown that results can be tricky. Therefore threshold and referent energy must be carefully chosen and experimentally evaluated.

The energy measure (*EM*) is introduced in the chapter. It was used to attach numerical value to an image. The calculated *EM* is then used to distinguish damaged component as a basis for computer vision algorithm decision making.

Further work could include:

- Automation of the process by SVM (Supported Vector Machines) and
- Finding the optimum level of decomposition and optimum exponent for the energy.

References

1. Haghighi, A.A.M.K.: Vibration-based damage detection and health monitoring of bridges. PhD Thesis, North Carolina State University (2010)
2. Wen, J., Choy, F.K., Xia, Z.: Wavelet denoising in electrical resistance based damage detection of carbon fiber composite materials. J. Mater. Sci. Res. doi: 10.5539/jmsr.v2n1p82 (2013)
3. Golmohamadi, M., Badri, H., Ebrahimi, A.: Damage diagnosis in bridges using wavelet. In: 2012 IACSIT Coimbatore Conferences. IACSIT Press, Singapore (2012)
4. Bagheri, A., Kourehli, S.: Damage detection of structures under earthquake excitation using discrete wavelet analysis. Asian J. Ci. Eng. (BHRC) **14**, 289–304 (2013)
5. Joo, D.J.: Damage detection and system identification using a wavelet energy based approach. PhD Thesis, Columbia University (2012)
6. Wang, W.J., McFadden, P.D.: Application of wavelets to gearbox vibration signals for fault detection. J. Sound Vib. **192**, 927–939 (1996)
7. Bajaba, N.S., Alnefaie, K.A.: Multiple damage detection in structures using wavelet transforms. Emirates J. Eng. Res. **10**, 35–40 (2005)
8. Bajaba, N.S.: Damage detection in plates by coupling modal analysis and wavelet transforms. In: ACE-X 2012, Istanbul, 2–4 July 2012
9. Buethe, I., Torres-Arredondo, M.A., Mujica, L.E., Rodellar, J., Fritzen, C.P.: Damage detection in piping systems using pattern recognition techniques. In: 6th European Workshop on Structural Health Monitoring, Dresden, Germany, July 3–6 2012, http://www.ewshm2012.com/Portals/98/BB/fr1d3.pdf. Accessed 10 Oct 2012
10. Reddy, D.M., Swarnamani, S.: Damage detection and identification in structures by spatial wavelet based approach. Int. J. Appl. Sci. Eng. **10**, 69–87 (2012)
11. Silva, R.S.Y.C., Bezerra, L.M., Brito, M.A.N.: Determination of damages in beams using wavelet transforms. In: Proceedings of the World Congress on Engineering, London, UK, 4–6 July 2012
12. Mizuno, Y., Fujino, Y.: Wavelet-based damage detection for shm under limited resources: computation, storage and transmission, pp. 529–536, Yokohama, Japan, 4–6 Dec 2006, http://iroha.scitech.lib.keio.ac.jp:8080/sigma/bitstream/handle/10721/2843/529-536_mizuno.pdf?sequence=1. Accessed 15 Sept 2012
13. Yu, L., Giurgiutiu, V.: Advanced signal processing for enhanced damage detection with piezoelectric wafer active sensors. Smart Struct. Sys. **1**, 185–215 (2005)

14. Thomas, J., Jeray, J., Kareem, A., Bowyer, K.: Efficacy of damage detection measures from satellite images. In: 11th American Conference on Wind Engineering, San Juan, Puerto Rico, 22–26 June 2009
15. Song, Y.Z., Bowen, C., Kim, H.A., Nassehi, A., Padget, J.: Non-invasive damage detection in composite beams using marker extraction and wavelets. In: Proceedings SPIE 7983, Nondestructive Characterization for Composite Materials, Aerospace Engineering, Civil Infrastructure, and Homeland Security, San Diego, CA, United States, 7–10 March 2011. doi:10.1117/12.880055
16. Uhl, T., Kohut, P., Holak, K., Krupiński, K.: Vision based condition assessment of structures. J. Phys. Conf. Ser. (2011). doi:10.1088/1742-6596/305/1/012043
17. Patsias, S, Staszewski, W.J.: Damage detection using optical measurements and wavelets. Struct. Health Monit. (2002). doi:10.1177/147592170200100102
18. Morlier, J., Salom, P., Bos, F.: New image processing tools for structural dynamic monitoring. Key Eng. Mat. (2007). www.scientific.net/KEM.347.239
19. Kohut, P., Holak, K., Uhl, T.: Monitoring of civil engineering structures supported by vision system. In: 6th European Workshop on Structural Health Monitoring, Dresden, Germany, 3–6 July 2012, http://www.ewshm2012.com/Portals/98/BB/p19.pdf. Accessed 23 July 2012
20. Lima, M.M., Amiri, G.G., Bagheri, A.: Wavelet-based method for damage detection of nonlinear structures. J. Civil. Eng. Urban **2**, 149–153 (2012)
21. Poudel, U.P., Fu, G., Ye, J.: Structural damage detection using digital video imaging and wavelet transformation. In: 2004 SEM X International Congress & Exposition on Experimental & Applied Mechanics. http://sem-proceedings.com/04s/sem.org-SEM-X-Int-Cong-s025p03-Structural-Damage-Detection-Using-Digital-Video-Imaging-Wavelet.pdf Accessed 13 Nov 2012
22. Papakostas, G.A., Karras, D.A., Mertzios, B.G., Boutalis, Y.S.: An efficient feature extraction methodology for computer vision applications using wavelet compressed zernike moments. ICGST Int. J. GV IP SI1, 5–15 (2005)
23. Livens, S.: Image analysis for material characterization. PhD Thesis, University of Antwerp, Belgium (1998)
24. Badashah, S.J., Subbaiah, P.: Surface Roughness Prediction with Denoising Using Wavelet Filter. IJAET Int. J. Adv. Eng. Technol. **3**, 168–177 (2012)
25. Yang, F., Abdi, H., Mitéran, J., Paindavoine, M.: A new image filtering technique combining a wavelet transform with a linear neural network: application to face recognition. Opt. Eng. (2000). doi:10.1117/1.1308492
26. Silveira, J., Ferreira, M.J., Santos, C., Martins, T.: Computer vision techniques applied to the quality control of ceramic plates. In: Proceedings of the IEEE International Conference on Industrial Technology, ICIT 2009, Gippsland, 10–13 Feb 2009, http://www.dei.uminho.pt/pessoas/jaa/asbg/wiki/uploads/Article_v1.pdf. Accessed 10 Nov 2012
27. Christopher, H., Walnut, D.F.: Fundamental papers in wavelet theory. Princeton University Press, London (2006)
28. Kuzmanić, I., Šoda, J., Antonić, R., Vujović, I., Beroš, S.: Monitoring of oil leakage from a ship propulsion system using IR camera and wavelet analysis for prevention of health and ecology risks and engine faults. Materialwiss Werkst (2009). doi:10.1002/mawe.200900424
29. Mallat, S.: A wavelet tour of signal processing, 3rd edn. Academic Press, New York (2009)
30. Kuzmanić, I., Vujović, I., Šoda, J.: Corrosion monitoring in marine environment using wavelet description. Key. Eng. Mat. (2011). doi:www.scientific.net/KEM.478.40
31. Vujović, I., Šoda, J., Kuzmanić, I.: Cutting-edge mathematical tools in processing and analysis of signals in marine and navy. Trans. Marit. Sci. (2012). doi:10.7225/toms.v01.n01.005
32. Kuzmanić, I., Beroš, S.M., Šoda, J., Vujović, I.: Pre-processing for image sequence visualization robust to illumination variations. In: Öchsner, A., Silva, L.F.M., Altenbach, H. (eds.) Design and Analysis of Materials and Engineering Structures. Springer-Verlag, London (2013)

33. Vujović, I.: Suppressing illumination variations in motion detection by wavelet transform. PhD Thesis, University of Split, Faculty of Electrical Engineering, Mechanical Engineering and Naval Architecture (2011)
34. Šoda, J., Beroš, S.M., Kuzmanić, I., Vujović, I.: Discontinuity detection in the vibration signal of turning machine. In: Öchsner, A., Altenbach, H. (eds.) Experimental and Numerical Investigation of Advanced Materials and Structures. Springer-Verlag, London (2013)
35. Šoda, J.: Wavelet transform based discontinuities detection of vibration signal. PhD Thesis, University of Split, Faculty of Electrical Engineering, Mechanical Engineering and Naval Architecture (2010)

The Fretting Fatigue Behavior of Bolted Assemblies

Ali Benhamena, Laïd Aminallah, Abdelghani Baltach, Abdelkrim Aid, Mohamed Benguediab, Abdelwaheb Amrouche and Noureddine Benseddiq

Abstract This study is a contribution to various scientific researches on fracture behavior of bolted joints. In our study, a three-dimensional finite element model (3D-FEM) is developed to predict the mechanical behavior of bolted joints. Fretting-fatigue tests and numerical simulations were conducted to study the effect of the clamping torque on the fatigue behavior and failure mode of bolted assemblies. The results obtained in numerical simulations are evaluated and compared with experimental results. The numerical model showed the capacity to simulate the behavior of the bolted assemblies under fatigue. Crack initiation changed from the edge of the hole to close the edge of the contact zone. A good correlation was found between the FEM simulations and the experimental results.

A. Benhamena (✉) · L. Aminallah · A. Baltach · A. Aid
Laboratory LPQ3M, University of Mascara, BP 763, Mascara, Algeria
e-mail: ali_benhamena@yahoo.fr

L. Aminallah
e-mail: abdaminn@yahoo.fr

A. Baltach
e-mail: baltachabdelghani@yahoo.fr

A. Aid
e-mail: aid_abdelkrim@yahoo.com

A. Benhamena · L. Aminallah · A. Aid · N. Benseddiq
Laboratory of Mechanic of Lille, University of Science and Technology of Lille, UMR CNRS, 8107 USTL, France
e-mail: noureddine.benseddiq@univ-lille1.fr

M. Benguediab
Department of Mechanical engineering, University of Sidi Bel Abbes, BP 89, Cité Ben M'hidi, 22000 Sidi Bel Abbes, Algeria
e-mail: benguediab_m@yahoo.fr

A. Amrouche
Laboratoire de Génie Civil et géo-Environnement LGCgE, EA4515, Université d'Artois, Faculté des Sciences Appliquées Béthune, Béthune, France
e-mail: abdelwaheb.amrouche@univ-artois.fr

A. Öchsner and H. Altenbach (eds.), *Design and Computation of Modern Engineering Materials*, Advanced Structured Materials 54,
DOI: 10.1007/978-3-319-07383-5_14,

Keywords Clamping force · Torque · Fretting fatigue · Crack initiation site · Stick · Slip · Adhesion

1 Introduction

Fretting damage is considered an important failure mechanism for a variety of mechanical components and assemblies. However, the analysis of practical fretting problems entails significant difficulties related to the characterization of the contact interactions [1–5]. Bolted joints in mechanical structures transmit a more important effort in various applications. According to Valtinat et al. [6] bolted joints have higher tensile and fatigue strengths than welded joints. The prediction of fracture and the reliability of such assembly in various practical applications are primordial given their impact on the economic plan and security. Fretting is caused by the oscillating movement with small amplitude that may occur between contacting surfaces subjected to vibration. The oscillations cause sliding to occur in a small region at the edge of contact, while the center of contact remains stuck together. The damage is related to cyclic loading and relative displacement at the interface in such assemblies. The amplitude of relative displacement (slip) and the contact forces (or contact pressure) are two significant factors, which control the fretting fatigue behavior. Lee and Mall [7] investigated the variation of fretting fatigue life with the increase of amplitude of relative displacement on Ti-6Al-4V material. They postulated that the amplitude of relative displacement increases from 0 to 50 μm (partial slip and mixed slip), the fretting fatigue life decreases. On the other hand, the increase of relative displacement from 50 to more than 100 μm (gross slip), induces the increase of fretting fatigue life. In the same context, several studies [8–11] have been carried out to investigate the damage of contact surface and the effect of fastening forces. All these authors observe that the fretting fatigue life decreases with the increase of contact forces. It is well known that stress distribution at the contact edge (i.e. the tangential stress and the compressive stress) plays a dominant role in influencing fretting fatigue strength [12–14]. From this point of view, in the previous study, a fretting fatigue strength prediction method based on the tangential stress range compressive stress range diagram has been developed and discussed for Ni-Cr-Mo-V steel [15]. Chakherlou et al. [16] conducted fatigue tests and numerical simulation to study the effect of cold expansion and bolt clamping on fretting fatigue behaviour of Al 2024-T3 in double shear lap joints. The results reveal that significant changes on the fatigue behaviour of the double shear lap joints at different load ranges. Depending on the degree of cold expansion and clamping torque. The references cited above show some contradictory tendencies of the fretting fatigue life. In service and under cyclic loading, the relative displacement at the interface is inevitable for bolted assemblies, in addition the effect of clamping force on the stress concentration close to the hole, the frictional stresses and the friction coefficient are not fully understood and represent a very active research field. This fact has led the authors to focus on the current study, in order to contribute in this way. We recall also, that the present study is an extension of previous studies [2]. Finite element analysis (FEA) is an important tool to design a practical mechanical component, such as, for example, pipelines. The three-dimensional finite element method 3D-FEM under consideration of non-linear models can predict well the behaviour of bolted assemblies (steel-aluminium) loaded in fatigue. Moreover, a number of the fatigue–fractured sections of the specimens are studied to help identify the location of crack initiation and also to examine the propagation regions.

2 Experimental and Numerical Model

2.1 Materials and Specimens

The details of the experimental fretting-fatigue test programme were previously reported in [2] and are just briefly explained here. Aluminium alloy A6xxxx from 6.0 mm thickness and high strength low alloy steel (HSLA355) of 3.0 mm thickness were used to produce (6.0 mm A6xxxx + 3.0 mm HSLA355) joints for this investigation (Fig. 1) and the fretting fatigue test method is shown in Fig. 2, full details of the test setup are given elsewhere [1, 2].

2.2 Finite Element Model

Fretting tests may be carried out in the partial slip regime. As highlighted by Hills et al. [17], it is very difficult to achieve a well controlled experiment using an external actuator due to the low displacement amplitudes. Although the relative displacement cannot be easily measured and its value is required, an analysis of the stress field at the process zone (contact zone) in partial slip regime. Finite element analysis (FEA) is an important tool to design practical mechanical joints, such as bolted assemblies. According to the dimensions of the structure, a three-dimensional model was generated using the commercial software ANSYS® (ANSYS 11 [18]) in order to determine and to perform the analyses of the stress field at the contact zone. The bolted assembly shown in Fig. 2 is symmetric in the longitudinal direction (*y*-direction). So, a half model with symmetry conditions was used in the finite element model in order to reduce the calculation time (Fig. 2a). Three-dimensional brick elements (SOLID45) are used for modelling of the bolted assembly. This element is defined by eight nodes and each having three degrees of freedom. In addition, a surface-to-surface contact element, which consists of contact elements (CONTAC173) and target surface elements (TARGE169), is used on the interfaces between all connected parts of the bolted assembly in order to

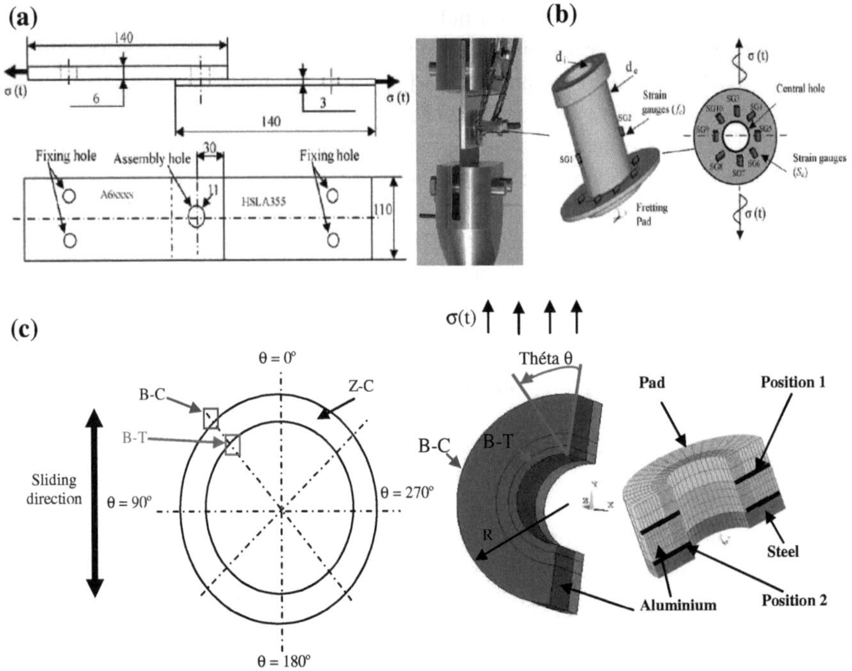

Fig. 1 **a** Specimen geometry (all dimensions in mm), **b** schematic drawing of central part of the specimen used in fretting fatigue tests and the strain gauges and **c** nomination of the Cartesian coordinates, different planes and positive θ direction at a half of bolted assemblies

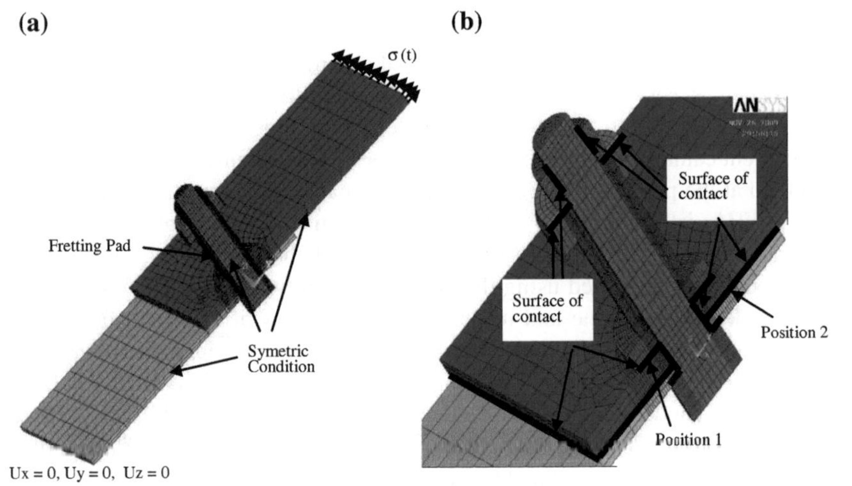

Fig. 2 **a** Geometrical and finite-element mesh of assembly, **b** modeling of contact surface (damaged surface: position 1 and position 2)

simulate numerically the contact problems. Friction between the contact surfaces at the connection is modeled using the classical Coulomb model, where the friction coefficient was set at 0.2. Various mesh schemes are tried to achieve convergence. The optimized model has 14,752 nodes, 17,020 elements and 190 contact elements. A two-step nonlinear analysis was performed in order to simulate the clamping force numerically, in the first step, the preload (clamping force) was modelled as a uniform negative pressure applied on the screw and a positive pressure applied over a ring of 10 mm inner radius and 16 mm outer radius, this ring represents the action of the nut (Fig. 2b). This phase represents the joining of the bolted assembly. After this operation, there is a relaxation of the tensile strain due to the deformability of the pad and the plates under the clamping force action, on attaining equilibrium and in the second step, this assembly was subjected to cyclic loading that generated a multi-axial stress fields at the contact zone. The bolt shank had a clearance of 1 mm due to the dimensional tolerance of the boltholes in the two plates (steel and aluminum) and pad. In practice the plates and pad are tightened by a torque applied to the bolts; this action was simulated by transferring the torque as axial strain in the bolt shank during the assembly process. Although during cyclic loading there is a possibility that the two plates or pad surface in boltholes could contact the surfaces of the bolt shank. However, the clearance was considered large enough for this action not to occur in order to simulate the fretting fatigue at position 1 and at position 2. The theory of incremental plasticity is introduced to model the material nonlinearity. The iterative Newton-Raphson method is used as an approach to solve nonlinear equations by finite elements.

3 Results and Discussions

3.1 Validation of the Finite Element Model

The finite element model is validated on the basis of the compressive axial strain measured by the strain gauges (SG1, SG2) glued at the pad during the fretting fatigue test. After that, the numerical model is applied to evaluate the stress distributions at the contact surface in bolted assemblies for various loading condition. Figure 3 shows the compressive axial strain during the simulation of fretting fatigue test under maximal cyclic loading equal to 70 MPa and a tightening torque value of 6 daN.m. We noticed in this case that the compressive strain is strongly concentrated at the interface between the pad and the aluminium plate. This phenomenon can be explained by the fact of strain incompatibility at the interface (position 1, see Fig. 1) and the frictional coefficient effect. Indeed, under this condition and in position at the pad, the compressive strain measured by the strain gauges (SG1, SG2) is 0.001085. At the same position, the compressive strain is calculated numerically and its value equal to 0.000954. So it can be seen according to Fig. 3 that the numerical results of the computational model are in good agreement with experimental results.

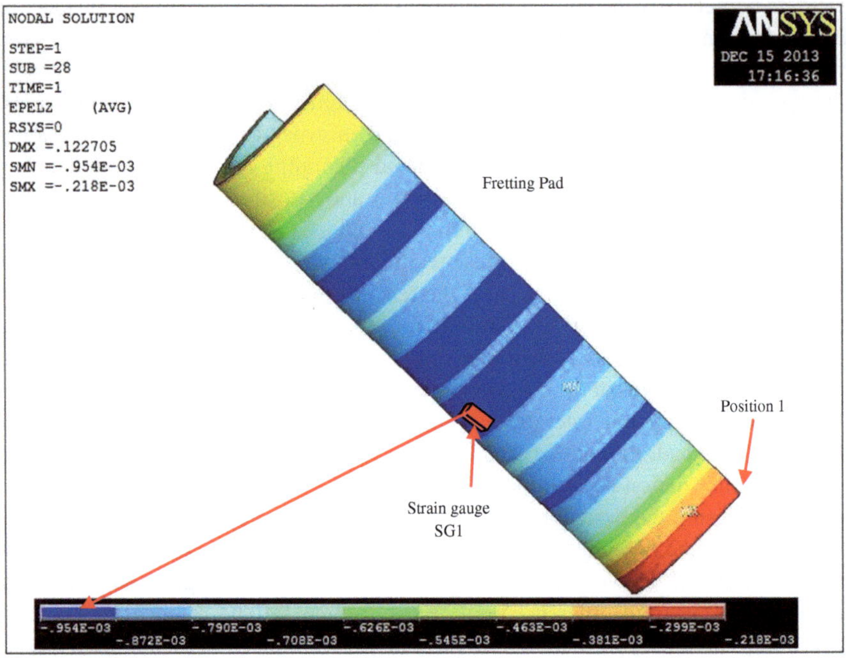

Fig. 3 Compressive strain at the pad under maximal cyclic loading $F_{max} = 12$ KN and torque 6 daN.m

These results were confirmed by the compressive strain (experimental and the numerical results) versus tightening torque presented in Fig. 4; this presents the clamping force variation according to the tightening torque. In this figure we observe that a linear trend was found between the compressive strains measured on the pad as a function of torque (clamped force). This confirms that the material of the pad remains in the elastic region, even for very high compression forces (contact force higher). We note also a fairly good agreement between the numerical and experimental results, this later establishing the confidence in the results of the finite element modeling for bolted assemblies.

3.2 Numerical Results and Analysis

Figure 5 shows the stress distribution along the longitudinal direction of cyclic loading (σ_y) for this loading condition. We observe that the stress distribution is not uniform; this is due to the anti-symmetry of various parameters: the loading conditions, the geometry and the heterogeneity of the behavior of plates. We also observe, on the one hand, that there is a high tensile local stress located at the head of the bolt ($\sigma_y = 435$ MPa), this is due to the interaction of the stress field

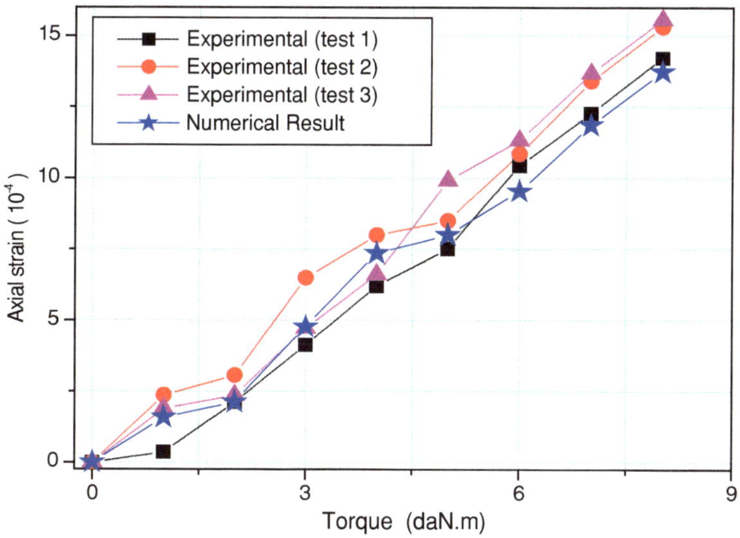

Fig. 4 Axial strain versus tightening torque

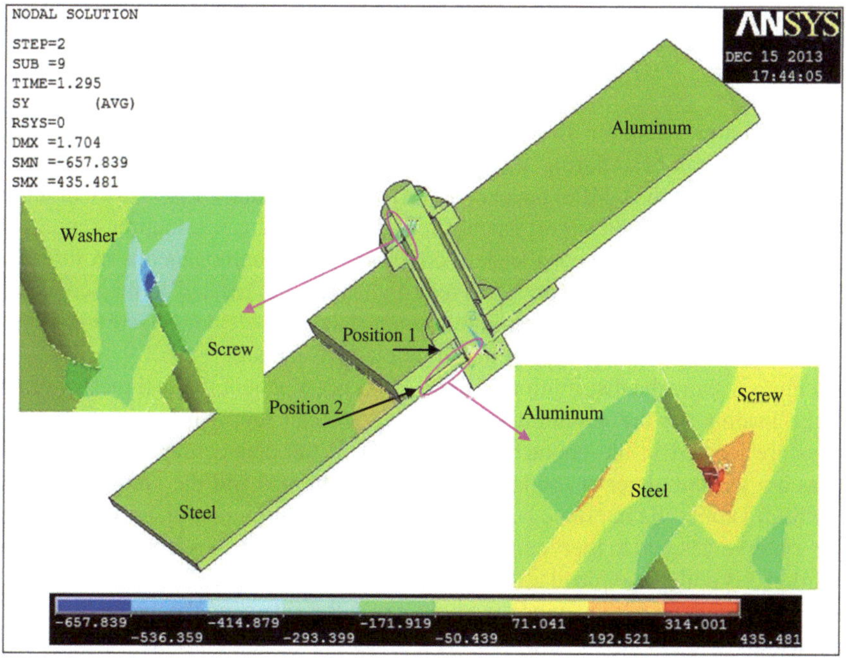

Fig. 5 The distribution of stress σ_y in the direction of cyclic loading of bolted assembly ($T = 6$ daN.m and $F_{max} = 12$ KN)

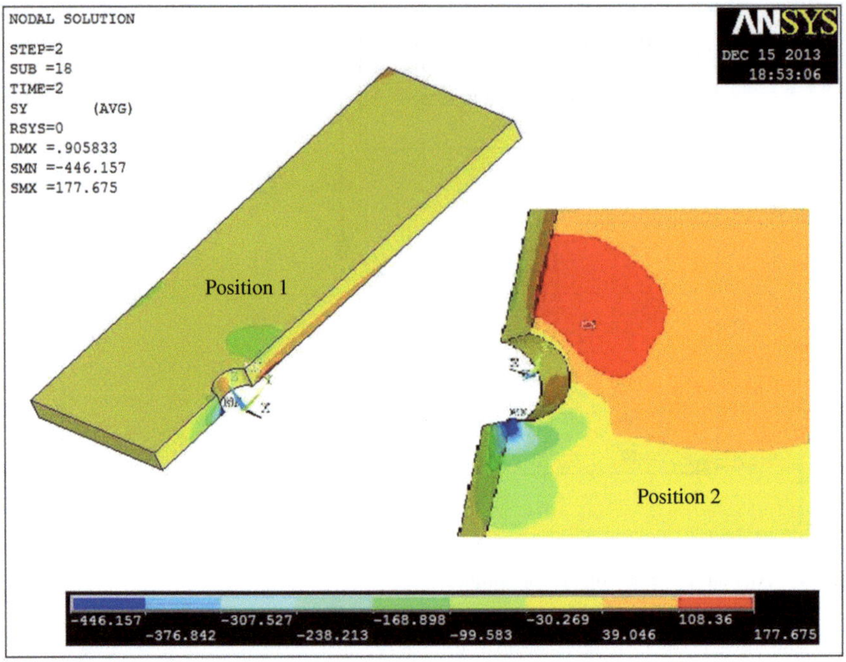

Fig. 6 The distribution of stress σ_y in the direction of cyclic loading of aluminium plate ($T = 6$ daN.m and $F_{max} = 12$ KN)

(tension-compression) in the contact surfaces at the interface between the steel plate and the bolt head during sliding. On the other hand, a very high compressive local stress ($\sigma_y = 657$ MPa) appears in the end of the contact areas between screw and nut. This phenomenon corresponds to a situation that can sometimes appear in such bolted assembly where the field constraint is not homogeneous in threaded elements. These constraints are relatively higher and are located in the screw-nut, which is reasonable when taking into account that these elements (screw and nut) are considered as rigid bodies in the finite element model.

Figure 6 show the distribution of the σ_y stress component in the direction of the cyclic loading for the aluminium plate only. We can see, that the most stressed areas are located in the contact interface near at the connecting element. These areas are located either at the interface between the pad and the aluminium plate (position 1) or at the interface between the steel and aluminium plates (position 2). We observe also a higher compression field located in position 1, while the position 2 is under combined effect (tension and compression effect), the tension is observed at the top of the connecting hole, i.e. at the side were the cyclic loading is applied and the compression is observed at the bottom of the hole.

The explanation for this phenomenon is the following: in the tightening phase, all contact surfaces in the aluminium plate are under compression. After the application of the load (cyclic loading) in direction perpendicular to the plane of application of the clamping force, a frictional stress is produced at the interface of the contact surfaces, in particular at the interface characterized by the position 2 (interface between the steel and aluminium plates). The presence of a large tensile stress in this position (position 2), leads to the risk of crack initiation in favourable directions at this position (position 2), while the existence of a compressive stress in such position in the contact surface significantly reduces the risk of crack initiation (for example at the near of the hole). We recall that the stress field of the compression and the tension is related by several factors: the coefficient of friction, sliding, sticking, secondary bending effect for this type of bolted assembly, geometrical parameter, etc. According to our experimental results [2] and the observations of Wagle et al. [19] for a bolted assembly under cyclic loading, the level of torque and the amplitude of the cyclic loading are two key variables which control the fretting fatigue life and the wear mechanism in bolted assemblies under fatigue. So, the intensity and the distribution of the stress field at the contact surface determine the risk of failure of bolted assemblies under fatigue.

3.2.1 Distribution of Stress at Position 1

Figure 7 shows the variation of the normal stresses (σ_x, σ_y and σ_z) and shear stresses (σ_{xy}, σ_{yz} and σ_{xz}) in Cartesian coordinates at the contact surface characterized by the radius *R* and angle theta θ (see Fig. 1) between aluminum plate and pad (position 1). At the edge of the contact area (BC), for position characterized by the angle θ less than 90° ($\theta < 90°$) along the (BC), all the normal stresses (σ_x, σ_y and σ_z) are negative values and for position characterized by the angle θ greater than 90° ($\theta > 90°$) the two components (σ_x and σ_y) are positive and the component (σ_z) is zero. This phenomenon can be explained by the fact that the aluminum plate is under complex loading (tension and compression) along the (BC) contact zone between aluminum plate and pad (position 1). We also observe that the value of shear stress (σ_{xy}, σ_{yz} and σ_{xz}) are very low values compared to the value of the normal stresses, therefore the failure mode is governed by the normal stresses. We note that the traction accelerates the risk of cracks in the contact area, while the compression results in the inverse problem. Through a comparison of the stresses evaluated along the edge of the central hole (BT) and the edge of the contact area (BC), we find that the intensity of the stress field in BC is higher than that in BT. Some stress component (σ_z) has significant negative values at the edge of the hole (BT). This is clearly due to the magnitude of torque which leads to produce an intense compressive stress field at the edge of the hole (BT) and therefore eliminates the risk of crack initiation in the contact zone, in particular at the edge of the hole (BT). It is obvious that the compressive stress significantly reduced the crack initiation and hence increasing the lifetime of such bolted assembly, while the traction accelerates the risk of the initiation of fatigue crack. So it is necessary to identify the critical position in the contact surface which leads to the failure of the bolted assembly from the analysis of the stress distribution at the interface between steel and aluminum plates (position 2) for which we observed the tension and compression in the contact area (position 2).

3.2.2 Distribution of Stress at Position 2

It is necessary to demonstrate the capacity of the finite element model to predict the characteristics of all the stresses at the interface in the contact surfaces for the bolted assembly. This information is not easily accessible from the experience especially in the interface between steel and aluminum plates (position 2). The numerical analysis is an alternative method to provide additional information for a better understanding for the case of bolted assemblies under fatigue. Figure 8 shows the evolution of each stress component in Cartesian coordinates at the contact zone characterized by the radius R and angle theta θ (see Fig. 1) at the interface between steel and aluminum plates (position 2). Figure 8 shows that the stress which triggers the initiation and propagation of the crack in the edge of the contact area (BC) is important for those on the edge of the hole (BT) in particular positions characterized by the angle θ less 90° ($\theta < 90°$) along the (BC). We also observe that the constraint (σ_z) in the direction of tightening is higher at the edge of the hole (BT) compared to the contact edge (BC). This is usually due to the high pressure generated in the contact surfaces by the elements of assembly (screw-nut) during clamping phase. The interaction between the tightening torque and the level of cyclic loading will produce a friction phenomenon at the interface of the contact surfaces which leads to cracking or degradation of the contact surfaces under a fatigue stress. Therefore, the finite element model is considered effective to predict the failure of the bolted joint. It is interesting to trace the distribution of stress (σ_y) in the direction of cyclic loading as a function of angle θ in the contact area (position 2) for different radius (R) in order to locate the point of crack initiation at the contact area.

Figure 9 illustrates the stress in the direction of cyclic loading (σ_y) as a function of the angle θ for three radius ($R = 6.5$, 8.0 and 12.5 mm). According to this Fig. 9a, we can see that the curves had two well differentiated parts. Whatever the radius in the contact area is (position 2), all constraints are negative for orientation characterized by the angle $\theta > 100°$, which means that there is lower risk of crack initiation in this portion at the contact surface. In positions between $0° \leq \theta \leq 43°$, the (σ_y) stress increases gradually and finally reaches the maximum value for positions close to an orientation $\theta = 43°$ for each radius in the contact area. Beyond this position ($\theta > 43°$) the stress decreases to achieve the null value as function of the angle θ and the radius (R) in the contact area. We note that there is a critical stress value corresponds to an orientation of $\theta = 43°$ at the edge of the contact zone characterized by the radius ($R = 12.5$ mm). We can therefore conclude that the initiation and the propagation of cracks in the interface between steel and aluminum plates (position 2) is in a favorable orientation characterized by the angle $\theta = 43°$ relative to the direction of cyclic loading. These results lead us to study the evolution of the stress according to the orientation ($\theta = 43°$) for different

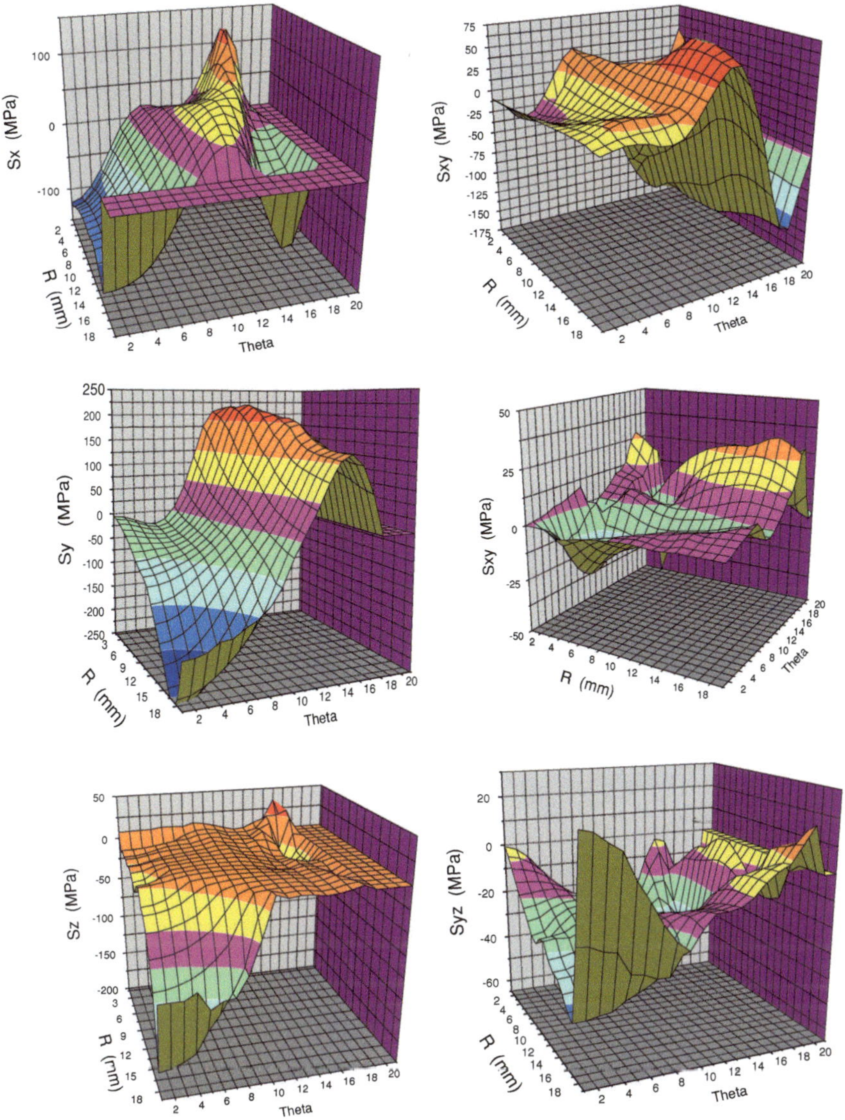

Fig. 7 The distribution of stress component along the BT and BC in contact zone (position 1)

radius in the contact zone and for a radius exceeding the limit of the contact area. Figure 9b shows the evolution of the stress (σ_y) at $\theta = 43°$ as a function of radius *R*. We note that the stress (σ_y) increases significantly with the radius. The maximum value is achieved for a radius *R* varied between 8.0 and 12.5 mm ($8 \leq R \leq 12.5$, where $R = 12.5$ is the radius of the edge of the contact surface at the interface between steel-aluminum: position 2), after the stress decreases for a higher radius ($R > 12.5$ mm: radius higher as the edge of the contact surface). So,

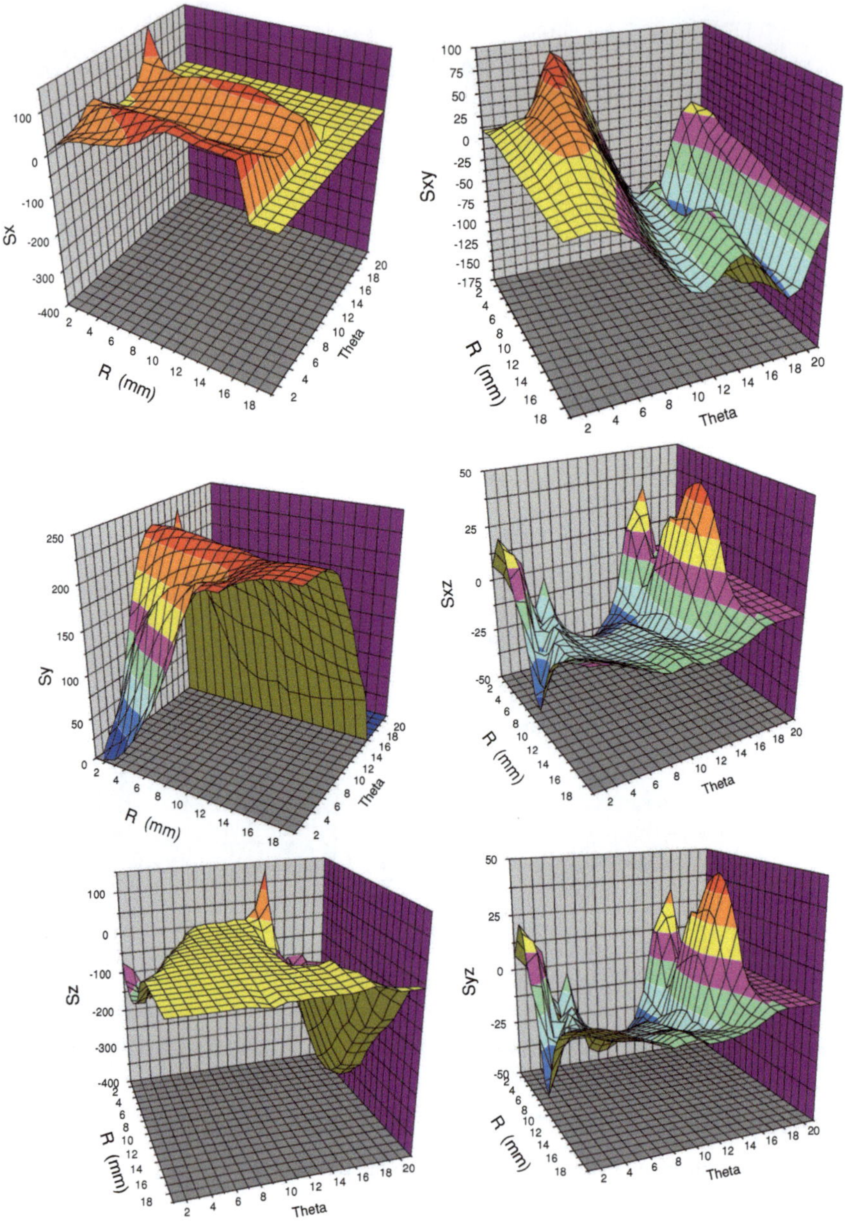

Fig. 8 The distribution of stress component along the BT and BC in contact zone (position 2)

our calculation at the interface steel-aluminum (position 2) has been limited to a radius not exceeding a value of 12.5 mm ($R \leq 12.5$ mm) although the size of the contact surface at the interface steel-aluminum (position 2) greatly exceeds the size seen in the numerical calculation.

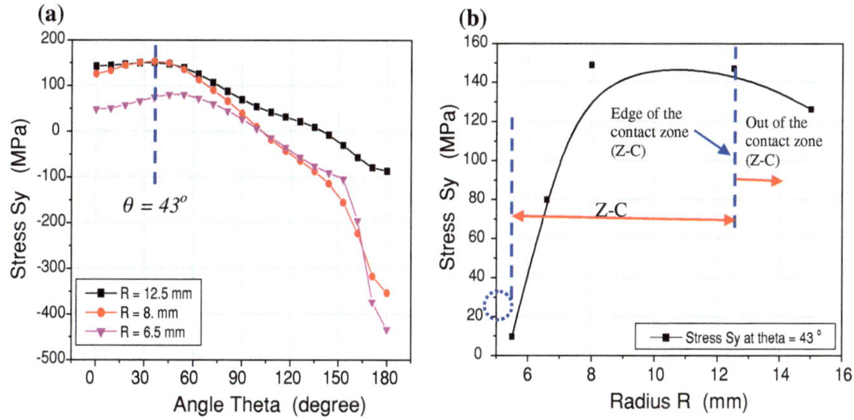

Fig. 9 The distribution of stress σ_y : **a** in contact zone for various radius (position 2), **b** an contact zone for angle $\theta = 45°$ for various radius (position 2)

We can now say that the size of the contact surface between the two plates (position 2) is strongly influenced by the intensity of the torque. For radius far from the center of the bolt ($R > 12.5$ mm) there is displacement with no friction in the contact surface. So, these surfaces at the interface of the plates are not in contact then no charge transfer occurs in these areas. For radius varied between 5.5 and 12.5 mm ($5.5 \le R \le 12.5$, where $R = 5.5$ mm is the radius of the hole), there are now two different conditions: adhesion (stick) and sliding (slip) in different levels of cyclic loading.

3.2.3 Contact Status at Position 1

It has been observed through literature review [20, 21] that the fretting contact zone is divided into two zones: adhesion zone (stick) and a sliding zone (slip). The knowledge of these two areas in a contact problem is no longer possible with the option of contact STATUS in ANSYS software. In addition, we recall also that the cracking phenomenon of the fretting fatigue occurs in the transitional phase stick-slip which is considered as the most critical phase according by many authors [20–24]. Figure 10 illustrates the contact STATUS at the interface between the pad and aluminum plate (position 1).

A first observation from the reading of this figure (Fig. 10) is that there are three different zones: red zone, characterizes the adhesion (Stick) where you can not connect the friction stress as a function of pressure according to Coulomb's law; orange zone, characterizes the slip where one can connect the friction stress as a function of pressure according to Coulomb's law; yellow zone, characterizes the non-contact in position 1. A concentration of stress between the sticking and the sliding zones which leads to the risk of crack initiation within the contact zone

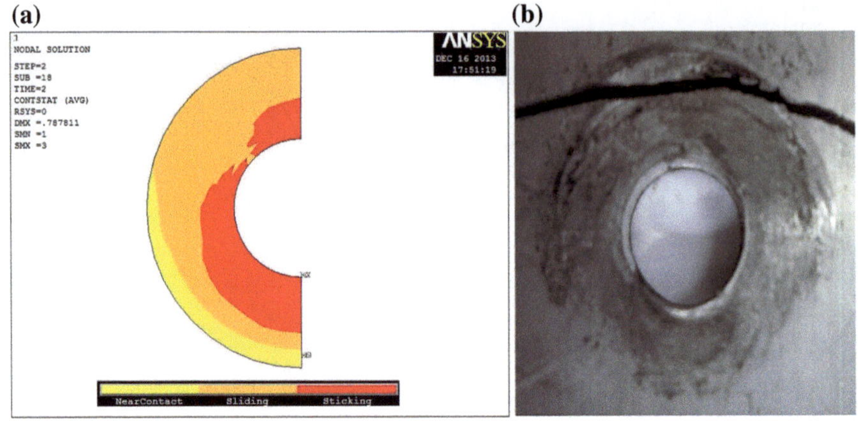

Fig. 10 **a** Status of the contact area (position 1) and **b** the experimental observation ($T = 6$ daN.m and $F_{max} = 12$ KN)

(position 1) in favorable positions. In conclusion, a transition from the sticking contact to the sliding contact characterized by an angle $\theta = 45°$.

To pinpoint the favorable orientation of the risk of crack initiation at the interface between the pad and aluminum plate (position 1), we present in Fig. 11 the evolution of the contact pressure, the friction stress and the sliding along the two paths (BT and BC) in the contact zone (position 1).

From Fig. 11, a first note is that the friction stress in positions characterized by the angle θ above 90° is zero and for angles $\theta \leq 90°$, the friction stress increases as function of angle θ to reach their optimum value at angle $\theta = 0°$ along the contour B–C. The same trend is observed for the evolution of slip and contact pressure, where the slip takes the maximum value at $\theta = 0°$. This indicates that the risk of crack initiation takes place in this portion in the zone of contact ($\theta \leq 90°$) during loading. We recall that the sliding at the edge of the hole (B-T) is negligible compared to the sliding at the edge of the contact zone (BC) due to the intensity of torque during the clamping phase.

3.2.4 Contact Status at Position 2

As previously mentioned, the torque and the level of cyclic loading have a significant effect on the distribution of contact STATUS (stick, slip and non-contact) in the contact zone, in particular at the interface between steel and aluminum plates (position 2).

Figure 12 presents the effect of the level of cyclic loading on the contact STATUS for a torque value equal to 8 Nm ($T = 8$ Nm). Three levels of cyclic loading are considered: 0 % of the cyclic loading (0 % F) is only in the tightening phase, 100 % of the cyclic load (100 % $F = F_{max}$) is the maximum value of the

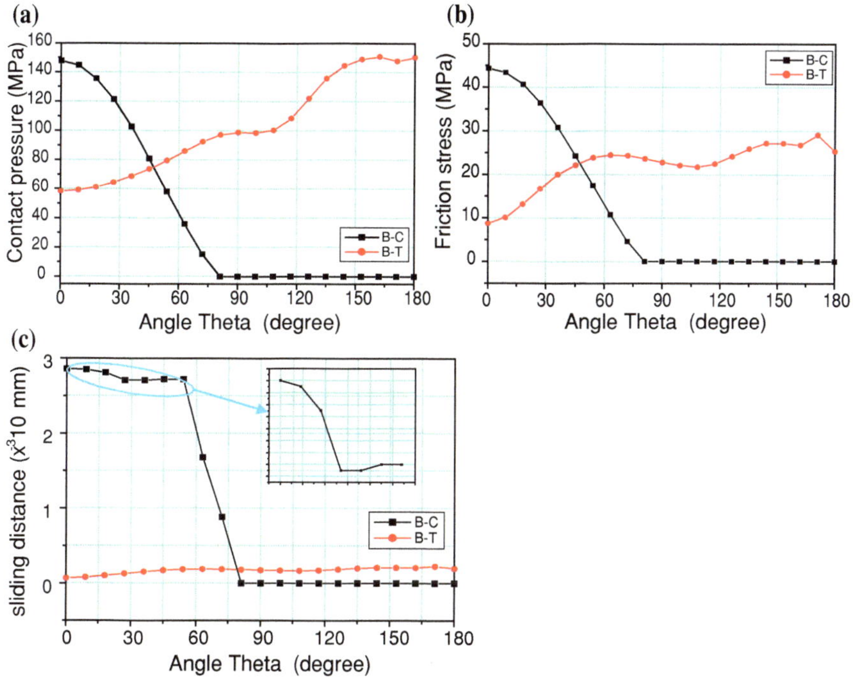

Fig. 11 The distribution of **a** contact pressure, **b** friction stress and **c** the sliding along the BT and BC in contact zone (position 1)

cyclic load and 50 % of F_{max} corresponding to an average load (0 % $F_{max} < F < 100$ % F_{max}). The analysis of Fig. 12 shows that, whatever the level of cyclic loading for this level of torque, the stick (adhesion) and slip zones were observed for a radius less than 12.5 mm ($R \leq 12.5$ mm) in the contact zone (position 2). For a radius greater than 12.5 mm (R > 12.5 mm), there is no-contact at the interface between steel and aluminum plates. So, the adhesion and slip zones completely disappear for $R > 12.5$ mm. We also see in this contact zone ($R \leq 12.5$ mm) that the adhesion zone decreased with increasing of the level of cyclic loading and her size decreases on 50 % of the maximum cyclic loading (50 % F_{max}). So, the transition from sticking (adhesion) to sliding is the most critical phase in fretting fatigue. Generally, the cracking occurs in this area.

One of the most important parameters which control the fretting fatigue life of bolted assembly is the sliding at the interface of the contact surfaces (position 2). Knowledge of the slip distribution is fundamental in the contact area in order to locate precisely the site of crack initiation at the interface between steel and aluminum plates (position 2). Figure 13 shows the distribution of slip for this loading condition. From Fig. 13 we observe that the value of the slip is even more important when approaching to the border of the contact zone ($R \rightarrow 12.5$), particularly in positions characterized by the angle θ less of 50° (θ < 50°) along the

0% F_{max}

50% F_{max}

100% F_{max}

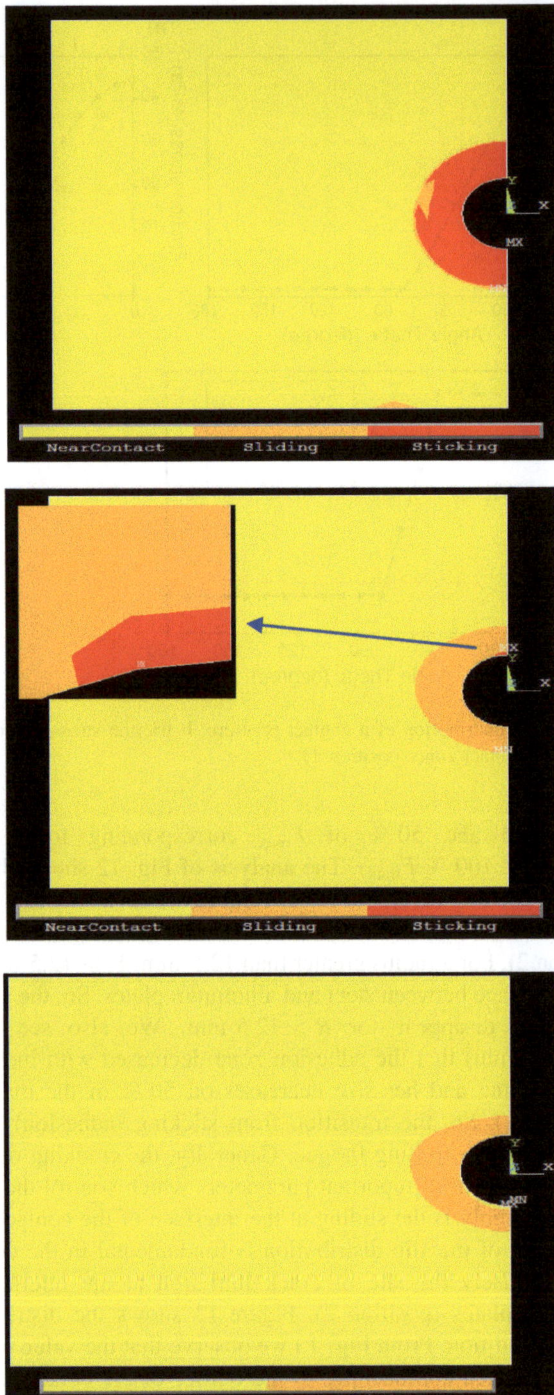

Fig. 12 Status of the contact area (position 2) versus to the cyclic loading for $T = 6$ daN.m

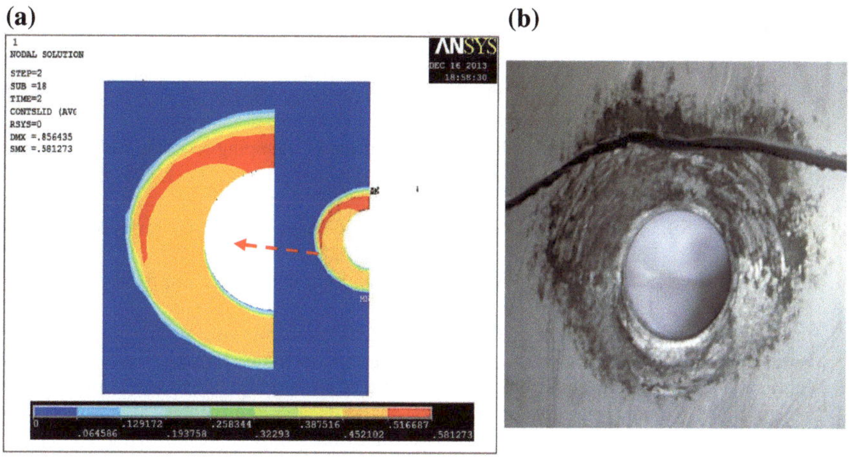

Fig. 13 **a** The distribution of sliding in the contact area (position 2) and **b** the experimental observation ($T = 6$ daN.m and $F_{max} = 12$ KN)

edge of the contact zone (BC). Therefore, through all the numerical results and experimental observations [2] we can now consider that the risk of crack initiation of any bolted assembly occurs in the edge of the contact zone (BC) in orientation inclined at an angle θ varied between 40 and 50° (40° ≤ θ ≤ 50°) relative to the direction of cyclic loading. The comparison between the experimental results and the status of the contact (Fig. 10b) leads to the same conclusion.

4 Conclusions

The aim of this study was to analyze the effect of tightening torque on bolted assemblies under fretting fatigue condition. From these experimental and numerical results, we can deduce the following conclusions:

- The developed finite element modelling approach to simulate clamping force (tightening torque) in the bolt of such bolted assembly was validated against the experimental results.
- The size of the adhesion and slip zones in the contact zone is related at the magnitude of tightening torque.
- Crack initiation changes from the edge of the hole to close the edge of contact zone.
- The magnitude of the clamping torque and the level of cyclic loading determine the mode failure of bolted assemblies.
- The crack initiation will be transferred to the edge of the contact zone at the interface between steel and aluminium plates (position 2) in a plane inclined at an angle θ varied between 40 and 50° relative to the sliding direction.

References

1. Guo, R., Duan, R.C., Mesmacque, G., et al.: Fretting fatigue behavior of riveted Al6XXX components. Mater. Sci. Eng. A **483–484**, 398–401 (2008)
2. Benhamena, A., Talha, A., Benseddiq, N., et al.: Effect of clamping force on fretting fatigue behaviour of bolted assemblies: Case of couple steel–aluminium. Mater. Sci. Eng. A **527**, 6413–6421 (2010)
3. Iyer, K., Hu, S.J., Brittman, F.L., et al.: Fatigue of single-and double-rivet self-piercing riveted lap joints. Fatigue. Fract. Eng. Mater. Struct. **28**, 997–1007 (2005)
4. Chen, Y.K., Han, L., Chrysanthou, A., et al.: Fretting wear in self-piercing riveted aluminium alloy sheet. Wear **255**, 1463–1470 (2003)
5. Han, L., Chrysanthou, A., Young, K.W., et al.: Characterization of fretting fatigue in self-piercing riveted aluminium alloy sheets. Fatigue. Fract. Eng. Mater. Struct. **29**, 646–654 (2006)
6. Valtinat, G., Hadrych, I., Huhn, H.: Strengthening of riveted and bolted steel constructions under fatigue loading by preloaded fasteners—experimental and theoretical investigations. In: Proceedings of the International Conference on Connections in Steel Structures IV, AISC and ECCS, Roanoke, USA (2000)
7. Lee, H., Mall, S.: Fretting behavior of shot peened Ti-6Al-4V under slip controlled mode. Wear **260**, 642–651 (2005)
8. Lee, H., Mall, S.: Investigation into effects and interaction of various fretting fatigue variables under slip-controlled mode. Tribol. Int **39**, 1213–1219 (2006)
9. Hutsona, A., Lee, H., Mall, S.: Effect of dissimilar metals on fretting fatigue behavior of Ti–6Al–4V. Tribol. Int **39**, 1187–1196 (2006)
10. Rajasekaran, R., Nowell, D.: Fretting fatigue in dovetail blade roots: experiment and analysis. Tribol. Int **39**, 1277–1285 (2006)
11. Sadeler, R.: Influence of contact pressure on fretting fatigue behaviour of AA 2014 alloy with dissimilar mating material. Fatigue. Fract. Eng. Mater. Struct. **29**, 1039–1044 (2006)
12. Xu, J.Q., Mutoh, Y.: Stress field near the contact edge in fretting fatigue tests. JSME Int. J. Ser. A **45–4**, 510–516 (2002)
13. Wright, G.P., O'Conner, J.J.: Finite-Element Analysis of alternating axial loading of an elastic plate between two elastic rectangular blocks with finite friction. Int. J. Eng. Sci. **9**, 555–570 (1971)
14. Nagata, K., Fukakura, J.: Effect of contact materials on fretting fatigue strength of 3.5Ni-Cr-Mo-V rotor steel and life prediction method. Trans. Jpn. Soc. Mech. Eng. **58–553**, 1561–1568 (1992)
15. Jayaprakash, M., Mutoh, Y., Asai, K., et al.: Effect of contact pad rigidity on fretting fatigue behavior of NiCrMoV turbine steel. Int. J. Fatigue. **32**, 1788–1794 (2010)
16. Chakherlou, T.N., Shakouri, M., Akbari, A., et al.: Effect of cold expansion and bolt clamping on fretting fatigue behaviour of Al 2024-T3 in double shear lap joints. Eng. Fail. Anal. **25**, 29–41 (2012)
17. Hills, D.A., Nowell, D.: Mechanics offretting fatigue. Kluwer Academic Publishers, Dordrecht (1994)
18. Swanson Analysis Systems Inc. ANSYS, User's guide for revision11, ANSYS 11 Documentation, ANSYS Element reference, Element, library. SOLID45/CONTA173 (2007)
19. Wagle, S., Kato, H.: Ultrasonic detection of fretting fatigue damage at bolt joints of aluminium alloy plates. Int. J. Fatigue. **31**, 1378–1385 (2009)
20. Cattaneo, C.: Sul Contato di Due Corpo Elastici, Accademia dei Lincei, Rendiconti, Series 6–27:342–348, 434–436 and 474–478 (1938)
21. Mindlin, R.D.: Compliance of elastic bodies in contact. J. Appl. Mech. Trans. ASME **16**, 259–268 (1949)
22. Waterhouse, R.B.: Fretting fatigue. Applied Science Publishers, London (1981)
23. Vingsbo, O., Soderber, S.: On fretting maps. Wear **126**, 131–147 (1988)
24. Munoz, S., Proudhon, H., Dominguez, J., et al.: Prediction of the crack extension under fretting wear loading conditions. Int. J. Fatigue. **28**, 1769–1779 (2006)

FPGA-Realization of Inverse Kinematics Control IP for Articulated and SCARA Robot

Ying-Shieh Kung, Bui Thi Hai Linh, Ming-Kuang Wu, Feng-Chi Lee and Wen-Chuan Chen

Abstract The computation of kinematics/inverse kinematics in robot manipulators consumes much CPU time and certainly slows down the motion speed of robots. To solve this problem, this paper proposes a hardware implementation of inverse kinematics control IP (Intelligent properties) for robot manipulators based on the FPGA technology. One robot is an articulated type robot manipulator and the other one a SCARA type robot manipulator. In this chapter, the formulations of inverse kinematics for articulated type and SCARA-type robot manipulator are firstly derived. Then, VHSIC Hardware Description Language (VHDL) will be adopted to describe the aforementioned circuit behavior. The design of the VHDL code will apply the finite state machine (FSM) to reduce the hardware resource usage. Under this design, the executing time for computing the inverse kinematics algorithm needs only several micro-seconds. Finally, the simulation results of the inverse kinematic for two type robot manipulators will be compared using the Matlab and Quartus II software in order to demonstrate the effectiveness and correctness of the proposed FPGA-realization of inverse kinematics.

Y.-S. Kung (✉) · B. T. H. Linh
Department of Electrical Engineering, Southern Taiwan University of Science and Technology, Tainan 710, Taiwan, ROC
e-mail: kung@mail.stust.edu.tw

B. T. H. Linh
e-mail: hailinh311@gmail.com

M.-K. Wu
Department of Mechanical Engineering, Southern Taiwan University of Science and Technology, Tainan 710, Taiwan, ROC
e-mail: w2m@mail.stust.edu.tw

F.-C. Lee · W.-C. Chen
Mechanical and System Research Laboratories, Industrial Technology Research Institute Hsinchu, Hsinchu, Taiwan, ROC
e-mail: lifengchi@itri.org.tw

W.-C. Chen
e-mail: JasonChen@itri.org.tw

A. Öchsner and H. Altenbach (eds.), *Design and Computation of Modern Engineering Materials*, Advanced Structured Materials 54,
DOI: 10.1007/978-3-319-07383-5_15,

Keywords FPGA · VHDL · Robot manipulator · Inverse kinematics

1 Introduction

Robotic control is currently an exciting and highly challenging research focus. Several solutions for implementing the control architecture for robots have been proposed [1–4]. Kabuka et al. [1] applied two high-performance floating-point signal processors and a set of dedicated motion controllers to build a control system for a six-joint robot arm. Yasuda [2] adopted a PC-based microcomputer and several PIC microcomputers to construct a distributed motion controller for mobile robots. Li et al. [3] utilized an FPGA to implement an autonomous fuzzy behaviour control in a mobile robot. Oh et al. [4] presented a DSP and an FPGA to design the overall hardware system in controlling the motion of biped robots. Kung et al. [5] employed the hardware/software co-design technology to the motion control of an articulated robot arm. However, these methods did not provide the solution for the computation of the complicated formulation in FPGA, such as inverse kinematics, and they also did not consider how to reduce the resource usage in FPGA without loss of system performance.

Inverse kinematics is a major problem in the control of robot manipulators. The inverse kinematics problem has a wide range of applications in robotics. Most of our high level problem solving about the physical world is posed in Cartesian space. To calculate the inverse kinematics in real time, it is necessary for the robot to control algorithms. Solving the inverse kinematics is computationally expensive and generally takes very long time in the real time control of robot manipulators. The robot kinematics comprises forward kinematics and inverse kinematics. Given the sets of joint angles, calculation the position and orientation of the end-effector of the robot manipulator is called the forward kinematics. The forward kinematics problem is straightforward and there is no complexity deriving the equations. Tasks to be performed by a robot manipulator are in the Cartesian space, whereas actuators work in joint space. The Cartesian space includes orientation matrix and position vector, and the joint space is represented by joint angles. The conversion of the position and orientation of a robot manipulator end-effector from the Cartesian space to the joint space is called the inverse kinematics problem [6]. This chapter studies and implements the inverse kinematics. Moreover, the FPGA-realization of the inverse kinematics for two types of robot manipulators is presented.

2 Inverse Kinematics of Two Types Robot Manipulators

This section describes the inverse kinematics regarding two types of robot manipulators, in which one is an articulated type robot manipulator, and the other one is a SCARA type robot manipulator.

2.1 Articulated Type Robot Manipulator

Figure 1a shows the link coordinate system of the five-axis articulated robot manipulator using the Denavit-Hartenberg convention [7]. The inverse kinematics of the the articulated robot manipulator will transform the coordinates of robot manipulator from the Cartesian space R^3 (x,y,z) to the joint space R^5 $(\theta_1, \theta_2, \theta_3, \theta_4, \theta_5)$. The computational procedure for the articulated robot is summarized as follows.

$$Step1: \quad \theta_1 = \theta_5 = a\tan 2(y, x) \tag{1}$$

$$Step2: \quad b = \sqrt{x^2 + y^2} \tag{2}$$

$$Step3: \quad \theta_3 = \cos^{-1}\left(\frac{b^2 + (d_1 - d_5 - z)^2 - a_2^2 - a_3^2}{2a_2a_3}\right) \tag{3}$$

$$Step4: \quad S_2 = (a_2 + a_3\cos\theta_3)(d_1 - d_5 - z) - a_3 b\sin\theta_3 \tag{4}$$

$$Step5: \quad C_2 = (a_2 + a_3\cos\theta_3)b + a_3\sin\theta_3(d_1 - d_5 - z) \tag{5}$$

$$Step6: \quad \theta_2 = a\tan 2(S_2, C_2) \tag{6}$$

$$Step7: \quad \theta_4 = -\theta_2 - \theta_3 \tag{7}$$

where a_1, a_5, d_1, d_5 are the Denavit-Hartenberg parameters.

2.2 SCARA Type Robot Manipulator

The inverse kinematics problem of the SCARA robot manipulator [8] is that we calculate the joint angles $\theta_1, \theta_2, d_3, \theta_4$ when the *(n, o, a)* and *p(x,y,z)* are given. The link coordinate system of the SCARA robot manipulator is shown in Fig. 1b. This four axis arm is in R-R-P-R configuration. The $\theta_1, \theta_2, \theta_4$ are the revolute joint angle variables and d_3 is the prismatic joint variable. The computational procedure for the SCARA robot is summarized as follows:

$$Step1: \quad C_2 = \frac{x^2 + y^2 - a_1^2 - a_2^2}{2a_1a_2} \tag{8}$$

$$Step2: \quad S_2 = \sqrt{1 - C_2^2} \tag{9}$$

$$Step3: \quad \theta_2 = \pm\cos^{-1}\left(\frac{x^2 + y^2 - a_1^2 - a_2^2}{2a_1a_2}\right) \tag{10}$$

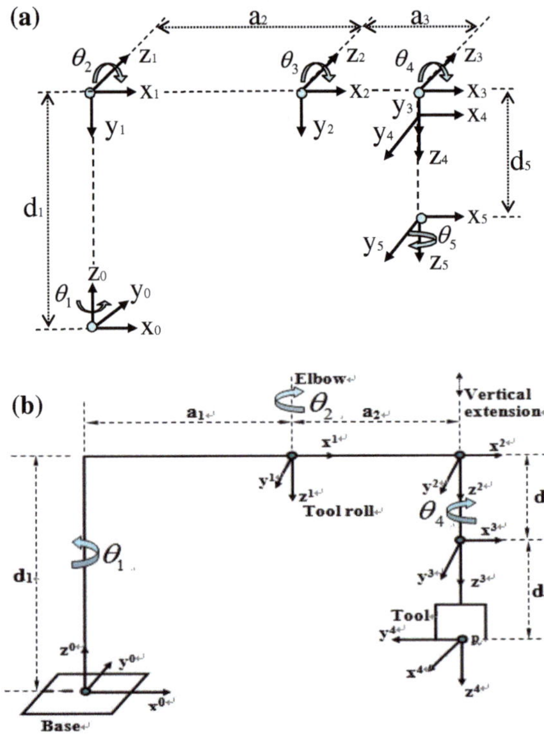

Fig. 1 The link coordinate system of **a** a five-axis articulated robot and **b** a four-axis SCARA robot

$$Step4: \quad S_1 = \frac{a_2S_2x + (a_1 + a_2C_2)y}{(a_2S_2)^2 + (a_1 + a_2C_2)^2} \tag{11}$$

$$Step5: \quad C_1 = \frac{(a_1 + a_2C_2)x - a_2S_2y}{(a_2S_2)^2 + (a_1 + a_2C_2)^2} \tag{12}$$

$$Step6: \quad \theta_1 = a\tan 2(S_1, C_1) \tag{13}$$

$$Step7: \quad d_3 = d_1 - z - d_4 \tag{14}$$

$$Step8: \quad \theta_3 = 0 \tag{15}$$

$$Step9: \quad C_4 = n_xC_{12} + n_yS_{12} \tag{16}$$

$$Step10: \quad S_4 = -n_xS_{12} + n_yC_{12} \tag{17}$$

$$Step11: \quad \theta_4 = a\tan 2(S_4, C_4) \tag{18}$$

where a_1, a_2, d_1, d_4 are the Denavit-Hartenberg parameters, and C_x, S_y, C_{xy}, S_{xy}respectively represents Cosθ_x, Sinθ_y, Cos$(\theta_x + \theta_y)$, Sin$(\theta_x + \theta_y)$.

3 FPGA Implementation of Inverse Kinematics for Two Types of Robot Manipulations

The implementation of the inverse kinematics applies the finite state machine method which can save computational resources in the case of FPGA. The FPGA chip adopted herein is an Altera DE2 Cyclone II EP2C35F672, which has 33,216 LEs, 105 M4 K RAM blocks 4,83,840 total RAM bits, 35 embedded multipliers, 4 PLLs, 475 user I/O pins, fine Line BGA 672-pin package, and a Nios II embedded processor which has a 32-bit configurable CPU core, 16 M byte Flash memory, 1 M byte SRAM and 16 M byte SDRAM, is used.

3.1 Hardware Design of Inverse Kinematics for Articulated Type Robot Manipulator

The hardware implementation of the inverse kinematic is designed by the finite state machine method, and it is shown in Fig. 2 which shows that 42 steps are needed to perform the overall computation of the inverse kinematics in (1)–(7). The circuit herein needs 2 multipliers, 1 divider, 2 adders, 1 square root function, 1 component for atan2 function, 1 component for acos function, 2 look-up-tables for sin and cos function and some comparator. From the simulation results, to realize the inverse kinematics of the articulated type robot manipulator, 840 ns (20 ns/step* 42 steps) computation time is needed and it uses 6,033 LEs resource. The computation for the inverse kinematics takes 5.803 ms for a C program by the Nios II processor and this computation time is about 6,908 times slower than the computation time on the FPGA-hardware.

3.2 Hardware Design of Inverse Kinematic for SCARA Type Robot Manipulator

Here, the finite state machine method is also adopted in designing the hardware implementation of the inverse kinematic for the SCARA type robot manipulator it is shown Fig. 3 that there are 55 steps to perform the inverse kinematics in (8)–(18), and the circuit needs 2 multipliers, 1 dividers, 2 adders, 1 square root function, 1 component for acos function, 1 component for atan2 function, 2 look-up-tables for sin and cos function. From the simulation results, it needs 1,100 ns (20 ns/step* 55 steps) computation time to realize the inverse kinematic and it uses 7,623 LEs resource. The computation for the inverse kinematics takes 4.203 ms for a C program by the Nios II processor and this computation time is about 3,820 times slower than the computation time on FPGA-hardware.

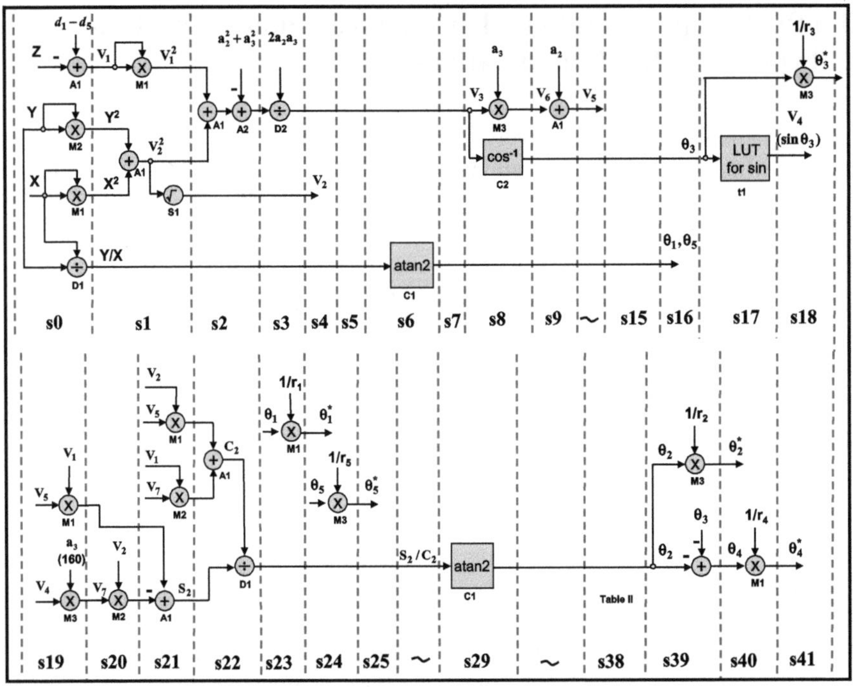

Fig. 2 Circuit of computing the inverse kinematics using finite state machine for articulated type robot manipulator

4 Simulation Results

4.1 Results of Articulated Type Robot Manipulator

The simulation results in Matlab are compared with implementation results in Quartus II which is shown in Fig. 4. The position is given as x = 200 mm, y = 100 mm, and z = 300 mm, then we get the $\theta_1, \theta_2, \theta_3, \theta_4, \theta_5$ in degree which is converted from the pulse based on the parameters of the robot manipulator. The results show that the outputs of the inverse kinematic for the articulated type robot manipulator in FPGA are almost the same as the calculation in Matlab, and the error is only within $10^{-3}-10^{-2}$ degree.

4.2 Results of SCARA Type Robot Manipulator

The simulation results in Matlab are compared with the implementation results in Quartus II which is shown in Fig. 5. The position is given as x = 200 mm, y = 100 mm and z = 300 mm, as well as $n_x = 0$, $n_y = 1$. Then we get the

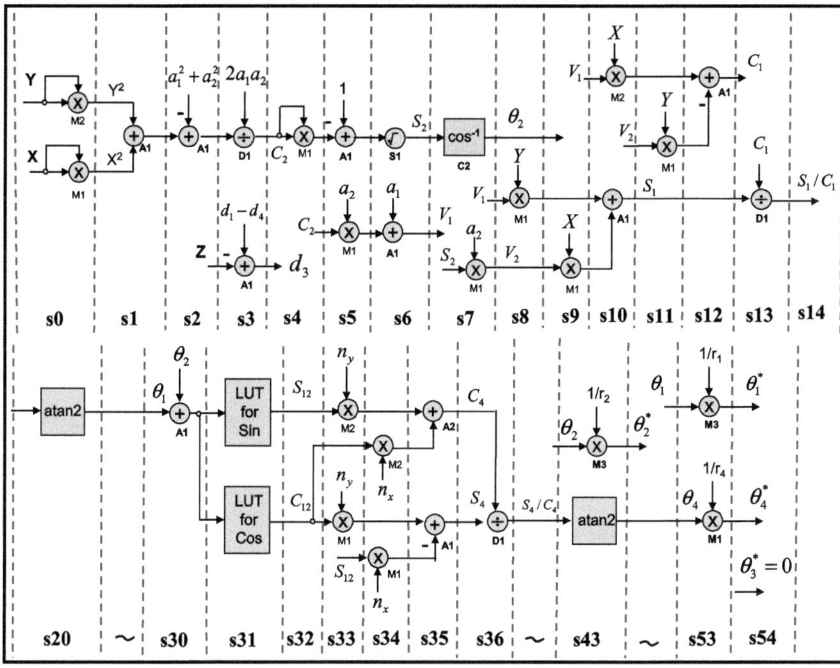

Fig. 3 Circuit of computing the inverse kinematics for SCARA type robot manipulator

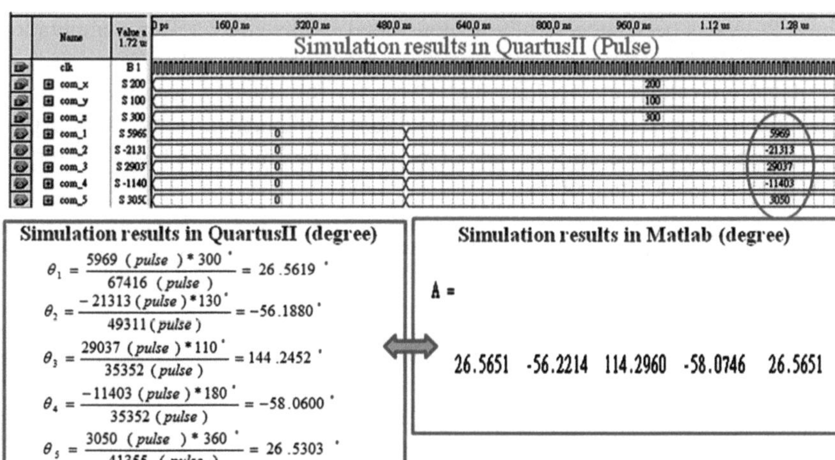

Fig. 4 Simulation results in Matlab and in Quartus II for articulated type robot manipulator

$\theta_1, \theta_2, \theta_4$ in degree which is converted from the pulse based on the parameters of the robot manipulator and d_3 is the distant in mm. The results also show that the outputs of the inverse kinematic for the SCARA type robot manipulator in FPGA is very good approximation with the calculation in Matlab.

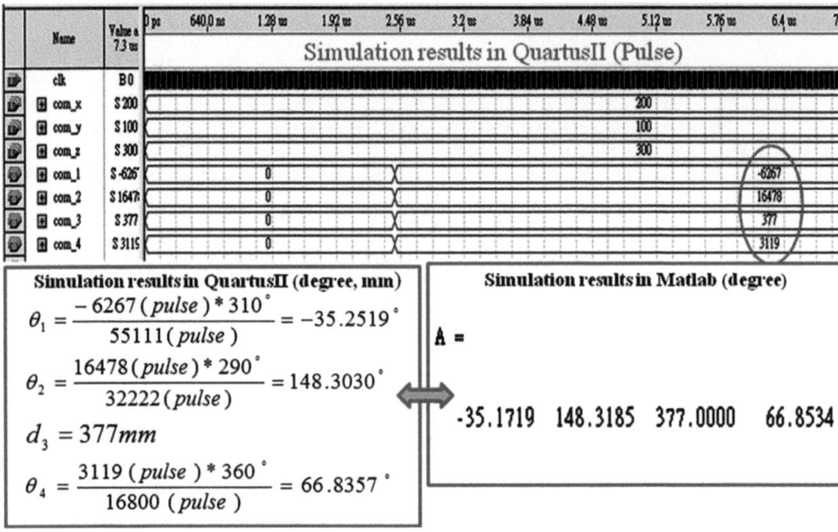

Fig. 5 Simulation results in Matlab and in Quartus II for SCARA type robot manipulator

5 Conclusions

The implementation of the inverse kinematics for two types of robot manipulators using FPGA has been successfully demonstrated in this chapter. The computation time to implement the inverse kinematics in hardware is only 840 ns for the articulated type robot manipulator and 1.1 μs for the SCARA type robot manipulator. Otherwise, the same formulas which are implemented in the Nios II processor need 5.60 ms for the articulated type robot manipulator and 5.83 ms for the SCARA type robot manipulator.

Acknowledgments This research is supported by Industrial Technology Research Institute.

References

1. Kabuka, M., Glaskowsky, P., Miranda, J., et al.: Microcontroller-based architecture for control of a six joints robot arm. IEEE Trans. Ind. Electron. **35**, 217–221 (1988)
2. Yasuda, G.: Microcontroller implementation for distributed motion control of mobile robots. Proceedings of International workshop on Advanced Motion Control, 114–119 (2000). doi:10.1109/AMC.2000.862840
3. Li, T.S., Chang, S.J., Chen, Y.X., et al.: Implementation of human-like driving skill by autonomous fuzzy behavior control on an FPGA-based car-like mobile robot. IEEE Trans. Ind. Electron. **50**, 867–880 (2003)
4. Oh, S.N., Kim, K.I., Lim, S., et al.: Motion control of biped robots using single-chip drive. IEEE Int. Conf. Robot **2**, 2461–2465 (2003)

5. Kung, Y.S., Shu, G.S.: FPGA-based motion control IC for Robot arm. Proceedings of the IEEE International Conference on Industrial Technology, 1397–1402 (2005). doi:10.1109/ICIT.2005.1600854
6. Kucuk, S., Bingul, Z.: The inverse kinematic solution of industrial Robot manipulators. Proceedings of the IEEE International Conference of Mechatronics, 274–279 (2004). doi:10.1109/ICMECH.2004.1364451
7. Kung, Y.S., Tseng, K.H., Chen, C.S., Sze, H.Z., Wang, A.P.: FPGA-Implementation of inverse kinematic and Servo controller for Robot manipulator. In: Robotics and Biomimetics, 2006 ROBIO'06 IEEE International Conference, 1163–1168 (2006). doi:10.1109/ROBIO.2006.340093
8. Alshamasin, M.S., Ionescu, F., Al-Kasasbeh, R.T.: Modeling and simulation of a SCARA robot using solid dynamics and verification by MATLAB/Simulink. Int. Model. Ident. Control **15**(1), 28–38 (2012). doi:10.1504/IJMIC.2012.043938

Development of Mathematical Models for Dissimilar Welding Pool Geometries

Ezzeddin Anawa and Abdulghani Olabi

Abstract This study presents the design of experiment methodology (DOE) for the development of mathematical models for predicting and optimizing fusion zone dimensions of aluminum/titanium welded components by using CO_2 laser welding (LW) process. The implemented DOE methodology used the Taguchi approach and orthogonal array to mathematical models to predict and an optimize fusion zone dimensions. These dimensions are the welding penetration (L_1), welding width at the surface (W_1) and cross section area of melted zone (A_1) in the aluminum plate. The laser power, welding speed, and defocusing distance ranges were experimentally determined with the objective of producing a welded joint with adequate penetration, minimum fusion zone size and acceptable welding profile. The fusion zone area and the shape of weld profile of the dissimilar aluminum/titanium were evaluated based on the selected laser welding parameters. Taguchi approach was used as a statistical design of experiment (DOE) technique for optimizing the selected welding parameters in terms of the minimizing area of melted zone (A_1), welding width at the surface (W_1) and maximizing the penetration (L_1). The results showed that, for the dissimilar titanium–aluminum overlap welds, the prediction of fusion depth and width that were obtained by the developed models are in good agreement with the results obtained experimentally. Since this methodology does not require complicated or excessive computation, it is especially useful for the actual welding process applications. It is also provided a robust approach to adaptive welding as well as to stabilize weld quality.

Keywords Dissimilar welding · Taguchi method · Penetration · Fusion zone

E. Anawa (✉)
Industrial and Manufacturing Engineering, Benghazi University, Benghazi, Libya
e-mail: ezzeddin.anawa@uob.edu.ly

A. Olabi
School of Mechanical and Manufacturing Engineering, Dublin City University, Dublin, Ireland
e-mail: Abdul.Olabi@uws.ac.uk

A. Öchsner and H. Altenbach (eds.), *Design and Computation of Modern Engineering Materials*, Advanced Structured Materials 54,
DOI: 10.1007/978-3-319-07383-5_16,

1 Introduction

Welding quality is strongly characterized by the weld penetration and the weld pool geometry [1]. Due to that the weld pool geometry plays an important role in determining the mechanical properties of the welded joints. Therefore, the selection of the welding process parameters is essential for obtaining optimal weld pool geometries. The important problem to be solved in welding engineering is to develop a model for determining the optimal process parameters.

A mathematical model for weld heat sources based on a Gaussian distribution of power density in space was presented by Goldak et al. [2]. In particular a double ellipsoidal geometry was proposed so that the size and shape of the heat source could be easily changed to model both the shallow penetration arc welding processes and the deeper penetration laser and electron beam processes.

Hsu and Rubinsky [3] have investigated a about a two-dimensional, quasi-stationary finite element numerical model to study the fluid flow and the heat transfer phenomena which occur during constant travel speed, keyhole plasma arc welding of metal plates.

Based on the computation model for quasi-steady heat transfer problems of welding with the boundary element method, Hang and Okada [4] had developed a computer program that used for the computation of thermal cycles at heat affected zones with gas shielded metal arc welding "GMAW" on medium thickness plates.

In a study by Zhang et al. [5], a polar coordinate model was proposed to characterize the weld pool geometrically. The identification of its parameters involves complicated non-linear optimization which cannot be done in real time using conventional algorithms. A neural network ANN was therefore proposed to identify the parameters in real time. Weld pool geometry was computed via numerical solution of a boundary integral equation used as a model for the autogenous full penetration welding of pure materials by Yeh and Brush [6]. Wahab and Painter [7] measured the full 3-dimensional weld pool shape for the GMAW process, and to study the use of this information within numerical models.

A computational modelling of welding phenomena within a versatile numerical framework was presented by Taylor et al. [8]. Fusion zone area and shape were evaluated by Gunaraj and Murugan [9] as a function of the selected submerged arc welding (SAW) parameters. Response surface methodology (RSM) was used as statistical design of experiment (DOE) technique for optimizing the selected welding parameters in terms of minimizing the fusion zone. In this study the modified Taguchi method was adopted to solve the optimal weld bead geometry with four the-smaller-the-better quality characteristics.

2 Experimental Procedure

Two sheets of Al 6082 T6 and Ti G2 with dimensions of 160 × 80 × 1 mm were selected to be joined by CO_2 laser welding.

2.1 Materials Specifications

The chemical and mechanical properties of materials selected are presented in Tables 1 and 2.

2.2 Sample Preparation

A small sample was cut from each welded plate perpendicular to the welding line to study the weld pool shape and geometry. The titanium side of the weldment was etched in reagent consisting of (10 ml HF and 5 ml HNO_3 in 85 ml of water), and the rest of the regions of the weldment were etched with Keller's reagent (1 % HF, 1.5 % HCl, 2.5 % HNO_3 and H_2O solution).

2.3 Welding Parameters Selection and Experiment Design

The selected welding parameters for these dissimilar materials are: Laser power, welding speed and focus point position. Table 3 shows the welding input variables and experiment design levels. The welding experiments were carried out in the Mechanical School workshop following the Taguchi designed matrix in random order generated by the Design Expert software, as presented in Table 4. The welding pool geometry, mechanical destructive tests (tensile shear strength) and cost per meter welded calculations were carried out in the joined specimens and the results are presented in Table 4. Each presented result in Table 4 in each column is an average of at least of three readings.

The measured results of welding pool area for each sample are presented in Table 4. Figure 1a–d shows the effect of the welding parameters and the variation on the total weld pool (fusion area) 'A_1' at aluminum plate only, welding widths at the specimen surface of aluminium 'W_1' and welding widths at the penetration of welding depth in aluminum plate 'L_1' of some experiments selected from Table 4.

Table 1 Chemical composition of the Ti G2 and the Al 6082 (wt%)

Material	C	Fe	H	N	O	Ti
Ti G2	0.1	0.3	0.015	0.03	0.25	99.2

Material	Si	Fe	Cu	Mn	Mg	Cr	Zn	Ni	Ti	Al
Al 6082	1.05	0.5	0.106	0.934	1.05	0.262	0.223	0.008	0.114	Bal.

Table 2 Mechanical properties of the welded materials

Grade	Tensile strength, [MPa]	Yield strength, [MPa]	Elongation %	Hardness [HB], max	Elastic modulus, [GPa]
Al 6082	210	140	11	94	70
Ti G2	344	275	20	14.5	105

Table 3 Process parameters and design levels used

Variables	Code	Unit	Level 1	Level 2	Level 3	Level 4
Laser power	*P*	kW	0.675	0.850	1.025	1.200
Welding speed	*S*	mm/sec	120	140	160	180
Focus	*F*	mm	−1.0	−0.67	−0.33	0.00

Table 4 Welding input variables, experiment design levels and the welding pool geometry calculations

Std	Run	*P*, kW	*S*, m/sec	*F*, mm	W_1, μm	L_1, μm	A_1, μm
16	1	1.200	180	−1	644	451	155.2
7	2	0.850	160	0.00	392	362	114.1
15	3	1.200	160	−0.67	619	354	171.8
5	4	0.850	120	−0.67	317	267	71.3
6	5	0.850	140	−1.00	341	257	67.7
1	6	0.675	120	−1.00	508	362	131.8
12	7	1.025	180	−0.67	527	312	88.6
14	8	1.200	140	−0.33	449	407	144.3
8	9	0.850	180	−0.33	394	215	33.7
11	10	1.025	160	−1.00	252	491	122.7
3	11	0.675	160	−0.33	597	165	63.3
4	12	0.675	180	0.00	478	251	78.1
13	13	1.200	120	0.00	474	281	155.2
9	14	1.025	120	−0.33	422	297	31.2
10	15	1.025	140	0.00	248	280	24.8
2	16	0.675	140	−0.67	552	272	103.9

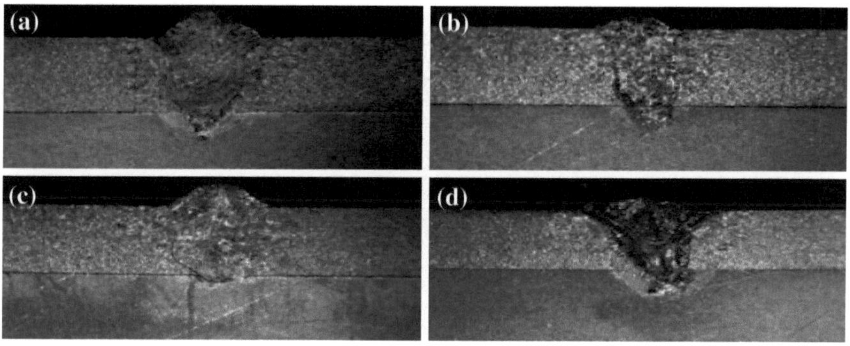

Fig. 1 **a** Welding pool specimen 1, **b** Welding pool specimen 4, **c** Welding pool specimen 8 and **d** Welding pool specimen 14. Effect of the welding parameters and the variation on the total weld pool dimensions

3 Results and Discussion

The fusion zone dimensions in the Al plate were measured by using the transverse sectioned specimens, optical microscope and image analysis software. The measured responses are listed in the same Table 4. Design Expert 7 software was used for analysing the responses. The fit summary output indicates that the quadratic models which are developed are statistically significant for the prediction of the responses (W_1 and L_1); therefore, they will be used for further analysis. It has been seen from the achieved results that the welding pool geometry, shape and penetration are controlled by the rate of heat input, which is a function of laser power and welding speed. The focusing position has also a strong effect on the responses.

3.1 Analysis of Variance

The test for significance of the regression model and the test for significance on individual model coefficients were performed using the Design Expert 7 software. The backward elimination regression method was applied and exhibited in ANOVA Tables 5 and 6 for the suggested reduced quadratic models. Tables 5 and 6 summarize the analysis of variances of the responses and show the significant models. The same tables also show the other adequacy measures R^2, adjusted R^2 and adequacy precisions. All adequacy measures were close to 1, which is reasonable and indicates an adequate model. The adequate precision compares the range of the predicted value at the design points to the average predicted error. In this study the values of adequate precision for the W_1 and L_1 are significantly greater than 4. The adequate precision ratio above 4 indicates adequate model discrimination. The developed quadratic models in terms of coded factors and actual values are exhibited in Eqs. 1–4.

Table 5 ANOVA for response 'W_1'

Source	Sum of squares	df	Mean square	F_v value	p-value prob. > F_v	
Model	169764	5	33952.8	6.0685	0.0078	Significant
P	312.05	1	312.05	0.0558	0.8181	
S	19096.2	1	19096.2	3.4131	0.0944	
F	4681.8	1	4681.8	0.8368	0.3818	
P^2	127449	1	127449	22.779	0.0008	
F^2	18225	1	18225	3.2574	0.1013	
Residual	55949.7	10	5594.97			
Cor. Total	225713.8	15				
$R^2 = 0.7521$			Adeq. Precision = 6.799			
Adj. $R^2 = 0.6282$						

Table 6 ANOVA for response 'L_1'

Source	Sum of squares	df	Mean square	F_v value	p-value prob. > F_v	
Model	69298.46	5	13859.69	3.34876	0.00491	Significant
P	32320.8	1	32320.8	7.809308	0.0190	
S	616.05	1	616.05	0.148849	0.7077	
F	1459.354	1	1459.354	0.352607	0.5658	
PS	3385.314	1	3385.314	0.817955	0.3870	
F^2	12432.25	1	12432.25	3.003863	0.1137	
Residual	41387.5	10	4138.75			
Cor. Total	110686	15				
$R^2 = 0.6261$			Adeq. Precision = 6.842			
Adj. $R^2 = 0.4391$						

Final Equation in Terms of Coded Factors:

$$W_1 = 381.50 + 5.92P + 46.35S - 22.95F + 200.81P^2 - 75.94F^2 \quad (1)$$

Final Equation in Terms of Actual Factors:

$$W_1 = 2591.09 - 5441.71P + 0.09S - 349.65F + 2914.29P^2 - 303.75F^2 \quad (2)$$

Final Equation in Terms of Coded Factors:

$$L_1 = 279.16 + 60.30P + 8.32S - 21.60F + 44.13PS + 62.72F^2 \quad (3)$$

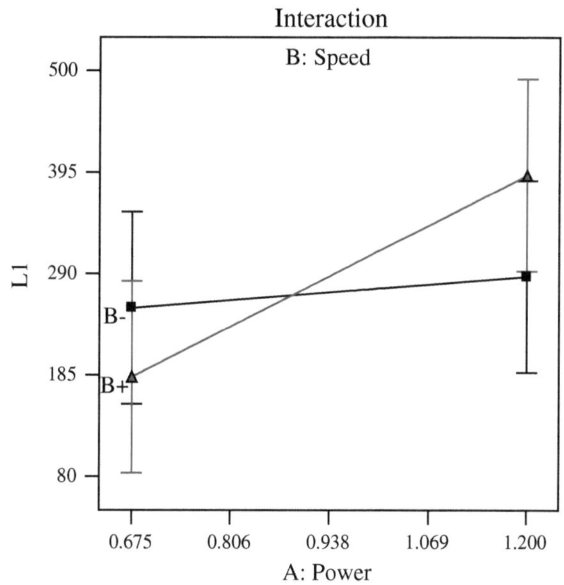

Fig. 2 Interactions between the welding parameters (*P*, *S*) with respect to the depth of penetration response at $F = -0.5$ mm

Final Equation in Terms of Actual Factors:

$$L_1 = 851.34 - 610.87P - 0.30S + 207.68F + 0.34PS + 250.88F^2 \quad (4)$$

For the response W_1 of the developed model, the analysis of variance indicates that the welding speed '*S*' and focus position '*F*' are the stronger welding parameters affecting the responses. The focus position '*F*' has a greater affect on the response W_1 than the laser power '*P*'. The W_1 model indicates that the studied parameters (*S*, *F*) significantly affect the response. For the response L_1 of the developed model, the analysis of variance indicates that the laser power '*P*' and focus position '*F*' are the stronger welding parameters affecting the responses. Focus position '*F*' has a greater affect on the response W_1 than welding speed '*S*'. The L_1 model indicates that the studied parameters (*P*, *F*) significantly affect the response. The L_1 model indicates that the welding parameters have interactions between *P* and *S* exhibited in Fig. 2. The figure exhibits the interaction of the welding speed with the laser power at focus position $F = -0.5$ mm.

3.2 Model Validation

The aim of this step is to predict and verify the improvement of the response using the optimal levels of the welding process parameters. Figures 3 and 4 show the relationship between the actual and predicted values of W_1 and L_1, respectively.

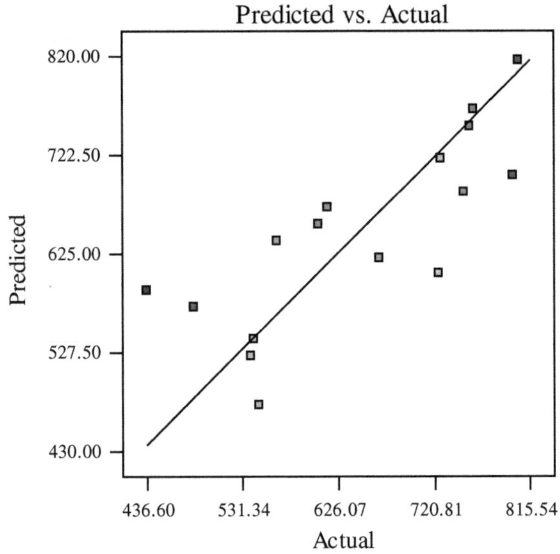

Fig. 3 The predicted values of the L_1 versus actual measured values

These figures indicate that the developed models are adequate because the residuals in prediction of each response are negligible, since the residuals tend to be close to the diagonal line.

Furthermore, to verify the satisfactoriness of the developed models, three confirmations experiments were carried out using new test conditions at different parameters conditions, obtained using the software and the developed mathematical models. The values of W_1 and L_1 for validation experiments were calculated using the software. Table 7 summarizes the experimental conditions, the actual experimental values, the predicted values and the percentages of absolute errors. It could be concluded that the models developed can predict the responses with a very small errors. W_1 and L_1 were greatly improved through this optimization.

3.3 *Effect of the Parameters on Responses*

The reason for predicting the welding pool geometry is to develop a model which would include the optimization step.

- Welding Pool Width at the Work Piece Surface (W_1).

The results and the model obtained for the response indicate that the S and F are the most important factors affecting the W_1 value. An increase in S leads to a

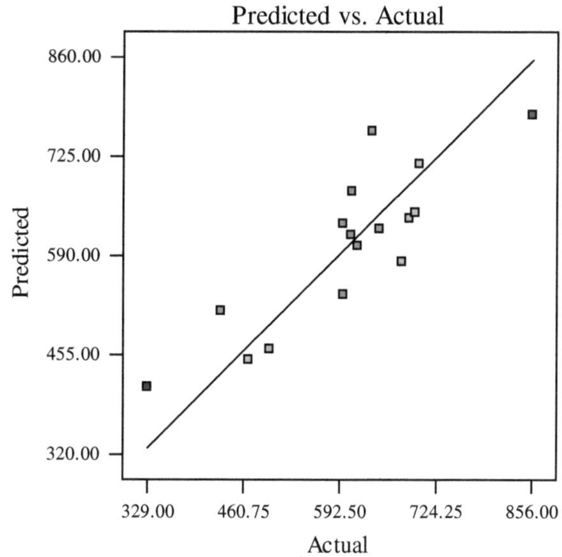

Fig. 4 The predicted values of W_1 versus actual penetration depth of measured values

Table 7 Confirmation experiments of the responses (W_1 and L_1)

Exp. no	P, kW	S, mm/sec	F, mm	W_1, mm		$\|E\|$ %	L_1, mm		$\|E\|$ %
				Act.	Pred.		Act.	Pred.	
1	1.134	120	0.00	382	408	6.37	299	327	8.56
2	1.134	150	0.00	501	399	25.56	323	365	11.51
3	0.871	180	−1.00	388	386	0.52	401	345	16.23

Act = Actual; Pred. = Predicted

decrease in W_1 and the increase of F leads to an increase in W_1. This is due to the fact that the laser beam is traveling at high speed over the welding line when S is increased. Therefore, the heat input decreases leading to less volume of the base metal being melted, consequently the width of the welded zone decreases. Moreover, a defocused beam, which is a wider laser beam, results in spreading the laser power over a wide area. Therefore, a wide area of the base metal will be melted leading to an increase in W_1 or vice versa. The result shows also that P contributes a secondary effect in the response width dimensions. Increasing P results in a slight increase in W_1, due to the increase in the power density. Figure 5 shows contour plots for the effect of the parameters on the W_1 width. Figure 5 illustrates the relationship between S and P with their impact on the welding pool width at the surface of the aluminum plate (W_1) at $F = -0.5$ mm.

- Welding Pool Width at the Middle of the Work Piece (L_1)

The results and the model obtained for the response indicate that the P and F are the most important factors affecting the W_1 value. An increase in P leads to

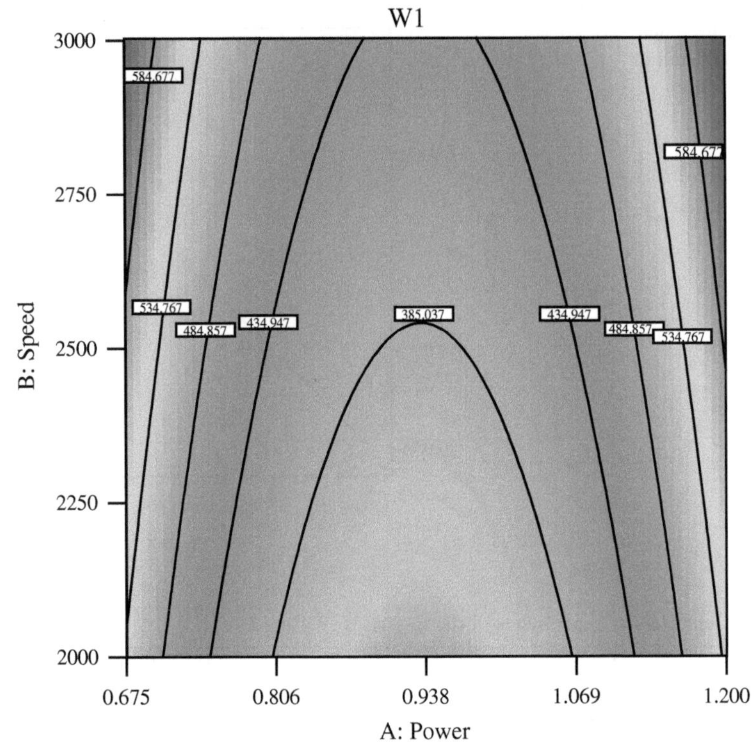

Fig. 5 Contour graphs of the effect of *P*, *S* parameters at $F = -0.5$ mm on the response L_l

an increase in L_1 and the increase in *F* leads to increase in L_l. This is due to the fact that an increase in the amount of laser power *P* that is transfer to the work piece leads to an increase in the response. Therefore, the heat input increases leading to an increase in the amount that the base metal melts, consequently the penetration of the welded zone increases. Moreover, a defocused beam, which is in a wider laser beam, results in spreading the laser power over a wide area. Therefore, a wide area of the base metal will be melted leading to an increase in L_l or vice versa. The result also shows that S contributes in a secondary effect in the response width dimensions. S is inversely proportional to the heat input. Increasing *S* results in a slight decrease in L_1; this is due to the decrease in the heat input. Figure 6 shows contour plots for the effect of the process parameters on the L_1 width. Figure 6 illustrates the relationship between *S* and *P* with their impact on the welding pool depth of the aluminum plate (L_l) at $F = -0.5$ mm.

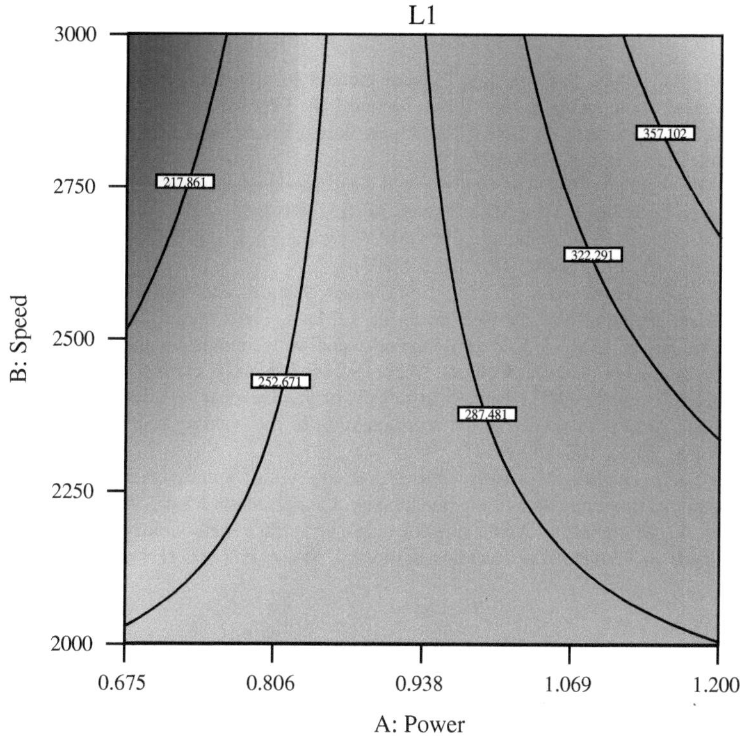

Fig. 6 Contour graphs exhibiting the effect of *P*, *S* parameters at $F = -0.5$ mm on the response W_l

4 Conclusions

Laser welding of nonferrous with nonferrous dissimilar material has been studied experimentally and analyzed statistically and the following points are concluded:

1. The dissimilar joint between aluminum alloy Al 6082 and titanium G2 alloys were successfully welded by CO_2 laser welding with a single pass and without filler material using the overlap joint design.
2. Laser welding is a very successful method for joining dissimilar nonferrous metals.
3. The models developed can satisfactorily predict the responses within the studied domain.
4. Appling of a DOE inspired by the Taguchi technique, best operating parameters were achieved and then develop models to control the welding parameters.

Acknowledgments The authors wish to thankful Mechanical and Manufacturing School in Dublin City University and Benghazi University for their support and assistance.

References

1. Anawa, E.M., Olabi, A.G.: Using Taguchi method to optimize welding pool of dissimilar laser-welded components. J. Opt. Laser Technol. **40**, 379–388 (2008)
2. Goldak, J., Chakravarti, A., Bibby, M.: A new finite element model for welding heat sources. J. Metall. Trans. B **15B**, 299–305 (1984)
3. Hsu, Y.F., Rubinsky, B.: Two-dimensional heat transfer study on the keyhole plasma arc welding process. Int. J. Heat Mass Transf. **31**(7), 1409–1421 (1988)
4. Hang, M., Okada, A.: Computation of GMAW welding heat transfer with boundary element method. J. Adv. Eng. Softw. **16**(1), 1–5 (1993)
5. Zhang, Y.M., Kovacevic, R., Li, L.: Characterization and real-time measurement of geometrical appearance of the weld pool. Int. J. Mach. Tools Manuf. **36–7**, 799–816 (1996)
6. Yeh, J.Y., Brush, L.N.: A boundary integral equation technique for the calculation of weld pool shapes in thin plates. J. Comput. Mater. Sci. **6–1**, 92–102 (1996)
7. Wahab, M.A., Painter, M.J.: Numerical models of gas metal arc welds using experimentally determined weld pool shapes as the representation of the welding heat source. Int. J. Press. Vessels Pip. **73**(2), 153–159 (1997)
8. Taylor, G.A., Hughes, M., Strusevich, N., et al.: Finite volume methods applied to the computational modeling of welding phenomena. J. Appl. Math. Model. **26**(2), 311–322 (2002)
9. Gunaraj, V., Murugan N.: Application of response surface methodology for predicting weld bead quality in submerged arc welding of pipes. J. Mater. Process. Technol. **88**(1–3), 266–275 (1999)

Efficient Multi-objective Optimization for Gas Turbine Discs

Francisco Javier Garcia-Revillo, Jesús R. Jimenez-Octavio, Cristina Sanchez-Rebollo and Alexis Cantizano

Abstract An original multi-objective optimization strategy for aeronautical gas turbine discs is presented in this chapter. A sensitivity analysis together with an optimal design of experiments (DoE) are accomplished prior applying a multi-objective genetic algorithm (MOGA) based on a Kriging surrogate model. Fatigue life prediction and total geometry mass are taken as optimization objectives in this work, whose results can be validated against laboratory tests but going further experimental conditions. To begin with, (i) a first case study considers laboratory test conditions, basically centrifugal loads due to the blades and the own mass of the disc. However, (ii) a second case study considers real conditions such as the centrifugal loads, airflow forces and thermal loads as well as taking into account different typical materials for this application. Results show that the proposed methodology successfully leads to the optimal geometric parameters both for laboratory and real conditions. The main conclusions and qualitative differences are finally summarized, highlighting the suitability of A718Plus among other materials and the *Cob width* parameter as the most critical one.

Keywords Multi-objective optimization · Turbine disc · Design of experiments · Kriging model · Genetic algorithm

---

F. J. Garcia-Revillo · J. R. Jimenez-Octavio (✉) · A. Cantizano
Department of Mechanical Engineering, Universidad Pontificia Comillas, c/Alberto Aguilera 25, 28015 Madrid, Spain
e-mail: jesus.jimenez@upcomillas.es

F. J. Garcia-Revillo
e-mail: javier.garcia-revillo@gea.com

A. Cantizano
e-mail: alexis.cantizano@upcomillas.es

C. Sanchez-Rebollo
Institute for Research in Technology, Universidad Pontificia Comillas, c/Santa Cruz de Marcenado 26, Madrid 28015, Spain
e-mail: cristina.sanchez@iit.upcomillas.es

A. Öchsner and H. Altenbach (eds.), *Design and Computation of Modern Engineering Materials*, Advanced Structured Materials 54,
DOI: 10.1007/978-3-319-07383-5_17,

1 Introduction

Rotating discs are one of the most critical parts of any rotating machinery, thus the optimal design of turbine discs becomes indeed essential for aeroengines [1]. The failure of a disk is absolutely unacceptable due to the hazardous and catastrophic effects that would suffer the engine, airframe, ship or installation if this happens. In order to prevent this fact and achieve an acceptably low probability of individual failure, the discs have very strict design criteria, where the material is carefully selected and stress, thermal, creep, vibration and fatigue analysis have to be satisfied more strictly every day [2].

From the pioneer analytic solutions for rotating discs [3, 4], the significant development in computational continuum mechanics boosts the suitability of much more powerful but computationally expensive techniques like the finite element method (FEM), which really fits from its very beginning the more realistic simulation and design of turbine discs [5–9]. The traditional design process was based on a trial and error method leaded by a group of experts. Despite the computing capabilities which allow FE simulations, optimal designs were subjected to the same iterative process: searching for the objectives of design but satisfying allowable stress, temperature and fatigue life.

Metaheuristic algorithms are proven to be suitable to tackle multi-objective nonlinear optimization problems, but due to the complexity of gas turbine discs, this may be computationally expensive and almost unapproachable when classic genetic algorithms (GA) are used [10–12]. Shape optimization of the discs used to be the goal when minimum weight designs are intended [13, 14]. However, not only one objective but also two or more can be assumed to lead the optimization process, which requires complementary techniques and sensitivity analysis [15] to simplify the methodology. Thus, simple surrogate models are able to construct an approximate response surface from an initial budget of FE based simulations, which allows optimizing via GA not so costly. Definitively, if the design of experiments (DoE) and the surrogate model are well fitted, a huge amount of FE simulations can be saved within the optimization process. The Kriging method is one of the most extended surrogate models for multidisciplinary optimization [16–18], which has been previously applied to the optimization of turbine discs together with GA, see [19, 20]. In order to save computational costs, even FE analysis can be completely avoided in optimization runs by the coupling of the GA with simplified response surface equations [21].

Furthermore, the selection of the base material is undoubtedly an essential variable of design. Both the mechanical properties and the economic costs are key factors which lead to the optimal design. Nowadays, turbine discs are typically formed from nickel alloy forgings [22]. Due to the high nickel content in its chemistry composition, these nickel-based super alloys present an exceptional combination of mechanical properties like toughness, high temperature strength or resistance to degradation in oxidizing and corrosive environments. They are used widely in aerospace applications and, currently, 40–50 % of the total weight of an aircraft engine is typically constituted by nickel-based super alloys. More specifically, it is applied in the parts that make up the turbine and the combustion chamber, where the highest operation temperatures are reached and highest loads are suffered.

Avoiding economic criteria, 3 nickel alloys are analyzed in this paper in order to include the material selection in the optimal design. The first one is A718 [23], a nickel based super alloy used extensively in aircraft applications. This alloy has been used for turbine discs for more than 20 years because of its exceptional high temperature mechanical properties. It has excellent strength, hot and cold workability, very good weldability and moderate cost [24]. On the other hand, Waspaloy [25] shows a better weldability and an improved thermal stability but higher economic costs. Likewise, A718Plus [26] is an improved alloy compared with the previous ones. It offers a full 55 °C temperature advantage over A718, maintaining many of the features of it, like good weldability, good workability (even better than Waspaloy), improved thermal stability and moderate cost [27, 28].

Therefore, the goal of this chapter is to state a robust and flexible optimization methodology applied to turbine discs based on low-fatigue life prediction. Basically finding the Pareto front with the optimal solutions which allow the decision maker to quantify the trade-offs in satisfying any different objectives. The optimization objectives herein proposed are the minimization of the geometry mass and the maximization of the fatigue life. The proposal gathers some of the abovementioned techniques, both for the numerical calculation and the optimization process, which provide the backbone of this methodology. A convenient design of experiments and sensitivity analysis prior applying a surrogate model and a multi-objective genetic algorithm is originally applied in this paper, saving huge CPU time. Once any feasible optimal candidate is computed via FEM, computational time becomes a barrier, so the whole method is mainly focused on CPU time saving.

In order to prove the flexibility of the methodology herein proposed, two different cases have been studied: (i) a laboratory based case, whose laboratory test conditions allow the results to be easily checked in a test bench, and (ii) a more realistic case considering flight conditions. This second case includes a specific study on the base material of the rig disc, obtaining promising results. Geometry, loads and boundary conditions are close to real ones, but not exactly due to confidential reasons. Despite this, it is not essential when a methodology is presented, both the degree of complexity of the models and the accuracy of the results are obviously improvable at the expense of the CPU time.

After this brief introduction and review of the state of the art, the paper is organized as follows. Section 2 provides a description of the turbine disc model, details of the cases of study and certain calculation and simulation basis. Section 3 presents the optimization methodology together with its application to the turbine disc model. Section 4 applies the original methodology to the geometry optimization of a test rig turbine disc, used in industry to test the fatigue life of high pressure turbine discs made of A718Plus. Section 5 is an in-depth examination of real operating conditions. Furthermore, the set of materials previously described are herein analyzed. Finally Sect. 5 summarizes the main conclusions of the chapter and highlights the most relevant contributions.

2 Description of the Turbine Disc Model

According to Timoshenko et al. in [3], the most critical part of a rotating disc with hub is located in the inner radius of it. This is also applicable to real operating turbine discs, whose critical part is located also there, in the region commonly called bore. Much effort has been devoted to improve and optimize the bore geometry indeed, assuring that it can resist the typical high stresses of this application. Its geometry is shown in Fig. 1-left.

2.1 Cases of Study

Two different cases are analyzed. In the first case, laboratory isothermal test conditions are considered, that is, the behavior of the rig disc subjected to a Bore Spin Test is studied. The second case seeks to deepen in the methodology herein presented and how it can be applied to real flight conditions and not only to isothermal test conditions. These cases of study are described below.

2.1.1 Laboratory Test Conditions

The rig turbine disc is subjected to a Bore Spin Test carried out by the company Rolls-Royce Deutschland Ltd & Co KG. This test is intended to prove bore lives of the BR725 HP Turbine Stage 1 and 2 as well BR710 HPT Stage 1 (both modern Rolls-Royce Jet Engines) using the material Allvac®718PlusTM (A718Plus). The test is also part of a generic spin test program within the frame of the A718Plus material development, and full information about it can be found in [29].

This test is performed by means of spinning the disc at isothermal conditions. Thus, the disc is assembled to a backing disc driven by a shaft. The assembly including both parts is shown in Fig. 1-right.

The assembly has to be cycled from minimum speed to maximum speed and again to minimum speed, with the enough dwell time (2–5 s) at each speed to ensure stable conditions. These cycles are made in two increments: the first one with 25,000 cycles and the second one with 35,000 cycles. After each increment, the disc must be crack tested and also some parts of the disc like the bore, rim or drive arm must be re-measured. These tests conditions are summarized in Table 1.

Results obtained from this test guarantee that the disc would resist real flight speed conditions because the maximum rig speed is much higher than the real aero-engines spin speed (around 15,000 rpm).

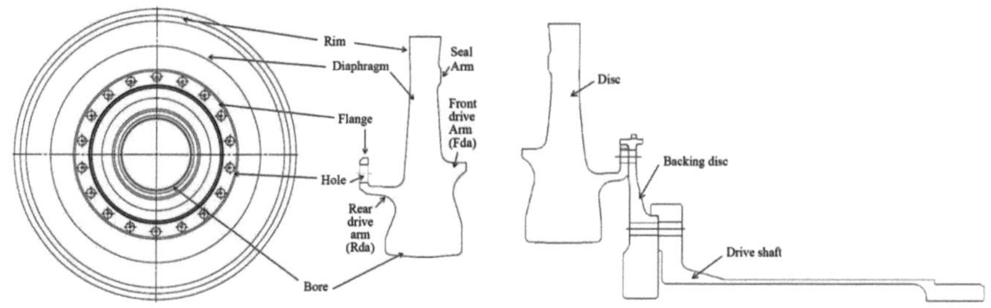

Fig. 1 Turbine disc geometry (*left*) and test assembly including (*right*)

Table 1 Test conditions

Maximum rig speed	21,500 rpm ± 25 rpm
Minimum rig speed	500 rpm ± 50 rpm
Test temperature	430 °C (isothermal) ± 15 °C
Increments	2
First increment	25,000 rig cycles
Second increment	35,000 rig cycles
Total number	60,000 rig cycles

2.1.2 Flight Conditions

A more realistic environment which is closer to real flight conditions is also analyzed. In this study, three different super alloys (A718, Waspaloy and A718Plus) are compared.

Real temperatures are applied to the disc model and a maximum speed of 1,600 rad/s, which is the maximum taking off speed. Furthermore, the surrogate modeling method is applied to the three proposed materials to check the influence of the material in the shape and mechanical behavior of the disc.

In this second case, the geometry used for the optimization is the same of the previous one. It must be highlighted that a real turbine disc has some features that are not included in this model, like the seals, front and rear drive arms or the multi-lobed roots (commonly named "fir-trees") used to locate the blades in the rim of the disc. These features can be observed in Fig. 2, where a real high pressure turbine disc is represented. All these elements have an influence on the stress and the temperature distribution of the disc, but these type of rig turbine discs are mainly used for the evaluation of the bore region. Therefore, the results are focused on that part of the model.

2.2 Basic Equations for Stress Calculation

Classic models of axisymmetric discs state the radial and hoop displacements as $\delta = [u, v]^T$, while the strain $\varepsilon = [\varepsilon_r, \varepsilon_\theta, \varepsilon_z, \gamma_{rz}]^T$ and the stress $\sigma = [\sigma_r, \sigma_\theta, \sigma_z, \tau_{rz}]^T$

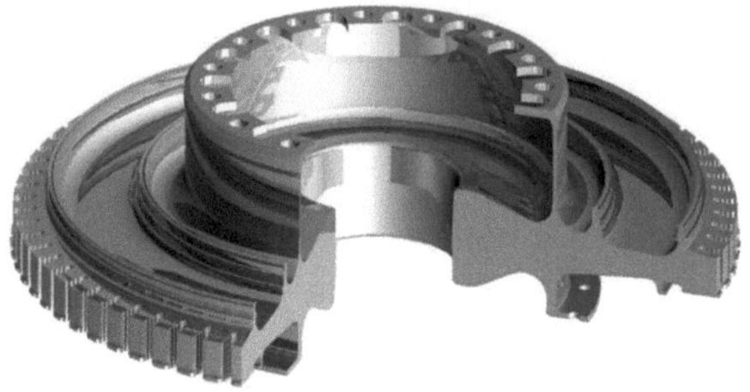

Fig. 2 High pressure turbine disc geometry

comprise radial, hoop, axial and shearing components. Therefore, the relation between strains and displacements in a turbine disc can be expressed as:

$$[\varepsilon_r, \varepsilon_\theta, \varepsilon_z, \varepsilon_{rz}]^T = \left[\frac{\partial u}{\partial r}, \frac{\partial v}{\partial z}, \frac{u}{r}, \frac{\partial v}{\partial r} + \frac{\partial u}{\partial z}\right]^T$$

While the stress-strain relationship in the disc, assuming thermal and elastic isotropy, is:

$$\begin{bmatrix}\sigma_r\\ \sigma_\theta\\ \sigma_z\\ \tau_{rz}\end{bmatrix} = \frac{E(1-v)}{(1+v)(1-2v)}\begin{bmatrix}1 & \frac{v}{1-v} & \frac{v}{1-v} & 0\\ \frac{v}{1-v} & 1 & \frac{v}{1-v} & 0\\ \frac{v}{1-v} & \frac{v}{1-v} & 1 & 0\\ 0 & 0 & 0 & \frac{1-2v}{2(1-v)}\end{bmatrix}\left(\begin{bmatrix}\varepsilon_r\\ \varepsilon_\theta\\ \varepsilon_z\\ \gamma_{rz}\end{bmatrix} - \begin{bmatrix}\alpha\Delta T\\ \alpha\Delta T\\ \alpha\Delta T\\ 0\end{bmatrix}\right)$$

where E, v and α denote the Young's modulus, Poisson's ratio and thermal expansion coefficient of the material, respectively. Finally T denotes the temperature field in the turbine disc.

The easiest suitable disc shape to analyze without numerical methods is a "coin shape" of constant thickness t, with a hole in the middle, inner radius r_1 and outer radius r_2, as is shown in Fig. 3-left. In this kind of models, the blade disc rim stress denoted by P_{rim} may be given by the centrifugal pull of the complete blade set divided by the disc perimeter and the thickness.

Assuming tensile plane stress without any thermal load for the sake of clarity and applying equilibrium equations on a differential piece of disc, thanks to Hooke's law in polar coordinates and certain manipulations it is possible to obtain the following generic solutions of the differential equation for the radial σ_r and hoop σ_θ stresses:

$$\sigma_r = \frac{E}{1-\mu^2}\left(\frac{\partial u}{\partial r} + \mu\frac{u}{r}\right) = \frac{E}{1-\mu^2}\left(C_1(1+\mu) - C_2(1-\mu)\frac{1}{r^2}\right) - \frac{\rho\omega^2}{8}(3+\mu)r^2$$

$$\sigma_\theta = \frac{E}{1-\mu^2}\left(\frac{u}{r} + \mu\frac{\partial u}{\partial r}\right) = \frac{E}{1-\mu^2}\left(C_1(1+\mu) + C_2(1-\mu)\frac{1}{r^2}\right) - \frac{\rho\omega^2}{8}(1+3\mu)r^2$$

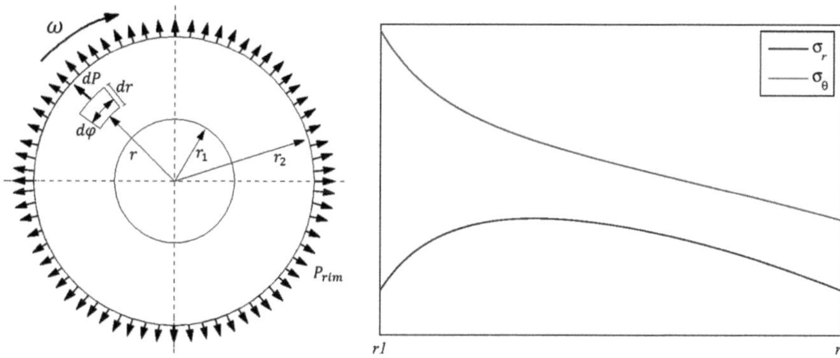

Fig. 3 Differential equilibrium (*left*) and principal stresses-radius (*right*)

Boundary conditions on the radial stress allow obtaining the integration constants and the final expressions of radial and hoop stresses are derived and plotted in Fig. 3-right. As it can be deduced from this figure, the most critical part of the disc is located in the position with the smallest radius and corresponds to the hoop stress. This part of the disc is called "*cob*" and it is strengthened in practice using a wider thickness at this point. Thus the critical hoop stress evaluated in the inner radius is:

$$\sigma_{\theta,\text{crit}} = \left(\frac{1-\mu}{4}\right) r_1^2 \rho \omega^2 + \left(\frac{3+\mu}{4}\right) r_2^2 \rho \omega^2 + \frac{2 P_{\text{rim}} r_2^2}{r_2^2 - r_1^2}$$

Therefore, the results obtained by this classical analysis lead the parameterization and finite element discretization of the computation models used within the methodology herein presented.

2.3 Boundary Conditions and Applied Loads

The dummy blades located in the rim of the rig disc are simulated as a distributed pressure in radial direction. Each dummy blade has a volume of 6,000 mm^3 and assuming they are made with the same material A718Plus, each dummy blade has an approximated weight of 49 g. And there are 36 dummy blades. Thus, the following simplified equation reflects the value of the distributed pressure in the rim face P_{rim},

$$P_{rim} = \frac{mN\omega^2}{2\pi V4}$$

where N is the number of blades with mass m, ω corresponds to the rotational speed of the disc (in this case fixed at 2251 rad/s) and finally *V4* is the *Rim width* parameter.

As the backing disc is also subjected to centrifugal forces and it would be deformed with the test rig disc, no restriction to the radial displacement of the flange of the disc is made. On the other hand, axial and transversal displacements remained to be fixed to simulate the transversal influence of the backing disc.

For the second case study, that is, for the flight conditions, different boundary conditions to those for the laboratory test conditions, collected in Table 1, need to be prescribed. In this case, three different temperatures are fixed [30]. It is important to highlight that these temperatures are evaluated in the hottest point of its working cycle; this is during the take-off. This point coincides with the most severe rotating situation due to the turbine disc is also spinning at its highest speed. Therefore, the temperature boundary conditions are:

- 800 °C in the rim zone. This temperature is the highest and it is caused by the hot gases that come from the combustion chamber.
- 700 °C in the front diaphragm zone. Here the cooling air that comes from the compressor is injected in this region in order to cool the blades. This airflow crosses the disc from this area to the rim through a duct that is located under the seals. This temperature is an approximation of the real temperature conditions a real disc can have.
- 600 °C in the bore zone. This part is the coolest of the disc and there are some secondary air flows that refresh this zone by convection.

These temperature gradients are another cause of stress in real turbine discs. This temperature difference is created because the disc does not heat up homogeneously, partly owing to the thick shape of the cob, which takes longer than the rim zone to heat up. Also the previous considerations of the cooling system affect the thermal temperature distribution along the section of the disc. For a typical turbine disc, about 20 % of the bore hoop stress is due to the blades; about 40 % is due to centrifugal.

Also, the rotational speed of the disc has changed to those imposed in the laboratory test conditions gathered in Table 1. This speed is fixed at 1,600 rad/s (15,278 rpm), which despite being slightly lower, it approximately appears during the take-off of the plane and becomes the most critical of all the flight.

2.4 Finite Element Modeling

The turbine disc has been modeled with the commercial software ANSYS Workbench as an axisymmetric geometry. A complete axisymmetric geometry using a two-dimensional section has been refused because of the holes for the bolts located in the flange of the disc, but it can be approached as a cyclic axisymmetry configuration. According to this symmetry approach, it is necessary to define the boundary conditions that indicate which faces corresponds to cyclical geometries repeated 18 times in a polar coordinates system, in order to extrapolate the cyclic behavior.

In Fig. 4, the mesh used to simulate the rig turbine disc is shown, with 42,395 nodes and 24,235 tetrahedral elements within each cyclic section. Special care has been taken with the bore region of the disc detailed in Fig. 4-left, which requires a refinement process to ensure the best possible results in the most critical region.

3 Optimization Methodology

The different steps of the methodology herein presented are described and shown in Fig. 5. In order to fulfill the global objectives which must lead the optimization process, the main parameters of the model which could have significant influence on the optimal design are initially chosen. A design of experiments (DoE) is carried out prior to obtain a response surface by means of the Kriging method, whose fitness quality is conveniently analyzed. These preliminary steps may result crucial for the minimization of the computational cost during the optimization process without giving up accuracy of the solution, which implies a simple iterative process for non-significant parameters removal. Once the design space is obtained, a sensitivity analysis is performed with mid-effects and Pareto charts and, finally, a multi-objective genetic algorithm (MOGA) is applied. With the results obtained, calculation of the fatigue life is carried out.

3.1 Model Parameters

From the detailed description of the rig disc geometry, a reduced set of 16 driving parameters have been taken into account in order to define the general geometry and more precisely its bore region. Thus the optimal design may be classified as sizing and configuration optimization because of the effects derived from the parameterization lead to both kind of changes in the geometry. In Fig. 6, the whole parameterization of the section of the model, with 68 parameters in total, is shown. The varying parameters are highlighted with red circles while the other parameters remain fixed.

The parameters subjected to variation are shown in Table 2. Their codification and name are related to the position they have, depending on the different parts where they are located (rim, bore, diaphragm, drive arm, flange and seal arm). Also identifications about the relative position on the part (rear or front) are provided. Furthermore, the upper and lower limits, as well as their base value are included.

The parameters coded with the letter *P* are not directly associated with any geometrical measurement. Due to negative dimensions have no sense in a

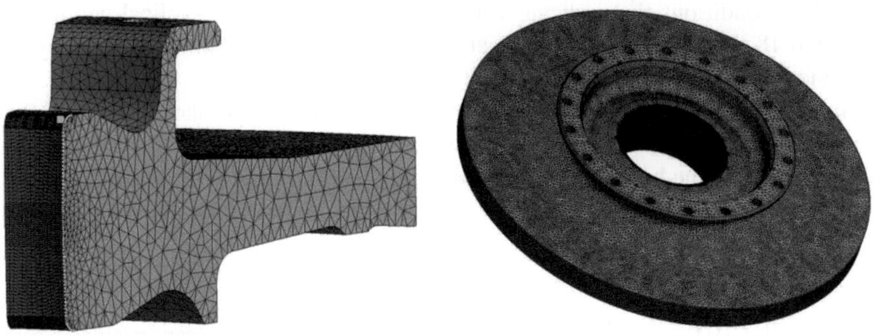

Fig. 4 Detail of the mesh in the bore (*left*) and whole mesh of the disc (*right*)

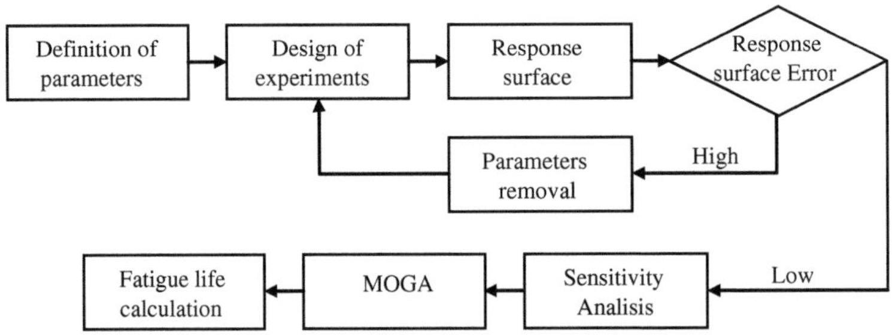

Fig. 5 Flowchart of the optimization methodology

parametric model but some of their basic ranges could include negative values (for example the diaphragm total angle orientation), certain auxiliary parameters that provide the negative varying capability have been added. They are the following:

$$\begin{cases} A83 = A82/2 + P78 \\ V88 = V86 - P84 \end{cases}$$

3.2 Design of Experiments

The sample method selected for the DoE is the Optimal Latin Hyper Cube technique [31]. In this study, each design point corresponds to a certain geometry configuration of the disc. The responses calculated are the two design objectives: geometry mass and critical stress in the bore region. In order to assure the best possible space, all the design explorations are done with 500 runs.

The stress criterion selected is the von Mises one, because it is proportional to the distortion energy. Therefore, it satisfies the property that two stress states with equal distortion energy have an equal von Mises stress. This is why it is considered

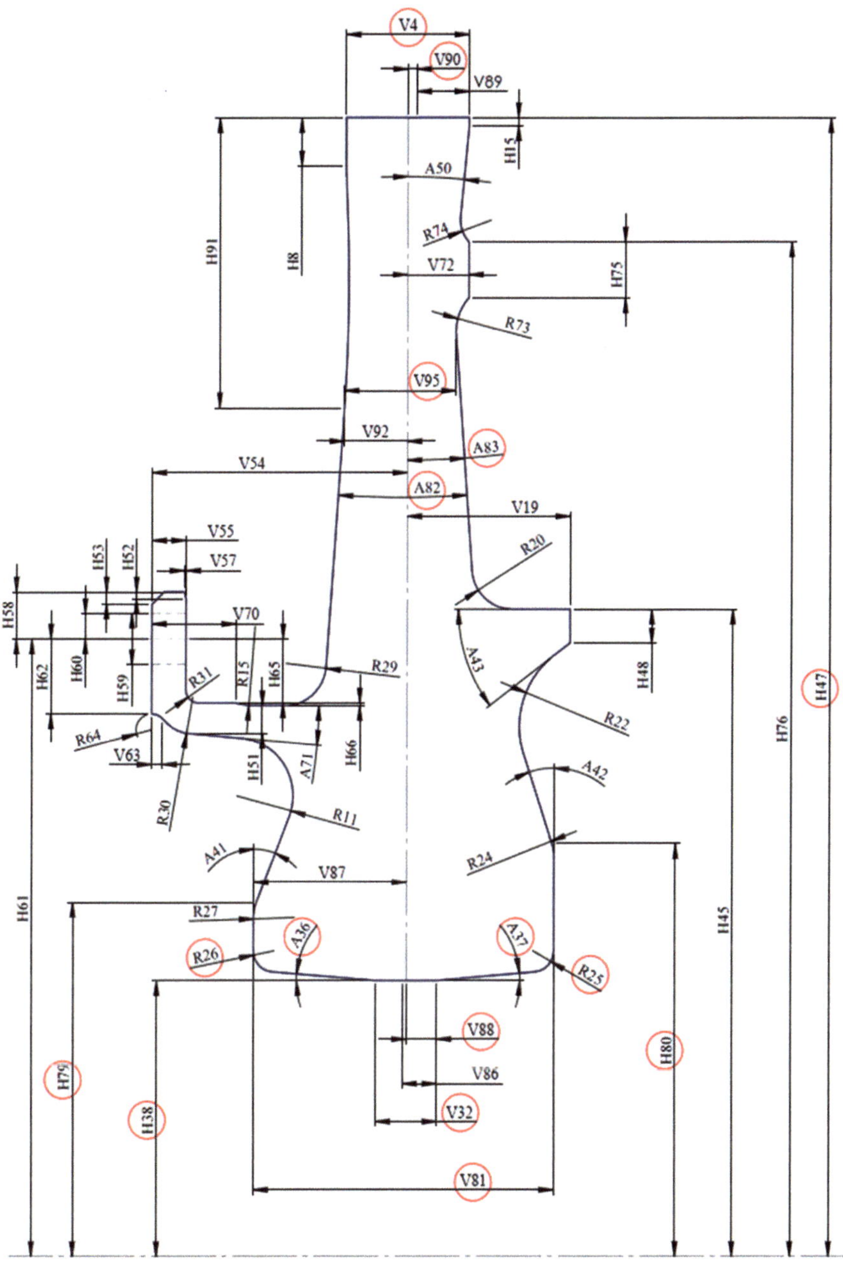

Fig. 6 Parameterization of the disc model

Table 2 Driving parameters of the disc model

Code	Parameter name	Base value	Units	L. limit	U. limit
A36	Cob rear inner angle	2	°	0.1	6
A37	Cob front inner angle	2	°	0.1	6
A82	Diaphragm total angle	8	°	1	10
H38	Cob bore rad.	52.5	mm	40	55
H47	Rim rad.	215.1	mm	213.6	221.6
H79	Cob rear outer dia.	68	mm	62	76
H80	Cob front outer dia.	76.6	mm	68	77
P78	Diaphragm angle aux	0	°	-0.5	1
P84	Cob center line offset	0.05	mm	-2	2
R25	Cob front inner fillet	4	mm	2	6
R26	Cob rear inner fillet	4	mm	2	6
V4	Rim width	24	mm	22.5	25.8
V32	Cob bore length	12	mm	3	15
V81	Cob width	58	mm	48	68
V90	Rim center line offset	1	mm	0	1.8
V95	Diaphragm width	20	mm	17.5	21

a very good measurement for ductile materials, such as metals, as the onset of yield for these materials does not depend on the hydrostatic component of the stress tensor and so it is a good average for the whole stress behavior of the disc.

3.3 *Evaluation of the Response Surface Error and Parameters Removal*

In order to analyze which are the most relevant parameters and which of them are statistically negligible, estimated parameters are reported for each response (critical stress and geometry mass). This analysis sorts all the different parameters, computing the estimates of each of them with the help of the design exploration data. These estimates can be easily compared using the t-ratio as a coefficient of the estimate and the standard error given by the distribution for each of the parameters. The result of this comparison is sorted by the absolute value of the t-ratio and with significance level of 0.05 and those non significant parameters are eliminated

3.4 *Sensitivity Analysis*

As it was previously mentioned, the use of Kriging surrogate models to calculate the regression-based interpolation of the response surface makes it possible to get better results for the approximation [32]. There have been used 34 refinement points in the Kriging algorithm in order to obtain a predicted relative error lower than 5 % in the surface, which can be considered good enough for this analysis. Also the calculation of the surface includes 5 verification points to confirm that the model really corresponds to the real behavior of the disc. Once the response surface has been calculated, it is possible to obtain results and conclusions.

The influence of each variable in the system is analyzed with main-effects charts for each of the responses, critical stress and geometry mass, for a better understanding of the whole system. Also, using a Pareto chart, all the variables are sorted by their influence on the response, providing as well quadratic and cross effects between parameters. This chart complements the conclusions that can be obtained from the main-effects chart allowing a deeper understanding of the behavior of the system.

3.5 Multi-objective Genetic Algorithm

Once the response surface has been calculated, a MOGA algorithm can be applied in order to obtain the optimal candidates of the problem [33]. Without loss of generality of the optimization methodology, two objectives lead the optimal designs in this work: the minimization of the geometry mass [34] and the maximization of fatigue life. Assuming simplified models, the second one may be considered inversely proportional to the critical stress, so the minimization of the stress becomes definitively the second optimization objective although the fatigue life is post-processed.

Pareto fronts are the most extended chart to depict the feasible region and the frontier of optimal design points when multi-objective optimization is performed. Once the process is leaded by antagonistic objectives, Pareto front becomes a very helpful tool for the decision maker due to its quantitative and qualitative representation of the non-dominated optimal candidates. The NSGA-II [35] multi-objective genetic algorithm has been applied due to its well known advantages when obtaining the Pareto fronts. This algorithm starts with 2,000 initial samples and each of its subsequent iterations continues with 1,000. The maximum allowable Pareto percentage was chosen as 55 %. This value can set the selection pressure of each one of the iterations. It is recommended to use values between 55 and 75 %. Finally, 50 iterations are set as the maximum allowable the algorithm can do.

4 Case Study I: Results for Laboratory Test Conditions

Regarding the laboratory test conditions detailed in Sect. 2.1, this section presents the results obtained step by step along the whole optimization process: (i) the evaluation of the response surface error and parameters removal, (ii) the sensitivity analysis, (iii) the results provided by the multi-objective genetic algorithm NSGA-II and, finally, (iv) the fatigue life calculation.

4.1 Evaluation of the Response Surface Error and Parameters Removal

In order to analyze which are the most relevant parameters and which of them are statistically negligible, estimated parameters are reported for each response (critical stress and geometry mass). Thus, these are sorted by the absolute value of the t-ratio, showing the most significant effects at the top. Also a bar chart is provided, which shows the t-ratio, with lines showing the 0.05 significance level. In each of these tables, Prob $> |t|$ column indicates the significance. Values with *-symbol beside them are p-values, i.e. the results which are statistically significant.

Initially, all the parameters are included in the model; therefore the first DoE explores 16 variables along 500 different design points. The results obtained in this analysis correspond to Tables 3 and 4. On the other hand, the created response surface has a global error of 129 %. This value is so high due to the elevated number of input parameters the system has. If no parameters removal would be wanted, more than 500 design points will be needed, but this becomes computationally unviable.

Comparing these two tables, it can be noted that the parameters *Cob rear inner fillet*, *Diaphragm angle aux*, *Cob front inner fillet*, *Cob center line offset* and *Rim center line offset* appear in both tables in the least relevant positions in the effect scale. This scale, corroborated by the p-values for each of the parameters allows depreciating these 5 parameters and removing them from the model, making them fixed at their base point value in further analysis. Also, it can be seen that the critical stress response only has in this analysis 2 statistically significant parameters, meanwhile the geometry mass has 7 parameters. This can be explained because the direct relationship that can be found between part dimensions and volume and hence also between dimensions and geometry mass.

This process is repeated until all the remaining parameters are relevant to explain the behavior of the system and the modeled response surface error is low enough to assure that it has been calculated properly. Tables 5 and 6 show the 6 parameters that left over after the iteration. In these two tables, it can be seen how all the remaining parameters have p-values statistically relevant for the geometry mass and 5 from 6 parameters in the critical stress. Also, the non-relevant parameter from this table has a p-value (0.069) really close to the significance level (0.05). Furthermore, it must be highlighted that all the remaining parameters correspond to the dimensions that roughly define the geometry (inner and outer diameters of the disc) or the widths of its most main parts (rim, cob and diaphragm). Moreover, the global error of the response surface calculated is 4.91 % which is low enough to assure correct and robust results.

According to these results, it can be concluded that the model has been cleaned from all the non-relevant parameters assuring a better optimization process as the model is simpler and also a better sensitivity analysis as all the parameters included in the model have relevance.

Table 3 Sorted parameter report for critical stress with 16 parameters

Parameter name	Estimate	Std error	t ratio	Prob > |t|
Cob width	−71429681.41	10821541,4	−6,6	<0.0001*
Cob rear outer dia.	−33141292.15	14758667.5	−2.25	0.0252*
Rim width	81782593.09	62510369.3	1.31	0.1914
Cob bore length	22385702.96	17168116.8	1.3	0.1929
Rim rad.	33203157.79	25853572.9	1.28	0.1997
Cob rear inner angle	−42823606.73	34869406.6	−1.23	0.22
Diaphragm total angle	−27738496.55	23160651.5	−1.2	0.2317
Cob front outer dia.	−24002510.8	22870002.7	−1.05	0.2945
Cob bore rad.	−12096135.83	13612280.5	−0.89	0.3747
Cob front inner angle	27210008.78	34902522.3	0.78	0.436
Rim center line offset	−74961854.05	114771615	−0.65	0.514
Cob center line offset	−22847704.25	51418346.2	−0.44	0.657
Cob front inner fillet	−22328215.39	51073482.2	−0.44	0.6622
Diaphragm angle aux	−58035858.21	137764042	−0.42	0.6738
Diaphragm width	9787156.128	58548133.5	0.17	0.8673
Cob rear inner fillet	−5.74E + 06	52142124.8	−0.11	0.9125

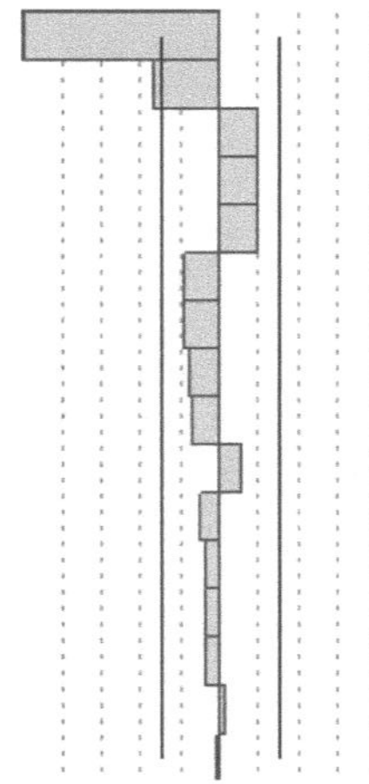

Table 4 Sorted parameter report for geometry mass with 16 parameters

Parameter name	Estimate	Std error	t ratio	Prob > |t|
Cob width	0.01126099	0.0006442	17.48	<0.0001*
Cob bore rad.	−0.00734693	0.00081033	−9.07	<0.0001*
Rim rad.	0.01329415	0.00153904	8.64	<0.0001*
Diaphragm width	0.02749039	0.00348532	7.89	<0.0001*
Diaphragm total angle	0.01074019	0.00137874	7.79	<0.0001*
Cob rear outer dia.	0.00592983	0.00087857	6.75	<0.0001*
Rim width	0.01431233	0.00372119	3.85	0.0001*
Cob front outer dia.	0.00262273	0.00136143	1.93	0.0547
Cob bore length	−0.00099901	0.001022	−0.98	0.3288
Cob front inner angle	−0.00126635	0.00207772	−0.61	0.5425
Cob rear inner fillet	0.00135929	0.00310398	0.44	0.6616
Cob center line offset	0.00048174	0.00306089	0.16	0.875
Diaphragm angle aux	0.0008143	0.00820099	0.1	0.9209
Cob rear inner angle	−0.0001941	0.00207575	−0.09	0.9255
Cob front inner fillet	−5.1942E-05	0.00304036	−0.02	0.9864
Rim center line offset	5.61E-06	0.00683226	0	0.9993

Table 5 Sorted parameter report for critical stress with 6 parameters

Parameter name	Estimate	Std Error	t Ratio	Prob > |t|
Cob width	−6440130.05	317733.92	−20.27	<0.0001*
Rim rad	8927804.42	789722.809	11.3	<0.0001*
Diaphragm total angle	−6450075.26	702999.878	−9.18	<0.0001*
Rim width	9831053.4	1917882.53	5.13	<0.0001*
Cob bore rad	2117015.63	421877.418	5.02	<0.0001*
Diaphragm width	3294740.53	1808070.66	1.82	0.069

Table 6 Sorted parameter report for geometry mass with 6 parameters

Parameter name	Estimate	Std Error	t ratio	Prob > |t|
Cob width	0.0115767	9.4105E-05	123.02	<0.0001*
Cob bore rad	−0.0079035	0.00012495	−63.25	<0.0001*
Rim rad	0.01396832	0.0002339	59.72	<0.0001*
Diaphragm width	0.03026036	0.00053551	56.51	<0.0001*
Diaphragm total angle	0.00967198	0.00020821	46.45	<0.0001*
Rim width	0.01487849	0.00056803	26.19	<0.0001*

4.2 Sensitivity Analysis

The sensitivity analysis herein presented uses a Kriging model to simulate the behavior of the system. A predicted relative error in the surface of 4.9162 % is

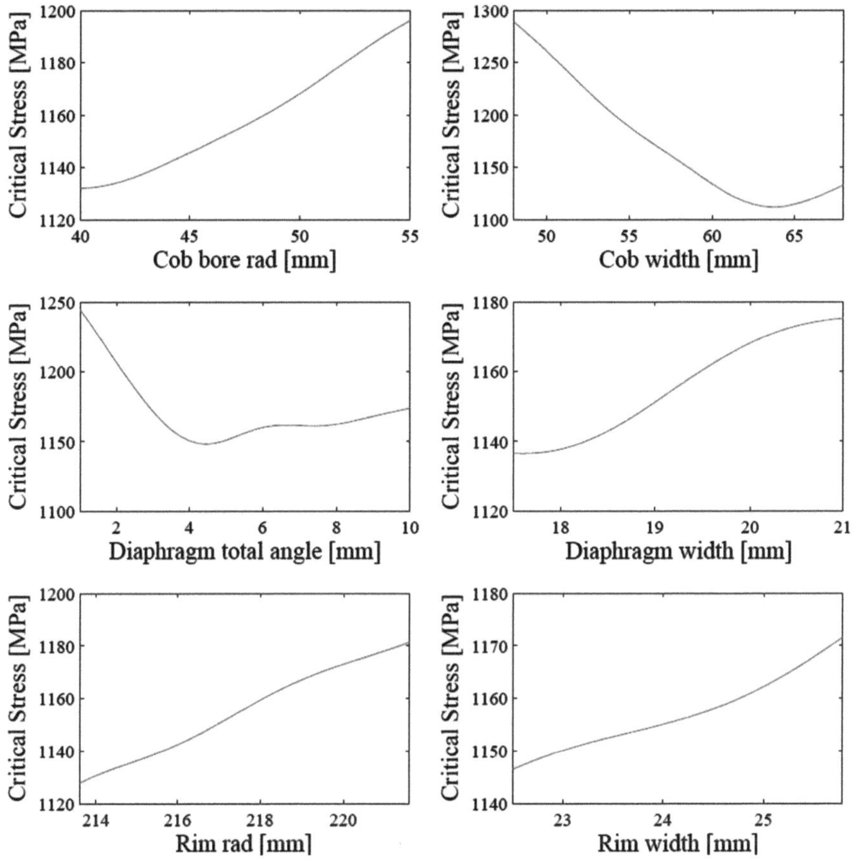

Fig. 7 Critical stress individual parameter responses of the disc

considered accurate enough for this analysis. So, once the response surface has been calculated, it is possible to obtain results.

The individual response of each of the parameters when the other parameters are fixed is firstly presented in Figs. 7 and 8. This allows obtaining a preliminary idea about how the system can be optimized according to the individual effect of each parameter.

In Fig. 7 the individual response to the critical stress of the disc of the 6 remaining parameters are shown. The represented responses model the individual behavior of each of the parameters supposing the other ones fixed at their medium value of each range of variation. It can be seen in this figure how the *Cob bore rad*, *Diaphragm width*, *Rim rad* and *Rim width* parameters have a strictly increasing shape, while the *Cob width* and *Diaphragm total angle* parameters have a minimum in the middle of their range of variation. These results are very useful to understand how the critical stress behaves with the variation of each parameter.

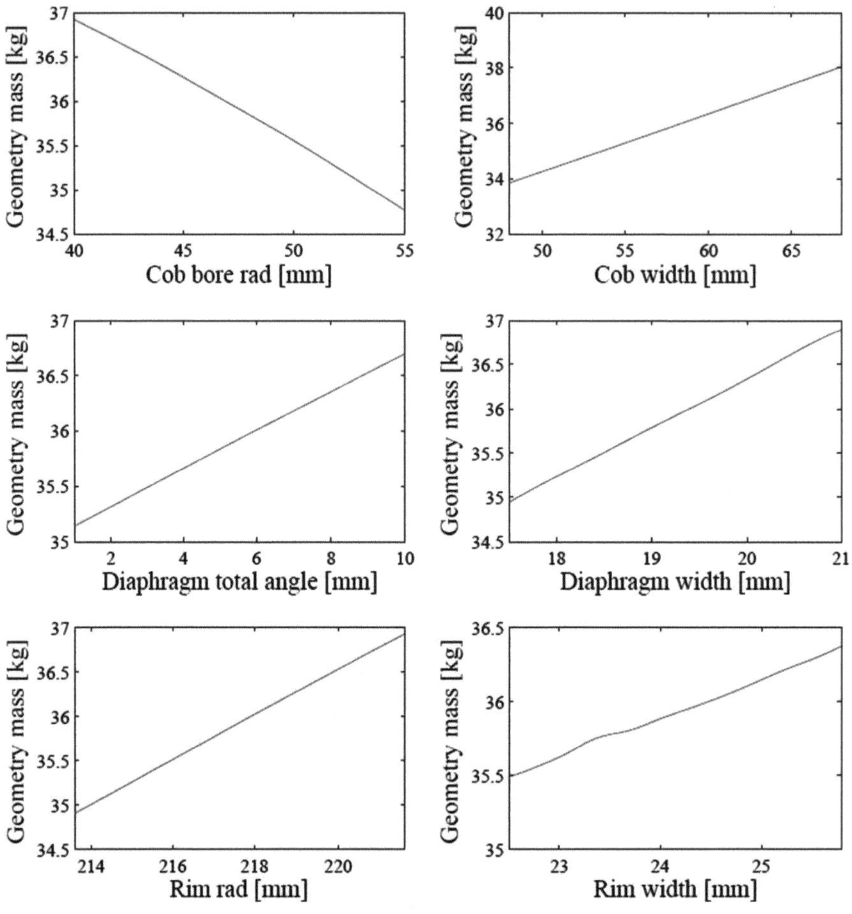

Fig. 8 Geometry individual parameter responses of the disc

Alternatively, the geometry mass responses of all the parameters are shown in Fig. 8. At first glance, it is possible to note that all of them have quasi-linear behavior in contrast to critical stress responses. This is explained because all the variation ranges of the parameters correspond to linear or quasi-linear dimensions in the model such as distances or angles, thus the influence on the volume and hence on the mass is linear and the responses show straight lines. Note the parameters related to radius, *Cob bore rad* and *Rim rad*, should obviously lead to a quadratic effect on the geometry mass unless the variation range is so small that it seems quasi-linear indeed. Furthermore, *Cob bore rad* is the only parameter with a decreasing shape. This coincides logically to the definition of this dimension that corresponds to the inner radius of the disc, and thus its increase causes directly mass growth.

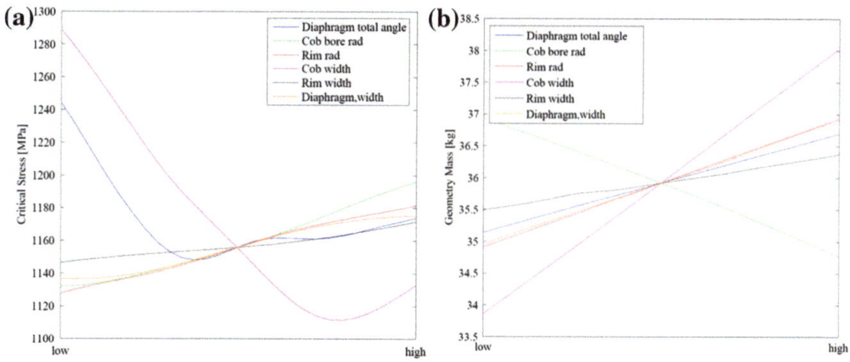

Fig. 9 Main-effects chart applied to critical stress and geometry mass

These results allow seeing separately the influence of each variable in the system but, a better understanding of the whole system can be reached using the main-effects charts for each of the responses (critical stress and geometry mass). Analyzing the main-effects chart it is possible to join all the previous responses and observe which are the most influential and the approximated consequences that would carry out varying one or other parameters.

In Fig. 9a, b can be found the main-effects chart applied to the critical stress and geometry mass of the disc, respectively. As it is shown, in both charts the most influential parameter is the *Cob width* of the disc. It must be reminded that all the plots cut each other in the middle point of their range because every parameter is evaluated remaining the others fixed at their medium value.

This analysis is extended by means of Pareto charts which show not only the influence of the variation of a parameter individually but also the combination of parameters or quadratic variation. Interesting results can be obtained from the response surface by means of these Pareto charts. Firstly, the one represented in Fig. 10 sorts the influence of the parameters in the critical stress response of the disc system. It only includes the effects that have greater influence than 1 %, so only 80 % of the whole response is represented in the chart. Thus the remaining 20 % corresponds to effects with negligible effects. It can be noted that *Cob width* is the parameter that affects to the critical stress the most. This corroborates what was deduced with the main-effects chart applied also to the critical stress.

In Fig. 11 the Pareto chart applied to the geometry mass response is provided. As well as in the previous chart, there are only effects represented with an influence greater than 1 %. In contrast to the critical stress Pareto chart, 84 % of the whole response can be explained with the simple parameters due to their quasi-linear behavior. Therefore, according to the ranking most part of the geometry mass can be explained just with simple terms.

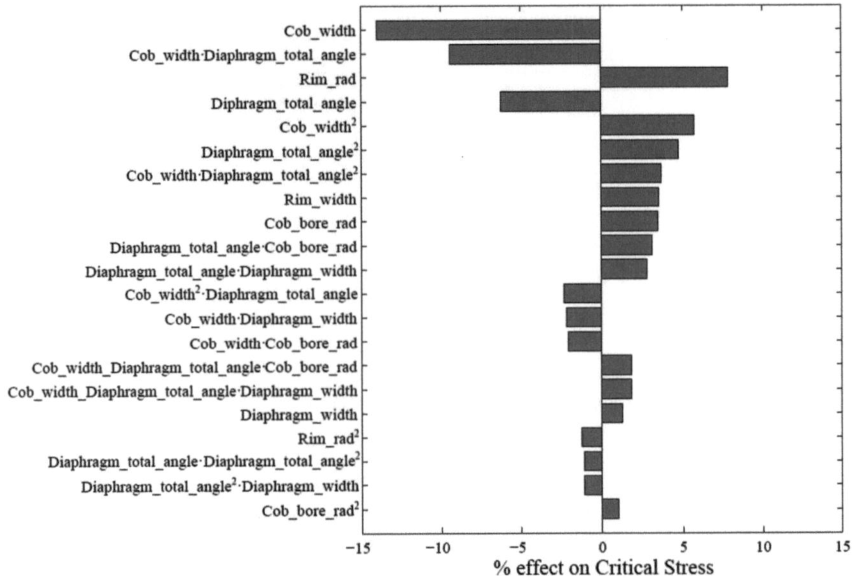

Fig. 10 Pareto chart applied to critical stress of the disc

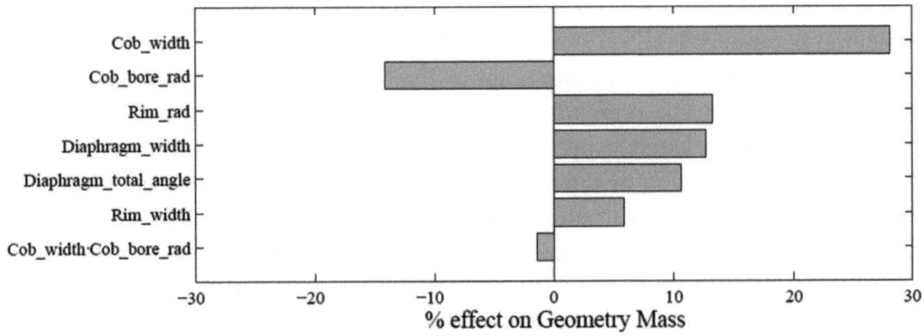

Fig. 11 Pareto chart applied to geometry mass of the disc

4.3 Multi-objective Optimization

Once the response surface has been calculated, it is possible to apply the MOGA algorithm in order to obtain the optimal candidates of the problem.

According to the reference values reported in Sect. 3.5, Fig. 12 shows the Pareto front obtained after the optimization calculation is done. In this chart, all possible candidates are shown gathered in a single Pareto front. As it was explained previously, all these design points are the set of non-dominated optimum values of the system.

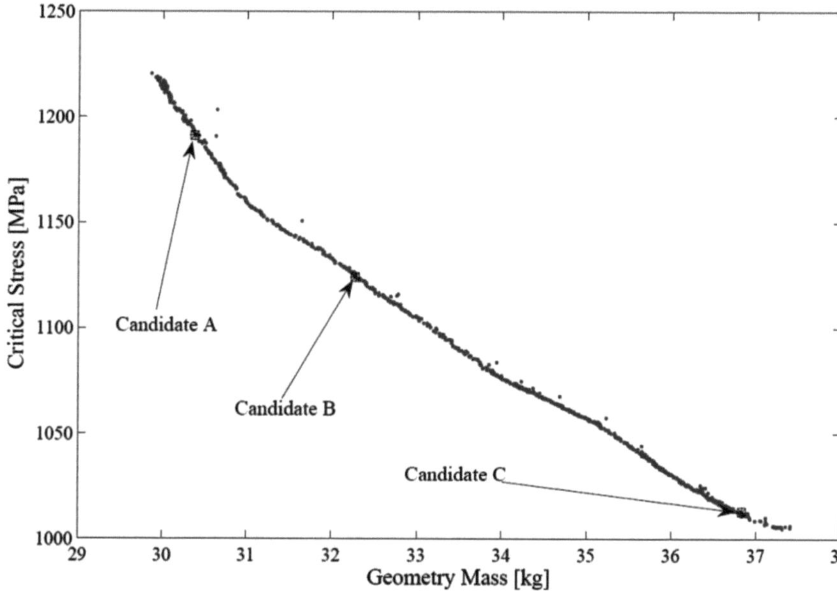

Fig. 12 Pareto front chart of the disc optimization with optimal candidates

Due to the shape of the front, which does not provide clearly any region of dominance, no point of the front could be selected a priori as a candidate suiting better both objectives. Therefore three candidates are selected from this Pareto front and they are proposed as Candidates A, B and C to the global optimum. On the one hand Candidate A has critical stress reduction priority while Candidate C has geometry mass reduction priority, and on the other hand, Candidate B is more balanced geometry mass reduction and critical stress reduction.

The following Table 7 shows the values of all the relevant driving parameters for each of the candidates proposed, strongly validating the conclusions obtained by means of the sensitivity analysis.

It can be observed how the most varying parameter is the *Cob width* of the disc. This corresponds to its key role in the Pareto charts (Figs. 10 and 11). On the other hand, *Rim width* parameter remains almost fixed in the three candidates, which can be also explained by the Pareto charts. The rest of the parameters follow similar tendencies, although they are not so strongly correlated with the previous sensitivity analysis as the *Rim width* and *Cob width*.

In Fig. 13 the different stress states of the candidates are shown in contrast to the base design point. In this figure, it can be appreciated how the critical stress zone remains being in the cob of the disc, which is consistent with the results obtained from the Pareto front. The shape of the stress distribution is symmetrical with respect to its middle line and also it must be highlighted the existence of a loaded zone in the fillet between the rear drive arm and the diaphragm, but the values of stress in this area are considerably lower than the ones obtained in the

Table 7 Parameter values fot the optimal candidates of the disc optimization

Parameter name	Units	Candidate A	Candidate B	Candidate C
Diaphragm total angle	°	1.19	4.47	9.34
Cob bore rad.	mm	54.70	54.73	40.47
Rim rad.	mm	213.60	213.61	213.62
Cob width	mm	50.63	57.93	65.33
Rim width	mm	22.51	22.52	22.51
Diaphragm width	mm	17.52	17.52	17.53

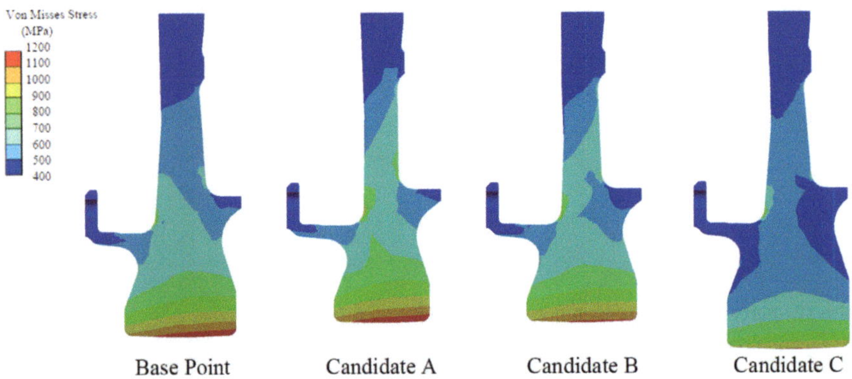

Fig. 13 Optimal candidates and base design point stress comparison

bore, thus this will never be a critical zone. Also it can be noted in this figure that Candidate A shows the thinnest shape, being the angle of the diaphragm and the width of the cob very small in contrast to the other candidates. On the contrary, Candidate C has larger volume and hence it is heavier too. Also it must be highlighted the smaller value of the radius of the cob, which is considerably lower than the other two candidates. On the other hand, Candidate B presents an intermediate value of cob width and diaphragm total angle which makes possible the balance between mass and stress as it was expected.

4.4 Fatigue Life Calculation

Once the stress distribution of the rig disc, material properties and load cycles are known, it is possible to calculate the fatigue life of each of the candidates. This purpose requires information about the curve alternating stress–life and uses the correction factors which take into account physical differences between the theoretical test specimen and the real part. The stress-life curve is created with and the actual designed part. According to [36], the correction factors are the following:

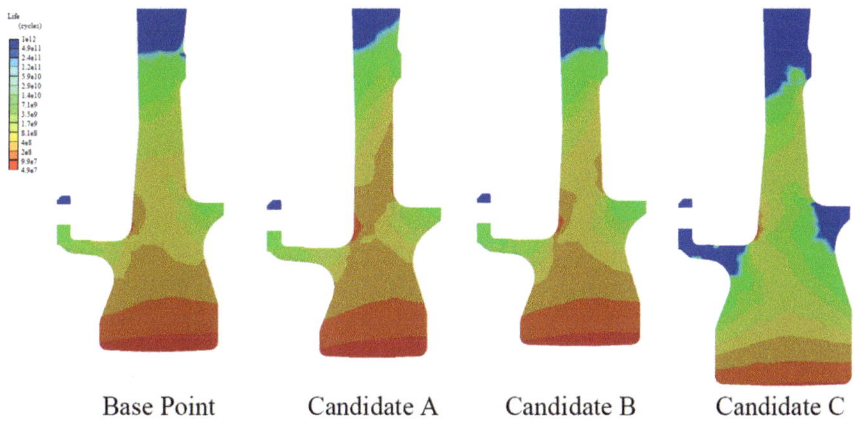

Fig. 14 Optimal candidates and base design point life comparison

$$
\begin{aligned}
C_{load} &= 1.00 \\
C_{size} &= 0.60 \\
C_{surf} &= 0.86 \\
C_{temp} &= 1.00 \\
C_{reliab} &= 0.90
\end{aligned}
$$

By means of these coefficients, the global fatigue coefficient which allows calculating the expected life of each one of the candidates, depicted in Fig. 14, is:

$$K_f = C_{load} C_{size} C_{surf} C_{temp} C_{reliab} = 0.46$$

The life shape of all the models is closely connected with the stress shape of Fig. 13, where the stress distribution is shown. Obviously the life of the disc is function of the alternating stress the part suffers, which lead to similar diagrams.

All the previous results (geometry mass, critical stress and fatigue life) of each one of the candidates and the base point are summarized in the following Table 8. In spite of the base geometry for the turbine disc, the loads and boundary conditions are not exactly real due to confidential reasons. It is noteworthy that the expected life for the three candidates are really close to real ones according to the feedback from industry.

Table 8 Optimization results

	Base point	Candidate A	Candidate B	Candidate C
Geometry Mass (kg)	35.38	30.40	32.29	36.82
Critical Stress (MPa)	1188.8	1193.2	1126.9	1020.2
Life (cycles)	51 328 000	49 527 000	61 035 000	85 583 000

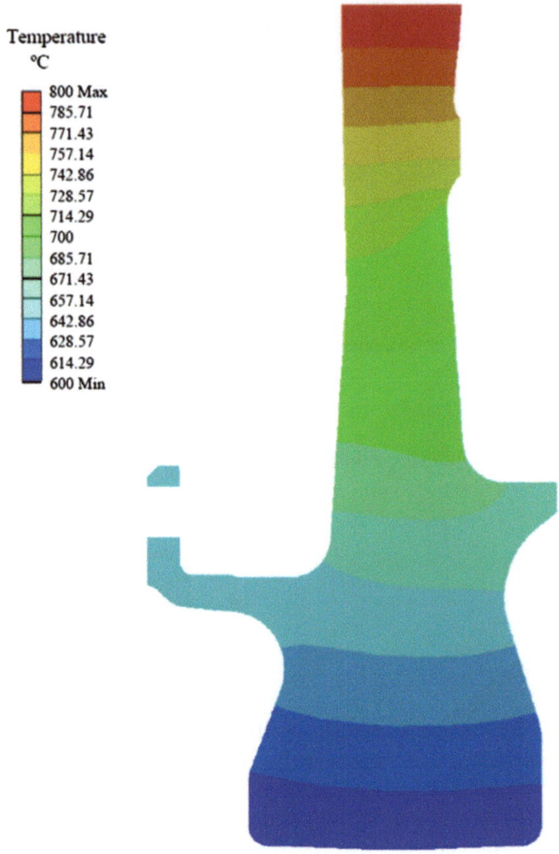

Fig. 15 Temperature distribution along the disc section

5 Case Study II: Results for Flight Conditions

This second case study seeks to deepen in how the methodology herein presented can be applied under real conditions and not only under isothermal test conditions. Therefore this part of the study looks to give a more realistic environment, closer to real flight conditions. In addition, this study intends to analyze the suitability of three typical turbine discs super alloys.

According to the flight boundary conditions pointed out in Sects. 2.1 and 2.3, herein the results obtained within the optimization process are presented. Once the most severe thermal distribution under the abovementioned conditions is reached, depicted in Fig. 15, the evaluation of the response surface error and parameters removal together with a sensitivity analysis has been carried out.

As well as the previous Case Study I, each one of the materials (Waspaloy, A718 and A718Plus) has been subjected to a design exploration process. The results obtained show that the relevant parameters are exactly the same as the ones from the Bore Spin Test conditions with A718Plus, see Tables 3 and 4. Therefore,

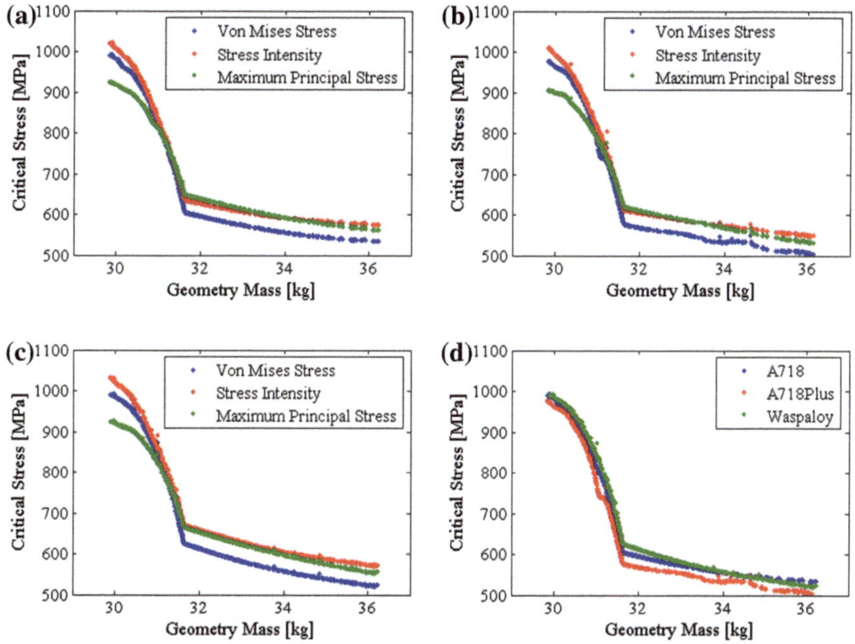

Fig. 16 Pareto fronts of material comparison. **a** A718 **b** Waspaloy **c** A718Plus **d** Von Mises Stress

the most relevant parameters to the critical stress and geometry mass of the disc are not case-dependent. Thus, each one of the material-based cases of study is evaluated with the final 6 critical parameters obtained for the Case Study I and gathered in Tables 5 and 6.

The response surface for each of the material-based cases is calculated using the Kriging method once again. The thermal influence is added as another input but maintaining 500 points of design in the DoE, so the global error of the response surface increases slightly to 5.5 % approximately for each case. In spite of this, this value is low enough to assure that the response surface is able to provide accurate and reliable results.

5.1 Multi-objective Optimization

After calculating the three response surfaces for each of the materials, it is possible to apply the MOGA optimization method to obtain the Pareto front with the optimal candidates. These Pareto fronts are shown in Fig. 16.

In this figure, each of the materials is evaluated separately, taking into account three different stress criteria. The shape of the diagram shows that for every criterion, there is a design point more likely to be the global optimum of the

Table 9 Parameters values for optimal candidates

Parameter name	Units	A718	Waspaloy	A718Plus
Diaphragm total angle	°	1.0023	1.0007	1.0023
Cob bore rad.	mm	40.0011	40.0001	40.0037
Rim rad.	mm	213.6020	213.6031	213.6020
Cob width	mm	48.0027	48.1307	48.0050
Rim width	mm	22.5008	22.5008	22.5008
Diaphragm width	mm	17.5000	17.5006	17.5009

system. This is the corner where the curves change their curvature. Also it is possible to see that in this point the most restrictive criterion is the von Mises equivalent stress. A preliminary way to compare which of the materials is the best option for a turbine disc is observing these Pareto fronts and comparing them for the same stress criterion. For that reason Fig. 16d presents a comparison among the three materials when the von Mises equivalent stress is one of the optimization criteria. As it can be seen the A718Plus curve remains at lower values of geometry mass and critical stress than Waspaloy and A718.

Thus the points located in the corner of the Pareto fronts for each of the materials are assumed as optimum candidates and their values are gathered in Table 9. Due to the geometry design is not material-dependent, the optimal geometry can be reached by means of any alloy. Therefore, the only way to choose the optimal material lies in the comparison of the different stress distributions keeping the geometry mass constant.

Figure 17 shows the stress distribution along the section of the disc for each of the materials in the optimum design point. Although results are quite similar, A718Plus leads to lower values of the von Mises equivalent stress in the bore, which is definitively the critical zone. This agrees that A718Plus is a refined alloy that improves the material properties of Waspaloy and A718, being more adequate for this application thanks to its consequent longer fatigue life.

5.2 Fatigue Life Calculation

The stress distribution allows calculating the fatigue behavior of the part using also the alternating stress—life curve and the correction factors previously mentioned. In this Case Study II, the only parameter that varies from the previous case is the C_{temp}, which swaps from 1.00 to 0.42, because it is evaluated taking into account the real environmental temperature approximated to 550 °C. According to previous considerations, the global fatigue coefficient is:

$$K_f = C_{load} C_{size} C_{surf} C_{temp} C_{reliab} = 0.19$$

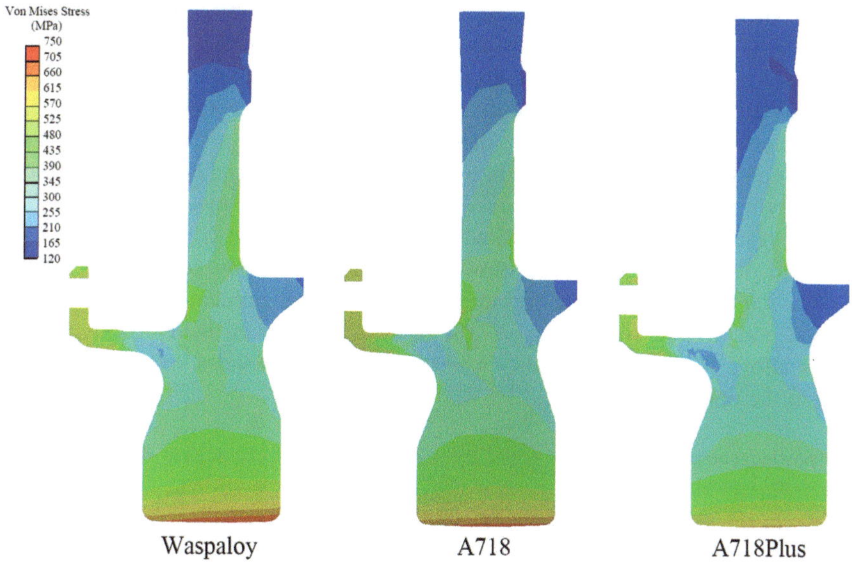

Fig. 17 Stress distribution comparison for the different materials

Table 10 Case study ii optimization results

	Waspaloy	A718	A718Plus
Geometry Mass (kg)	31.62	31.62	31.62
Critical Von Mises stress (MPa)	631.10	611. 94	581.67
Life (cycles)	∞	∞	∞

The results obtained in the fatigue life calculation reveal that all candidates present infinite life for the stress and conditions proposed in this case study. This corresponds to real design criteria that states that the critical parts cannot fail under any working circumstances due to the dangerous effects this could lead to.

All the previous results are summarized in Table 10, where the optimal design objectives obtained for each different material are collected. Once again A718Plus excels the other materials owing to its lower critical von Mises stress despite the fact that its geometry mass is exactly the same as the others.

6 Summary and Conclusions

This chapter presents a robust and flexible optimization methodology for turbine discs based on low-fatigue life prediction. Sizing and configuration optimal designs are performed according to two leading objectives: the minimization of the geometry mass and the maximization of fatigue life, which means definitively the minimization of the critical stress.

The methodology herein proposed iteratively gears an initial design of experiments with a sensitivity analysis and a surrogate model prior to run the multi-objective genetic algorithm. The greatest advantage of this procedure is the CPU time saving, but also it includes an accuracy improvement due to the non-significant parameters removal. The purging of non-feasible and non-relevant optimal solutions step by step becomes the main feature of this method and has revealed crucial to perform precise FEM computations of the thermo-mechanical model.

The results obtained in Case Study I, for laboratory test conditions, and Case Study II, for flight conditions, allow highlighting two main conclusions: (i) the *Cob width* parameter of the turbine disc is the most influential in the optimal design and (ii) the super alloy A718Plus evidences much better behavior than super alloys A718 and Waspaloy when realistic flight conditions are considered within the optimization process.

Acknowledgements The authors gratefully acknowledge the technological support of Rolls-Royce in Germany.

References

1. Witek, L.: Failure analysis of turbine disc of an aero engine. Eng. Fail. Anal. **13**(1), 9–17 (2006)
2. Rolls-Royce (2005) The Jet Engine
3. Timoshenko, S.P., Goodier, J.N.: Theory of Elasticity, 3rd edn. McGraw-Hill, New York (1970)
4. Leopold, W.: Centrifugal and thermal stresses in rotating disks. ASME J. Appl. Mech. **18**, 322–326 (1984)
5. Meguid, S.A., Kanth, P.S., Czekanski, A.: Finite element analysis of fir-tree region in turbine discs. Finite Elem. Anal. Des. **35**(4), 305–317 (2000)
6. Farshi, B., Jahed, H., Mehrabian, A.: Optimum design of inhomogeneous nonuniform rotating discs. Comput. Struct. **82**(9–10), 773–779 (2004)
7. Ranta, M.A.: On the optimum shape of a rotating disk of any isotropic material. Int. J. Solids Struct. **5**, 1247–1257 (1969)
8. Luchi, M.L., Poggialini, A., Persiani, F.: An interactive optimization procedure is applied to the design of gas turbine discs. Comput. Struct. **11**, 629–637 (1980)
9. Chem, J.M., Prager, W.: Optimal design of rotating disk for given displacement of edge. J Optimiz Theory App **6**, 161–170 (1970)
10. Genta, G., Bassani, D.: Use of genetic algorithms for the design of rotors. Meccanica **30**(6), 707–717 (1995)
11. Tiwari, B.R., Rao, R.: Optimum design of rolling element bearings using genetic algorithms. Mech. Mach. Theory **42**(2), 233–250 (2007)
12. Feng, F.Z., Kim, Y.H., Yang, B.: Applications of hybrid optimization techniques for model updating of rotor shafts. Struct Multidiscip O **32**, 65–75 (2006)
13. Silva, D.. Minimum weight design of disks using a frequency constraint. J Eng Ind **91**, 1091–1099 (1969)
14. Cheu, T.: Procedures for shape optimization of gas turbine. Comput. Struct. **54**, 1–4 (1990)
15. Shu-Yu, W., Yanbing, S., Gallagher, R.H.: Sensitivity analysis in shape optimization of continuum structures. Comput. Struct. **20**, 855–867 (1985)

16. Simpson, T.W., Mauery, T.M., Korte, J.J., et al.: Kriging models for global approximation in simulation-based multidisciplinary design optimization. AIAA J **39**(12), 2233–2241 (2001)
17. Huang, D., Allen, T.T., Notz, W.I., et al.: Sequential Kriging optimization using multiple-fidelity evaluations. Struct Multidiscip O **32**(5), 369–382 (2006)
18. Sakata, S., Ashida, F., Zako, M.: On applying Kriging-based approximate optimization to inaccurate data. Comput Methods Appl M **196**(13–16), 2055–2069 (2007)
19. Song, W., Keane, A.J.: An efficient evolutionary optimisation frame-work applied to turbine blade fir tree root local profiles. Struct Multidiscip O **29**, 382–390 (2005)
20. Huang, Z., Wang, C., Chen, J., et al.: Optimal design of aeroengineturbine disc based on Kriging surrogate models. Comput. Struct. **89**, 27–37 (2011)
21. Mohan, S.C., Maiti, D.K.: Structural optimization of rotating disk using response surface equation and genetic algorithm. Int J Comput Methods Eng Sci Mech **14**(2), 124–132 (2013)
22. Boyce, M.: Gas Turbine Engineering Handbook. Butterworth-Heinemann, Elsevier, Oxford (2006)
23. ATI Allvac (2012) ATI 718^{TM} Alloy
24. Pollock, T.M., Tin, S.: Nickel-based superalloys for advanced turbine engines: chemistry, microstructure, and properties. J. Propul. Power **22**(2), 361–374 (2006)
25. ATI Allvac (2011) ATI Waspaloy Alloy
26. ATI Allvac (2004) ATI 718Plus Alloy
27. Uzunonat, Y., Cemal, M., Cevik, S. et al.: Allvac 718Plus^{TM} superalloy for aircraft engine applications (2012)
28. Jeniski, R.A., Kennedy, R.L.: Development of ATI Allvac 718Plus Alloy and Applications (2011)
29. Rolls-Royce Deutschland Ltd & Co KG (2011) Test Requirement Document for Bores Spin Test SP.190 in A718Plus
30. Buschoff, T., Voigt, M., Chehab, E. et al.: Probabilistic Analysis of Stationary Gas Turbine Secondary Air Systems, GT2006-90261, Barcelona (2006)
31. Wang, G.: Adaptive response surface method using inherited Latin hypercube design points. J. Mech. Des. **125**(2), 210–220 (2003)
32. Forrester, A.I.J., Keane, A.J.: Recent advances in surrogate-based optimization. Prog Aerospace Sci **45**(1–3), 50–79 (2009)
33. Kesseler E, van Houten MH (2007) Multidisciplinary Optimisation of a Turbine Disc in a Virtual Engine Environment. 2nd European Conference for Aerospace Sciences (EUCASS), Amsterdam
34. Jahed, H., Farshi, B., Bidabadi, J.: Minimum weight design of inhomogeneous rotating discs. Int J Pres Ves Pip **82**(1), 35–41 (2005)
35. Deb, K., Pratap, A., Agarwal, S., et al.: A fast and elitist multiobjective genetic algorithm: NSGA-II. IEEE T Evolut Comput **6**(2), 182–197 (2002)
36. Norton, R.L. Machine design: an integrated approach. Vol. 3. New Jersey: Pearson Prentice Hall, (2006)

Composite Suspension Arm Optimization for the City Vehicle XAM 2.0

Massimiliana Carello and Andrea Giancarlo Airale

Abstract The use of composite materials is very important in the automotive field to meet the European emission and consumption standards set for 2020. The most important challenge is to apply composite materials in structural applications not only in racing vehicles or supercars, but also in mass-production vehicles. In this chapter is presented a real case study, that is the suspension wishbone arm (with convergence tie and pull-rod system) of the XAM 2.0 urban vehicle prototype, that has the particular characteristics that the front and rear, and left and right suspension system has the same geometry. The starting point was from an existing solution made in aluminum in the XAM urban vehicle to manufacture a composite one, in particular in carbon fiber. The first step was the development of a dynamic model of the vehicle to understand the suspension loads and behavior to define the suspension weight and stiffness targets with respect to the aluminum arm, because it was necessary to understand the tensile strain on the component to simplify and optimize the geometry. Once the wishbones external surfaces have been defined, a carbon fiber layer thickness and orientation optimization have been made to define the lamination lay-out. Generally, after the analysis of the composite thickness optimization result, it would be possible to build up a new CAD model that encounters the process constrains and would define the lamination process. The results of the final suspension in carbon fiber compared to aluminum one were a weight reduction of 5 % and an increasing of stiffness of 78 %. The final purpose of this work is not only to find the best suspension solution but to define an engineering methodology to design suspension in composite materials thanks to simulation and virtual analysis.

M. Carello · A. G. Airale (✉)
Politecnico di Torino - Department of Mechanics and Aerospace Engineering, Turin, Italy
e-mail: andrea.airale@polito.it

M. Carello
e-mail: massimiliana.carello@polito.it

A. Öchsner and H. Altenbach (eds.), *Design and Computation of Modern Engineering Materials*, Advanced Structured Materials 54,
DOI: 10.1007/978-3-319-07383-5_18,

Keywords Vehicle dynamics · Carbon fiber · Composite vehicle suspension · Structural analysis

1 Introduction

This chapter presents a carbon fibre suspension arm in which the target was to design the suspension system for a heavy quadricycle vehicle prototype XAM 2.0 from an existing suspension system designed and made for the vehicle XAM 1.0. The main important characteristics, to put evidence in the difference, of the two vehicles are shown in Table 1.

It is easy to note the difference in meaning of weight, power and performance, instead volumes, lay-out and suspension architecture of the two vehicles are the same. When XAM 1.0 was up-sizing, the suspension system design target was to maintain the same architecture, improving structural resistance, in particular increase stiffness to improve vehicle dynamics performance from XAM 1.0 to XAM 2.0 [1].

The vehicle XAM 2.0 was designed to have low consumption and has participated in the Future Car Challenge [2], where the consumption is expressed by the electric energy needed to run for 1 km (Wh/km) or in (L/100 km). For this kind of competition the most important key factors are: weight, aerodynamics resistance, friction of the wheels and bearing, drive-line and at last, but not least, hybrid power-train efficiency. The aim, for the chassis, mechanical subsystem and the body, is to reach the lowest weight maintaining the structural resistance [3]. Suspensions is one the most important but at the same time critical subsystem of the vehicle, so the second target was to decrease weight on suspension arms, to make a reduction to unsprung mass and improve vehicle dynamics, like the weight reduction on wheels [4]. So one of the solution is to design the upper and lower arm of XAM 1.0 suspension system in composite materials instead of aluminum (Fig. 1).

A lot of car maker and research centers are working on downsizing and lightweight on suspension using composite material, like glass fibre or carbon fibre:

- ZF: CFRP Front Mc Pherson System [5]
- ZF: GFRP Rear Spring System [5]
- GM: Chevrolet Corvette C6 GFRP Rear Spring [6]
- MAGNA STEYR: Aero Light Prototype CFRP Front and Rear Spring System [7]

The application of composite material on structures, in this case, in suspension systems is present-day and there is a lot of industrial interest [8], [9].

Table 1 XAM 1.0 and XAM 2.0 characteristics

Technical informations	XAM 1.0	XAM 2.0
Weight	197 kg	400 kg
Length	2,800 mm	2,800 mm
Height	1,280 mm	1,280 mm
Width	1,300 mm	1,300 mm
C_x	0.31	0.30
Maximum speed	30 km/h	80 km/h
Powertrain type	Parallel hybrid	Serial hybrid
Energy storage	Supercap	Li-Po batteries
Power	2 kW	15 kW
Chassis	Aluminum frame	Aluminum frame
Suspension	Double wishbone pull-rod	Double wishbone pull-rod
Road legal	No	Yes

Fig. 1 XAM 1.0 aluminum double wishbone pull-rod system starting point (*on the left*) and the upper arm to design it in composite materials (*on the right*)

2 Vehicle Dynamic Model

A vehicle dynamic model has been developed, through a multi-body software, to evaluate the loads applied on the suspension elements [10]. This multi-body model interacts with the vehicle dynamic model, developed in Altair Motion View software and allows to calculate the forces exchanged between tire and road, applied on the tire-road contact zone (Fig. 2).

As a consequence, from the applied loads and the maximum deformation tolerance obtained, a stiffness target for all the suspension system could be estimated. From the global stiffness target it would be possible to evaluate the different targets for each suspension element.

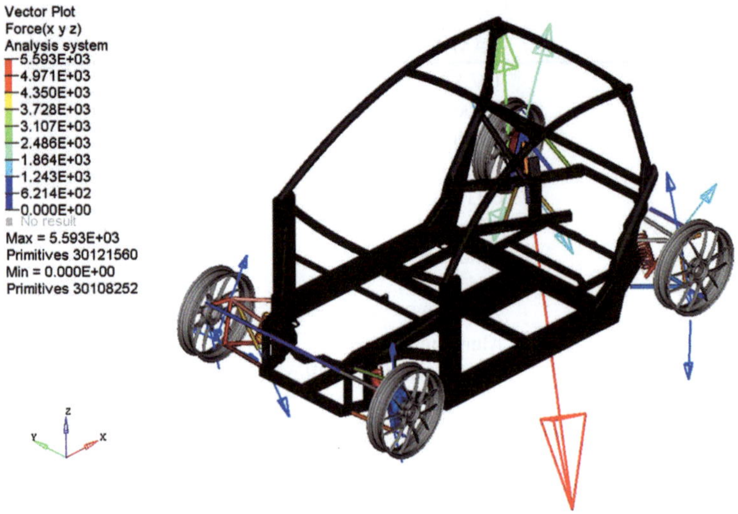

Fig. 2 Global multi-body model of XAM 2.0 (N)

Table 2 Suspension stiffness targets

Lower wishbone arm		
$\frac{kx}{m} = 132.0\text{N}/(\text{mm} \cdot \text{g})$	$\frac{ky}{m} = 338.4\text{N}/(\text{mm} \cdot \text{g})$	$\frac{kxy}{m} = 93.0568\text{N}/(\text{mm} \cdot \text{g})$
Upper wishbone arm		
$\frac{ky}{m} = 135.58\text{N}/(\text{mm} \cdot \text{g})$	$\frac{kz}{m} = 73.36\text{N}/(\text{mm} \cdot \text{g})$	$\frac{kyz}{m} = 66.37\text{N}/(\text{mm} \cdot \text{g})$

The double wishbone pull-rod suspension system presents a kinematics that loads mainly the lower arm on the *x* and *y* axis, and the upper arm on the *y* and *z* axis, due to the linkage with the pull-rod. Due to that, the stiffness of those components have been calculated in those directions. Under those assumptions the stiffness targets are shown in Table 2.

On the other hand, the design of the suspension has not been focused only on the component stiffness, but also on its ultimate strength. Through the developed multi-body model, different driving manoeuvers have been simulated to obtain the theoretical forces to which the suspension system must deal without damages or permanent deformations. Thanks to that, the maximum load that the component must resist under different conditions is known.

During the design phase, after the stiffness sizing of the component, the maximum static load will be checked. In the case that this verification highlights structural problems, the lamination or the geometry must be reviewed to achieve a positive static verification.

Point	Fx [N]	Fy [N]	Fz [N]
1	-40	810	37
2	35	-340	912
3	188	-515	-1131

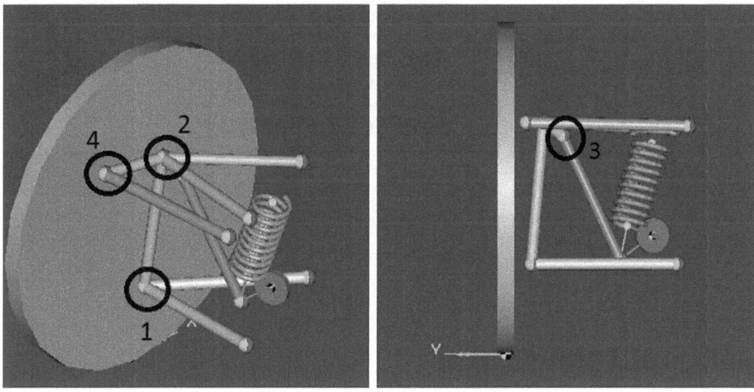

Fig. 3 Points analysis of the multi-body suspension model

In this way new components characterized by a extremely high stiffness/mass coefficient with respect to the equivalent components in aluminum alloy of the XAM 1.0 have been developed.

The target loads are obtained from the output of the step steer manoeuvre (Fig. 3).

Application Point	F_x [N]	F_y [N]	F_z [N]
1	–40	81	37
2	35	–340	912
3	188	–515	–1131

The Steep Steer Maneuver is used to evaluate the directional behavior and the vehicle stability. It consists on moving the steering wheel to a pre-fixed position instantaneously and keep that position for a period of time. Being defined those inputs:

- Vehicle mass: $m = 550$ kg (400 kg [vehicle curb mass] + 2.75 kg [passengers mass])
- Steering wheel input: $\delta = 0$ deg for $t < 0$ s, $\delta = 13$ deg for $t \geq 0$;
- Lateral acceleration: $a_y = 0.4$ g;
- Speed: V = 60 km/h;

Fig. 4 Surface design workflow

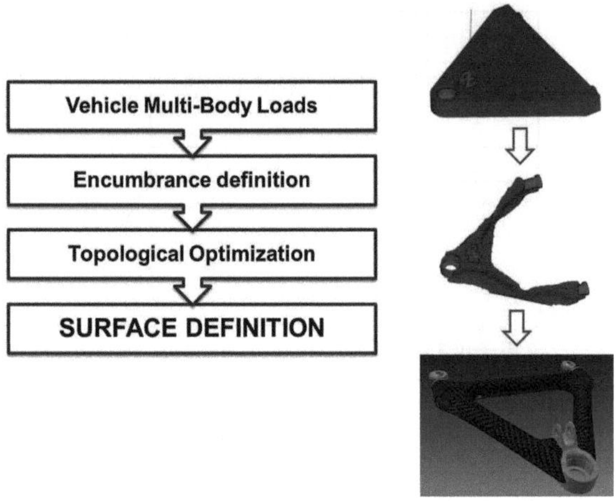

3 Topological Optimization

Before defining the component surfaces, it is necessary to understand the stress flow on the component to define and optimize the geometry. Thanks to the characteristics of the composites component design, it has been possible to generate complex shapes that has been matched with the mechanical requirements.

The first step of the design process was a topological optimization, through the software Altair Hypermesh with Optistruct solver, that solves topological optimization problems using the density method.

The results of the topological optimization highlights the areas where the load concentration is greater erasing the zones where the material is not influenced from the structural point of view.

Analyzing the optimization result, it is possible to obtain informations about the element concentration of the component considering a defined load condition which enables the geometry and the stiffness definition from the beginning of the design phase.

First of all it was necessary to enclose the maximum area available for the component optimization, that will be defined without interfere with the other elements of the vehicle. Then are evaluated the fixed volumes that will not be under study and will be reference points to apply the loads and constraints, in this case, those are the wishbone bushings for the chassis attachment and the metallic insert for hub assembly.

The different components are simulated with 3D elements to which an isotropic elastic material has been assigned. Figure 4 shows an example of the verified model results.

From the optimized model it is observed that the component present a greater concentration of the elements on the upper and lower planes of the optimized area close to the metallic insert. It is possible to observe also the presents of a well defined area that connects the two arms. Unconnected elements are present between the bushing elements. From the information obtained from this analysis it could be thought that the geometrical definition of the upper wishbone arm will be a V shape suspension with a Sect. 1 on the arms with a curated interface between the composite layers and metallic inserts.

4 Definition and Optimization of the Lamination

Once the wishbones external surfaces have been defined, a composite layer thickness and orientation optimization has been run to define the lamination layout. The optimization result shows a laminate thickness mapping that has been taken into account to define the lamination process by highlighting the areas that request a local reinforcement for stiffness reasons.

As a composite laminate is typically manufactured through a stacking and curing process, certain manufacturing requirements are necessary in order to limit undesired side effects emerging during this curing process[11]. For example, one typical constraint for carbon fiber reinforced composites is that plies of a given orientation cannot be stacked successively for more than 3 or 4 plies. This implies that a design concept that contains areas of predominantly single fiber orientation would never satisfy this requirement. Therefore, to achieve a manufacturable design concept, manufacturing requirements for the final product need to be keep into account during the concept design phase. For the particular constraint mentioned above, for instance, the design concept would offer enough alternative ply orientations to break the succession of plies of the same orientation if the percentage of each fiber orientation is controlled. In order to consider these needs, the following manufacturing constraints are made available for composite free-sizing:

- lower and upper bounds on the total thickness of the laminate,
- lower and upper bounds on the thickness of individual orientation,
- lower and upper bounds on the thickness percentage of individual orientation,
- constant thickness of individual orientation,
- thickness balancing between two given orientations.

The first step to set the optimization process was to define the material properties. In order to be conservative, the lowest characterized composite material properties are used, in our case a fabric of T300 Twill 2 $\times$ 2–200 g/m^2 have been selected.

The aim of the optimization process was to define the composite thickness that would satisfy a load requirement by limiting the displacement with the minimum mass. For that reason, the following control parameters are defined:

- x-displacement: in order to ensure a stiffness greater than the aluminum wishbone the maximum value that can be reached during the

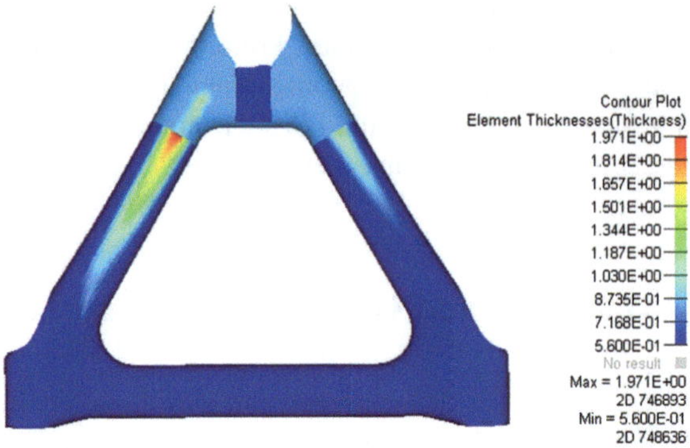

Fig. 5 Element thickness of the upper wishbone (mm)

optimization iteration was lower than the obtained during the aluminum stiffness calculation.

- y-displacement: in order to ensure a stiffness greater than the aluminum wishbone the maximum value that can be reached during the optimization iteration was lower than the obtained during the aluminum stiffness calculation.
- Mass: the purpose of the optimization process was to iterate with the laminate thickness by controlling the displacement of a reference node to obtain the minimum mass achievable.

The loads that will be the input of the lamination optimization calculation are an overestimation of those obtained in the step steer manoeuver. This assumption will lead to a stiffer results in the optimization process.

Once all the precedent steps are set it was possible to make the optimization simulations and evaluate the thickness map of the carbon fiber layer for each component.

The result obtained from the upper wishbone (Fig. 5) highlights that the most loaded areas are the upper and lower surfaces, especially at the transition between the metallic insert and the suspension arms. This result confirms that the production process thought previously during the surface generation, is also the best solution from the structural point of view, as unloaded areas will be split or glued.

Once the thickness optimization result model has been studied, the different lamination sequences have been developed. In order to generate the different plies and its correspondent on-plane development, the wishbone arm has been exported to *Catia* ambient to manage better the surfaces.

In order to achieve a good load transfer among the different layers and make a homogeneous lamination each composite orientation sequence is limited to a single consecutive ply. Under this assumption, the general lamination has the following sequence: 0, 45° and the reinforcement plies, if required.

The material selected to build up the suspension arms are a T300-2 $\times$ 2 Twill–200 g/m^2 for the general lamination and a high modulus unidirectional fiber M40 for the reinforcements.

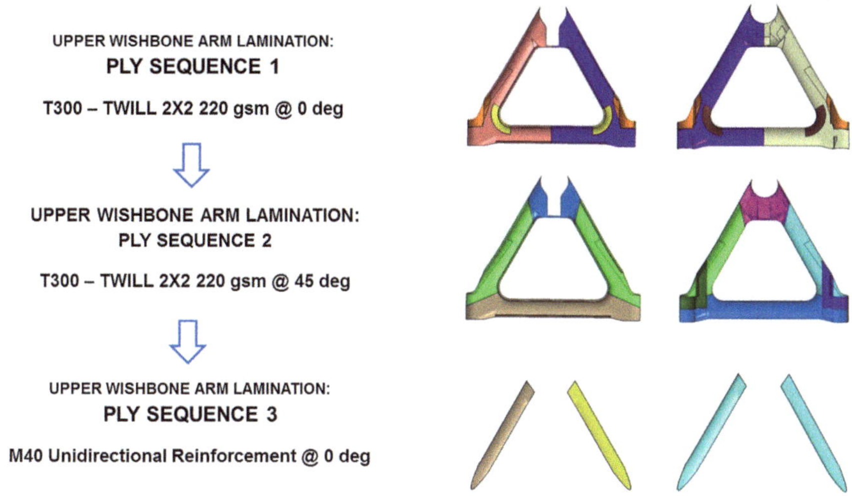

Fig. 6 Lamination sequence for the upper wishbone

While the surfaces of the wishbone arms were split in order to define the different plies that will compound the component, two main design aspects have been taken into account. Firstly, one of the objectives of the surface split is to follow as much as possible the 0° direction imposed on the FEM model with the real composite fabric minimizing the warpage of the composite layer when it is laminated.

Another important aspect regards the production process, it was always checked the on-plane development of all the surfaces that define the lamination to ensure that the plies designed could be obtained from the flat sheet of composite fabric. This point has facilitated the production process as the surfaces are perfectly defined for the automatic cutting machine and errors due to geometrical imprecision during lamination was minimized.

Figure 6 shows the first process sequence of the upper wishbone, the second one is the same just repeated.

5 Material Comparison and Static Verification

In order to conclude the lamination design it was necessary to ensure the performances of the pieces. For that reason a new FEM model of the wishbone arms has been designed taking into account the predefined lamination and the technical solution adopted for its production feasibility. Those modifications will reproduce as much as possible the final component.

For the components verification, the FEM result calculation must be changed, in this case the linear solver RADIOSS—Bulk Data Interface has been used.

Before analyzing the model and the results it would be convenient to overview how composites are simulated in FEM language using the RADIOSS solver.

The plates and shells that define the 2D elements can be made of layered composites in which several layers, also of different materials, (plies) are bonded together to form a cohesive structure. Typically, the plies are made of unidirectional fibers or of woven fabrics and they are joined together by a bonding medium (matrix). In RADIOSS—Bulk Data composite shells, the plies are assumed to be laid in layers parallel to the middle plane of the shell. Each layer may have a different thickness and different orientation of fiber directions.

Classical lamination theory is used to calculate effective stiffness and mass density of the composite shell. This is done automatically within the code using the properties of individual plies. The homogenized shell properties are then used in the analysis.

After the analysis, the stresses and strains in each layer and between the layers can be calculated from the overall shell stresses and strains. Then these results may be used to assess the failure indices of individual plies and of the bonding matrix.

Analysis of composite shells is very similar to the solution of standard shell elements. The primary difference is the use of the 2D property PCOMP property card, instead of PSHELL, to specify shell element properties. From the ply information specified on the PCOMP entry, RADIOSS—Bulk Data automatically calculates the effective properties of the shell element.

After the analysis, the available results include shell-type stresses as well as stresses, strains, and failure indices for individual plies and their bonding. These results are controlled by the result flags on the PCOMP or entry and the usual I/O control cards.

PCOMP defines the structure and properties of a composite lay-up which is then assigned to an element. The plies are only defined for that particular property and there is no relationship of plies that reach across several properties. Some remarks are given below regarding the specifics of composite analysis:

- The most typical material type used for composite plies is MAT8, which is a planar orthotropic material. The use of isotropic MAT1 (as the aluminum used for bushes and inserts) or general anisotropic MAT2 for ply properties is also supported. If MAT 1 or MAT2 would be integrated inside the lamination, the stress limits in tension, compression and shear must be defined in order to analyze the ply failure.
- While it is possible to specify ply angles relative to the element coordinate system, the results become strongly dependent upon the node numbering in individual elements. Thus, it is advisable to prescribe a material coordinate system for composite elements and specify ply angles relative to this system. The wishbone suspension arms present a relative simplified geometry, so the ply angles will be relative to the arms main direction. An accurate definition of the main direction must be addressed at the transition zones, where the arms are linked, to ensure

that no opposite directions will meet on two attached elements. A regular and smooth transition of main directions is also a request to simulate the wrapability of the composite fabric.

- Depending on the specific lay-up structure, the composite may be offset from the reference plane of the shell element, i.e. to have more material below than above the reference plane (or vice versa). In the case presented in this chapter, it was preferable to maintain the surfaces in contact with the mold as reference as it is a well-defined geometry. By considering the normal direction of the component surface in contact with the mold inwards, the lamination sequence will follow the production process. In this way it is easy to control and check the FEM model.
- Stress results for composites include both shell-type stresses and individual ply stresses. Importantly, shell-type stresses are calculated using homogenized properties and thus only represent the overall stress-state in the shell. To assess the actual stress-state in the composite, individual ply results need to be examined.

For the analysis of the results for composite shell elements, a number of composite-specific results are calculated. Due to the specific type of these results, some explanation is given below in place regarding their meaning.

- Ply stresses and strains: classical lamination theory assumes a two-dimensional stress-state in individual plies (so-called membrane state). The values of stresses and strains are calculated at the mid-plane of each ply, i.e. halfway between its upper and lower surface. For sufficiently thin plies, these values can be interpreted as representing uniform stress in the ply. Ply stresses and strains are calculated in coordinate systems aligned with ply material angles as specified on the PCOMP card. In particular, σ_1 corresponds to the primary ply direction, σ_2 is orthogonal to it, and σ_{12} represents in-plane shear stress.
- Inter-laminar stress: Inter-laminar bonding matrix usually has different material properties and stress-state than the individual plies.
- Failure indices: To facilitate prediction of potential failure of the laminate, failure indices are calculated for plies and bonding material. While there are several theories available for such calculations, their common feature is that failure indices are scaled relative to allowable stresses or strains, so that the value of a failure index lower than 1.0 indicates that the stress/strain is within the allowable limits (as specified on the material data card), and a failure index above 1.0 indicates that the allowable stress/strain has been exceeded. For this study the Tsai-Wu theory of ply failure has been adopted, whose index is calculated using the following equation:

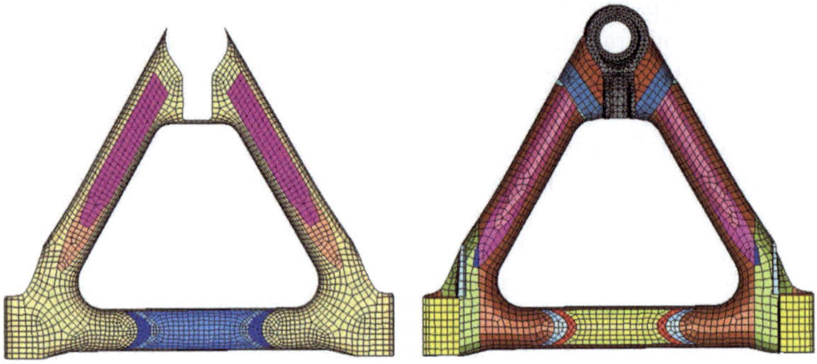

Fig. 7 Upper wishbone FEM model

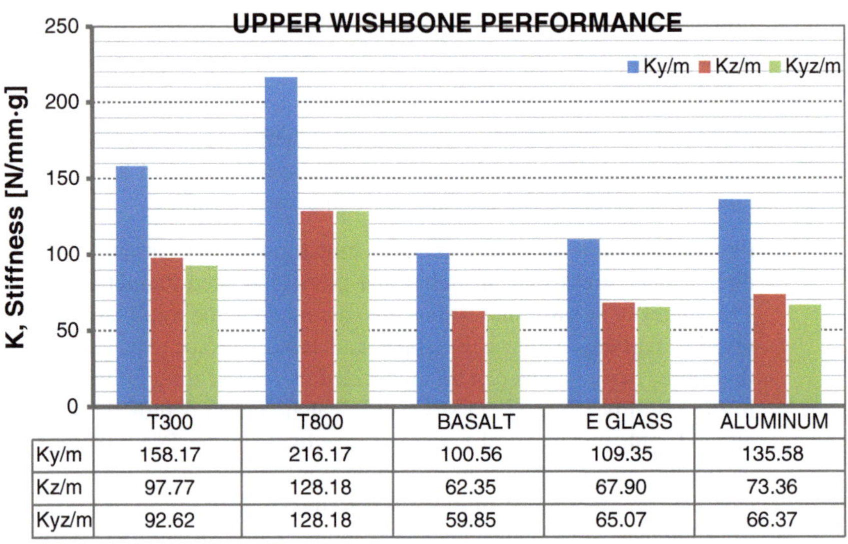

	T300	T800	BASALT	E GLASS	ALUMINUM
Ky/m	158.17	216.17	100.56	109.35	135.58
Kz/m	97.77	128.18	62.35	67.90	73.36
Kyz/m	92.62	128.18	59.85	65.07	66.37

Fig. 8 Performances comparison of upper wishbone in different materials

$$F = \left(\frac{1}{X_t} - \frac{1}{X_c}\right)\sigma_1 + \left(\frac{1}{Y_t} - \frac{1}{Y_c}\right)\sigma_2 + \frac{\sigma_1^2}{X_t X_c} + \frac{\sigma_2^2}{Y_t Y_c} + \frac{\tau_{12}^2}{S^2} + 2F_{12}\sigma_1\sigma_2 \quad (1)$$

Under those statements, new FEM models have been generated to evaluate the structural performances of the suspension wishbone arms. First of all the stiffness component has been calculated with the loads of the step steering maneuver, and after that the ply stresses, ply strain and failure index was calculated with the loads coming from the step collision maneuver.

Before the final simulations a comparison between different composite properties would be done in order to understand the benefits of using one material instead of another one. By maintaining the same properties and lamination

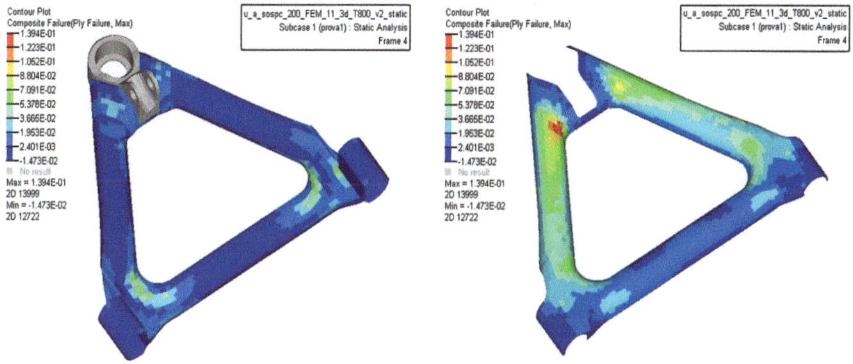

Fig. 9 Composite failure of the upper wishbone arm (*upper* and *lower* shell)

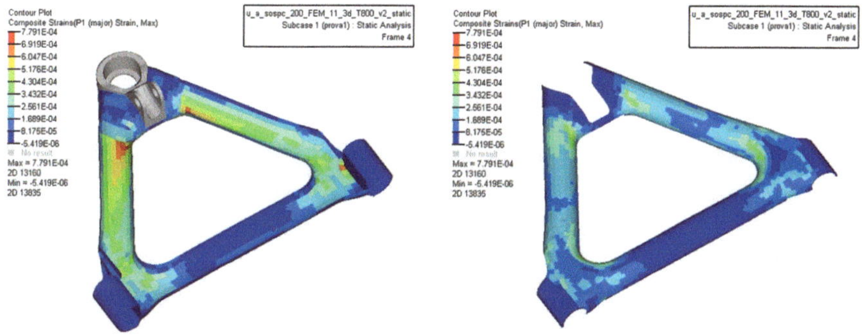

Fig. 10 Static strain of the upper wishbone arm (*upper* and *lower* shell)

sequence, only the material of the main fabric that defines the lamination has been changed (Fig. 7).

The materials considered are: carbon fibers T300, carbon fiber T800, Basalt and E-Glass all with the same epoxy resin. This comparison was done by comparing the component stiffness calculated with the step steer forces and after that their performances was evaluated with those of the previous upper wishbone arm done in aluminum (Fig. 8).

It is easy to verify that the configuration with the lamination with T800 presents the best performances. The lamination with T300 is the lighter and have better performances than the previous arm in aluminum, but its performances are below the T800 and this is not much heavier, so the solution that present the best compromise is the wishbone arm done in T800 fabrics.

Once the lamination has been decided a final static calculation can be performed to ensure the best mechanical characteristics of the component. For that purpose, a simulation where the step steer forces have been increased to simulate a fatal while driving. The results of those simulations will determine if the

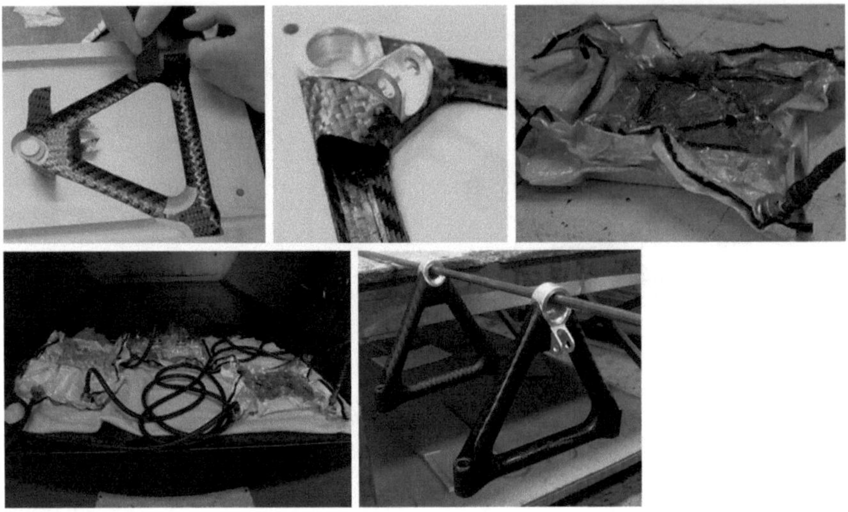

Fig. 11 Production sequence of upper wishbone

Fig. 12 The solution in aluminum on XAM 1.0 (*on the left*) and solution in carbon fiber on XAM 2.0 (*on the right*)

suspension components are able for its use at the race. In Figs. 9 and 10 and the results of the simulation are shown.

In the composite failure plot, it is possible to note that the component reacts properly to the load applied as the maximum failure value is 0.14, far below the maximum admissible 1. The strain and stress plots highlight the structural behavior of the component. The upper shell, that contains the metallic inserts reacts to the loads applied while the lower shell acts only as a cover.

After that verification it could be concluded that the lamination is ready to production. In Fig. 11 the production sequence of the upper wishbone is shown.

Table 3 Benefits in weight reduction

Material	Theoretical mass (g)	Real mass (g)
Aluminium	153	180
Carbon T800	146	171

Table 4 Benefits in stiffness increasing

Material	Ky/m (N/mm•g)	Kz/m (N/mm•g)	Kyz/m (N/mm•g)
Aluminium	135.58	73.36	66.37
Carbon T800	216.17	128.18	128.18

6 Conclusions

After the production the new system has been assembled to the XAM 2.0 urban prototype (Fig. 12) and made available to make track test and to participate in the Future Car Challenge 2012 in the UK.

The final solution does not present a relevant weight reduction, just a 5 % on the final wishbones, but in terms of stiffness increasing there was an average increase of 78 %, which gives the urban prototype a good dynamic performance take into account the target.

The evolution of this work will be to verify the reliability of the FEM model with experimental validation on the wishbones. Furthermore, we need to think about a different architecture of suspension to use better the composite properties, because the composite materials work in proper conditions if large and smooth surfaces are used, as the fibers could be aligned perfectly and the stress concentration points are reduced (Tables 3 and 4).

Acknowledgements The authors would like to thanks you all the H_2politO student team members (www.polito.it/h2polito) and all sponsors and partners (www.polito.it/h2polito/it/sponsors/xam-20.html), in particular here to the companies involved in composite field: Huntsman, Araldite, EXP compositi, Vaber, Impregnatex Compositi, Angeloni, Special Insert and Altair.

References

1. Carello, M., Airale, A.: Le sospensioni della XAM 2.0—The suspension of the XAM 2.0 vehicle. Compos. Mag. **VIII**(27), 36–41 (2013)
2. http://www.futurecarchallenge.com. Accessed Nov 25th 2011
3. Carello, M., Airale, A., Scattina, A.: Carbon fiber monocoque for a hydrogen prototype for low consumption challenge. In: Materials Science and Engineering Technology - Special

ACE-X Issue, vol. 42, n. 5, pp. 386–392, Wiley-VCH Verlag GmbH & Co., Weinheim (2011)
4. Carello, M., Scattina, A.: Structural design and experimental investigation of a carbon fibre wheel for low consumption vehicle. Book chapter in Material with complex behavior II, Springer, Berlin (2011)
5. http://www.zf.com. Accessed May 28th 2013
6. http://www.media.gm.com. Accessed June 12th 2013
7. http://www.magnasteyr.com. Accessed June 6th 2013
8. Nicolais, L., Meo, M., Milella, E.: Composite Materials—A Vision to the Future. Springer, Heidelberg (2011)
9. Richard, D.: Automotive suspension systems benefit from composites. J. Reinf. Plast. **47** (11), 18–21 (2003)
10. Noce, S.: Studio della dinamica del city vehicle XAM. Master of Science Thesis in Mechanical Engineering, Politecnico di Torino (2012)
11. Wiedemann, M., Sinapius, M.: Adaptive, Tolerant and Efficient Composite Structures. Springer, Berlin (2012)

Developing and Optimization Models for Multi-mechanical Properties of Dissimilar Laser Welding Joints

Ezzeddin Anawa and Abdulghani Olabi

Abstract Demands placed on dissimilar metals joints have increased from various viewpoints, such as, environmental concerns, energy saving, high performance, cost saving and so on. The aim of this manuscript is to predict and optimize laser welding of some economically important dissimilar materials in industry through applying a Taguchi orthogonal array design as a design of experiment (DOE) approach to design the experiments, develop mathematical models and optimize the welding operation. This was achieved by controlling selected welding parameters; laser power, welding speed and focus position, to relate the mechanical properties and weld bead geometry to the selected input welding parameters. The dissimilar materials studied in this work were low carbon steel and stainless steel. For the studied dissimilar welded material, mathematical models were developed to predict the required responses. Furthermore, the study of dissimilar were optimized by determining the best combinations of input process parameters in order to produce an excellent weld quality.

Keywords Design of experiments · Taguchi approach · Mechanical properties

1 Introduction

Laser welding is one of the most recent welding techniques available to industry. CO_2 laser beam welding with a continuous wave is a high energy density and low heat input process. The result of this is a small heat-affected zone (HAZ), which

E. Anawa (✉)
Industrial and Manufacturing Engineering, Benghazi University, Benghazi, Libya
e-mail: ezzeddin.anawa@uob.edu.ly

A. Olabi
School of Mechanical and Manufacturing Engineering, Dublin City University, Dublin 9, Ireland
e-mail: Abdul.Olabi@uws.ac.uk

A. Öchsner and H. Altenbach (eds.), *Design and Computation of Modern Engineering Materials*, Advanced Structured Materials 54,
DOI: 10.1007/978-3-319-07383-5_19, © Springer International Publishing Switzerland 2014

cools very rapidly with very little distortion, and a high depth-to-width ratio for the fusion zone [1].

Joining low carbon steel (white ferrite), with 316 stainless steel (black austenite) is known as black and white joints. These dissimilar joints are based on both technical and economical aspects, because they can provide satisfactory service performance and reasonable cost savings [2]. Joining ferritic low carbon steel is faced with the problem of coarse grains in the weld zone and heat affected zone of fusion welds and consequent low toughness and ductility due to the absence of phase transformation [3]. Joining ferritic low carbon steel to austenitic stainless steels is considered to be a major problem due to the difference in their thermal expansion coefficients, which may lead to crack formation at the interface, formation of hard zone close to the weld interface, relatively soft regions adjacent to the hard zone; large hardness difference of a hard and soft zones and the expected differences in the microstructure may lead to failures in service [4–6]. Recently, laser beam welding has received more attention for joining F/A [7, 8]. Hence, CO_2 laser welding may be a possible solution for this problem. Furthermore, the performance of the welded components in real life is sensitive to the service temperature. The welded components become very tough as the surrounding temperature decreases or vice versa. Hence, the most important mechanical property in this case is the impact strength [9]. Design of experiment (DOE) and statistical techniques are in massive use to optimize process parameters. Many researches were conducted to identify the optimal process input parameters [10–13].

In the present study, Taguchi's method is used to determine the influence of the laser welding input parameters (laser power, welding speed, gap between the jointed plates, and focus positions) on the some mechanical properties (tensile strength and impact strength). The emphasis is focused on the optimization of welding parameters combination, in order develop and to maximize both the tensile strength and impact strength.

2 Experimental Work

The materials selected to be joined by CO_2 laser welding and subjected to this study were stainless steel 316 and low carbon steel AISI 1008 cold drawing. The thicknesses of the used plates were 3 mm while the length and width of the plates was 160 × 80 mm.

Chemical composition and mechanical properties of the materials are exhibited in Tables 1 and 2. Butt joints were used to joint two abutting members lying approximately in the same plane. The plate's edges were well prepared to guarantee the full contact between the plates along the welding line during the butt-welding. A fixture was used to clamp the plates during the welding.

Pilot experiments of laser welding were carried out to determine the practical operating range of selected laser welding parameters [laser power (*P*), welding

Table 1 Chemical composition of the joined materials (wt%)

Material	C	Si	Mn	P	S	Cr	Ni	Nd	Mo	Fe
AISI 1008	0.093	0.027	0.21	0.001	0.005	0.043	0.065	0.024	0.006	Bal.
SST316	0.048	0.219	1.04	0.013	0.033	18.028	10.157	0.098	1.83	Bal.

Table 2 Mechanical properties of the materials

Grade	Tensile strength (MPa)	Yield strength (MPa)	Elongation (%)	Hardness brinell (HB), max	Elastic modulus (GPa)
AISI 1008	340	290	20	95	190
SST316	485	170	40	217	193

Table 3 The Taguchi design matrix with the actual values of the studied welding parameters and welding pool geometry, mechanical tests results and cost per meter welding calculations

Std.	Run	*P* (kW)	*S* (mm/min)	*F* (mm)	*G* (mm)	Impact St. (J)	Tensile St. (MPa)
1	9	1.00	200	−1.00	0.00	33	757
2	8	1.00	533	−0.67	0.05	30	643
3	10	1.00	867	−0.33	0.1.0	27	603
4	16	1.00	1200	0.00	0.15	16	559
5	11	1.15	200	−0.67	0.10	33	731
6	4	1.15	533	−1.00	0.15	31	659
7	1	1.15	867	0.00	0.00	26	617
8	3	1.15	1200	−0.33	0.05	4	544
9	15	1.30	200	−0.33	0.15	31	835
10	2	1.30	533	0.00	0.1	37	629
11	14	1.30	867	−1.00	0.05	35	692
12	13	1.30	1200	−0.67	0.00	7	613
13	5	1.45	200	0.00	0.05	31	842
14	6	1.45	533	−0.33	0.00	37	710
15	12	1.45	867	−0.67	0.15	16	646
16	7	1.45	1200	−1.00	0.10	12	570

speed (*S*), focus position (*F*) and gape between jointed plates (*G*)] and visual inspections for the joints were applied in order to produce acceptable quality. Argon gas was used as a shielding gas without any preheating or post heating treatments. Taguchi approach using L16 was carried out using the Design Expert 7 software. The welding input variables used are as presented in Table 3. The experiments were run in random order.

3 Results and Discussion

3.1 Development of a Mathematical Model for Tensile Strength

The experiments were carried out according to the design matrix given in Table 3. They were performed in random order to avoid any systematic error. The notched tensile strength 'NTS' samples were tested at room temperature 20 °C. Each tensile test result listed in Table 3 is an average of at least three or more samples.

3.1.1 Analysis of the Results

The raw data, the average tensile strength and the *S/N* ratio of the tensile test results are shown in Table 3. The average result of tensile strength and *S/N* ratio of the tensile test results are plotted for each of the welding parameters in Fig. 1. The average NTS tests appear to be mainly affected by the welding speed and laser power as shown in Table 4. The rank 1 in Table 4 indicates that speed has a stronger effect on the process followed by rank 2, power, which has also a strong effect, while rank 3 for gap parameter has less on the process. The minimum affect on the process, at rank 4, the focus position, has an insignificant effect on the process. To analyze the effects of the welding parameters in detail, analysis of variance (ANOVA) was conducted; these results are shown in Table 5.

In the ANOVA Table 5, the F_v is used to test the significance of a factor by comparing model variance with the residual (error) variance, which is calculated by dividing the model mean square by the model residual mean square. A high F_v value for a parameter means that the effect of the parameter on the characteristics is large. The result in Table 5 shows that the highest F_v value in the process was obtained for speed '*S*' equal to 58.61. The F_v value for the laser power '*P*' was equal to 5.54, which indicates that the laser power has a relatively low effect on the process. "Adequate precision" compares the range of the predicted values at the design points to the average prediction error. For this model it was equal to 15.506, as shown in Table 5. The same table also shows the other adequacy measures R^2 and Adjusted R^2. All the adequacy measures indicate that an adequate model has been obtained. The final mathematical models for predicting the tensile strength of dissimilar *F/A* joint in terms of coded factors and actual factors as determined by the Design Expert software are shown below in Eqs. (1) and (2).

Final Equation in Terms of Coded Factors:

$$\text{Tensile Strength} = 665.63 + 31.35P - 102.00S \tag{1}$$

Final Equation in Terms of Actual Factors:

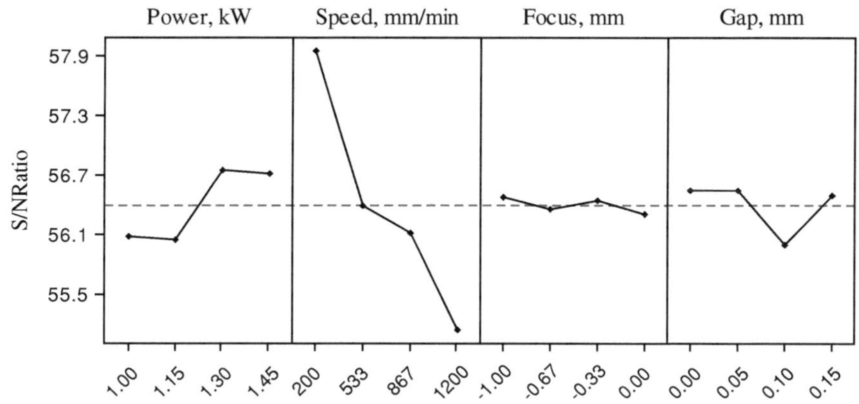

Fig. 1 The effect of the laser welding parameters on the tensile strength and S/N ratio

Table 4 Shows the tensile strength response for S/N Ratio

Levels	1	2	3	4	Delta	Rank
P	640.50	637.75	692.25	692.00	54.50	2
S	791.25	660.25	639.50	571.50	219.75	1
F	669.50	658.25	673.00	661.75	14.75	4
G	674.25	680.25	633.25	674.75	47.00	3

Table 5 ANOVA for selected factorial model

Source	Sum of squares	*df*	Mean square	F_v value	*p*-value Prob. > F_v	
Model	1.0E + 005	2	50608.1	32.07	<0.0001	Significant
P	8736.20	1	8736.2	5.54	0.0350	
S	92480.00	1	92480.0	58.61	<0.0001	
Residual	1.2E + 005	13	1577.8			
Cor. Total	43858	15				
$R^2 = 0.8315$				Adeq. Precision = 15.506		
Adj. $R^2 = 0.8056$						

$$\text{Tensile Strength} = 637.74 + 139.33P - 0.20S \tag{2}$$

Figure 2 shows the actual response versus the predicted response for NTS. From this figure, it can be seen that the model adequately describes the response within the limits of the factors being investigated herein, as the data points are close to the diagonal line. Furthermore, three extra confirmation experiments were carried out using different test conditions, which are presented in Table 6 along with the resulting percentage error. It can be noticed that the NTS value obtained

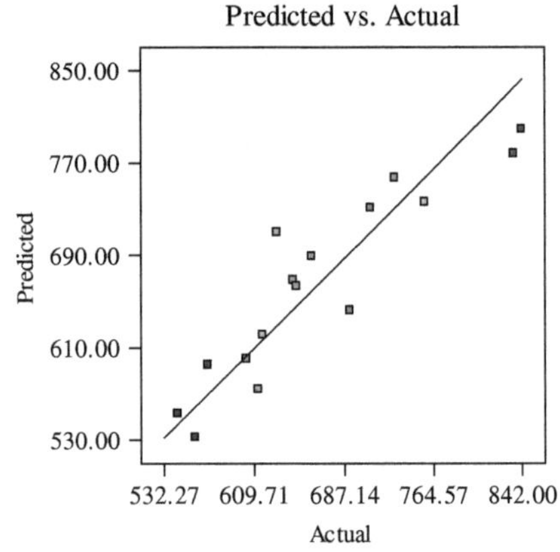

Fig. 2 Predicted versus actual for notched tensile strength NTS, MPa

Table 6 Confirmation experiments of the responses compared with model results

Exp. No	*P* (kW)	*S* (mm/min)	*F* (mm)	*G* (mm)	Tensile strength (MPa)		$\|E\|$ (%)
					Actual	Predicted	
1	1.30	304	–0.33	0.129	802	756	6.1
2	1.37	1119	–0.61	0.141	579	600	3.5
3	1.00	236	–0.04	0.146	754	729	3.4

after laser welding is greater than the base metals value, especially when compared to low carbon steel side.

3.1.2 Effect of Process Parameters on the Response

Welding Speed It can be seen that the welding speed is the most significant factor associated with the response, as shown in Fig. 1. The highest tensile strength value (842 MPa) was observed to be at a speed of 200 mm/min. It is evidence that by increasing welding speed with or without changing focus position the response would decrease. Figure 3 shows a 3D graph of the effect of *P* and *S* on the response at $F = -0.5$ mm and $G = 0.075$ mm.

Laser Power It is evident from the results that the laser power also has a strong effect on the tensile strength of the laser-welded joint, as shown in Fig. 1. It is clear that the higher welding speed laser power resulted in a higher response value, due to the fact that using high laser power increases the power density. This leads to more penetration resulting in an improved response.

Gap Parameter The result in Fig. 1 indicates that the stronger effect is at 0.1 mm.

Focus Point Position The results indicate that the focus point position has no obvious effect on the response within the parameter range domain applied.

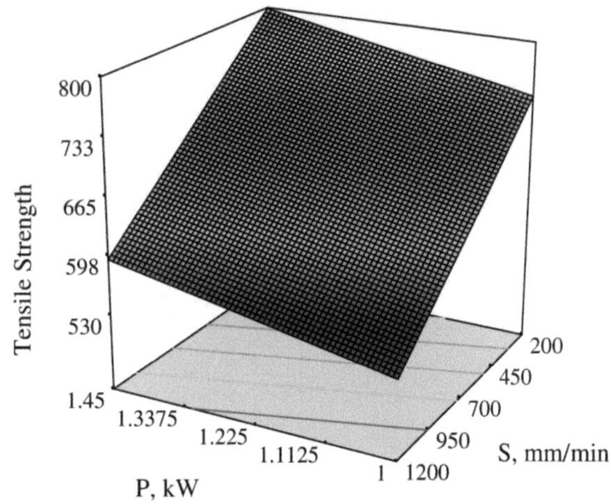

Fig. 3 3D graph of the effect of the effect of P and S on the tensile strength dissimilar components at $F = -0.5$ mm and $G = 0.075$ mm

3.2 Development of the Mathematical Model for Impact Strength

The experiments were carried out according to the design matrix given in Table 3. They were performed in random order to avoid any systematic error. The standard impact samples were tested at room temperature 20 °C. Each impact tested result listed in Table 3 was an average of at least three or more samples. From the impact test it was noted that fracture generally occurred in HAZ or base metal (low carbon steel side) which indicates that the produced joints were stronger than the low carbon steel plate.

The Taguchi method using the statistical software "Design-expert 7" was applied for designing the experiments, analyzing and optimizing the experimental data. Regardless of the category of the quality characteristic, a larger S/N ratio corresponds to a better quality characteristic. Therefore, the optimal level of the process parameters is the level with the highest S/N ratio. In this study, the S/N ratio was chosen according to the criterion the-bigger-the-better, in order to maximize the impact resistance. The Taguchi experiment results are presented in Table 7 was exhibited in Fig. 4. From the obtained result, it is obvious that the impact resistance is mainly affected by the welding speed and focal position, while the laser power and gap have less affect on the response as shown in Table 7 and exhibited in Fig. 4. The rank 1 in Table 7 indicates that the welding speed has a stronger effect on the process followed by rank 2 which indicates that the focal position parameter also has a strong effect on the process. Rank 3 in the Table indicates that the laser power parameter has less effect on the process. Rank 4 in the same Table indicates that the gap parameter has a minimal effect.

Furthermore, a statistical analysis of the variance (ANOVA) is performed and presented in Table 8 to see which process parameters are statistically significant. The optimal combination of the process parameters can then be predicted. The

Table 7 Impact strength response for *S/N* ratio

Levels	1	2	3	4	Delta	Rank
P	28.1969	25.2967	27.2425	26.7335	2.9002	3
S	30.0775	30.5636	28.0192	18.8093	11.7542	1
F	28.1666	25.2664	25.6220	28.4146	3.1482	2
G	26.7785	25.7119	27.9840	26.9952	2.2721	4

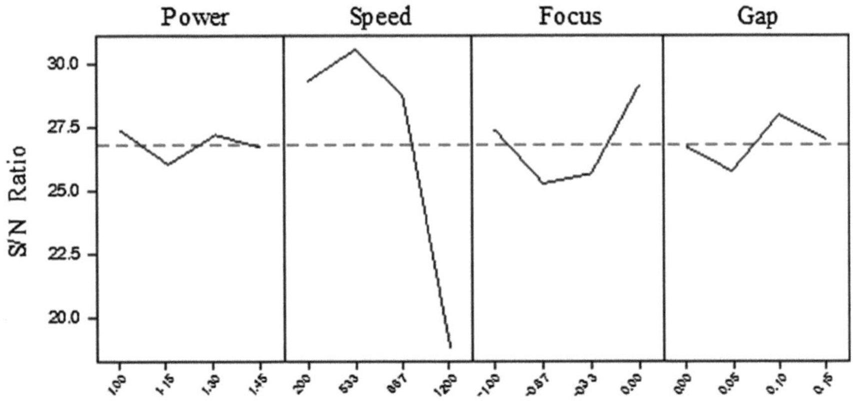

Fig. 4 Exhibited the effect of the laser welding parameters on the impact strength and *S/N* ratio

result shows that the highest F_v value obtained is 156.46 for the welding speed, but the focus and gap parameters F_v values were equal to 0.842 and 11.095 respectively which means that gap and focus have less effect on the process. The lowest F_v value was 0.360 for laser power this is due to effect of the gap parameter on the process. The same table shows also the other adequacy measures R^2, Adjusted R^2 and Adequate precision. All the adequacy measures indicate that an adequate model has been developed. The final mathematical models in terms of actual factors are shown below in Eqs. (3) and (4).

Final Equation in Terms of Coded Factors:

$$\begin{aligned}\text{Impact Strength} = {} & 29.44 + 0.43P - 16.19S - 0.64F - 2.73G - 7.58PF \\ & - 3.96PG + 3.90FG - 12.21S^2 + 5.02F^2. \end{aligned} \quad (3)$$

Final Equation in Terms of Actual Factors:

$$\begin{aligned}\text{Impact Strength} = {} & 48.73 - 14.17P + 0.04S + 93.50F + 303.29G - 67.36PF \\ & - 234.94PG + 103.88FG - 4.89\text{E} - 005S^2 + 20.069F^2 \end{aligned} \quad (4)$$

Table 8 ANOVA for impact strength response model

Source	Sum of squares	*df*	Mean square	F_v value	*p*-value prob. > F_v	
Model	1706.427	9	189.603	49.801	<0.0001	Significant
P	1.369	1	1.369045	0.360	0.5707	
S	595.688	1	595.6877	156.462	<0.0001	
F	3.206	1	3.205679	0.842	0.3942	
G	42.242	1	42.24206	11.095	0.0158	
PF	63.444	1	63.44368	16.664	0.0065	
PG	65.112	1	65.1117	17.102	0.0061	
FG	53.352	1	53.35217	14.013	0.0096	
S^2	384.933	1	384.933	101.106	<0.0001	
F^2	79.507	1	79.50665	20.883	0.0038	
Residual	22.843	6	3.807228			
Cor. Total	1729.270	15				
$R^2 = 0.9868$			Adeq. Precision = 22.691			
Adj. $R^2 = 0.9670$						

Table 9 Confirmation experiments of the responses compared with model results

No.	P	S	F	G	Impact strength		|*Error*| (%)
					Actual	Predicted	
1	1	906	0	0.15	39	37	4.9
2	142	963	–1.00	0	42	40	4.9
3	1	886	0	0.15	37	38	3.3

Finally, confirmation experiments were conducted to verify the optimal process parameters obtained from the design. They are presented in Table 9.

3.2.1 Model Validation

This step is to predict and verify the improvement of the response using the optimal level of the welding process parameters. Figure 5 shows the relationship between the actual and predicted values of impact strength. This figure indicates that the developed model is adequate because the residual in prediction of response is negligible, since the residuals tend to be close to the diagonal line. It could be concluded that the model developed can predict the response with a very small error. Furthermore, to verify the satisfactoriness of the developed models, three confirmation experiments were carried out using new test conditions at optimal parameters conditions, obtained using the design expert software. Table 7 summarizes the experiments conditions, the actual experimental values, the predicted values and the percentages of error. It could be concluded that the developed model can predict the response with a very small error. Impact strength was greatly improved through this optimization.

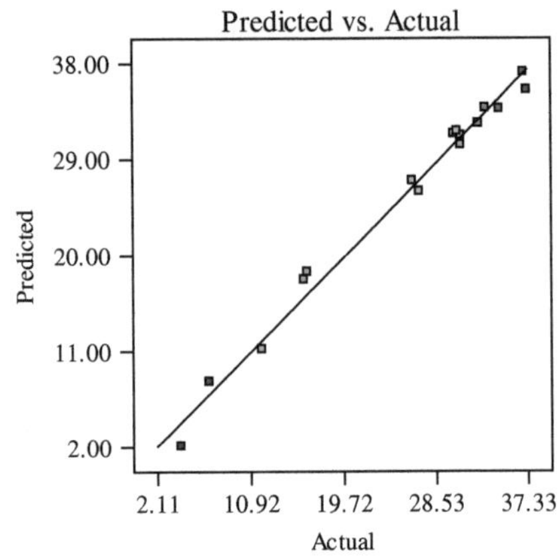

Fig. 5 Relationship between the actual and predicted values of the response

3.2.2 The Effect of the Parameters on the Response

The impact strength of dissimilar joints between stainless steel and low carbon steel was measured and plotted in 3D in Fig. 6; which shows that the welding speed parameter and laser power effect on the process at focus equal –0.5 mm and gap distance equal 0.75 mm.

Welding Speed The signal-to-noise (*S/N*) ratio analysis and ANOVA analysis indicates that the welding speed parameter has the most significant effect on the process. The increase in welding speed, leads to decreases in impact strength.

Focusing Position This factor has a strong effect on the response as it is indicated in *S/N* ratio analysis and ANOVA analysis. The focus parameter in this study interacts with the laser power parameter and influences its affect on process.

Laser Power In presence of the effect of focus and the gap parameters in the welding process the laser power parameter has an insignificant effect.

Gap Width Between the Welded Plates The presence of gap between jointed plates is an important factor to decrease the residual stresses resulting from heat input due to the welding process particularly when joining dissimilar materials. In this study, signal-to-noise (*S/N*) ratio analysis and ANOVA analysis indicates that the gap parameter has a strong effect on the process.

To compare the effect of all the considered welding parameters on the impact strength at a midpoint point position in the design space, a perturbation plotted is exhibited in Fig. 7. The response is plotted by changing only one factor over its range while holding of the other factors constant.

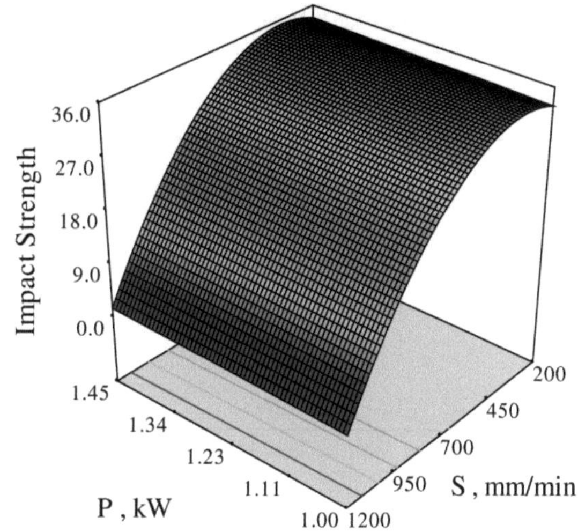

Fig. 6 Effect of welding parameters on the impact strength

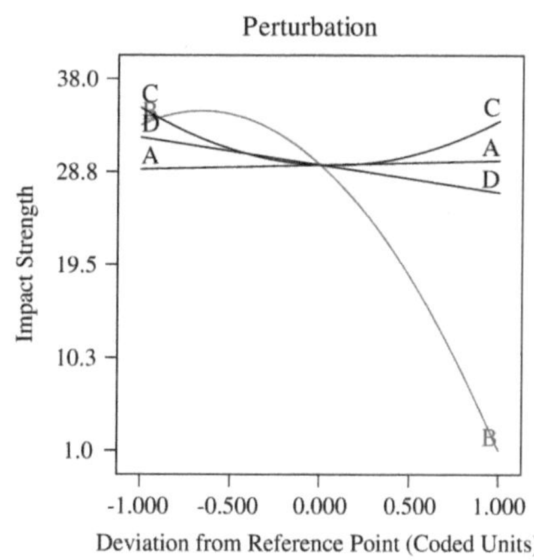

Fig. 7 Perturbation plots exhibiting the effect of welding parameters on the tensile strength, where: A = power, B = speed, C = focus and D = gap

4 Models (Multiple–Response) Optimization

The numerical multiple-response optimization criterion is to reach maximum tensile strength and maximum impact strength with minimizing laser power and maximizing welding speed while focus position was kept in range.

In first optimization criteria all the parameters received the same importance (+++) and same weight (1) as per the software default. In the second criteria a different weight was assigned for each parameter as presented in Table 10, while the importance for each parameter was kept same as (+++). The importance was changed for welding parameters in the third criteria while the weight was the same as per the soft ware default. All the decided welding optimization criteria and the resultant optimizations are presented in Table 10. The result presented in the

Table 10 Three optimization criteria with the optimization results using numerical multiple-response

Welding parameters	*P*		*S*		*F*		*G*		Impact resistance		Tensile strength		*D**.
Goal	Min.		Max.		In range		In range		Max.		Max.		
Criteria	Wt.*	Imp.*	Wt.	Imp.	Wt.	Imp.	Wt.	Imp.	Wt.	Imp.	Wt.	Imp.	
First criteria	1	+++	1	+++	–	–	–	–	1	+++	1	+++	
Result	1.01		897		0.00		0.15		37.3		595		0.686
Second criteria	0.5	+++	5	+++	–	–	–	–	5	+++	5	+++	
Result	1.388		848		–1.00		0.00		44		657		0.351
Third criteria	1	+	1	+++	–	–	–	–	1	+++++	1	+++++	
Result	1.34		943		–1.00		0.00		37.3		633		0.671

Imp.* = Important, Wt* . = Weight, *D.** = Desirabili

Table 10 at each criterion is selected from one of ten or more different optimum result calculated by the software.

The effect of changing the criteria on the optimization result is obvious in Table 10. For example by applying in the third criteria the tensile strength value will reach up to 633 MPa while if applying the second criteria it will be around 657 MPa. If the target is only to maximize the tensile strength regardless of the other parameters then the response value will be greater than the received values and this is true for each response individually optimized.

Impact strength was affected by changing the optimization criteria; its value is between (37–44 J) depending on its assigned weight and importance it received in each criterion. Also, the welding speed is relatively high, in range of (848–943 mm/min), in the optimization criteria which leads to increased production rate.

5 Conclusion

Ferrite/Austenite joints are a popular dissimilar metal combination in many applications. Therefore, exploitation of new processes for producing them is of interest to several industrial sectors. The following points can be concluded from this study:

1. By means of a Design of Experiment inspired by the Taguchi approach, it is possible to achieve the best operating parameter window and then develop models to control the welding parameters.
2. Successful welding can be achieved on butt stainless steel—low carbon steel joints using CO_2 laser welding. Using laser welding improves the mechanical properties and produce a narrow HAZ.

3. All the models developed for the considered stainless steel—low carbon steel joints can adequately predict the response within the factors domain.
4. Focus position and the gap between the jointed plates have a strong effect on impact strength while laser power has an insignificant effect on the impact strength.
5. The laser welded joint of F/A is stronger than both base metals.

Acknowledgement The authors wish to thankful Mr. Johnson Martin and Benghazi University for their support and assistance.

References

1. Zacharia, T., David, S.A., Vitek, J.M, Debroy, T.: Metall. Trans. **20A**, 1125 (1989)
2. Sun, Z.: The application of electron beam welding for the joining of dissimilar metals. Int. J Pres. Pip. **68**, 153–160 (1996)
3. Satyanarayana, V.V., Reddy, G.M., Mohandes, T.: Dissimilar metal friction welding of austenitic–ferritic stainless steels. J. Mater. Process. Technol. **160**, 128–137 (2005)
4. Allabhakshi, S., Reddy, G., Ramarao, V., Babu, C., Ramachandran, C.: National Welding Conference,India. Indian Institute of Welding, Chennai, 8 (2002)
5. Omar, A.: Effects of welding parameters on hard zone formation at dissimilar metal welds. Weld. J. **77–2**, 86–93 (1998)
6. Welding Hand Book (2004) 7th ed. Am. Weld. Soc. **4**, 93–128
7. Mai, T.A.: Characterisation of dissimilar joints in laser welding of steel–kovar, copper–steel and copper–aluminum. Mater. Sci. Eng. A **374**, 224–233 (2004)
8. Li, Z., Fontana, G.: Autogenous laser welding of stainless steel to free-cutting steel for the manufacture of hydraulic valves. J. Mater. Process. Technol. **74**, 174–182 (1994)
9. Anawa, E.M., Olabi, A.G.: Effect of laser welding conditions on toughness of dissimilar welded component. J. Appl. Mech. Mater. **5–6**, 375–380 (2006)
10. Tarng, Y.S., Juang, S.C., Chang, C.H.: The use of grey-based Taguchi methods to determine submerged arc welding process parameters in hardfacing. J. Mater. Process. Technol. **128**, 1–6 (2002)
11. Anawa, E.M., Olabi, A.G., Hashmi, M.S.J.: Application of taguchi method to optimize dissimilar laser welded components. In: presented at 23rd International Manufacturing Conference. Belfast, UK (2006)
12. Anawa, E.M., Olabi, A.G.: Using Taguchi method to optimize welding pool of dissimilar laser-welded components. J. Opt. Laser Technol. **40**, 379–388 (2008)
13. Lee, H.K., Han, H.S., Son, K.J., Hong, S.B.: Optimization of Nd YAG laser welding parameters for sealing small titanium tube ends. J. Mater. Sci. Eng. A **415**, 149–155 (2006)

The Effect of Strain Hardening in Stainless Steels Submitted to Nitriding Treatment

Mariana Zicari di Monte, Jan Vatavuk, Antonio Augusto Couto and Nelson Batista de Lima

Abstract The present study shows an evaluation of the effects of strain hardening on the kinetics of the nitriding thermochemical treatment in alloyed ferritic stainless steels AISI 430 and AISI 439 for different degrees of plastic deformation. The processes of application of the gas nitriding, in salt baths and plasma were performed on these alloys. The layer thickness was larger for the process of nitriding gas. Note also that the nitrogen concentration reduces with the depth of the nitrided layer as it comes near to the base metal. On the other hand, carbon is pushed inside of the sample, promoting decarburization in the region enriched in nitrogen and an accumulation in the regions just below to the nitride layer. In general, note that the ferritic matrix (Fe–Cr) and the formation of chromium nitride (CrN) in samples of AISI 430 and AISI 439 which suffered plasma nitriding and gas. In addition to this compound, it was possible to identify the presence of Fe_4N and Fe_3N. In the samples of AISI 430 and AISI 439 nitrided in salt baths, it can be observed that the nitrided layer consists in the ferritic matrix (Fe–Cr) with predominance of precipitates $(FeCr)_2N$ and without presence of peaks of CrN.

A. A. Couto · N. B. de Lima
IPEN-CNEN/SP, Rua da Consolação, 930, São Paulo, SP 01302-907, Brazil
e-mail: nblima@ipen.br

M. Z. di Monte · J. Vatavuk · A. A. Couto (✉)
Mackenzie Presbyterian University, Rua da Consolação, 930,
São Paulo, SP 01302-907, Brazil
e-mail: acouto@ipen.br

M. Z. di Monte
e-mail: Mariana.Monte@jtekt.com

J. Vatavuk
e-mail: janvatavuk@uol.com.br

A. Öchsner and H. Altenbach (eds.), *Design and Computation of Modern Engineering Materials*, Advanced Structured Materials 54,
DOI: 10.1007/978-3-319-07383-5_20, © Springer International Publishing Switzerland 2014

1 Introduction

The Stainless steels are known for their corrosion resistance. However, they have low hardness around 150 HV, preventing its application in situations that require wear resistance [1]. Thus, studies have been done looking for alternatives for changing the surface properties of these steels without compromising other properties such as corrosion resistance. An alternative that has been studied is the thermochemical process known as nitriding. The three main methods of nitriding are: gas nitriding, salt bath and plasma. The nitriding process, regardless of its method of process, increases the hardness of these steels by promoting higher wear resistance due to the formation of precipitates nitrides [2].

The nitriding offers as main advantage of their low application temperature in the range of 500–590 °C [3, 4], that no phase change is taking place during the process. Furthermore, the nitriding process does not require high cooling rates, resulting in minimal dimensional variation and little change of the mechanical properties of the core part [5]. The ferrous material in the nitriding thermochemical treatment will produce distinct regions, from the surface and in the depth order: a zone called "white layer" composed of micro constituents ε ($Fe_{2\text{-}3}N$) and γ'(Fe_4N), present only in some processes of nitriding, diffusion zone, transition zone (between the diffusion zone and the core material) and core material [6]. While for the alloy steels, the composite layers contain in addition to iron nitrides ε and γ', nitrides of alloying elements, such as Cr_2N and CrN.

The Stainless steels have been studied extensively as well as the nitriding processes. However, the effects of modifications of these steels by work hardening and the influence on the kinetics of formation of the nitrided layer are not sufficiently exploited. Therefore, this study aims to investigate the effects of the degree of plastic deformation in the process of nitriding of stainless steel with crystal structure predominantly ferritic.

2 Materials and Methods

In the present study, there were used rolled sheets and annealed ferritic stainless steels AISI 430 and AISI 439. Table 1 shows the nominal chemical composition performed by atomic absorption spectrophotometry on a sample of each steel used in this work. The stabilization of AISI439 with Ti and Nb allows being welded without the occurrence of embrittlement. The chromium content of the steel AISI 439 is larger than the AISI 430. Tensile specimens were prepared and deformed in a universal testing machine MTS 810. The test specimens were submitted to different degrees of cold plastic deformation: 0 (annealed), 5, 10, 20 and 30 %. Each stainless steel plate with 1.5 mm of thickness was shared into four samples as Fig. 1, with one of the samples was maintained and the others were nitrided.

Table 1 Chemical composition of ferritic stainless steel AISI 430 and AISI 439, nominal and obtained by Atomic Absorption Spectrophotometry

AISI		%C	%Mn	%Si	%P	%S	%Cr	%Ni	%Ti
Nominal chemical composition									
430	min	–	–	–	–	–	16.0	–	
	max	0.12	1.00	1.00	0.04	0.03	18.0	0.75	
439	min	–	–	–	–	–	17.0	–	
	max	0.07	1.00	1.00	0.04	0.03	19.0	0.5	
Chemical composition obtained by atomic absorption spectrometry									
430			0.22				16.8	0.21	
439			0.12				18.2	0.18	0.10

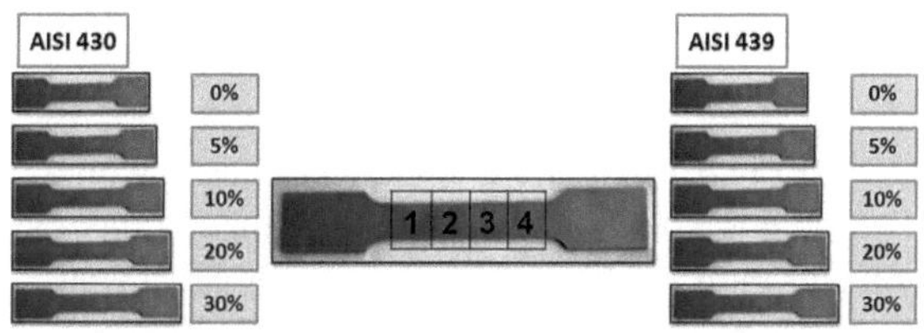

Fig. 1 Scheme of section of the test specimen for obtaining samples to be nitrided

An undeformed sample was also submitted to nitriding along with the deformed samples.

The specimens after undergoing plastic deformation were submitted to three different nitriding processes. Samples were nitrided under the following conditions:

Samples 1: Preserved for the initial state before nitriding;
Samples 2: Preserved for the initial state before nitriding;
Samples 3: Plasma nitriding at 480 °C for 10 h;
Samples 4: Liquid nitriding at 540 °C for 2 h.

For the characterization of the nitrided samples, the following analysis techniques were used to: microhardness profile, characterized by optical microscopy and scanning electron, thickness measurement of the nitrided layer, curves of nitrogen and carbon, determining the phases identified by X-ray diffraction and bending until fracture.

The preparation of metallographic specimens consisted of conventional techniques of sanding, polishing and etching with 5 % of Nital. The Vickers microhardness profile was achieved for each 20 μm from the surface of the sample. The thickness of the nitrided layer was determined by measuring at three different regions along the nitrided layer. The semi quantitative analysis of nitrogen and carbon from the surface of the sample was obtained in a scanning electron microscope (SEM) coupled to a detection system for energy dispersive X-ray spectroscopy (EDS) and wavelength dispersive X-ray spectroscopy (WDS). The analysis of the phases

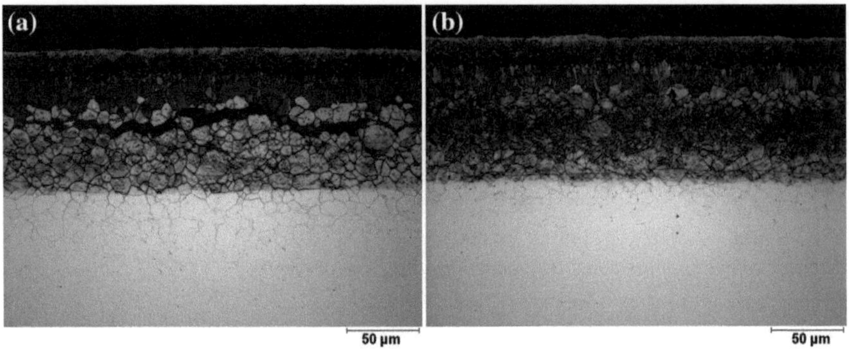

Fig. 2 Micrographs of ferritic stainless steels samples gas nitrided AISI 430 **a** without plastic deformation and **b** 20 % of plastic deformation

identified by X-ray diffraction was performed by using kα radiation of copper ($\lambda = 0.1542$ nm), graphite monochromator, with a range of 20–80° and a speed of 1°/min. For a better understanding of the fracture mechanisms, bending tests of the samples until fracture and subsequently SEM analyses was carried out.

3 Results and Discussion

By optical microscopy were analyzed the microstructure of alloys samples AISI 430 and AISI 439 without going through the process of nitriding. This analysis did not observe the presence of cracks in the original samples and also in samples with plastic deformation. Figures 2, 3 and 4 correspond respectively to the micrographs of representative samples of AISI 430 stainless steel nitrided in liquid, gas and plasma obtained by light microscopy. No significant difference was observed in the nitrided layer due to the degree of plastic deformation. Due to that, micrographs of samples are shown for each steel and nitriding process without plastic deformation and with 20 % of plastic deformation. Likewise that was shown for the steel AISI 430, Figs. 5, 6 and 7 display the micrographs of representative samples of AISI 439 stainless steel nitrided in liquid, gas and plasma and achieved by light microscopy.

In the samples of ferritic stainless steels AISI 430 alloy, it is observed that the grains are shown in the lower range from 5 to 10 μm. In the samples of ferritic stainless steels AISI 439 alloy, it is noted that the grains presented to be higher than those of alloy AISI 430. The microhardness of the materials after strain hardening and before nitriding showed values of 150 HV to 220 HV in the degree of plastic deformation of 30 %. The response of stainless steel to the nitriding process is very relevant, since it is starting from steel with about 150–220 HV, the hardness achieved in the nitride layer can reach values of the order 1,000–1,100 HV.

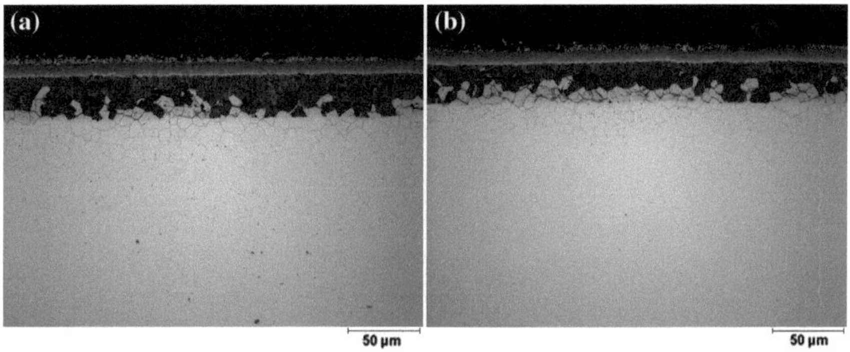

Fig. 3 Micrographs of ferritic stainless steels samples liquid nitrided AISI 430 **a** without plastic deformation and **b** 20 % of plastic deformation

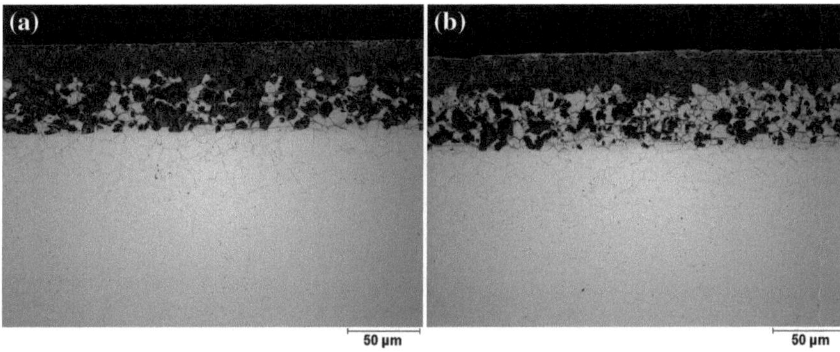

Fig. 4 Micrographs of ferritic stainless steels samples plasma nitrided AISI 430 **a** without plastic deformation and **b** 20 % of plastic deformation

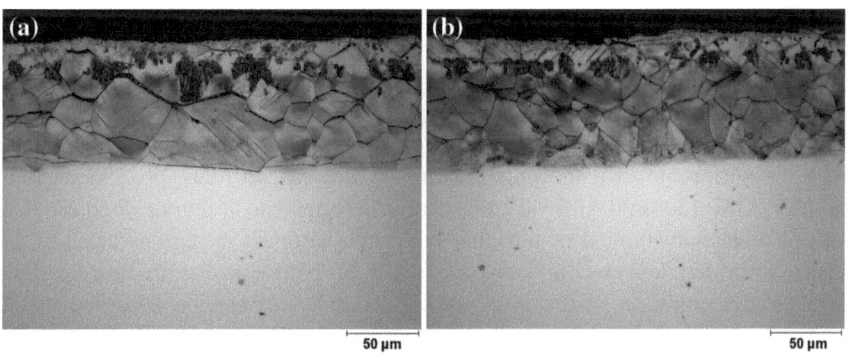

Fig. 5 Micrographs of ferritic stainless steels samples gas nitrided AISI 439 **a** without plastic deformation and **b** 20 % of plastic deformation

Fig. 6 Micrographs of ferritic stainless steels samples liquid nitrided AISI 439 **a** without plastic deformation and **b** 20 % of plastic deformation

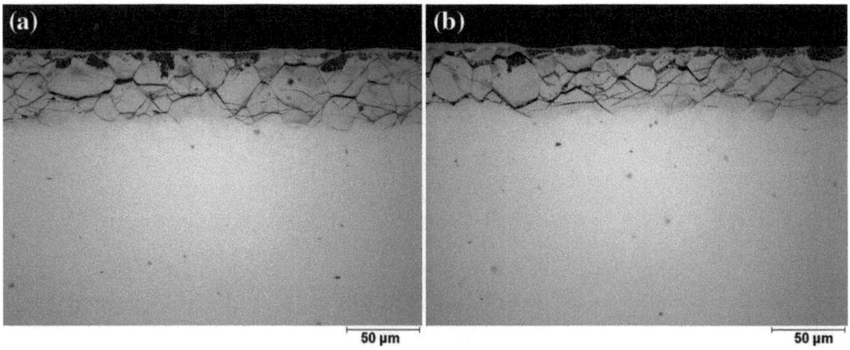

Fig. 7 Micrographs of ferritic stainless steels samples plasma nitrided AISI 439 **a** without plastic deformation and **b** 20 % of plastic deformation

This hardness value is in the same range of magnitude found in some tool steels. The optical micrographs of AISI 430 showed a clearer area and another darker in the nitrided layer. The microhardness measurements on samples of AISI 430 stainless steel nitrided after the nitriding process indicated that the lighter area is slightly greater than the hardness of the darker region.

The graphs of Figs. 8 and 9 show the results of microhardness profiles obtained from the surface toward the core relating to the gas nitriding process of steel samples AISI 430 and AISI 439 plastically deformed to different degrees. It is noticed a behavior typical of nitriding with high hardness in regions close to the surface (nitrided layer) with subsequent fall of hardness due to decreased of nitrogen concentration in the layer. A similar behavior can be observed in the process of salt bath nitriding and plasma as presented in the graphs of microhardness profiles in Figs. 10, 11, 12 and 13.

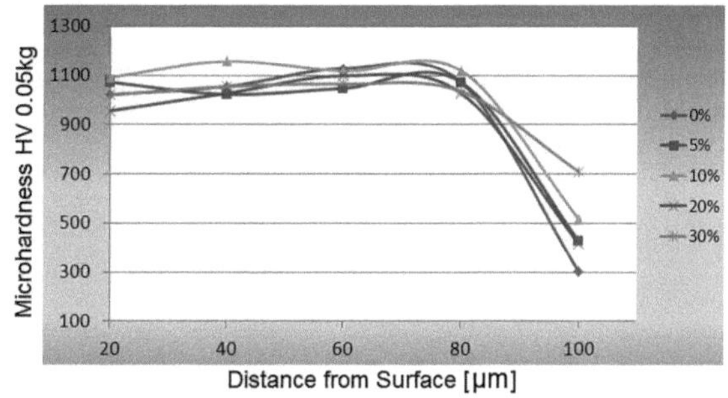

Fig. 8 Microhardness profile of AISI 430 previously strain-hardened and gas nitrided

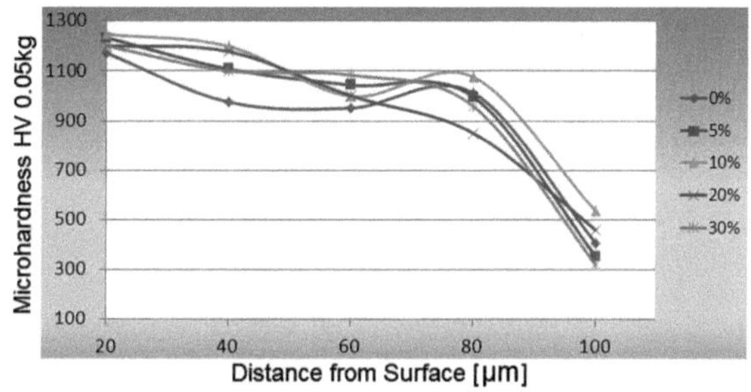

Fig. 9 Microhardness profile of AISI 439 previously strain-hardened and gas nitrided

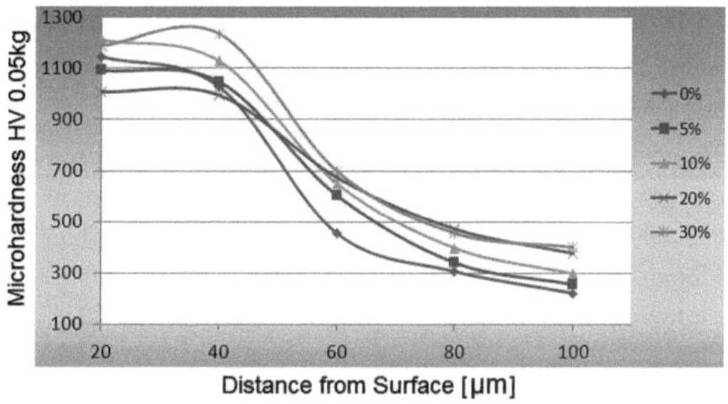

Fig. 10 Microhardness profile of AISI 430 previously strain-hardened and salt bath nitrided

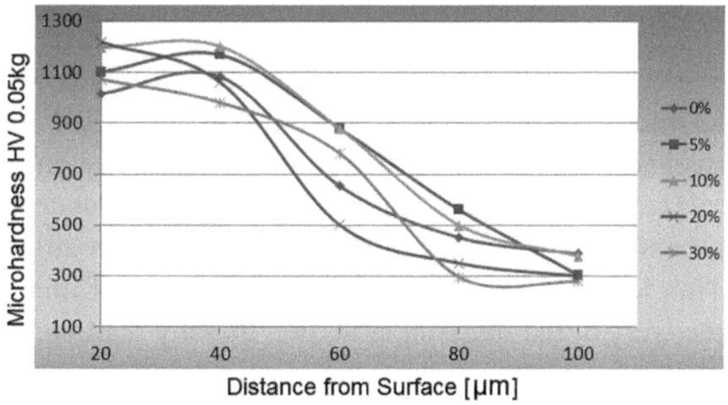

Fig. 11 Microhardness profile of AISI 439 previously strain-hardened and salt bath nitrided

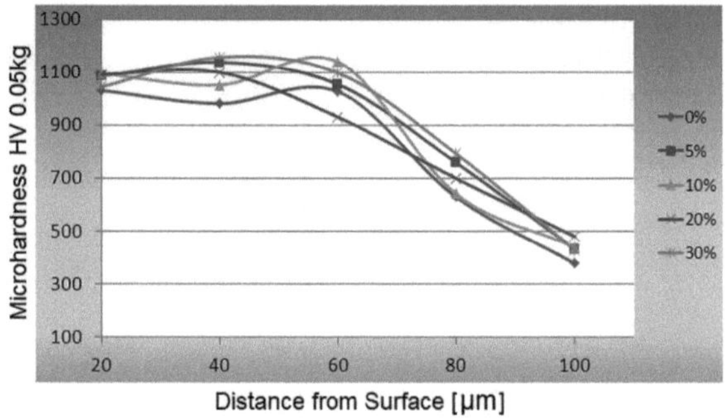

Fig. 12 Microhardness profile of AISI 430 previously strain-hardened and plasma nitrided

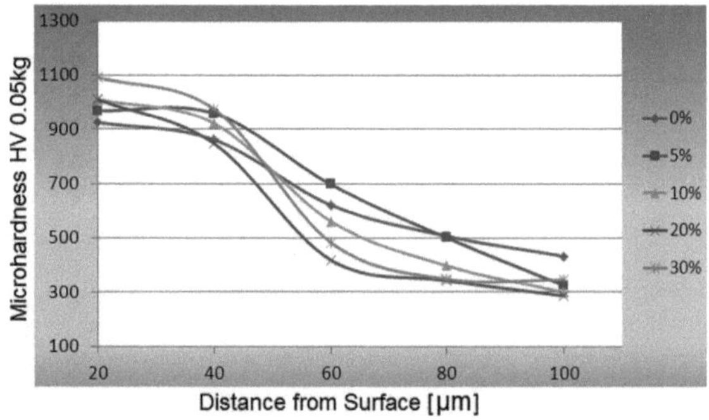

Fig. 13 Microhardness profile of AISI 439 previously strain-hardened and plasma nitrided

Table 2 Nitrided layer thickness due to different nitriding processes of samples of steel AISI 430 and AISI 439 previously deformed plastically

Nitriding-AISI	0 %	5 %	10 %	20 %	30 %
Gas-430	$93.9^{+/-0.96}$	$95.8^{+/-1.54}$	$96.8^{+/-0.19}$	$93.4^{+/-0.19}$	$92.8^{+/-0.96}$
Gas-439	$83.9^{+/-0.96}$	$84.4^{+/-0.96}$	$81.7^{+/-1.67}$	$84.0^{+/-1.15}$	$83.4^{+/-0.30}$
Liquid-430	$37.2^{+/-0.96}$	$39.5^{+/-0.85}$	$35.6^{+/-0.96}$	$35.6^{+/-0.96}$	$35.1^{+/-0.19}$
Liquid-439	$39.9^{+/-0.19}$	$48.9^{+/-0.96}$	$49.4^{+/-0.96}$	$49.4^{+/-0.96}$	$48.3^{+/-1.67}$
Plasma-430	$56.4^{+/-0.38}$	$56.1^{+/-0.96}$	$53.1^{+/-0.38}$	$61.8^{+/-0.19}$	$59.4^{+/-0.96}$
Plasma-439	$47.2^{+/-0.96}$	$45.6^{+/-0.96}$	$43.9^{+/-0.26}$	$42.8^{+/-0.96}$	$45.4^{+/-0.96}$

The differences between the three nitriding processes are related to the thicknesses of nitrided layers, as seen from Table 2. The gas nitrided samples, both ferritic stainless steel type AISI 430 and AISI 439, showed the highest layer thickness in relation to other nitriding processes due to the used higher temperature (570 °C). The results obtained indicate that the cold plastic deformation does not influence the hardness of the nitrided layer on any of the three processes studied. This fact can be related to the temperature of nitriding processes that provides a recovery of the material structure, thereby rearranging and annihilating most of the dislocations generated by the plastic deformation.

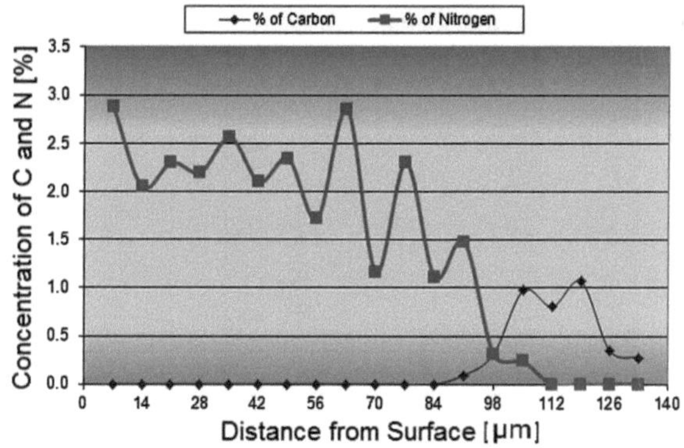

Fig. 14 Concentration profile of Nitrogen and Carbon in relation to the depth of the gas nitrided layer of steel AISI 430

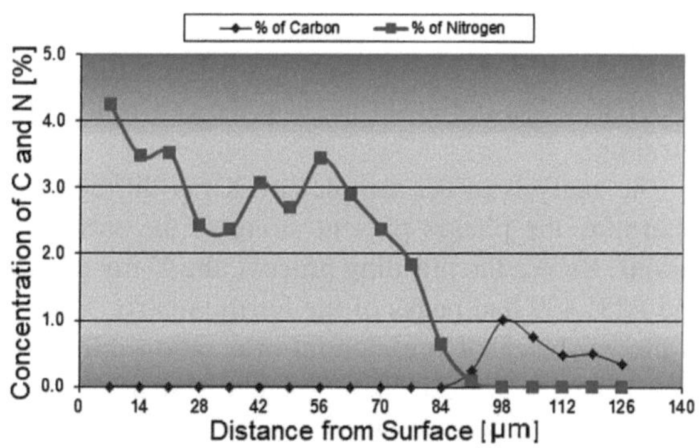

Fig. 15 Concentration profile of Nitrogen and Carbon in relation to the depth of the gas nitrided layer of steel AISI 439

The concentration profiles of nitrogen and carbon from the surface of the gas nitrided samples of steels AISI 430 and AISI 439 were determined by an X-ray spectrometer for wavelength coupled to a scanning electron microscope and they are presented in Figs. 14 and 15. The nitrided layer of ferritic stainless steel AISI 430 gas nitrided has an average content of N around 1.8 % reaching the range of carbon content of 1 % and the sample of ferritic stainless steel AISI 439 gas nitrided presents an average content of N around 2.5 % with a peak for carbon content of approximately 1 %.

The steels of this study have a maximum carbon content of 0.07 % by weight for the alloy AISI 430 and 0.12 % for the AISI 439 and the gas and plasma nitriding processes are free of carbon with the exception of salt bath nitriding process. Thus, it can be concluded that probably a decarburization in the diffusion region with the nitrogen occurred and those carbon atoms will diffuse to regions below to the nitrided layer, creating a region with a higher concentration of carbon. The continuous diffusion of nitrogen and residual stresses transformation are responsible for redistribution of carbon whereas carbon atoms, initially in the matrix, diffuse to regions free of stresses, toward the front of nitriding, leading to

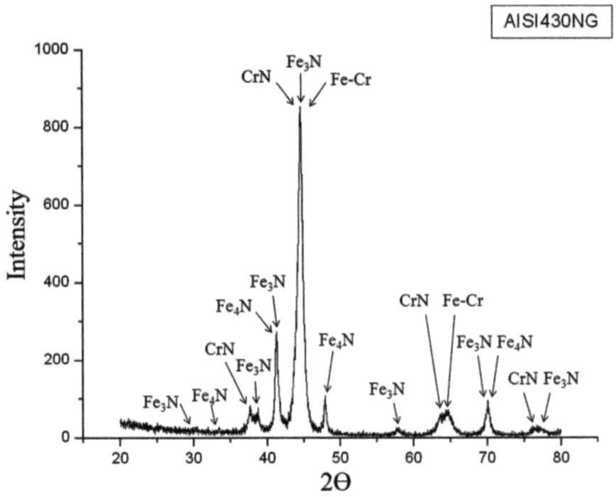

Fig. 16 X-ray diffractogram of steel AISI 430 gas nitrided

decarburization of the surface and the generation of a region rich in carbon before nitriding.

The analysis of the samples by X-ray diffraction was performed in an effort to determine the phases present along to the surfaces nitrided under different conditions. Before the nitriding process, the X-ray diffractograms of samples AISI 430 and AISI 439 had peaks of the ferritic matrix. Other aspect of such diffractograms is the existence of preferential orientation differentiated between the two steels. This behavior is most likely related to this crystallographic texture derived from different final processes to obtain these plates: cold rolling for steel AISI 430 and hot rolling for steel AISI 439. Figures 16, 17, 18, 19 and 21 show X-ray diffractograms of samples of ferritic stainless steel AISI 430 and AISI 439 after nitriding, identifying the present phases.

In general, the spectra of X-ray diffraction indicate the presence of the phase α (Fe–Cr) for all conditions analyzed in this study, as well as the generation of chromium nitride (CrN) for the samples that underwent plasma and gas nitriding processes. Figures 16 and 17 show the spectra of X-ray diffraction where it is possible to identify the ferrite matrix (Fe–Cr) with precipitates of chromium nitride (CrN) and the presence of peaks Fe_4N (γ'-face centered cubic) for the steels AISI 430 and AISI 439 and Fe_3N (ε-hexagonal compact) only for the steel AISI 430. Comparing the results of X-ray diffraction of the steels AISI 430 and AISI 439 gas nitrided (Figs. 16 and 17) with the respective micrographs observed by optical microscopy (Figs. 4 and 5), it can be noticed that the steel AISI 439 does not display significant presence of the dark phase. The diffractogram presents a ferritic matrix (Fe–Cr) with precipitates rich in chromium nitride (CrN) and presence of peaks of Fe_4N precipitates. Unlike that observed in steel AISI 439, the AISI 430 shows a predominance of both the dark phase as clear, with the diffractogram showing the presence of peaks of Fe_4N and Fe_3N precipitates. Therefore, it is understood that the dark region would be composed of Fe_3N

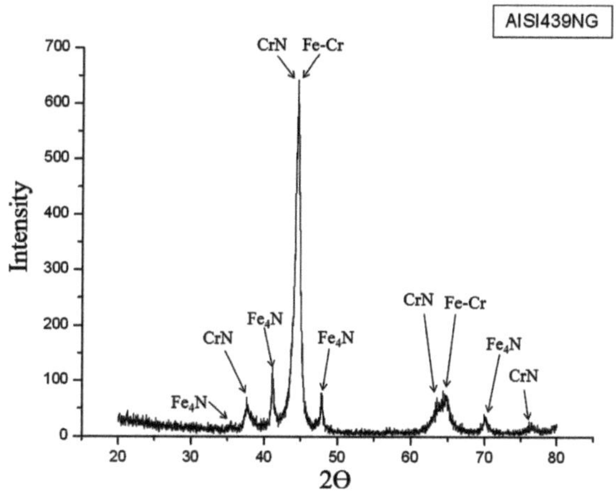

Fig. 17 X-ray diffractogram of steel AISI 439 gas nitrided

precipitates, since only the diffractogram of the AISI 430 stainless steel showed the presence of this Fe_3N precipitate.

Figures 18 and 19 show the phases identified by X-ray diffraction of the sample nitrided in a bath of salts. It can be noted that the nitride layer consists of a ferritic matrix distribution (Fe–Cr) with a predominance of $(FeCr)_2N$ precipitates and peaks without the presence of chromium nitride (CrN). Figures 20 and 21 show the phases identified by X-ray diffraction in plasma nitrided samples. In these diffractograms, it can be observed that the nitrided layer consists of a ferritic matrix distribution (Fe–Cr) and presence of peaks Fe_4N, Fe_3N and CrN.

During the nitriding process, the nitrogen from the nitriding medium penetrates in the steel by diffusion, combining with the present alloying elements, especially chromium in the case of stainless steels. Depending on the nitrogen activity of the nitriding medium, the generation of a nitride-based layer can occur which is followed by a diffusion layer composed by the matrix and nitride precipitates. The presence of chromium is attributed to a better response of nitriding in terms of the hardness levels reached due to the nitrides based on this alloying element. The compressive residual stresses are not related directly to the chromium content, generally being a common observation in the nitrided layers. The nitriding occurs with volume expansion and the core prevents this expansion by tensile residual stresses which produce compressive residual stresses at the surface [7]. The impoverishment of chromium in solid solution causes a reduction in corrosion resistance of stainless steels [8].

It is observed in some papers that the precipitation of chromium nitride (CrN) in the ferrous matrix constitutes the dark phase in the micrograph of austenitic stainless steels [9]. However, in this study microhardness analyzes were performed in metallography of steel AISI 430 after the process of gas, liquid and plasma nitriding processes and the results showed microhardness values in the order of 1,000 HV both in the clear region as well as in the dark of the studied micrographs. Thus, it is concluded that the chromium nitride can be present both in the

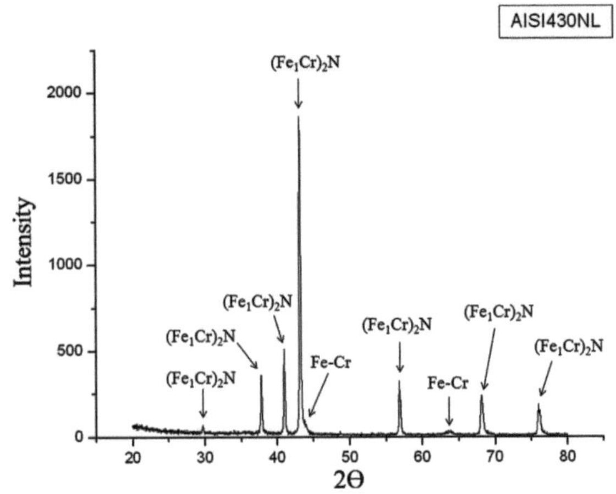

Fig. 18 X-ray diffractogram of steel AISI 430 liquid nitrided

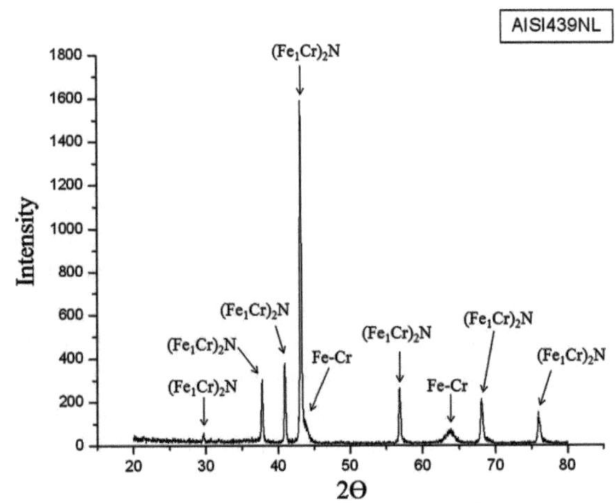

Fig. 19 X-ray diffractogram of steel AISI 439 liquid nitrided

clear region as dark of the nitrided layer and it cannot be distinguished that the CrN precipitate in the ferrous matrix constitutes the dark region of the nitrided layer.

With the purpose of analyzing the mechanism of fracture found in ferritic stainless steels AISI 430 and AISI 439 under different nitriding processes, the fracture surfaces of samples submitted to bending were observed by scanning electron microscopy (SEM). The samples analyzed were those of ferritic stainless steels AISI 430 and AISI 439 gas and plasma nitrided with 5 % degree of plastic deformation. Figure 22 shows fractography of sample of AISI 430 stainless steel gas nitrided with 5 % plastic deformation.

The arrow in Fig. 22a shows the location of the sample where the nitrided layer was observed. It can be observed in Fig. 22c the presence of cleavage and in Fig. 22d the region from the interface of the nitrided layer and the substrate with

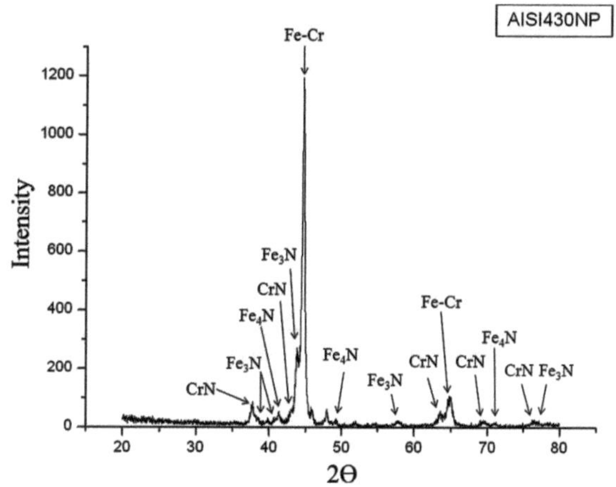

Fig. 20 X-ray diffractogram of steel AISI 430 plasma nitrided

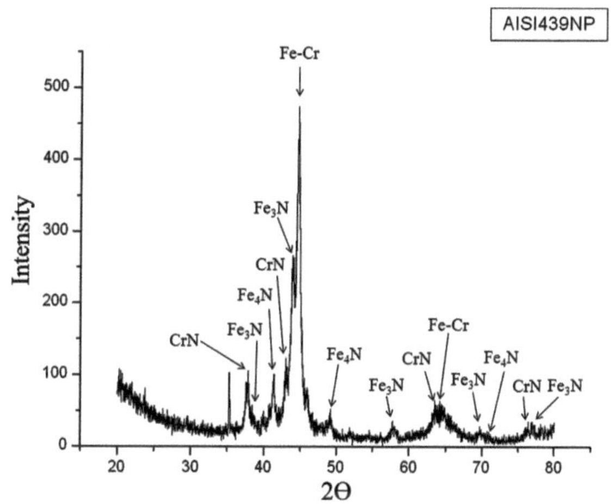

Fig. 21 X-ray diffractogram of steel AISI 439 plasma nitrided

generation of microvoids. In can be observed in this figure that the fracture showed predominance of cleaving mechanism close to the surface followed by plastic fracture characterized by tearing microvoids from the interface between the nitride layer and the substrate. These observations had already been observed by Vatavuk [10, 11] in AISI 304L. In the fractography of the nitrided layer is also observed the intergranular fracture from cracks parallel to the sample surface, as shown in Fig. 23. The cracks from the nitriding process had both intergranular and transgranular character, as can be observed by SEM in the micrograph of Fig. 24 of the nitrided sample before fracture.

According to Wulpi [12], in most of the cases, a metal fracture predominantly occurs by a failure mode, whether it is ductile or brittle however not being unique. According to the author, it can be found in the analysis of fracture, cleavage regions and regions of intergranular fracture, as well as, it is also possible to find

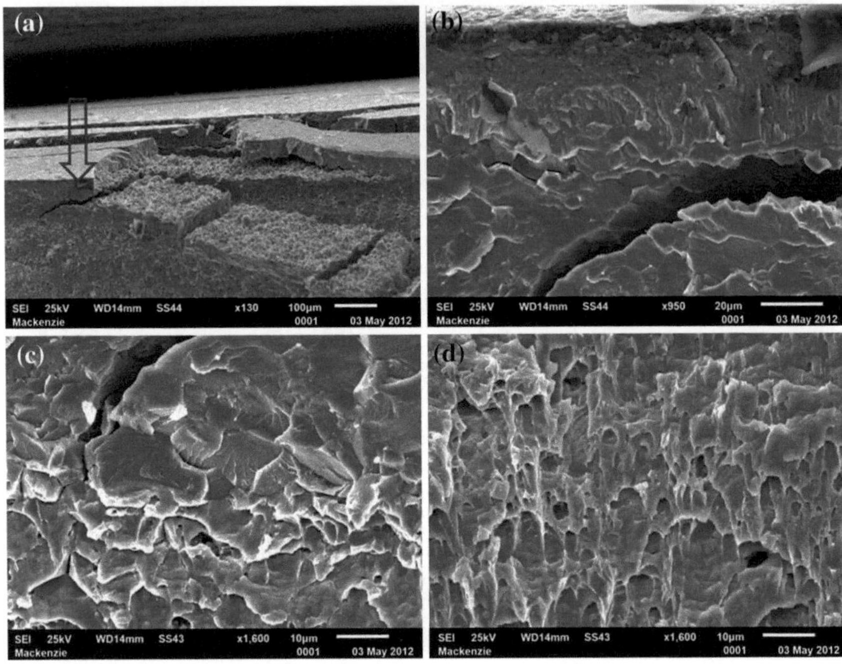

Fig. 22 Fractography of gas nitrided layer of steel AISI 430 **a** Overview. **b** Detail of the nitrided layer. **c** Detail showing the presence of transgranular fracture. **d** Region from the interface of the nitrided layer and substrate with generation of microvoids

some regions having cleavage and shear (ductile fracture) in varying proportions. This combination of different modes of fracture depends basically on the chemical composition, the stress state, the environment, possible imperfections or crystal orientation of the grains of the material.

Figure 25 shows micrograph of the fracture of steel AISI 439 with 5 % plastic deformation gas nitrided, indicating in general, the presence of cleavage (Fig. 25b). In the interface region of the nitrided layer and substrate (Fig. 25c), it is observed a change in the mechanism of intergranular fracture for the generation of microvoids (dimples) in the region of the substrate. In the AISI 5160 carbonitrided, quenched and tempered, it is noted the same behavior, predominantly intergranular fracture in the surface layer with carbon content of 0.9 %, while the core with formation of microvoids in the substrate. It is considered that with higher carbon content, that the hardness values increase with higher carbon content and it can produce higher stress than the cohesive force of the grain boundaries [12]. In Fig. 26, it is observed the fractography of AISI 439 with a 5 % degree of plastic strain nitrided by plasma. In Fig. 26c, it is observed the transgranular fracture with presence of river of cleavage that converge towards the facet and in Fig. 26d note the generation of microvoids in the region of the substrate.

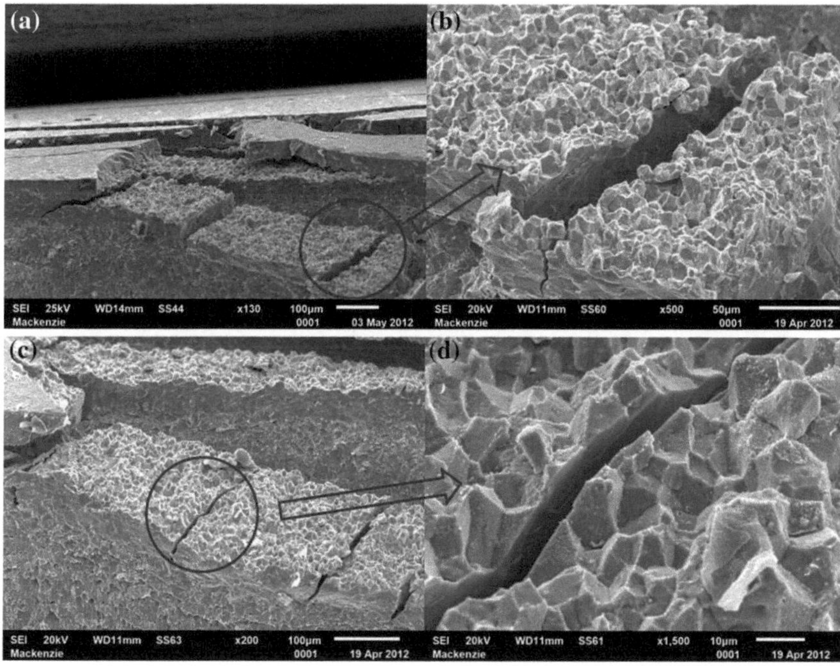

Fig. 23 Fractography of gas nitrided layer of steel AISI 430 indicating the presence of intergranular fracture

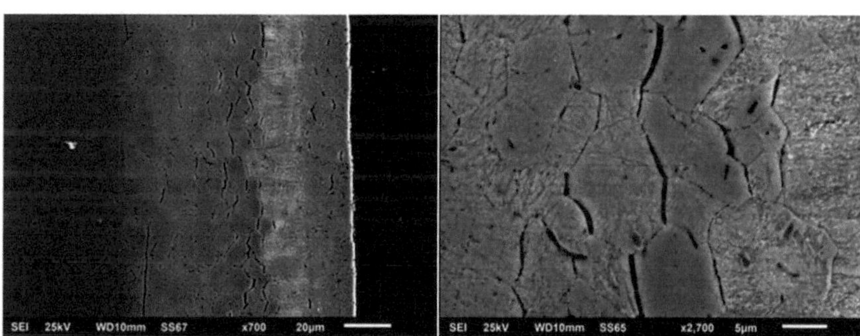

Fig. 24 Aspect of cracks before fracture of ferritic stainless steel AISI 430 gas nitrided

In this study, either the gas nitrided samples as plasma showed a predominance of transgranular fracture in the nitrided layer and generation of microvoids in the substrate. It is known that fracture by cleavage occurs along crystalline structures specifically in the body-centered cubic (CCC) and hexagonal compact (HC). In the case of micromechanisms, in general, the ductile fracture presents microvoids, since

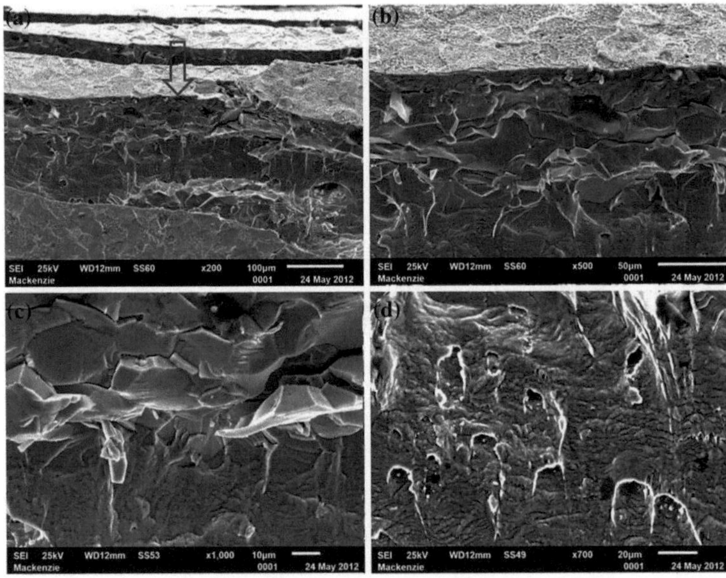

Fig. 25 Fractography of steel AISI 439 gas nitrided **a** Overview of the sample. **b** Presence of cleavage. **c** Region from the interface of the nitrided layer and substrate. **d** Formation of microvoids in the region of the substrate

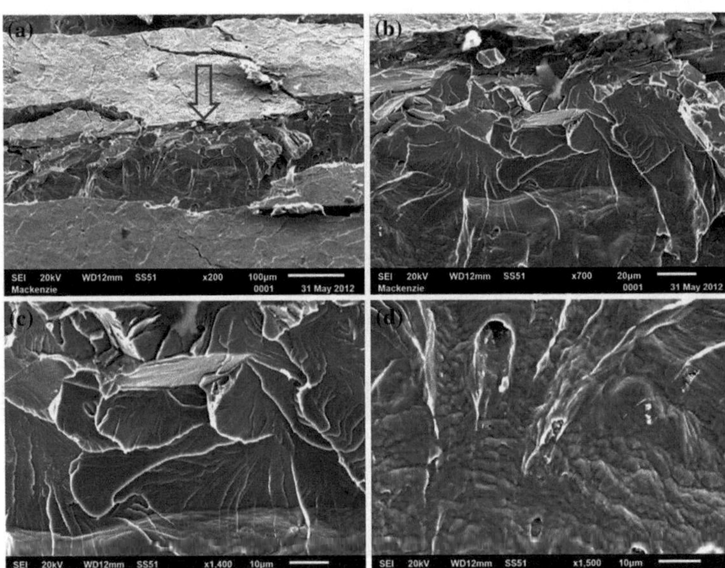

Fig. 26 Fractography of AISI 439 plasma nitride

the brittle fracture is more strongly associated with cleavage, when the matrix is ferrite, quasi-cleavage when composed by tempered martensite, bainite or intergranular character when present grain boundary embrittlement elements [7].

4 Conclusions

The study of nitriding of ferritic stainless steel AISI 430 and AISI 439 allowed to concluding that:

- The steel AISI 430 after the nitriding process presented a clearer region and another darker one in the nitrided layer. It can be noted that the clearer region has hardness slightly higher than the dark region.
- The thickness values were random and showed no influence of strain hardening on the kinetics of diffusion of nitrogen.
- The hardness values of the nitrided layer were in the range of 1,000–1,100 HV in all conditions for both steels: AISI 430 and AISI 439.
- The nitrided layer is constituted by a ferrous matrix rich in nitride (diffusion layer), with nitrogen content around 1.8 % for the AISI 430 and 2.5 % for the AISI 439.
- The nitrogen content is decreasing, especially when approaching of the substrate. During nitriding, the carbon steel is diffused to the threshold between the nitrided layer and the substrate, with accumulation in this region.
- In general, it is observed the presence of the ferritic matrix (Fe–Cr) and the formation of chromium nitride (CrN) in samples of AISI 430 and AISI 439 that suffered plasma and gas nitriding. Besides this compound, it was possible to identify the presence of Fe_4N and Fe_3N. In the samples of AISI 430 and AISI 439 nitrided in salt baths, it can be observed that the nitride layer consists of ferritic matrix (Fe–Cr) with a predominance of $(FeCr)_2N$ precipitates and without the presence of peaks of chromium nitride (CrN).
- In the fracture surface of AISI 430 and AISI 439 after nitriding, it was noted predominantly transgranular fracture in the nitrided layer and the presence of ductile fracture with formation of microvoids in the substrate.

Acknowledgments The authors acknowledge the Mackenzie Research Fund—Mackpesquisa for supporting the granting of partial scholarship for the master's student Mariana Zicari Di Monte.

References

1. Rolinski, E.: Effect of Plasma Nitriding Temperature on Surface Properties of Austenitic Stainless Steel. Surf. Eng. **3**, 35–40 (1987)
2. ASM Metals Handbook. Surface Engineering, vol. 5 (1991)

3. Knerr, C.H., Rose, T.C., Filkowski, J.H.: Gas nitriding, Heat Treating. In: ASM Handbook, vol. 4, pp. 387–409. ASM International, Columbus (1991)
4. Thelning, K.: (1975) Steel and Its Heat Treatment. McGraw-Hill, New York
5. Gavriljuk, V.G.: (1996) Nitrogen in iron and steel. ISIJ Int. **36**(7), 738–745
6. Hoffmann, F.T., Mayr, P.: Nitriding and Nitrocarburizing. Friction, Lubrification and Wear Technology. In: ASM Handbook, vol. 18, pp. 878–883, ASM International,Columbus (1992)
7. Vatavuk, J.: Aulas elaboradas para o Programa de Mestrado Profissional em Engenharia de Materiais. Universidade Presbiteriana Mackenzie. São Paulo: Editora Mackenzie (2011)
8. Vatavuk, J., Vendramin, J.C.: Efeito da adição de cobalto na resistência à corrosão de aços inoxidáveis de composição base Fe-17Cr-0,8C. Natal: Anais do CBECIMAT (2002)
9. Oliveira, A.M., Casteletti, L.C.: Nitretação e carbonitretação por plasma em aços inoxidáveis e suas influências nas resistências à corrosão e ao desgaste. Tese de Doutorado São Carlos (2005)
10. Vatavuk, J.: Análise Topográfica de Trinca de Corrosão sob Tensão em aço AISI 304L: Estudo de caso. Revista Mackenzie de Engenharia e Computação, vol. 4, pp. 151–163. São Paulo: Editora Mackenzie (2001)
11. Vatavuk, J.: Steel Failures due to Tempering and Isothermal Heat Treatment: Failure Analysis of Heat Treated Steel Components. ASM International, Columbus (2008)
12. Wulpi, D.J.: Understanding How Components Fail, 2nd edn. ASM International, Columbus (1999)

Mathematical Modeling of Afterglow Decay Curves

Yih-Ping Huang, Chi-Yang Tsai and Yung-Chieh Huang

Abstract It is popular to use multiple single exponential equations to simulate an afterglow decay curve from experimental tests. This chapter shows methods to check generated models, including an examination of the definition for associated parameters used in equations, quotes of physical properties for experimental data, and views profiles of individual components. To solve multiple exponential equations, we transform non-linear equations into linear forms and propose a scheme to search for an appropriate solution. Examples cited from various resources are provided to demonstrate the validity of the five guide lines proposed in this chapter. Several unsolved problems in simulation are also indicated and discussed.

Keywords Afterglow · Decay curve · Least square · Modeling

Y.-P. Huang
Department of Civil Engineering, Feng Chia University,
Taichung 407, Taiwan, R.O.C
e-mail: yphuang@fcu.edu.tw

C.-Y. Tsai (✉)
Graduate Institute of Dentistry, College of Oral Medicine,
Taipei Medical University, Taipei, Taiwan, ROC
e-mail: cytsai@tmu.edu.tw

C.-Y. Tsai
Department of Dentistry, Taipei Medical University Hospital,
Taipei, Taiwan, ROC

Y.-C. Huang
Department of Pediatrics, Taichung Veterans General Hospital,
Taichung 407, Taiwan, ROC
e-mail: huang1985john@pchome.com.tw

A. Öchsner and H. Altenbach (eds.), *Design and Computation of Modern Engineering Materials*, Advanced Structured Materials 54,
DOI: 10.1007/978-3-319-07383-5_21,

1 Introduction

The afterglow luminescent behavior is interpolated by the trap depths phenomenon with unfixed trap depths that are continuous with a wide distribution range [1]. However, in order to simplify the problem, several types of traps with different trap depths are present within a crystal lattice. These traps are correlated with the behavior of the curve. Decay studies carried out on strontium aluminates related phosphors by many researchers reveal that they are successfully fitted from the use of empirical multiple exponential equations [2]. The different values for decay times (τ_i) in an equation suggests that the decay process for afterglow consists of initial decay, intermediate decay, and slow decay according to Eq. (1):

$$I = I_0 + \sum_{i=1}^{n} \alpha_i \exp(\frac{-t}{\tau_i}) \tag{1}$$

where I is the intensity, t is the time, I_0 and α_i and τ_i (i = 1, 2, 3) are the constants. When only one exponential term is applied, as $n = 1$, it is a single exponential equation; and $n = 2$ for two terms as double exponential equation; and $n = 3$ for three terms as triple exponential equation. τ_i is called the decay time with this value usually treated as indicators that correlate with associate phosphor decay behaviors [1–6]. α_i is the time-invariant constant, very few researchers mentioned about it. For example, Chang [2] points out that it is a popular concept that better afterglow properties are accomplished with larger values for slow decay times (τ_2 or τ_3) in equations, i.e. it is a slower decay speed.

Among the three possible exponential equations described in Eq. (1) for $n = 1$–3, there are doubles and triples that prevail in use over single exponential equation. Huang [3] claims that although it is common to use the double exponential equation fitting, the underlying reason is still not explicit. There is another way to fit afterglow decay profiles, Sharma [4, 5] used a peeling-off procedure for numerical processes, which split the ln I (the log function of the light intensity) versus t (duration time) curves into minimum number of straight lines as well as calculates decay constants for an equation. Yuan [6] depicted a decay curve profile along with the three exponential equations simulation, with $n = 1$–3 in Eq. (1) to show that triples is most suited.

In numerical modeling of luminescence decay curve, a frequently encountered problem is the number of exponential terms for an equation that is unknown beforehand. However, it can be decided with computer programs based on some predefined criteria. The implication to answer this problem to evaluate calculated parameters in an equation and examine them is still relevant. This chapter examines the calculated parameters in an equation through the use of related definitions, physical properties, and graphical displays. In other words, this chapter attempts to find a better choice between double and triple exponential equations from a comparison of the accuracy of numerical operation, and also discusses the meaning contained for those calculated constants.

2 Guidelines

This chapter proposes two hypotheses to explain mathematical models to simulate experimental afterglow decay curve data, as:

(1) Based on a set of experimental data, acceptable multiple single exponential equations are fitted by positive exponential sums with high accuracy as well as minimum terms [7, 8].
(2) Those exponential components in an equation stand for the initial, intermediate, and long-term decay respectively; some may use another three names as fast, medium and slow components.

To apply the above hypotheses with Eq. (1), the following five guidelines are made:

(1) The optimal multiple exponential terms of a model possesses an acceptable accuracy and minimum terms.
(2) Those constants in Eq. (1) as I_0, α_i and τ_i are all positive.
(3) The relationships as $\tau_1 < \tau_2 < \tau_3$ is defined in hypothesis (2) and implies $\alpha_1 > \alpha_2 > \alpha_3$. The α_i values represent the initial intensity of these components. Figure 1 shows the generated decay curve as the upper one. The fast exponential term is the Curve 1 that dominates the initial portion of the decay curve. Curve 2 serves as the medium term and Curve 3 forms the major part of the final decay curve. The right hand side of the figure shows the tabulated parameters.
(4) Initial light intensity is formulated as in Eq. (2). For a decay curve, this value is fixed and can be estimated from experimental test data. Equation (2) should always be the same regardless of how many exponential terms are used in an equation or how to process the numerical simulation.

$$I_{initial} = I_0 + \sum_{i=1}^{n} \alpha_i \tag{2}$$

(5) The use of different terms in an equation in simulation may result in different parameters. As many researchers use the calculated τ_i to interpolate some phosphor behaviors, this affect should be taken into account.

3 Numerical Process

With powerful software and hardware available today, the computational efficiency as well as CPU time is not an issue. We coded computer programs with Fortran and are linked with the IMSL mathematical library [9], which benefited us by its versatility for linear least-squares computation requirements. Equation (3) is a matrix expression for a set of data using this mathematical formula.

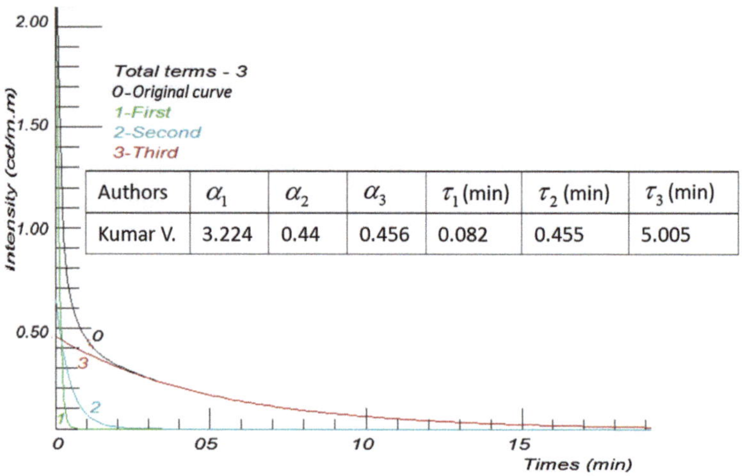

Authors	α_1	α_2	α_3	τ_1 (min)	τ_2 (min)	τ_3 (min)
Kumar V.	3.224	0.44	0.456	0.082	0.455	5.005

Fig. 1 The data is cited from Kumar [10], Curve 0 is the original curve and its three components in the equation is marked as 1, 2 and 3 in the figure

$$[A]_{n\times m}\{X\}_{m\times 1}=\{B\}_{n\times 1} \tag{3}$$

In which A is the coefficient matrix, m is the number of components in the equation, n is the data point number, and X are the parameters of the components of the equation and B is the known data. Using the IMSL mathematical library, Eq. (3) can be solved with a single line of Fortran code, as:

$$X = (A \cdot tx \cdot A) \cdot ix \cdot (A \cdot tx \cdot B) \tag{4}$$

In which X, A, and B are the matrix as in Eq. (3). $\lceil .tx. \rfloor$ is a mathematical operator that takes the transpose matrix on the left hand side and multiplies the right hand side matrix. $\lceil .ix. \rfloor$ is a mathematical operator that takes the inverse of the left hand side matrix and then multiplies the right hand side matrix.

Equation (1) is a non-linear equation and three types of parameters have to be calculated as α_i, τ_i and I_0. The process procedures used in this chapter are described as follows:

1. Predefines τ_i values, and lets $\tau_i < \tau_{i+1}$ as in Eq. (1), then the processed exponential equation becomes a linear equation.
2. Calculates α_i and I_0 by using the least square method as in Eq. (3) by adopting experimental data. Also, the associated deviation (D) is derived by using Eq. (5) for the generated curve.

$$D = \sqrt{\frac{\sum_{1}^{n} [y_i - f(x_i)]^2}{n}} \qquad (5)$$

D is for deviation and defined as the sum of the squares for the offsets of the experimental data y_i from the computed value $f(x_i)$ with Eq. (1).

3. Selects a set of optimal parameters for α_i and τ_i. The computer program sequentially fits models as in Eq. (1) to the experimental data involving one, two, and three exponential models, and terminates the process when an acceptable model is found. The choice of the correct model is based on meeting the minimum deviation and the predefined guidelines as described in the previous section.

4 Exhausted Search

A solve scheme is proposed in our chapter to execute the least squares method. This search process is to set predefined ranges for τ_i values with a constant increment and keeps $\tau_i < \tau_{i+1}$. For example, the search range for τ_1 is from 1 to 1.0 with 0.001 interval, i.e., 1,000 elements in the range. For τ_2 is τ_1 +0.1 to 3.0 with 0.001 interval; and τ_3 is τ_2 +0.1 to 10.0 with 0.002 interval. The relationship of $\tau_1 < \tau_2 < \tau_3$ is retained in this process.

This is a straightforward method that searches through a predefined range for τ_i's and records down the calculated D (deviation), I_0 and α_i. Then, one can choose the pair α_i and τ_i that satisfies with the minimum D and the required guidelines. Obviously, as one employs a wider search range or smaller increment, a dramatic response will pop up in regard to CPU processing time as well as memory size. The search ranges as well as increment are two key parameters in the program operation that may lead to different results. Experience is required to select suitable values for these parameters in a case, and it is important to deal with the evaluation of confidence intervals for the determined parameters.

5 Examples

Five well documented decay curves cited from five published articles are referred in this section. Figure 1 displays different curve profiles that are generated from the parameters recorded in Kumar [10] as tabulated in the figure including the Curve 0 which represents the original curve, and Curves 1 to 3 are the three individual components used in the equation. Curve 1 forms a major part at the initial stage, then Curve 2, and finally the curve is dominated by Curve 3.

The parameters in Fig. 1 are provided to generate data points along a curve to create parameters in models formed by single, double, and triple exponential equations in this chapter. The profiles of the original curve and the three new generated models are illustrated in Fig. 2. It shows that triple exponential equation fits well, Curve 1 is the original curve and Curve 4 is triples with only 0.00078 in D value as Eq. (5); the single exponential equation is bad as Curve 2, and the double exponential equation is barely ok for use as Curve 3. The calculated α_i and τ_i are different from the use of different terms in simulation. The slow component as τ_3 in triple case is larger than τ_2 in the double case, and the τ_1 in the single case is the smallest among those three slow components.

Curve 1 in Fig. 3 is generated from Yadav's data [11] that is described by a double exponential equation. Three simulated curves are included in the figure that are created by single, double, and triple exponential models. Again, the single exponential equation (Curve 2) seems unable to trace the original curve well. The double (Curve 3) and triple (Curve 4) exponential models work quite well. However, as one examines the triple case, the parameter α_2 is close to α_3 and also for τ_2 and τ_3 that can be interpolated as the medium component is similar to the slow one; in other words, these two components may combine as one. In other words, the double is enough in this case.

Figure 4 is an interesting case described by Paulose [12] that uses a triple exponential equation (Curve 0). It is obvious that this curve is dominated by the slow component (as Curve 3) alone all the way from the very beginning to the end. The relationship among the decay times as $\tau_1 < \tau_2 < \tau_3$ is retained; however, another relationship among the time-invariant constants $\alpha_1 > \alpha_2 > \alpha_3$ is not existing as shown in the tabulation in the figure; the three components in the data are plotted as Curve 1, 2 and 3. It indicates that the Curve 1 (fast) as well as Curve 2 (medium) have both a minor affect in the curve description. This may violet the generally accepted hypothesis that the exponential components in an equation stand for the initial, intermediate, and long-term decay respectively. We use the data to generate the associated double and triple equations as in the tabulation as named My-2 and My-3 in the figure, both of them can retain the relationship among the decay times as well as time-invariant constants as proposed in this chapter. In this case we recommend that using the triple one is more appropriated due to the accuracy reason.

Figure 5 is depicted by Wu [13] and tabulates the generated data in the lower right hand side. He mentions that the sequence of initial intensity from strong to weak is S1 > S3 > S0 > S2 > S4 and remains this sequence during the whole measured time. It implies that the afterglow duration of the samples may also conform to this sequence. He also suggests that a double exponential equation is appropriated for these cases. There are two questions existing as:

(1) The use of single term can model these problems quite well. As tabulates in the upper right hand side of that figure, those five cases can be modeled by a single term equation within 0.4 % of accuracy, and there is no significantly advantage of using double term in modeling.

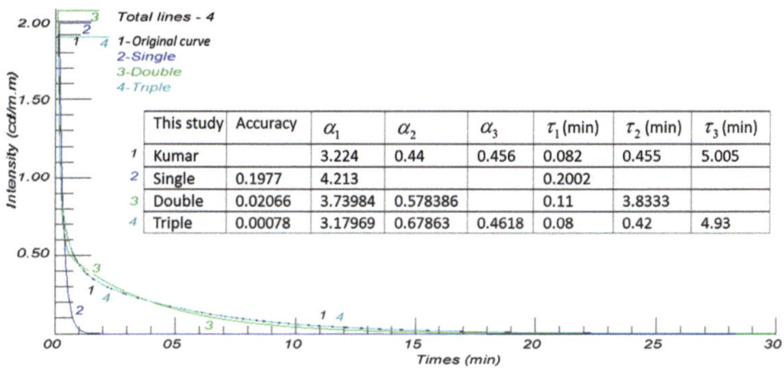

	This study	Accuracy	α_1	α_2	α_3	τ_1 (min)	τ_2 (min)	τ_3 (min)
1	Kumar		3.224	0.44	0.456	0.082	0.455	5.005
2	Single	0.1977	4.213			0.2002		
3	Double	0.02066	3.73984	0.578386		0.11	3.8333	
4	Triple	0.00078	3.17969	0.67863	0.4618	0.08	0.42	4.93

Fig. 2 The original curve by Kumar is displayed as line 1, and the three new generated curves are the triples as line 4, the doubles as line 3, and the single as line 2

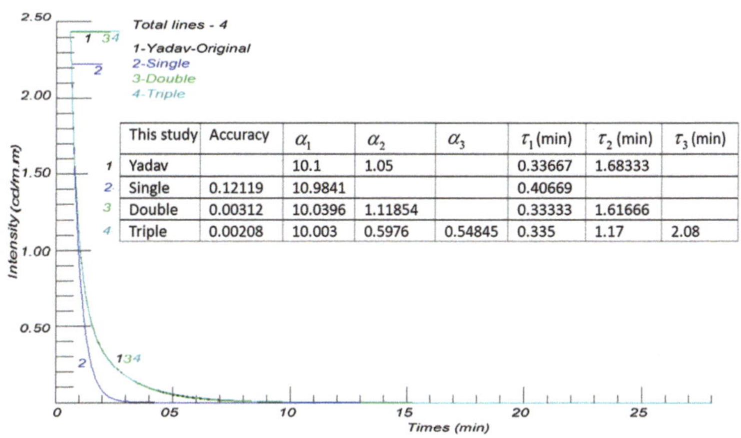

	This study	Accuracy	α_1	α_2	α_3	τ_1 (min)	τ_2 (min)	τ_3 (min)
1	Yadav		10.1	1.05		0.33667	1.68333	
2	Single	0.12119	10.9841			0.40669		
3	Double	0.00312	10.0396	1.11854		0.33333	1.61666	
4	Triple	0.00208	10.003	0.5976	0.54845	0.335	1.17	2.08

Fig. 3 Curve 1 is generated from Yadav's data [11], curves 2 to 4 represent the use of single, double and triple terms in simulation

(2) The relationship of $\alpha_1 > \alpha_2$ should be retained in order to guarantee that the fast component is in charge of the initial intensity of a decay curve.

Those calculated τ_1 values by using single term simulation are very close to the associated calculated τ_2 values by using the double term process. The relationships between the luminous behaviors and τ values claimed by Wu [13] are still valid for those τ_1 or α_1 values by using the single term equation calculated in this chapter.

Sharma [3] adopted a triple exponential equation without I_0 term to fit the experimental hyperbolic decay curve by using a peeling-off procedure, that is to find the minimum number of straight lines on the Ln I versus t curve plot. The result is shown in Fig. 6. The tabulation in the figure presents other two sets of parameters that are modeled by double and triple terms in this chapter, respectively. The accuracy for the double term case is quite good (0.365 %) and its

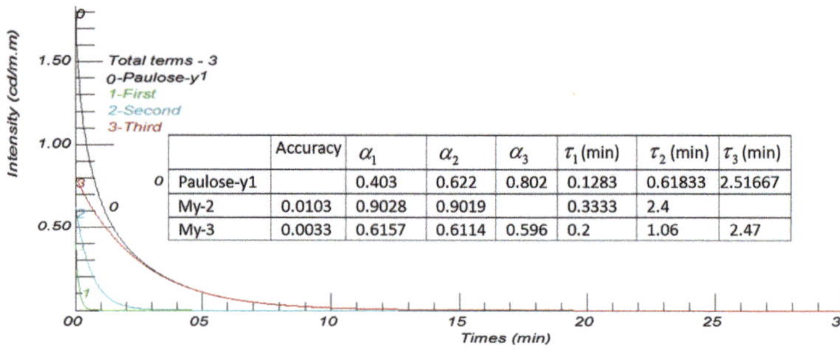

	Accuracy	α_1	α_2	α_3	τ_1 (min)	τ_2 (min)	τ_3 (min)
0 Paulose-y1		0.403	0.622	0.802	0.1283	0.61833	2.51667
My-2	0.0103	0.9028	0.9019		0.3333	2.4	
My-3	0.0033	0.6157	0.6114	0.596	0.2	1.06	2.47

Fig. 4 Curve 0 is displayed by using Paulose's data [12], its associated components are shown as Curve 1, 2 and 3. My-2 and My-3 represent the created parameters for using double and triple exponential equations respectively in this chapter

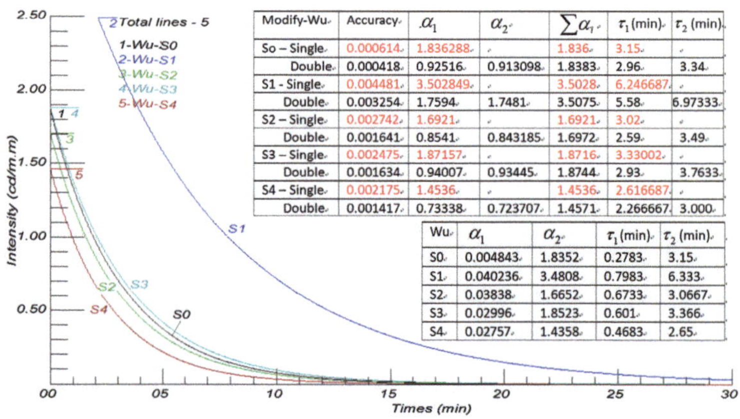

Modify-Wu	Accuracy	α_1	α_2	$\sum\alpha_i$	τ_1 (min)	τ_2 (min)
So – Single	0.000614	1.836288		1.836	3.15	
Double	0.000418	0.92516	0.913098	1.8383	2.96	3.34
S1 - Single	0.004481	3.502849		3.5028	6.246687	
Double	0.003254	1.7594	1.7481	3.5075	5.58	6.97333
S2 – Single	0.002742	1.6921		1.6921	3.02	
Double	0.001641	0.8541	0.843185	1.6972	2.59	3.49
S3 – Single	0.002475	1.87157		1.8716	3.33002	
Double	0.001634	0.94007	0.93445	1.8744	2.93	3.7633
S4 – Single	0.002175	1.4536		1.4536	2.616687	
Double	0.001417	0.73338	0.723707	1.4571	2.266667	3.000

Wu	α_1	α_2	τ_1 (min)	τ_2 (min)
S0	0.004843	1.8352	0.2783	3.15
S1	0.040236	3.4808	0.7983	6.333
S2	0.03838	1.6652	0.6733	3.0667
S3	0.02996	1.8523	0.601	3.366
S4	0.02757	1.4358	0.4683	2.65

Fig. 5 Wu [13] uses double terms to simulate S0 to S4 curves as recorded in the lower right hand side. This chapter regenerates the associated parameters by using single and double terms in an equation as tabulated in the upper right hand side

individual components are plotted as Curve 1 and Curve 2; even higher accuracy (0.0241 %) can be found by using triple term simulation. It may be confusing about how to select an appropriate terms in an equation, how to set the checking boundary is worth further chapter.

It can be found that those guidelines set in this chapter can be held for all of the above examples and discussed as following:

(1) The optimal multiple exponential terms of a model possesses appropriate accuracy and minimum terms. We suggest that the accuracy in a modeling can be set higher than 0.3 %, say 0.5 %, because the luminance meter we have in this chapter can read 1 mcd/m.m ± 1 mcd/m.m, that is the accuracy can reach 0.3 % cd/m.m. Figures 5 and 6 are not in agreement with this statement in regards to use the minimum terms in an equation.

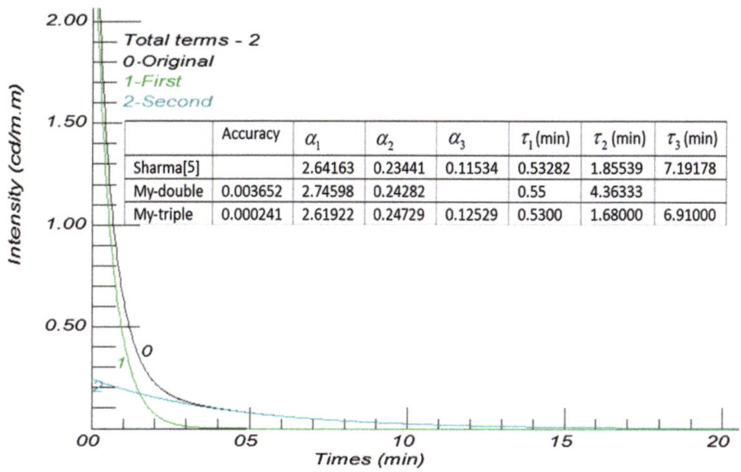

	Accuracy	α_1	α_2	α_3	τ_1 (min)	τ_2 (min)	τ_3 (min)
Sharma[5]		2.64163	0.23441	0.11534	0.53282	1.85539	7.19178
My-double	0.003652	2.74598	0.24282		0.55	4.36333	
My-triple	0.000241	2.61922	0.24729	0.12529	0.5300	1.68000	6.91000

Fig. 6 Curve 0 is obtained from Sharma [5]. My-double and My-triple are data regenerated in this chapter by using double and triple terms in an equation. Curves 1 and 2 are the associated components of My-double case that uses double terms in an equation

(2) All parameters in an equation are positive values, that is I_0, α_i and τ_i are all positives. This argument is valid for all the cases in this chapter.

(3) The relationship of $\tau_1 < \tau_2 < \tau_3$ and $\alpha_1 > \alpha_2 > \alpha_3$ is valid in an equation for a simulation. This statement is debatable; Figs. 2, 4 and 5 are not coincided with it.

(4) The initial intensity is constant as described in Eq. (3). As an examination through Figs. 2, 3 , 4, 5, and 6 shows, one can find that the initial intensity is the same for using different terms to simulate a case.

(5) For all the cases present in this chapter, it indicates that the use of the lower number of terms in an equation may turn out the smaller τ_{i} but larger α_i values in the slow component (the last term).

6 Conclusion

The measured decay curve of phosphor is often fitted by using multiple single exponential equations. The problem is how to choose suitable number of exponential terms, and to decide appropriate calculated parameters in the equation. We look into some physical properties for decay curves as well as the definition of applying a curve model. To solve the nonlinear exponential equation, we show a method to turn the equation into a linear form. A solve scheme is proposed to search for the best fit solution. Five guidelines are proposed in this chapter to help to choose appropriate equations for a model. We realized that it is a complex problem to deal with a decay curve, and some of the statements proposed in this chapter are still debatable and may not be convincible from different perspective. Thus, more solid physical mechanism is needed for further chapter.

Acknowledgements The work described in this chapter consists of part of research projects sponsored by the National Science Council, Taiwan, R.O.C. as NSC101-2221-E-035-023, and NSC102-2221-E-035-049 whose support are greatly appreciated.

References

1. Zhao, Z., Wang, Y.: The synthesis and afterglow luminescence properties of a novel red afterglow phosphor: ZrO_2: Sm^{3+}, Sn^{4+}". J. Lumin. **132**, 2842–2846 (2012)
2. Chang, C., Li, W., Huang, X., et al.: Photoluminescence and afterglow behavior of Eu^{2+}, Dy^{3+} and Eu^{3+}, Dy^{3+} in $Sr_3Al_2O_6$ matrix. J. Lumin. **130**, 347–350 (2010)
3. Huang, P., Zhang, Q., Cui, C., Li, J.: Influence of excitation wavelengths on luminescent properties of $Sr_3Al_2O_6$: Eu^{2+}, Dy^{3+} Phosphors prepared by sol-gel-combustion processing. Opt. Mater. **33**, 1252–1257 (2011)
4. Sharma, S.K., Pitale, S.S., Malik, M.M., et al.: Spectral and defect analysis of Cu-doped combustion synthesized new $SrAl_4O_7$ phosphor. J. Lumin. **130**, 240–248 (2010)
5. Sharma, S.K., Pitale, S.S., Malik, M.M., et al.: Luminescence studies on the blue-green emitting $Sr_4Al_{14}O_{25}$: Ce^{3+}phosphor synthesized through solution combustion route. J. Lumin. **129**, 140–147 (2009)
6. Yuan, S., Yang, Y., Fang, B., Chen, G.: Effects of doping ions on afterglow properties of Y_2O_2S: *Eu* phosphors. Opt. Mater. **30**, 535–538 (2007)
7. Holmstrom, K., Petersson, J.: A review of the parameter estimation problem of fitting positive exponential sums to empirical data. Appl. Math. Comput. **126**, 31–61 (2002)
8. Jukie, D., Scitovski, R.: Existence of optimal solution for exponential model by least squares. J. Comput. Appl. Math. **78**, 317–328 (1997)
9. Chapman, S.J.: Fortran 95/2003 for Scientists and Engineers, 3rd edn. McGraw-Hill, New York (2007)
10. Kumar, V., Pitale, S.S., Mishra, V., et al.: Luminescence investigations of Ce^{3+} doped CaS nanophosphors. J. Alloy. Compd. **492**, L8–L12 (2010)
11. Yadav, P.J., Joshi, C.P., Moharil, S.V.: Persistent luminescence in $Ca_8Zn(Sio_4)_4Cl_2$:Eu^{2+}". J. Lumin. **1332**, 2799–2801 (2012)
12. Paulose, P.I., Joseph, J., Warrier, M.R., et al.: Relaxation kinetics of Sm: Ce-doped CaS phosphors. J. Lumin. **127**, 583–588 (2007)
13. Wu, H., Hu, Y., Kang, F., et al.: Observation on long afterglow of Tb^{3+} in $CaWO_4$. Mater. Res. Bull. **46**, 2489–2493 (2011)

Comparative Study for Removal of the Methyl Red by Two Illites Clays

Reda Marouf, Fatima Ouadjenia, Faiza Zahaf and Jacques Schott

Abstract In the present work, we were interested to characterize two clays from Mascara deposits in western Algeria, and compare their behaviour to eliminate the Methyl Red (MR) dye from aqueous solution. XRD, BET, FTIR, were used to investigate the natural illites clays. XRD analysis identified illite as the major clay mineral. The ability of the materials to remove MR from aqueous solution at 293, 303, 313 and 323 K was investigated. The results indicated that the adsorption was affected by *pH* values. The isotherms adsorption of dye was very well described by the Langmuir model. The adsorption reaction follows the pseudo-second-order kinetic. Enthalpy ($\Delta H°$), entropy ($\Delta S°$) and Gibbs free energy ($\Delta G°$) were calculated and the results indicated that the adsorption of Methyl Red in the clays was a spontaneous and endothermic process.

Keywords Adsorption · Illite · Kinetic · Methyl red · Thermodynamic

R. Marouf (✉) · F. Ouadjenia · F. Zahaf
Laboratoire de Matériaux, Applications et Environnement, Université de Mascara, Mascara, Algérie
e-mail: reda_marouf@yahoo.fr

F. Ouadjenia
e-mail: fatouadj2@yahoo.fr

F. Zahaf
e-mail: zahaf.faiza@yahoo.fr

J. Schott
Laboratoire Géosciences Environnement Toulouse, CNRS (UMR 5563)-OMP-Université Paul-Sabatier, Toulouse, France
e-mail: jacques.schott@get-obs.mip.fr

A. Öchsner and H. Altenbach (eds.), *Design and Computation of Modern Engineering Materials*, Advanced Structured Materials 54,
DOI: 10.1007/978-3-319-07383-5_22, © Springer International Publishing Switzerland 2014

1 Introduction

The industrial activities present a source of significant pollution and contribute to the deterioration of the environment and human health. These industrial activities are such as textile, agro alimentary, pharmaceutical, and paper [1]. Due to the high toxicity of dyes in natural water, they can cause several diseases in humans, like dizziness, irritations and headaches [2].

The toxicity and mass production of dyes leads to the necessity of treatment. Various techniques including coagulation [3], precipitation [4], adsorption [5] and photocatalytic degradation [6], were used for removal dyes from wastewater. The adsorption in aqueous solutions has been found to be an efficient and economical process for removing the pollutants as the dyes. Many adsorbents have been tested such as activated carbon [7], silica [8], chitosan [9], zeolite [10] and fly ash [11]. However these adsorbents are expensive due to high costs of manufacturing, or they have a weak capacity of adsorption.

Clays have been accepted as one of the low cost adsorbents for removal of dye from wastewater. Among the mineral clays studied, illite has an important consideration as an adsorbent. The general structure of illite is formed by alternation of two tetrahedral silica sheets and an octahedral alumina sheet [12]. The negative charge of the clay is compensated by exchangeable K^+ cations. The wide usefulness of clay minerals is essentially due their high specific area, and a variety of surface and structural properties [13].

In the present chapter we report the very high adsorption potential of the natural clay soil of Algerian origin for Methyl Red. To the best Knowledge of the authors, there are no works concerning the adsorption of this dye by the clay available in literature. However to characterize the material, we use the XRD, BET, FTIR and chemical analysis techniques such as cationic exchange capacity (CEC) and point of zero charge (PZC). The effects of various adsorption conditions such as pH and temperature have been investigated. To fit the adsorption experimental data, we used Langmuir and Freundlich isotherm models. Kinetic and thermodynamic parameters ($\Delta H°$, $\Delta S°$ and $\Delta G°$) were also calculated to describe the adsorption mechanism.

2 Materials and Methods

2.1 Adsorbate

The dye Methyl Red (MR) is a colored indicator acid, C.I. Acid Red 2, with the chemical formula: $C_{15}H_{15}N_3O_2$. MR is a red crystalline powder with a molecular weight of 269.30, which was supplied by Merck Chemicals and is used without further purification. The chemical structure of methyl red is shown in Fig. 1. MR is pink in aqueous solution of pH under 4.2, yellow in pH over 6.2 and orange in between.

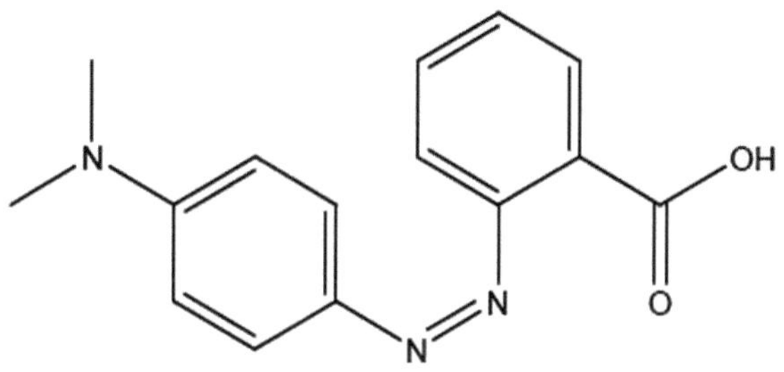

Fig. 1 Molecular structure of Methyl Red

2.2 Adsorbent

The samples of clays were obtained from deposit of Zahana (C–Z) and silt Chorfa dam (C–C) (Mascara region), in the form of fine powder of brick red color for C–Z and yellowish brown for C–C. Before using, the materials were washed with dionized water Milli-Q, dried at 95 °C for 24 h, sieved and ground to 80 μm. The raw clay compositions are presented in Table 1.

X-ray analyses were performed using INEL CPS 120 diffractometer employing cobalt Ka radiation ($\lambda = 0.178$ nm). The specific surface area and porosity data were determined by adsorption of nitrogen via an ASAP 2010 instrument. The Fourier Transformed Infrared (FTIR) spectrums are performed by a Nicolet 5700 spectrometer, using KBr pellets in the range of 4000–400 cm^{-1}. The point of zero charge of raw clay was determined by the solid addition method [14]. The values of finding $pH_{(PZC)}$ are approximately 6.4 and 7.83 for C–Z and C–C, respectively. This means that in the values of pH lower than the $pH_{(PZC)}$ the surface electric charge of the material will be positive, whereas for the values of pH over the $pH_{(PZC)}$ the electric charge becomes negative.

The determination of the cation exchange capacity (CEC) of clay was performed through a conductimetric titration method. This method is based on the saturation of the sample in $BaCl_2$ solution [15]. The CEC values are 94.22 and 96 meq/100 g for C–Z and C–C, respectively.

2.3 Adsorption Experiment

A methyl red solution was prepared in the range of initial concentrations 10–110 mg/L. For each experiment, 20 ml of dye solution was added to 0.2 g of the raw clays. The suspensions were shaken at room temperature (22 ± 2 °C) for 3 h. The pH was adjusted at 3–4 by the addition of 0.1 N NaOH or 0.1 N HNO_3 solutions. When the adsorption procedure completed, the mixture was centrifuged to get a supernatant liquid. The residual concentrations of MR were detected using UV-vis spectrophotometer (UV- mini 1240). The dye concentration retained by the adsorbent was calculated using the following relation:

Table 1 Chemical composition of raw clays (wt%)

Element	SiO_2	CaO	MgO	Al_2O_3	Fe_2O_3	Weight loss
Zahana clay	36.44	24.90	01.45	07.80	03.59	24.21
Chorfa clay	29.37	18.22	01.33	09.12	03.84	36.46

$$q_e = C_0 - C_e \cdot \frac{V}{m} \tag{1}$$

where, q_e is the equilibrium adsorption capacity, C_0 is the initial dye concentration, C_e is the equilibrium dye concentration, V is the volume of solution containing adsorbate and m is the mass of the adsorbent.

2.4 Isotherms and Models Fitting

The Langmuir sorption isotherm has been widely used to characterize the adsorption phenomena from solutions. The isotherm is valid for monolayer adsorption onto a surface containing a finite number of identical sites. The form of Langmuir isotherm can be represented by the following equation [16, 17]:

$$q_e = Q_0 \cdot \frac{K_L C_e}{1 + K_L C_e} \tag{2}$$

Equation (2) can be represented by linear form:

$$\frac{C_e}{q_e} = \frac{C_e}{Q_0} + \frac{1}{K_L Q_0} \tag{3}$$

where Q_0 is the maximum adsorption capacity, is the amount of MR at complete monolayer coverage (mg/g), and K_L (L/mg) is a constant that relates to the heat of adsorption.

The Freundlich isotherm can represent properly the sorption data at low and intermediate concentrations on heterogeneous surfaces [18]. The model is expressed as:

$$q_e = K_F C_e^{\frac{1}{n}} \tag{4}$$

Equation (4) can be expressed in linear form:

$$\ln q_e = \ln k_F + \frac{1}{n} \ln C_e \tag{5}$$

K_F and n are the Freundlich constants, indicating the capacity and intensity of adsorption, respectively.

2.5 Kinetic Studies

In order to examine the mechanism of adsorption process, we use the two simplest kinetic models, i.e. pseudo-first-order and pseudo-second-order model. The linear form of the pseudo-first order rate equation is given as [19]:

$$\ln q_e - q_t = \ln q_e - k_1 t \tag{6}$$

where q_t is the amount of adsorbate adsorbed (mg/g) at time t, k_1 is the rate constant (min^{-1}). The values of k_1 were calculated by plots ln (q_e − q_t) versus t from different concentrations.

The pseudo-second-order kinetic model is expressed as [20]:

$$\frac{t}{q_t} = \frac{1}{k_2 q_e^2} + \frac{t}{q_e} \tag{7}$$

where k_2 is the rate constant for second-order adsorption (g/mg.min). If the plot t/q_t versus t shows a linear relationship, the second-order kinetic model is applicable.

2.6 Thermodynamic Studies

To calculate thermodynamic parameters, the adsorption of MR onto clay material was carried out in 303, 313 and 323 K, and using the equation [21, 22]:

$$\ln k_d = \frac{\Delta S^\circ}{R} - \frac{\Delta H^\circ}{RT} \tag{8}$$

where ΔH°, ΔS°, and T are the enthalpy, entropy and temperature in Kelvin, respectively, and R is the gas constant. The plot of Ln K_d versus $1/T$ yields straight lines with the slope and intercept giving values of ΔH° and ΔS°.

The distribution coefficient (K_d) is calculated from the following equation [23]:

$$k_d = \frac{C_0 - C_e}{C_e} \cdot \frac{V}{m} \tag{9}$$

The Gibbs free energy, ΔG°, of a specific adsorption is represented by following equation:

$$\Delta G^\circ = \Delta H^\circ - T\Delta S^\circ \tag{10}$$

3 Results and Discussions

3.1 Characterization of Adsorbent

The result of chemical composition investigation indicates the presence of silica, lime and alumina as major constituents, as well as iron and magnesium oxides.

The XRD patterns of the natural samples are given in Fig. 2. The two materials contain clay minerals, calcite and quartz as main components, with dolomite as accessory mineral. Illite is the dominant clay mineral, the kaolinite is a minority.

The FTIR spectrums of raw clays are illustrated in Fig. 3. The most characteristic peaks of our materials are those in 1436–1032 cm^{-1}, which is corresponding to the absorbance bands of the carbonates (CO_3^{2-}) and the liaison Si–O, respectively. The bands observed at 873, 527 and 473 cm^{-1} are attributed to group Al-OH, vibration bond of MgO and Si–OM (M: Mg, Al or Fe) deformation, respectively [24, 25]. The spectrum also displays two bands at 3619 cm^{-1} and 3421 cm^{-1} which are assigned to the stretching vibration of octahedral OH groups of Al_2OH [26].

The specific surface areas were found to be 48.80 for C–Z and 65.57 m^2/g for C–C. It is noted that these values are almost the same or higher compared with the surface area of the other materials such as kaolinite [27], montmorillonite [28] and china clay [29].

3.2 Effect of pH

Effect of *pH* was studied in the *pH* range 2–7, with initial concentration of methyl red fixed at 50 mg/L and temperature of 23 °C. It should be noted that the wavelength absorption of the methyl red is not fixed, but changes according to the *pH*. The maximum wavelength obtained for each desired pH value is shown in Table 2 [30].

Figure 4 depicts the effect of pH on the adsorbed amount of MR onto the clays C–Z and C–C. The high adsorption was obtained at *pH* values of 3–4. At low pH values, the low adsorption observation was explained due to increase in positive charge (protons) density on the surface sites and thus, electrostatic repulsion occurred between the dye and the edge groups with positive charge ($Si–OH^{2+}$) on the surface. In an alkaline medium, the electrostatic repulsion becomes weak thus resulting in an increase MR adsorption.

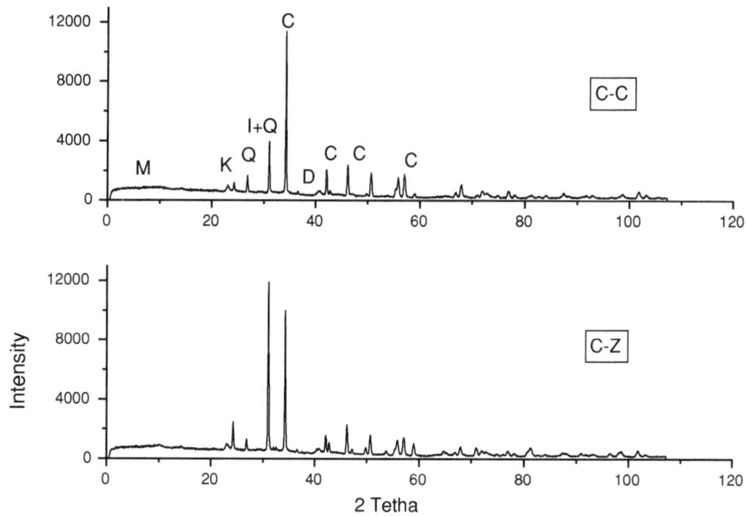

Fig. 2 X-ray diffractometers analysis for the raw clays. Illite (*I*), Quartz (*Q*), Calcite (*C*), Kaolinite (*K*), Dolomite (*D*) and Montmorillonite (*M*)

3.3 Adsorption Isotherm Models

The relation between the amount adsorbed of MR and its equilibrium concentration in aqueous solution is shown in Fig. 5. It can be seen that when the equilibrium concentration of MR increases, the amount adsorbed increases. Using the classification of Giles et al. [31], the experimental isotherms obtained are of type S. Figures 6 and 7 show that the isotherms of adsorption at various temperatures have the same shape. It is also clear that the adsorbed amount of MR has been affected by the temperature, where she (22.15 and 8.15 mg/g for C–Z and C–C, respectively) is more significant at 50 °C. We remark that the adsorbed amount in the case of clay Zahana is more significant than the clay Chorfa.

Table 3 shows that the experimental data of isotherms are suitably described by the Langmuir model, because the value of R^2 is very high (0.969 for C–Z and 0.997 for C–C) at ambient temperature for the two materials. Indeed the equation has a satisfactory representation of the adsorption isotherms of the clays with the methyl red.

The maximum adsorption capacity for acid dye calculated in this chapter is almost similar or more than reported values in the literature. For example, the adsorption of Acid Blue 62 on sepiolite follows the Freundlich isotherm model with an adsorption capacity of 20.9 mg/g [32]. The adsorption capacity of anionic Acid blue 193 and Acid Black 1 on fly ash has been shown to be 10.93 and 10.33 mg/g, respectively, by Sun et al. [33].

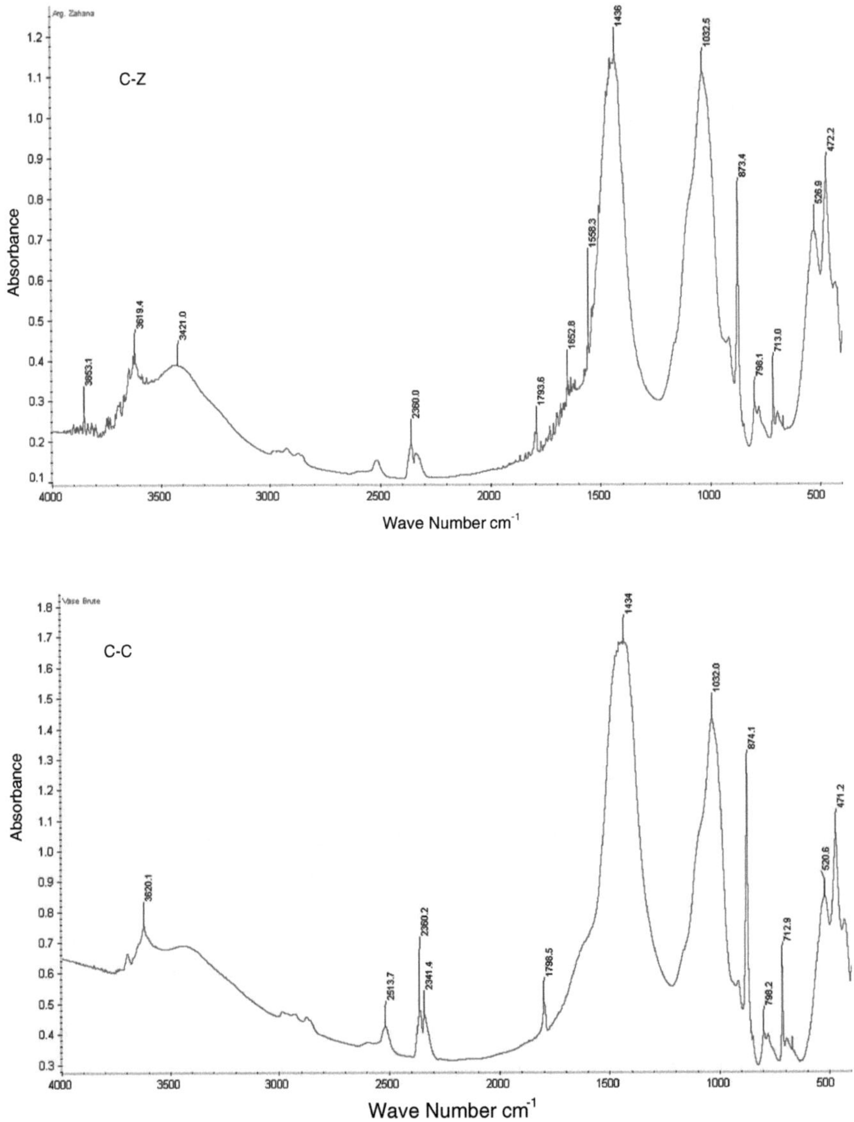

Fig. 3 FT-IR spectrums of the raw C–Z and C–C

3.4 Adsorption Kinetics

The results presented in Table 4 allow us to notice that there is a slight difference between the correlation coefficients R^2 of the kinetics of the pseudo-first-order and that of the pseudo-second-order. Except that in the case of the reaction of the second order R^2 is almost equal to unity. However, the adsorption of MR in

Table 2 Wavelength of MR solution at different pH

pH	Wavelength (nm)
2	527
3	525
4	433
5	403
6	407
7	420

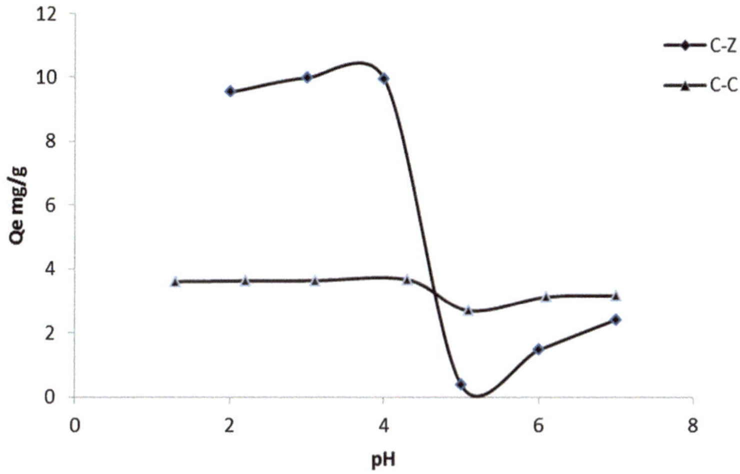

Fig. 4 Effect of pH on adsorbed amount of MR onto the clays C–Z and C–C

aqueous solution by Zahana clay and Chorfa clay follows the pseudo-second-order kinetics. Salleh et al. [34] reported that the adsorption of anionic dye onto agricultural solid wastes follows the pseudo-second-order model. Anbia and Salehi investigated the adsorption of several acid dyes such as Acid Blue 113, Acid Blue 114, Acid Green 14 and Acid Yellow 127, using mesoporous material hexamine functionalized SBA-3, and they found the results indicating the second-order nature of adsorption process [35].

3.5 *Adsorption Thermodynamics*

Thermodynamic parameters for the adsorption of MR are shown in Table 5. According to Eq. (9), $\Delta H°$ and $\Delta S°$ can be calculated from the slope and intercept of the plot of Ln K_d versus $1/T$, respectively. The values of the standard enthalpy were found in the range of 04.08–12.58 kJ/mol for Zahana clay and were all positive. But in the case of the Chorfa clay, Eq. (9) is valid only for of initials

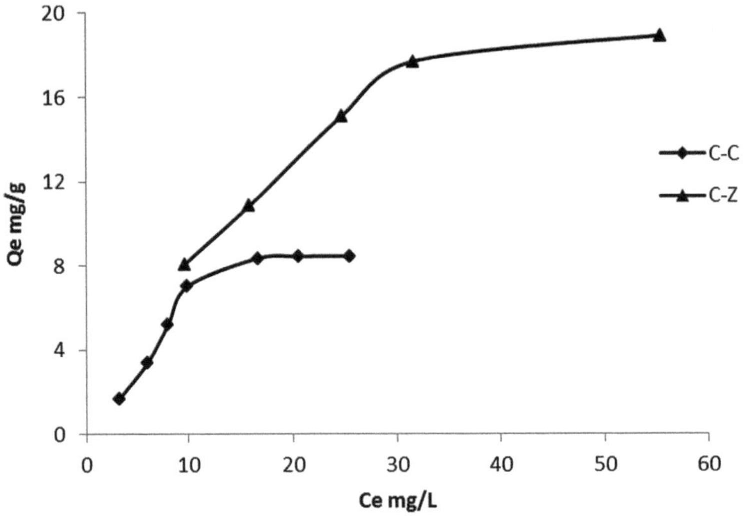

Fig. 5 Methyl Red adsorption isotherms on C–C and C–Z clays at ambient temperature

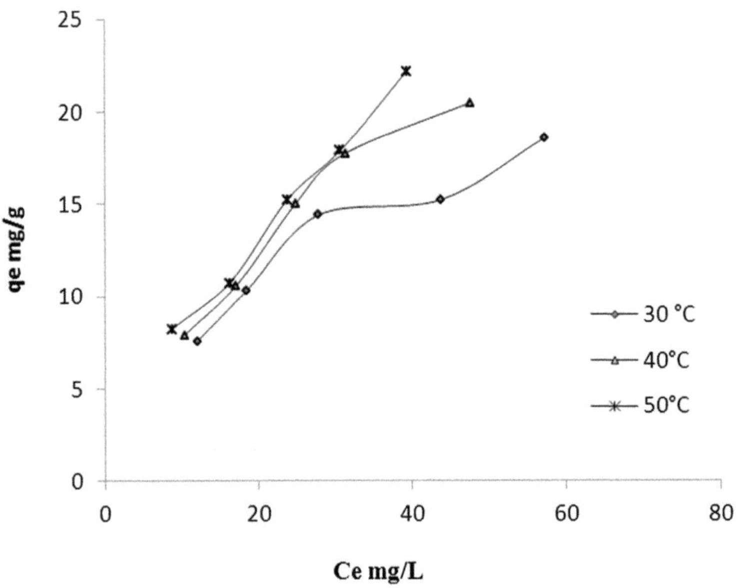

Fig. 6 Adsorption isotherms of MR on Zahana clay at temperatures 30, 40 and 50°C

concentrations 105 and 110 ppm, where the values of enthalpy were 0.873 and 1.101 kJ/mol. So we can calculate the thermodynamic parameters for C–C only at high initials concentrations. This suggests that the adsorption reaction of methyl red on the two clays is endothermic and physical in nature. The positive value of

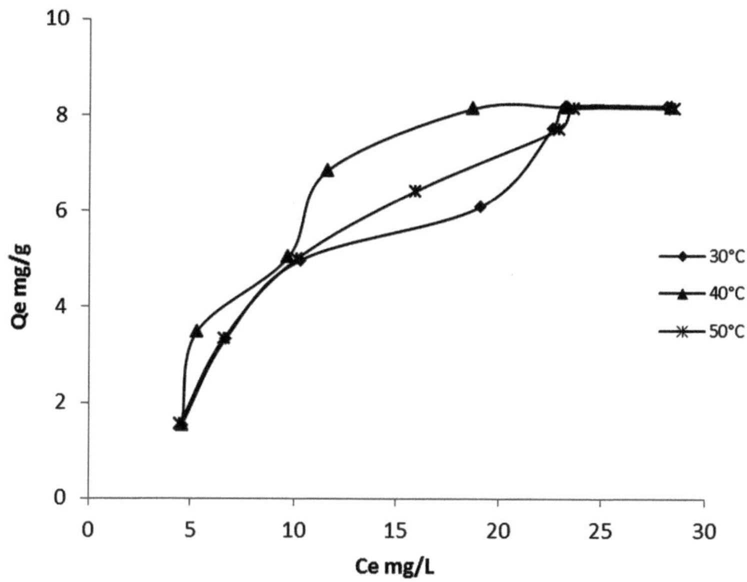

Fig. 7 Adsorption isotherms of MR on Chorfa clay at temperatures 30, 40 and 50°C

Table 3 Langmuir and Freundlich isotherm constants for MR adsorption

Sample	Langmuir				Freundlich		
	T K	K_L (L/g)	Q_0 (mg/g)	R^2	K_F	n	R^2
Zahana clay	296	0.047	27.03	0.969	0.105	1.93	0.933
Chorfa clay	296	0.111	08.62	0.997	2.985	1.31	0.989

Table 4 Kinetic models for MR adsorption

	T (K)	Pseudo-first-order		Pseudo-second-order	
		K_1 (min^{-1})	R^2	K_2 (g/mg min)	R^2
Chorfa clay	296	0.093	0.989	1.536	1
Zahana clay	296	0.037	0.952	0.124	0.999

$\Delta S°$ shows increased disorder at the solid-solution interface during the adsorption of dye. For both materials the maximal values of the free energy are almost identical. The negative values of Gibbs free energy change, $\Delta G°$, show that the adsorption process for the clay sample is spontaneous and the degree of spontaneity of the reaction increases with increasing temperature.

Table 5 Thermodynamic parameters for MR adsorption

Sample	C_0 (mg/L)	ΔH (kj/mol)	ΔS (j/mol K)	ΔG (kj/mol)			
				303 K	313 K	323 K	R^2
Zahana clay	50	12.58	46.66	−1.56	−3.07	−3.57	0.987
	70	04.08	18.10	−1.41	−1.55	−1.73	0.992
	100	04.99	20.82	−1.32	−1.58	−1.79	0.958
	120	09.63	34.94	−0.96	−1.31	−1.66	0.854
	170	09.71	34.71	−0.81	−1.15	−1.50	0.992
Chorfa clay	105	1.101	12.05	−2.55	−2.67	−2.79	0.947
	110	0.873	12.92	−3.03	−3.16	−3.28	0.994

4 Conclusion

XRD analysis shows that the clay material origin of Zahana or Chorfa deposits contains some minerals such as calcite, quartz and clays. The nature of these clays is illite, with the existence of very small quantity of the kaolinite. The present chapter shows that the clays can be used as an adsorbent for the removal of methyl red from aqueous solutions. The Langmuir adsorption isotherm was found to have the best fit to the experimental data at ambient temperature, suggesting monolayer adsorption on a homogeneous surface with identical energetic sites. The adsorption kinetics can be predicted by the pseudo-second-order model. The adsorption of MR on clay was physical in nature and endothermic. The negative values of free energy change revealed that the adsorption process is spontaneous and the degree of spontaneity increases with increasing temperature.

It is concluded that the both clays have similar physicochemical properties, only difference is the specific area, where C–C has the highest value. By cons, it is has the lowest value of adsorbed amount of RM, compared to the C–Z. So it is sure that the specific area factor is not decisive on the adsorption phenomenon.

References

1. Gupta, S., Bhattacharyya, P.K., Pamulapati, J., Mourou, G.: Optical properties of high quality InGaAs/InAlAs multiple quantum wells. J. Appl. Phys. **69**, 3219–3225 (1991)
2. Walker, G.M., Weatherley, L.R.: Adsorption of acid dyes on to granular activated carbon in fixed bed. Water Res. **31**, 2093–2101 (1997)
3. Aksu, Z.: Application of biosorption for the removal of organic pollutants: a review. Process Biochem. **40**, 997–1026 (2005)
4. Crini, G.: Non-conventional low cost adsorbents for dye removal: a review. Bioresour. Technol. **60**, 67–75 (2006)
5. Doğan, M.M., Özdemir, Y., Alkan, M.: Adsorption kinetics and mechanism of cationic methyl violet and methylene blue dyes onto sepiolite. Dyes Pigment. **75**, 701–713 (2007)
6. Sahoo, C., Gupta, A.K., Pali, A.: Photocatalytic degradation of Methyl Red dye in aqueous solutions under UV irradiation using Ag^+ doped TiO_2. Desalination **181**, 91–100 (2005)

7. Wang, S., Zhu, Z.H.: Effects of acidic treatment of activated carbons on dye adsorption. Dyes Pigment. **75**, 306–314 (2007)
8. Donia, A.M., Atia, A.A., Al-amrani, W.A., El-Nahas, A.M.: Effect of structural properties of acid dyes on their adsorption behaviour from aqueous solutions by amine modified silica. J. Hazard. Mater. **161**, 1544–1550 (2009)
9. Monvisade, P., Siriphannon, P.: Chitosan intercalated montmorillonite: preparation, characterization and cationic dye adsorption. Appl. Clay Sci. **42**, 427–431 (2009)
10. Akgül, M., Karabakan, A.: Prometed dye adsorption performance over desilicated natural zeolite. Micropor. Mesopor. Mater. **145**, 157–164 (2011)
11. Suna, D., Zhanga, X., Wub, Y., Liu, X.: Adsorption of anionic dyes from aqueous solution on fly ash. J. Hazard. Mater. **181**, 335–342 (2010)
12. Gu, X., Evans Les, J.: Modeling the adsorption of Cd(II), Cu(II), Ni(II), Pb(II), and Zn(II) onto Fithian illite. J. Colloid Interface Sci. **307**, 317–325 (2007)
13. Lin, S.H., Juang, R.S., Wang, Y.H.: Adsorption of acid dye from water onto pristine and acid-activated clay in fixed bed. J. Hazard. Mater. **113**, 197–202 (2004)
14. Srivastava, V.C., Mall, I.D., Mishra, I.M.: Characterization of mesoporous rice husk ash (RHA) and adsorption kinetics of metal ions from aqueous solution onto RHA. J. Hazard. Mater. **134**, 257–267 (2006)
15. Navia, R.: Environmental use of volcanic soil as natural adsorption material. PhD Thesis, University of Leoben, Austria (2004)
16. Langmuir, I.: The adsorption of gases on plane surfaces of glass, mica and platinum. J. Am. Chem. Soc. **40**, 1361–1403 (1918)
17. Sawalha, M.F., Peralta-Videa, J.R., Romero-Gonzalez, J., Gardea-Torresdey, J.L.: Biosorption of Cd(II), Cr(III), and Cr(VI) by saltbush (Atriplex canescens) biomass: thermodynamic and isotherm studies. J. Colloid Interface Sci. **300**, 100–104 (2006)
18. Freundlich, H.M.F.: Über die adsorption in lösungen. Z. Phys. Chem. **57**, 385–470 (1906)
19. Lagergren, S., Svenska, B.K.: Zur theorie der sogenannten adsorption geloester stoffe. Veternskapsakad Handlingar **4**, 1–39 (1898)
20. Ho, Y.S., McKay, G.: Sorption of dye from aqueous solution by peat. Chem. Eng. J. **70**, 115–124 (1998)
21. Chen, C.L., Wang, X.K.: Adsorption of Ni(II) from aqueous solution using oxidized multiwall carbon nanotubes. Ind. Eng. Chem. Res. **45**, 9144–9149 (2006)
22. Fan, Q.H., Shao, D.D., Hu, J., Wu, W.S., Wang, X.K.: Comparison of Ni^{2+} sorption to bare and ACT-graftattapulgites: effect of pH, temperature and foreign ions. Surf. Sci. **602**, 778–785 (2008)
23. Wang, X.K., Chen, C.L., Hu, W.P., Ding, A.P., Xu, D., Zhou, X.: Sorption of ^{243}Am(III) to multiwall carbon nanotubes. Environ. Sci. Technol. **39**, 2856–2860 (2005)
24. Sales, J.A.A., Airoldi, C.: Epoxide silylant agent ethylenediamine reaction product anchored on silica gel–thermodynamics of cation–nitrogen interaction at solid/liquid interface. J. Non-Cryst. Solids **330**, 142–149 (2003)
25. Lin, J., Siddiqui, J.A., Ottenbrite, R.M.: Surface modification of inorganic oxide particles with silane coupling agent and organic dyes. Polym. Adv. Technol. **12**, 285–292 (2001)
26. Post, J.L., Borer, L.: Physical properties of selected illites, beidellites and mixed-layer illite–beidellites from southwestern Idaho, and their infrared spectra. Appl. Clay Sci. **22**, 77–91 (2002)
27. Sarı, A., Tuzen, M., Citak, D., Soylak, M.: Equilibrium, kinetic and thermodynamic studies of adsorption of Pb(II) from aqueous solution onto Turkish kaolinite clay. J. Hazard. Mater. **149**, 283–291 (2007)
28. Bhattacharyya, K.G., Gupta, V.K., Sen, S.: Inffluence of acid activation on adsorption of Ni(II) and Cu(II) on kaolinite and montmorillonite. Chem. Eng. J. **136**, 1–13 (2008)
29. Sharma, Y.C.: Thermodynamics of removal of cadmium by adsorption on an indigenous clay. Chem. Eng. J. **145**, 64–68 (2008)
30. Rosemal, M., Mas Haris, H., Sathasivam, K.: The removal of Methyl Red from aqueous solutions using banana pseudostem fibers. Am. J. Appl. Sci. **6**, 1690–1700 (2009)

31. Giles, C.H., MacEwan, T.H., Nakhwa, S.N., Smith, D.: Studies in adsorption. Part XI. A system of classification of solution adsorption isotherms, and its use in diagnosis of adsorption mechanisms and in measurement of specific surface areas of solids. J. Chem. Soc. **786**, 3973–3993 (1960)
32. Alkan, M., Çelikçapa, S., Demirbaş, Ö., Doğan, M.: Removal of reactive blue 221 and acid blue 62 anionic dyes from aqueous solutions by sepiolite. Dyes Pigment. **65**, 251–259 (2005)
33. Sun, D., Zhang, X., Wu, Y., Liu, X.: Adsorption of anionic dyes from aqueous solution on fly ash. J. Hazard. Mater. **181**, 335–342 (2010)
34. Salleh, M.A.M., Mahmoud, D.K., Abdul Karim, W.A.W., Idris, A.: Cationic and anionic dye adsorption by agricultural solid wastes: a comprehensive review. Desalination **280**, 1–13 (2011)
35. Anbia, M., Salehi, S.: Removal of acid dyes from aqueous media by adsorption onto amino-functionalized nanoporous silica SBA-3. Dyes Pigment. **94**, 1–9 (2012)

Synthesis and Characterization of Nano Ti-50%Al by Mechanical Alloying

Jinan B. Al-Dabbagh, Rozman Mohd Tahar, Mahadzir Ishak and Siti Aisyah Harun

Abstract In this chapter, powder metallurgy process of TiAl nano alloys were performed via mechanical alloying (MA) of Ti-50%Al powder using planetary ball milling equipment. The characteristics of the powder samples including the compositions and microstructure changes were investigated by using X-ray diffraction and field emission scanning electron microscopy (FESEM) coupled with energy-dispersive X-ray spectroscopy (EDX). Estimation by using Scherrer equation indicated that MA performed at different duration up to 15 h have successfully refined the crystallite size from 89.51 nm of Al to 28.29, and 67.6 nm of Ti down to minimum of 17.17 nm. Longer MA duration also exhibits a better effect on the thermal behaviour of Ti-50%Al powders and micro-hardness value which is gradually increased along with MA duration.

Keywords Tial alloys · Mechanical alloying · Nanostructured materials · Thermal properties

J. B. Al-Dabbagh (✉) · R. M. Tahar · S. A. Harun
Faculty of Industrial Sciences and Technology, University Malaysia Pahang, 23600 Gambang, Kuantan, Malaysia
e-mail: jdabbagh@ump.edu.my

R. M. Tahar
e-mail: iroz80@yahoo.com

S. A. Harun
e-mail: aishahh@ump.edu.my

M. Ishak
Faculty of Mechanical Engineering, University Malaysia Pahang, 26600 Pekan, Malaysia
e-mail: mahadzir@ump.edu.my

A. Öchsner and H. Altenbach (eds.), *Design and Computation of Modern Engineering Materials*, Advanced Structured Materials 54,
DOI: 10.1007/978-3-319-07383-5_23,

1 Introduction

Over the past decades, inter-metallic titanium aluminides, particularly γ-TiAl based alloys, have gained a great deal of attention in numerous structural, non-structural and functional applications in different engineering fields [1, 2]. Equipped with low density, high specific strength to weight ratio, good oxidation and corrosion resistance at elevated temperature [3, 4], γ-TiAl based alloys are considered as a very promising material for potential replacement of Nickel based super-alloys and conventional titanium alloys in high-temperature structure applications especially in aerospace, automotive and power turbine market [5, 6]. However, the development and the actual utilization of γ-TiAl based for structural applications are plagued by poor ductility and fracture toughness at room temperature [7, 8]. The ductility of the γ-based TiAl alloy is very sensitive to the microstructures. For example, the duplex (α2/γ) structure is significantly more ductile at room temperature, compared to lamellar or single equaxed γ structure [9].

New age strategies in the development of a novel class of materials are focused to control the microstructure to achieve a set of desired properties [10]. Advancement in micro alloying, composition modification, grain refinement to nano-meter size and refining near-gamma grains and the lamellar colonies through heat treatment is a viable method to improve ductility as well as mechanical strength. One of the most promising methods to synthesize materials which can produce ultrafine, homogenous and manipulable microstructures is the mechanical alloying [11, 12]. Hence, the production of an ultrafine and homogenous powder is predicted to overcome the scattering in mechanical properties due to the segregation in the composition of TiAl alloys manufactured by conventional casting routes [13].

The purpose of this work is to synthesize and evaluate the formation of Ti–Al nano alloys compounds during MA process of elemental Ti and Al powders. The effect of subsequent heat treatment to powder processed up to 15 h was also studied. In addition, the thermal and mechanical properties of the produced alloys were investigated.

2 Experimental

The MA processes were carried out using a Retsch PM 100 planetary ball mill for duration varying from 5 to 15 h. Elemental powders of Ti (99.5 %)-100 mesh and Al (99.97 %) were mixed together to form a composition of Ti-50Al50 (at.%). For each experiment, 5 g of the powder mixture were poured into a tungsten carbide (WC) jar (250 ml). Tungsten carbide balls (ø10 mm) were used as a milling media with the ball-to-powder weight ratio of approximately up to 20:1. Small amount of Hexane was added as process control agent (PCA) to prevent excessive

Table 1 MA parameters and conditions for Ti-50%Al powders

Parameters	Conditions
Milling type	Planetary ball mill (Retsch PM 100)
Milling jar	Tungsten Carbide, WC (250 ml)
Grinding balls	Tungsten Carbide, WC (ø10 mm)
Starting powder	Ti, 100 mesh (99.5 % purity), Al (99.97 % purity)
Rotation speed	Up to 300 rpm
Milling duration	Up to 15 h
Ball-to-powder mass ratio	Up to 20:1
Process control agent	Hexane
Environment	Ar (99.9 % purity)

Table 2 MA group for Ti-50%Al powders

Group	Rotation speed (rpm)	PCA's	Ball to powder mass ratio
A	200	–	10:1
B	300	Hexane (50 wt%)	10:1
C	400	Hexane (50 wt%)	10:1
D	300	Hexane (50 wt%)	20:1
E	300	Hexane (25 wt%)	20:1

agglomeration of the powders to the milling tools. The jar then was air tight sealed and back-filled with pure Argon (99.9 %) where the pressure in the jar was kept at 0.1 MPa. The rotation speed was set at 300 rpm with interval time at every 5 min. The milling was interrupted at selected 5, 8, 10, 12 and 15 h and a small amount of powder was removed for characterizations.

The MA process parameters and conditions are as shown in Table 1 below.

The Ti-50%Al powders samples were mechanically alloyed under different parameters and conditions as indicated in Table 2.

The surface morphology and microstructure of the processed powder were characterized by using a Zeiss Evo 50 scanning electron microscope (SEM) at an accelerating voltage of 10 kV. X-ray measurements were applied to the samples to identify the powder component, phase transformation and structural changes of their crystal structure with a Rigaku Miniflex X-ray diffractometer, using Cu Kα radiation ($\lambda = 1.54062$ Å). Step-scanning has been carried out from 20 to 80θ with a counting time of 5 s every 0.02θ. The crystallite size of the milled powders was determined from X-ray line broadening by using the Scherrer equation;

$$D = 0.9\ \lambda/\beta\ cos\ \theta,$$

where, D is the mean crystallite size, λ is the CuK_α wave length of X-ray, θ is the diffraction angle and β is the full width at half maximum (FWHM) of the XRD peaks.

To study the thermal properties of the Ti-50%Al powders as well as the transformation of the powder products during heating, the samples were heated in dynamic vacuum atmosphere up to 850 °C at a rate of 10 °C/min in a Linseis P75 Platinum Series dilatometer. The results obtained including the value of differential thermal analysis (DTA) and coefficient of thermal expansion (CTE), using Al_2O_3 as reference material. The powder was then analyzed with XRD again to investigate the powder transformation after heating.

Hardness tests were performed using a Matsuzawa MMT-X7 micro hardness test machine according to ASTM E 92-82 standard Method under a load of 0.02 kg (VH20) for 10 s. Prior indentation, Ti-50%Al powders were formed into ø12 mm pellets by pressing the material in a special mould at 100 Psi for 10 s, hot mounted in a polyester resin, cured for 8 days, polished and dried following powder metallurgical processing method.

3 Result and Discussion

3.1 Morphological Evolution of Ti-50%Al Powder

The morphology of the MA-ed powder at different milling duration was investigated by SEM. For the initial Ti-50%Al powder mixture (0 h), Ti particles appear in irregular shapes and various sizes as the initial Ti powder was in 100 mesh size, but Al particles are mainly in much smaller size with irregular shape. For the 2 h initial stage of MA with the absences of PCA (A samples), it appeared that the powder particles underwent a repeated cold welding, fracturing and re-welding resulting in the formation of rounded shaped beads and deformed particles. Severe agglomeration of the powder to the balls and milling jar were observed due to ductility of Al. After 4 h of milling, these beads and particles then evolved and disintegrate to dull and more uniform flake shaped particles. On further milling to 6 h, the MA process has increased the number of particles fracturing and refining, these flaky particles disintegrate more as uniform smaller size flakes and particles. But the major drawback of dry milling was severe sticking of the powder to the balls and milling jar during the process due to ductility of Al which were hard to be removed.

On the other hand, with an addition of Hexane as a process control agent in B, C, D & E group, the sticking of the powder to milling tool appeared to be very minimum and the formation of beads were not observed. The use of Hexane has proven to be an effective means to minimize agglomeration and to optimize the milling yield as the amount of agglomerated powder to the balls and milling jar was almost negligible. Only small size flaky particles were observed in the initial stages 2–4 h. Further milling up to 8 h, leads by increased deformation and work hardening, these flakes then turn into a finer and relatively smaller structure, and by 15 h, equiaxed particles with a homogeneous structure were obtained. The FESEM images of B samples after 8 h of milling are shown in Fig. 1.

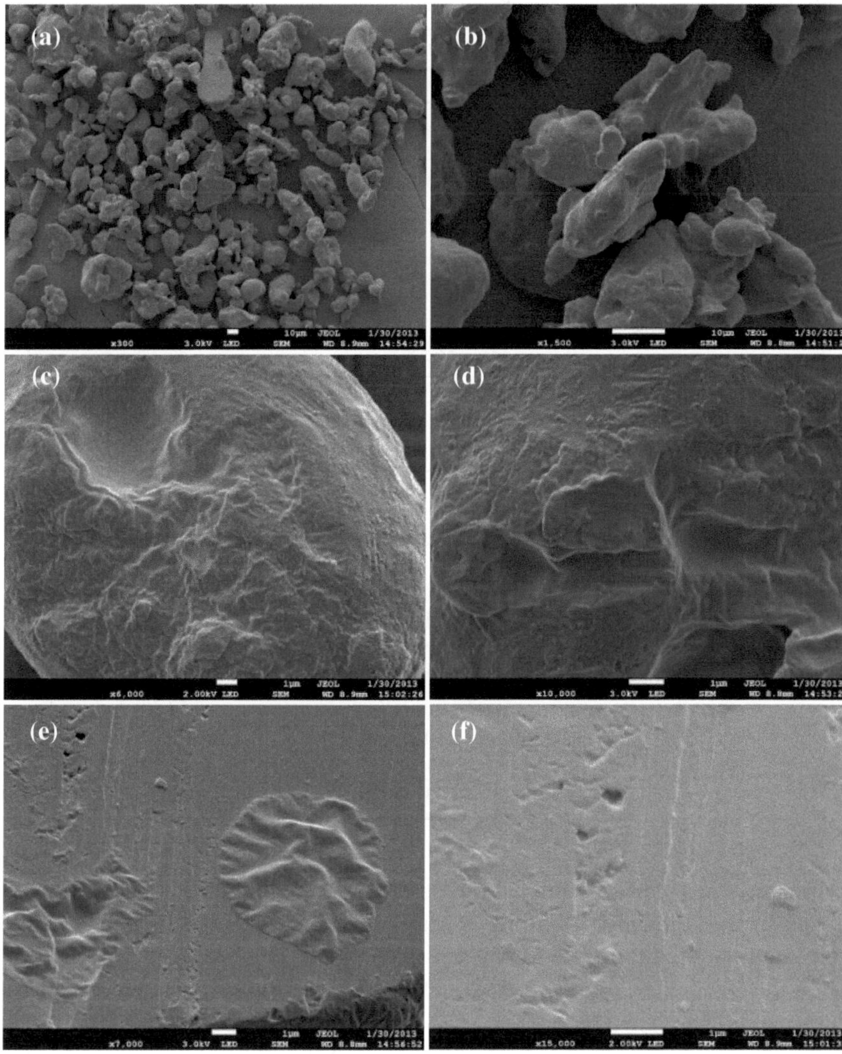

Fig. 1 FESEM images of B samples after 8 h of milling. **a** and **b** powder particles, **c** and **d** particles morphology, **e** and **f** particle surface

3.2 Crystallite Size Evolution of Ti-50%Al Powder

The investigation of the Ti-50%Al powders transformations for different milling duration up to 15 h was followed by XRD. The XRD patterns of all sample groups exhibit the reflections of a well defined Ti and Al spectrums. In general, the Ti and Al peaks are weakened and broadened with increasing milling time to form a solid solution. The broadening of Ti and Al peaks, suggests an increase of strain in the

Table 3 Crystallite size evolution of Ti at various milling duration

Milling duration (h)	Crystallite size (nm)				
	A	B	C	D	E
0			67.6		
2	22.55				
4	17.17	59.12	59.89		
5					48.41
6	18.56	57.20	49.85		
8		58.70		68.36	
10		65.61		44.91	69.3
12		58.13		57.97	
15		56.58		44.95	42.48

Table 4 Crystallite size evolution of Al at various milling duration

Milling duration (h)	Crystallite size (nm)				
	A	B	C	D	E
0			89.51		
2	40.33				
4	38.85	60.35	69.27		
5					53.39
6	28.29	62.24	55.61		
8		60.91		57.30	
10		57.10		48.92	62.68
12		62.31		48.74	
15		56.03		48.25	50.41

internal crystallite or a decrease in the effective crystallite size, or both. This is proven by the estimation from the Scherrer equation made by the data obtained from the Full Width at Half Maximum (FWHM) of the XRD pattern. The evolution in crystallite size of Ti and Al for respective samples group are calculated and listed in Tables 3 and 4.

Amongst of all the sample, A group which was dry milled (200 rpm, 10:1 bpr) exhibits a sharp intensity reduction and peak broadening (Fig. 2) were observed due to the great decrease in crystallite size from 67.6 nm of Ti and 89.51 nm of Al in the initial Ti-50%Al powder mixture to 18.76 and 28.29 nm respectively. Even though the result of this group shows a progressive decrease in crystallite size and disappearance of Ti and Al spectrums, it does not exhibit any new peaks neither inter-metallic nor of any other compound. This result was in a good agreement with the EDX analysis which identified some un-reacted particles as shown in Fig. 3.

In contrast, even after 15 h of milling, the XRD results only show a gradual intensity decrease and peak broadening (Figs. 4, 5, 6, 7). The crystallite size refinements were also slower and less progressive compared to A the group. For

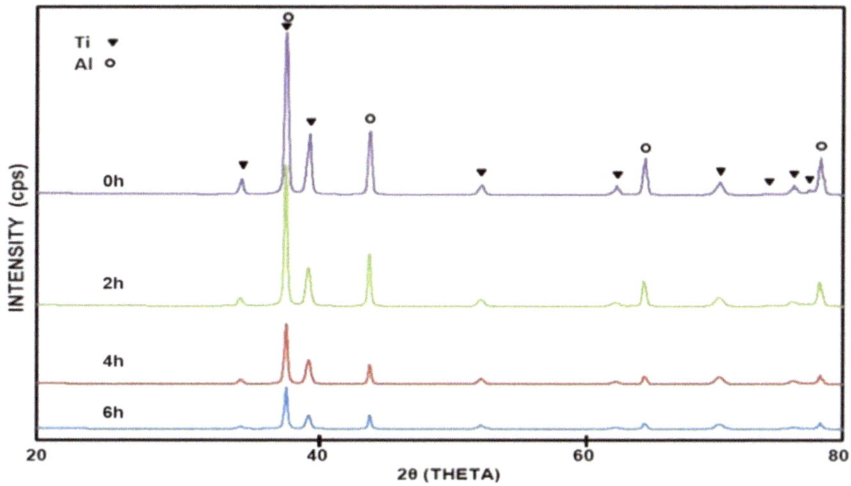

Fig. 2 X-ray diffraction spectrum of A group (Dry, 200 rpm, 10:1) at various milling duration

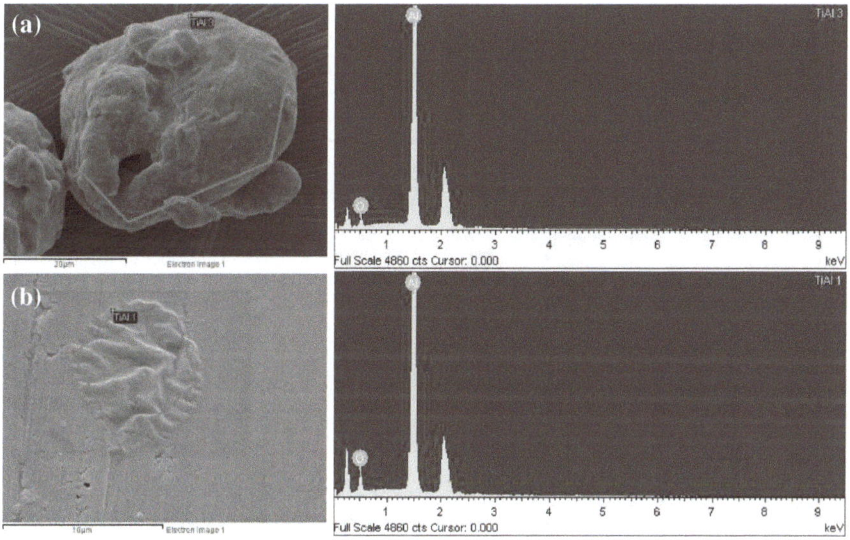

Fig. 3 FESEM images and EDX spectrums of unreact powder particles of B samples after 8 h of milling. **a** Al particles and **b** Al patch on Ti particle

instance, the B group (300 rpm, 10:1 bpr, 50 wt% Hexane) exhibits a crystallite size decreased of Ti to only 57.20 nm, and Al to only 62.24 nm after 6 h of milling. Even after prolonged milling of up to 15 h, the size only decreased to 56.58 nm of Ti and 56.03 nm of Al, respectively. Whereas for C samples which were mechanically alloyed under higher rotation speed (400 rpm, 10:1 bpr, 50

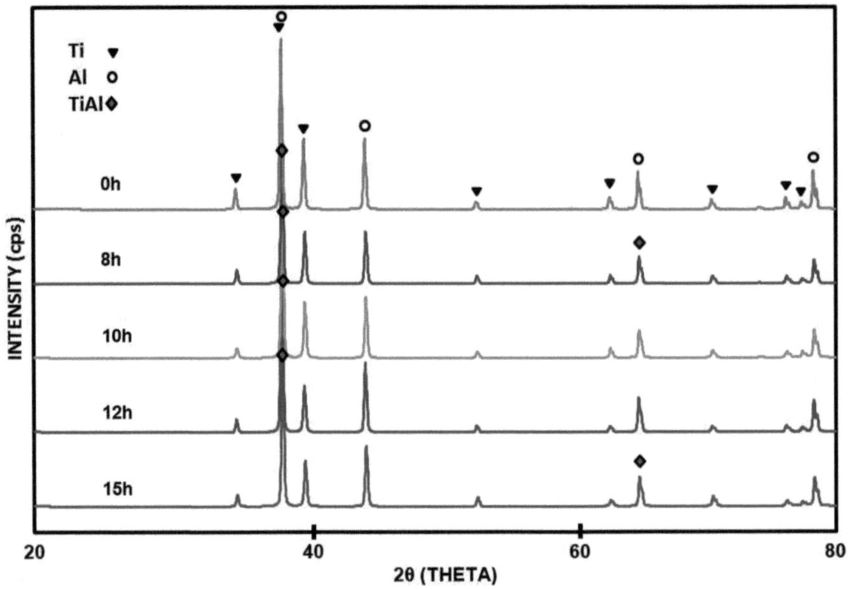

Fig. 4 X-ray diffraction spectrum of B group (Wet 50 wt% Hexane, 300 rpm, 10:1) at various milling duration

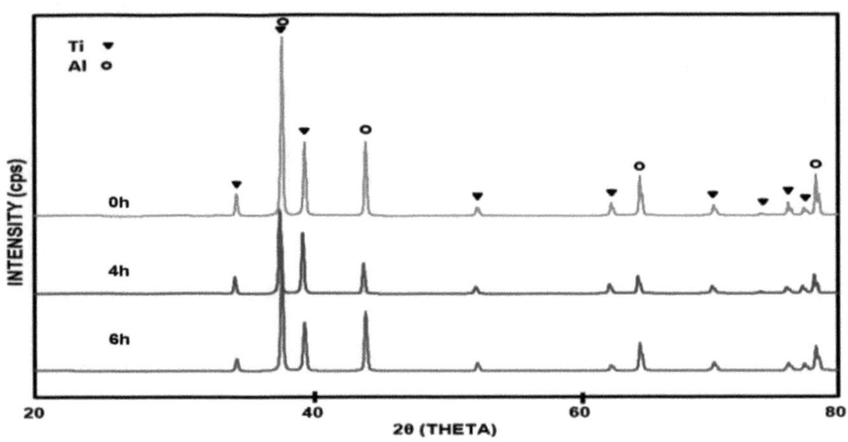

Fig. 5 X-ray diffraction spectrum of C group (Wet 50 wt% Hexane, 400 rpm, 10:1) at various milling duration

wt% Hexane), the crystallite size reduction was slightly better as it was decreased to 49.85 nm for Ti, and 55.61 nm for Al after 6 h. This occurred by increasing the rotation speed, the likelihood of collision between the ball to the powder and the milling jar has also increased.

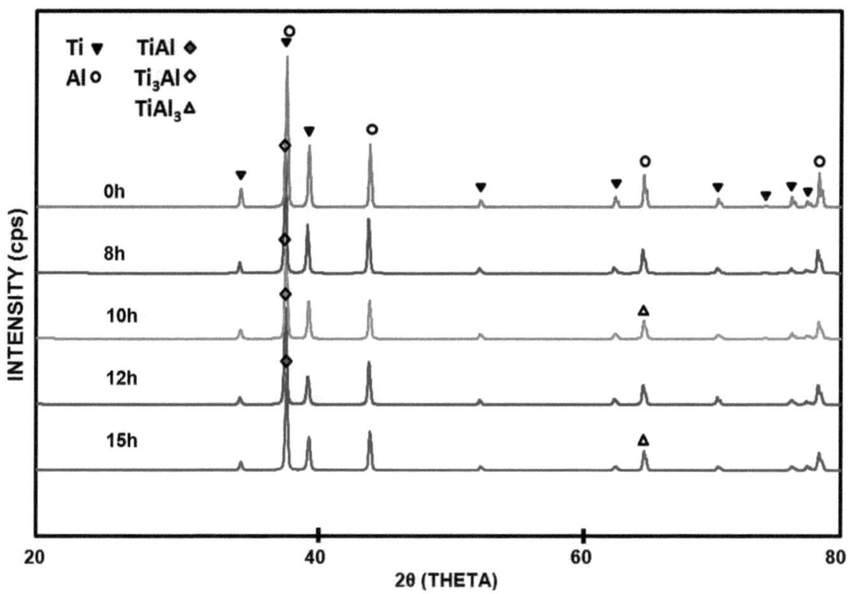

Fig. 6 X-ray diffraction spectrum of D group (Wet 50 wt% Hexane, 300 rpm, 20:1) at various milling duration

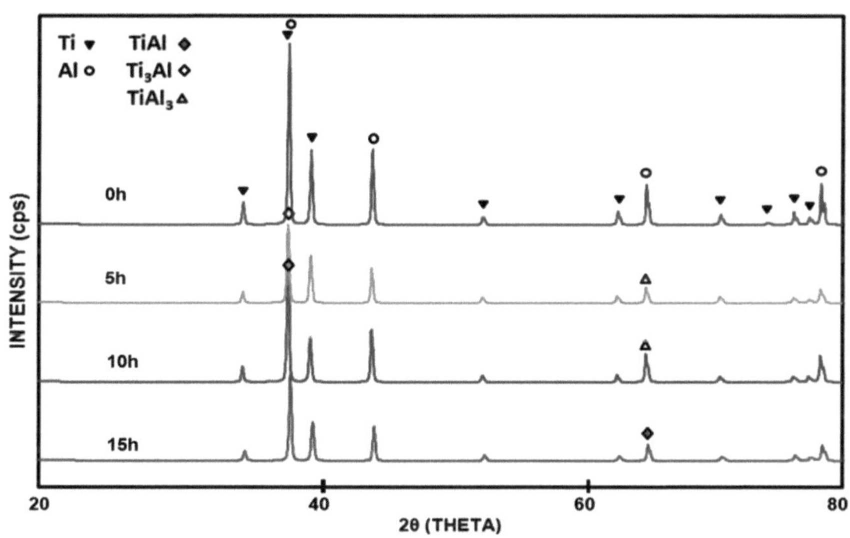

Fig. 7 X-ray diffraction spectrum of A group (Wet 25 wt% Hexane, 300 rpm, 20:1) at various milling duration

The use of 20:1 ball to powder ratio in D group (300 rpm, 20:1 bpr, 50 wt% Hexane), somehow exhibits a better refinement of crystallite size compared to B group, as after 15 h of milling, it was decreased to 44.95 nm for Ti and 48.25 nm of Al respectively. This result has also suggested that increase of the ball to powder weight ratio has accelerated the MA process as the surface area for ball to powder impact has doubled. In the case of E group (300 rpm, 20:1 bpr, 25 wt% Hexane) which was mechanically alloyed by using less amount of PCA for 15 h, more progressive refinement was observed as the Ti crystallite size was decreased to 42.28 nm whilst Al was reduced to 50.41 nm. The result obtained shows that the amount of Hexane used during milling has also played an important role in grain refinement. In addition, all above results has also confirmed that the use of Hexane has delayed the alloying process as partial of the kinetic energy during milling were absorbed by PCA instead of the powder particles.

The nature of crystallite size reduction in MA is by the continuous cold welding, fracturing, re-welding and re-fracturing, resulting in the breaking of the powder particles and therefore the grain boundaries of the materials. In general, fracturing of ductile materials was not easy and it was difficult to obtain a very small crystallite size. As predicted, the Ti with less ductility (hcp structure) has a smaller size than that of very ductile Al (fcc structure) in the final powder product. These phenomena may occur due to crystallite size of the initial powder feed with 89.51 nm of Al, and 67.6 nm of Ti. But the decreased rate of Ti crystallite size was much slower than Al, which is in good agreement with the ductile-brittle material behaviour.

The observation of variation in crystallite size of Ti and Al in Ti-50%Al powder samples shows that the structures of the powder were strongly dependent on the mechanical alloying process conditions and parameters. It can be deduced from the result obtained that dry milling without an addition of PCA's is proven to be the most effective means in reducing the Ti-50%Al crystallite size, despite the fact that, milling with higher rotation speed, higher ball to powder ratio and addition of less PCA resulted in better crystallite refinement.

3.3 Thermal Stability of Ti-50%Al Powder

Thermal analysis was performed by heating the powder sample in an alumina adapter in dynamic vacuum atmosphere up to 850 °C at a heating rate of 10 °C/min. As observed from the DTA spectrums of selected B samples, gas release occurs in the temperature range of 70–315 °C. All spectrums also show that the DTA peak appeared between 665–740 °C. The minimum onset temperature for this sample was at 681.2 °C for 8 h MA-ed powder which corresponds to the melting of the Al phase, and maximum temperature of 689.50 °C for 10 h MA-end powder (Fig. 8). The maximum point of exothermic peak was observed to appear within the temperature range of 697.9–702.8 °C and between 56.3-65.743 μV. This is associated with the transformation of the meta-stable solid solution into the equilibrium TiAl

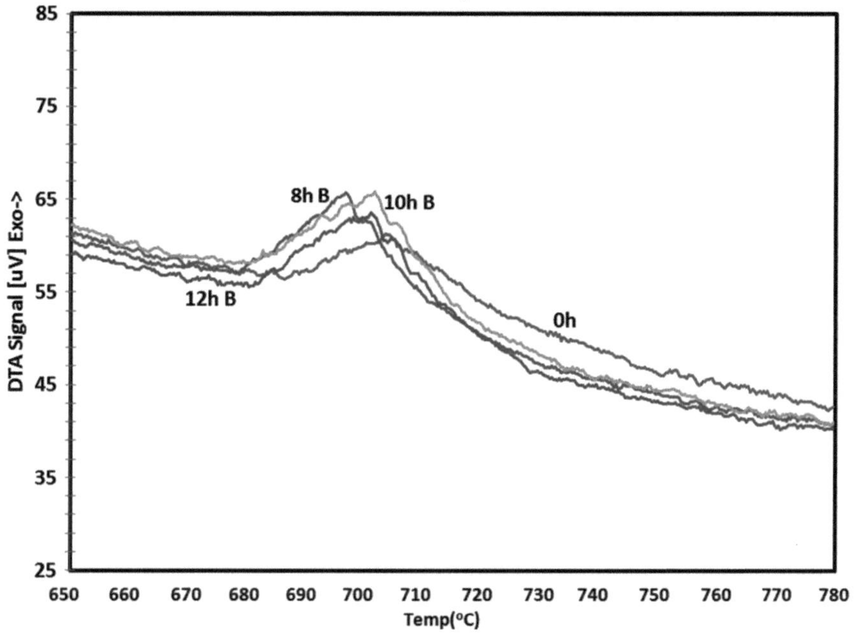

Fig. 8 DTA thermogram for Ti-50%Al powders B group at various milling duration up to 15 h

Table 5 DTA peak measurement data for selected Ti-50%Al powder at various milling duration

Samples	Milling duration (h)	On set Temp (°C)	Off set	Point of reaction	Heat changes μVS
Initial	0	675.80	697.10	678.50	1369.22
B	8	681.20	704.40	685.70	1983.40
B	10	689.50	708.80	696.80	2142.01
B	12	687.00	711.70	691.60	2305.43
D	12	709.50	739.90	719.80	2147.09
D	15	708.70	750.20	717.40	2437.96
E	10	698.4	739.1	701.7	2125.285
E	15	696.3	738.3	702.1	2463.848

phase. Accordingly, the maximum, onset, offset and reaction point of all samples increased gradually by milling duration over initial powder value as seen in Table 5.

The increased mechano-chemical reaction temperature reflects a progressive intermixing or formation of nano-crystallite Ti(Al) solid solution as well as crystallite size refinement. As only single TiAl phase appears to form in these sample group, it is suggested that the reduction in crystallite size was the main reason in the shifting of DTA peak to a higher temperature. On the other hand, it is revealed that progressive intermixing and multiphase nano-crystallite Ti(Al) solid solution formation in D and E samples, has resulted in the shifting of exothermic

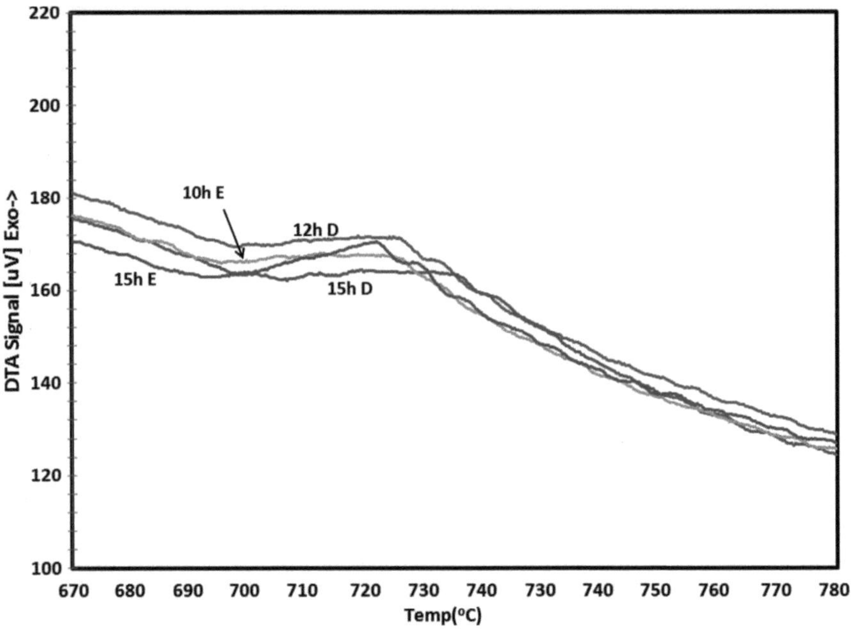

Fig. 9 DTA thermogram for selected Ti-50%Al powders D and E group at various milling duration

peaks to even higher temperatures compared to B samples. As shown in Fig. 9, the DTA traces exhibit better thermal properties as DTA peak for both samples appear to co-exist in the higher temperature range of 670–795 °C with minimum onset temperature at 696.3 °C for 15 h(E) MA-ed powder and maximum temperature of 709.50 °C for 12 h(D) MA-ed powder.

Maximum points of exothermic peak were observed to appear within 713.1–726.1 °C temperature range and between 163.6–171.4 μV. As multiple TiAl phases appear to form in these sample groups, it is well explained the inconsistency in DTA peak value. Whereas the increasing of heat changes suggested the energy consumed to induce the mechano-chemical reaction is caused by the formation of nano-crystallite Ti(Al) solid solution as well as crystallite size refinement. This DTA observation further confirmed that the use of higher ball to powder ratio from 10:1 to 20:1 has accelerated the MA process as it resulted in better thermal behaviour.

As shown in Figs. 10 and 11, it was found that the pattern of α-alpha (CTE) under controlled heating up to 850 °C exhibits two different groups of spectrums as DTA result. B samples which were milled with 10:1 ratio, exhibit a higher peak in a narrow peak range with CTE value range between 9.4357–10.9467 E-6/K, while D and E samples which were milled with 20:1 ratio have a lower peak but wider peak range and lower CTE value range between 6.8785–9.7642 E-6/K.

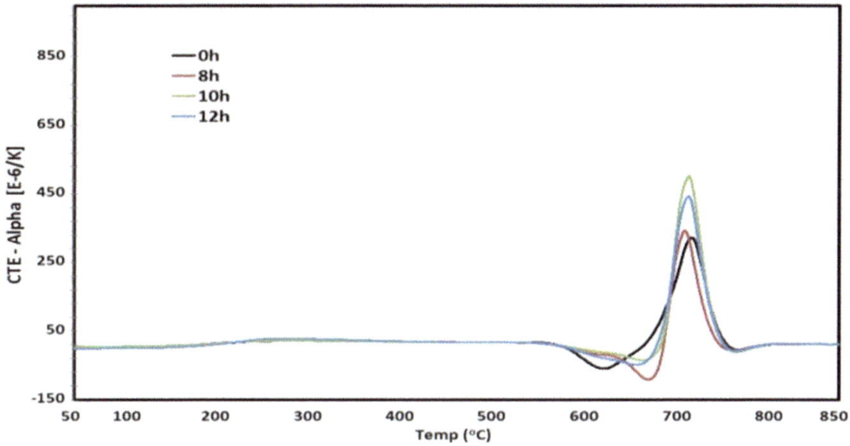

Fig. 10 CTE thermogram of selected Ti-50%Al powders at various milling duration of B group

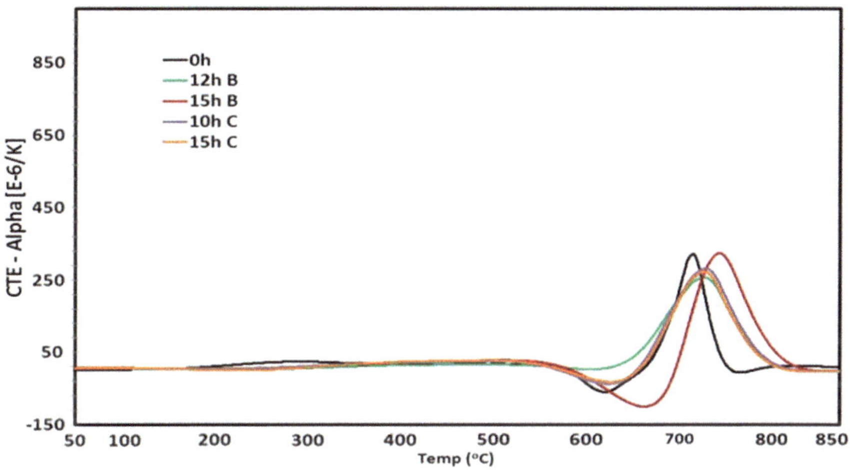

Fig. 11 CTE thermogram of selected Ti-50%Al powders at various milling duration of D & E group

3.4 Micro-Hardness of Ti-50%Al Powder

Micro-hardness test were performed on selected samples at various milling duration. The test result for each sample is the average value of at least 10 successive indentations. As shown in Fig. 12, the micro hardness values of D samples have gradually increased over milling duration. After 12 h of MA, the micro hardness value is 2 times higher than the initial powder mixture (0 h). The increased Ti-50%Al hardness is not only due to increased fineness of the Ti–Al

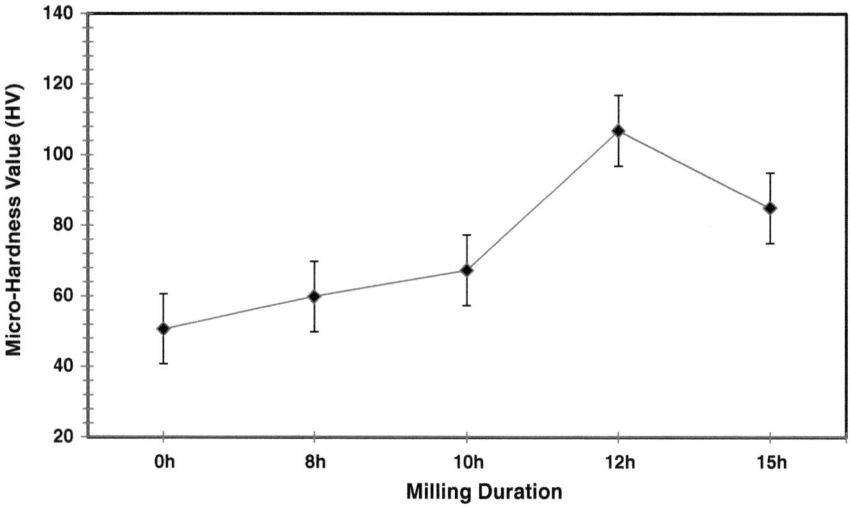

Fig. 12 Vickers Micro-hardness (Hv) of Ti-50%Al D powder samples at various milling duration

powder microstructure but also due to the formation of formation Ti(Al) solid solutions.

4 Conclusion

The results of this paper show that the milling parameter plays an important role in the efficiencies of the MA process;

i. Mechanical alloying (MA) of elemental Ti and Al powders promoted the formation TiAl alloys. Longer milling duration and less addition of hexane resulted in a better formation of TiAl alloys as Ti(Al) solid solution was formed after 5 h of milling.
ii. Dry milling without an addition of PCA's led to a dramatic decrease in the crystallite size of powder product but the agglomeration effect resulted in low milling yield. In contrast, with an addition of hexane, even though effectively minimize the agglomeration problem, proved to delay the MA process.
iii. Higher rotation energy, less addition of PCA and higher ball to powder weight ratio could accelerate the crystallite size reduction.
iv. Longer milling duration exhibits a better effect on the thermal behavior of Ti-50%Al powders as the reaction temperature has increased to 719.80 °C compared with 678.50 °C of the initial powder mixture. As in the DTA results, the observation on α-alpha value shows that higher ball to powder weight ratio used in MA also resulted in better thermal behavior.

v. The micro-hardness value of the MA-ed powders systematically increased by milling duration with a maximum value of 106.93 Hv for 12 h MA-ed powder as a result of grain refinement and the formation of new phases.

It can be conclude that the physical and thermal behaviour of Ti-50%Al powder in the early stage of MA, is determined by the morphology and microstructure of powder samples. By further MA in the intermediate and final stage, progressive intermixing between the Ti–Al plays a vital role in determining the changes in the physical and thermal behaviour.

Acknowledgement This work was supported by UMP Research Grant under research project UMP RDU 100370.

References

1. Bhattacharya, P., Bellon, P., Averback, R.S., Hales, S., et al.: Nanocrystalline TiAl powders synthesized by high-energy ball milling: effects of milling para-meters on yield and contamination. J. Alloy. Compd. **368**, 187–96 (2004)
2. Forouzanmehr, N., Karimzadeh, F., Enayati, M.H., et al.: Study on solid-state reactions of nanocrystalline TiAl synthesized by mechanical alloying. J. Alloy. Compd. **471**, 93–97 (2009)
3. Goral, M., Swadzba, L., Moskal, G., Jarczyk, G., Aguilar, J., et al.: Diffusion aluminide coatings for TiAl intermetallic turbine blades. Intermetallics **19**, 744–747 (2011)
4. Yamaguchi, M., Inui, H., Ito, K., et al.: High-temperature structural intermetallics. Acta Mater. **48**, 307–322 (2000)
5. Kothari, K., Radhakrishnan, R., Wereley, N.M., et al.: Advances in gamma titanium aluminides and their manufacturing techniques. Prog. Aerosp. Sci. **55**, 1–16 (2012)
6. GE-Aviation:Genx.GE.[Online] (2007). http://www.Geae.Com/Engines/Commercial/Genx/S
7. Hu, D.: Effect of composition on grain refinement in TiAl-based alloys. Intermetallics **9**, 1037–1043 (2001)
8. Cuevas, F.G., Cintas, J., Montes, J.M., Gallardo, J.M., et al.: Al–Ti powder produced through mechanical alloying for different times. J. Mater. Sci. **41**, 8339–8346 (2006)
9. Bin, F.W., Wen, L.X., Fei, S.H., Feng, D.Y., et al.: Characterization of Ti-50%Al composite powder synthesized by high energy ball milling. Trans. Non-Ferrous Met. Soc. China **21**, 333–337 (2011)
10. Farhang, M.R., Kamali, A.R., Samani, M.N., et al.: Effects of mechanical alloying on the characteristics of a nanocrystalline Ti–50 at.%Al during hot pressing consolidation. Mater. Sci. Eng., B **168**, 136–141 (2010)
11. Wen, L.X., Fei, S.H., Bin, F.W., Feng, D.Y., et al.: Structure and morphology of Ti-Al composite powders treated by mechanical alloying. Trans. Non-Ferrous Met. Soc. China **21**, 338–341 (2011)
12. Gabbitas, B., Cao, P., Raynova, S., Zhang, D., et al.: Microstructural evolution during mechanical milling of Ti/Al powder mixture and production of inter-metallic TiAl cathode target. J. Mater. Sci. **47**, 1234–1243 (2012)
13. Fadeeva, V.I., Leonov, A.V., Szewczak, E., Matyja, H., et al.: Structural defects and thermal stability of Ti(Al) solid solution obtained by mechanical alloying. Mater. Sci. Eng., A **242**, 230–234 (1998)

Equilibrium, Langmuir Isotherms and Thermodynamic Studies for Adsorption of Cu(II) on Natural Clay

Mustapha Djebbar, Fatiha Djafri and Mohammed Bouchekara

Abstract We have studied the pH and the temperature effects on copper ions adsorption on natural and treated clays from ENOF Chemical Ltd., Research Company, Algeria. Treated clay was studied to improve the adsorption capacity. X-Ray diffraction identified montmorillonite and kaolinite as major clay minerals. The langmuir adsorption model was used for the mathematical description of the adsorption equilibrium and the equilibrium data adhered very well to this model. The treated and natural clay had the monolayer adsorption capacity equal to 15.40 and 12.22 mg.g^{-1} respectively at pH value of 6.5 and 20 °C and the adsorption isotherms could be fitted with Langmuir isotherms, and the coefficients indicated favorable adsorption of Cu(II) on the clays. Determination of the thermodynamic parameters, *H, S* and *G* showed that the adsorption process was spontaneous and exothermic accompanied by decrease in entropy and Gibbs energy. Results of this study will be useful for future scale up for using this material as a low-cost adsorbent for the removal of Cu(II) from wastewater.

Keywords Activated clay · Sorption isotherms · Thermodynamics · Montmorillonite adsorption

M. Djebbar (✉)
Laboratory for Materials, Applications and Environment, Faculty of Science and Technology, University of Mascara, BP 763 Route de Mamounia Mascara, Mascara, Algeria
e-mail: djebbar_mustapha@yahoo.fr

F. Djafri
Laboratory for Chemical and Materials, Faculty of Science and Technology, University of Mascara, University of Oran, BP 1524 El Menouar Oran, Oran, Algeria
e-mail: djafri-fatiha@yahoo.fr

M. Bouchekara
Faculty of Science and Technology, University of Mascara, BP 763 Route de Mamounia Mascara, Mascara, Algeria
e-mail: Bouchekara_mouhamed@yahoo.fr

A. Öchsner and H. Altenbach (eds.), *Design and Computation of Modern Engineering Materials*, Advanced Structured Materials 54,
DOI: 10.1007/978-3-319-07383-5_24,

1 Introduction

The removal of heavy metals from water and wastewater is important in terms of protecting public health and environment [1]. Many industrial activities such as metal plating, industry fertilizer, mining operations, metallurgy, manufacturing batteries and dyeing in textile industries introduce heavy metals into the environment via their waste effluents [2]. Precipitation, ion-exchange, ultrafiltration, membrane separation and adsorption are the usual methods for the removal of heavy metal ions from aqueous solutions [3].

Due to its simplicity and easy operational conditions, adsorption is a widely-used process. In this study, we report the adsorption of Cu(II) from aqueous solutions on natural and activated clays.

In continuation of previous works carried out in this field, in the present work, a more efficient method for modification of Clay was studied. One of the objectives of this study was to evaluate the effect of this modification on the capability and mechanism of Cu(II) adsorption from water by the produced adsorbent.

2 Materials and Methods

2.1 Preparation of Na-Montmorillonite

The natural clays were washed several times with distilled and deionised water and were completely dispersed in water. After 7 h at rest, the dispersion was centrifuged for 1 h at 2,400 rpm. The size of the clay particles obtained was 2 mL.

These clay particles were dispersed in water and heated at 75 °C in the presence of a solution composed of the sodium salts of bicarbonate (1 M), citrate (0.3 M), and chloride (2 M) [4, 5]. The purpose of this operation was to eliminate inorganic and organic compounds, aluminium found in the inter-layer spaces and various free captions. Carbonates were removed by treatment with HCl (0.5 M) and chloride was eliminated after several washings. The organic matter was eliminated completely by treatment with H_2O_2 (30 % v/v) at 70 °C. The purified clay was dried at 110 °C, and then saturated with sodium (Na^+). To ensure complete transformation into the sodium form, all samples were washed several times with a NaOH solution (1 M), 40 g of Na-montmorillonite was treated with 400 mL of 5 N sulphuric acid (analytical grade) at 90 °C for 3.5 h [6] in a stirred glass reaction vessel with reflux. After the acid treatment, the sample was filtered and washed with distilled water until they were free of SO_4^{2-}. The samples were dried at 60 °C for 12 h and ground to pass through a 0.074 mm sieve. The Activated clay acid was characterized using FTIR, DRX and DTA/TG techniques.

3 Results and Discussions

3.1 X-ray Diffraction

The final product of clay was separated by centrifugation, then washed and dried at 60 °C and studied and X-ray diffraction patterns (XRD) of powder samples were recorded at room temperature under air conditions on a Siemens D-5000 instrument, using Cu Kα radiation ($\lambda = 1.5406$ Å) and TG/DTA techniques and FT-IR spectroscopy.

Results of the X-ray diffraction analysis for natural and activated clay are shown in Fig. 1. It clearly shows that the d-spacing of clay increased from 12.93 Å, $2\theta = 6.83°$ to 16.50 Å, $2\theta = 5.35°$ which could be attributed to the natural and activated clay. Quartz (reflection at $d = 13.92$ Å, $2\theta = 21.07, 26.34°$) and calcite (reflection at $d = 4.42$ and 2.49 Å, $2\theta = 20.05°, 35.97°$) are the major impurities. The reflection at $d = 7.07$ Å, $2\theta = 12.5°$ is characteristic of kaolinite. Montmorillonite and Kaolinite as major clay minerals were identified by XRD (Fig. 1).

3.2 Infrared Spectroscopy Study

FT-IR spectra of the natural and activated samples are in the region of 4000–400 cm^{-1} in fact, IR techniques have been used by many researchers for the identification of natural clay minerals [7] as shown in Fig. 2. The characteristic vibrations of hydroxyl groups, the silicates anions and the octahedral captions are present in the IR spectra of the studied samples. The peak positions agree with the values given in literature [8]. In the region of 3,700–3,200 cm^{-1}, a number of sharp peaks at 3,625 and 3,460 cm^{-1} are observed. In the region above 3,000 cm^{-1}, the wavenumber contains information about the silanols.

The characteristic vibration peaks of montmorillonite are at 3,625 cm^{-1} (O–H stretching), 1,040 cm^{-1} (Si–O stretching), 628 cm^{-1} (Al–O–Si stretching), 525 cm^{-1} (Si–O–Al bending) and 467 cm^{-1} (Si–O–Si bending). The small peaks in the region of 1,650–1,500 cm^{-1} are due to O–H stretching vibrations from H–O–H, which implies that there is a presence of little amount of adsorbed water in the clay samples. The peak at 914 cm^{-1} and the weak band at 440 cm^{-1} were assigned to O–Si–O asymmetric stretching; the doublet at 850 and 790 cm^{-1} indicates the presence of quartz. The characteristic bands of montmorillonite-Na were observed at 1,039 and 467 cm^{-1} [8]. The high intensity of the peak appearing at 1,039 cm^{-1} is an indication of the large amount of this mineral in the sample.

3.3 Thermo-Gravimetric Analysis

The thermal analysis diagrams for *natural* and *activated* clay are shown in Figs. 3 and 4.

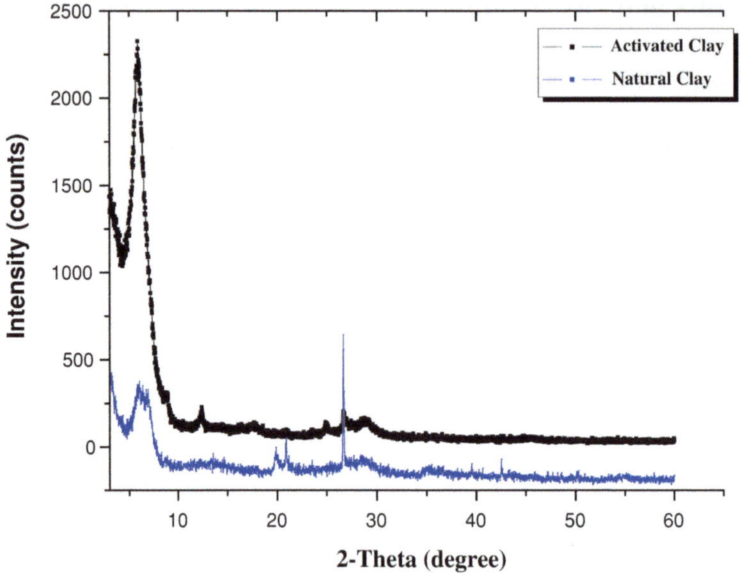

Fig. 1 XRD pattern for natural clay and activated clay

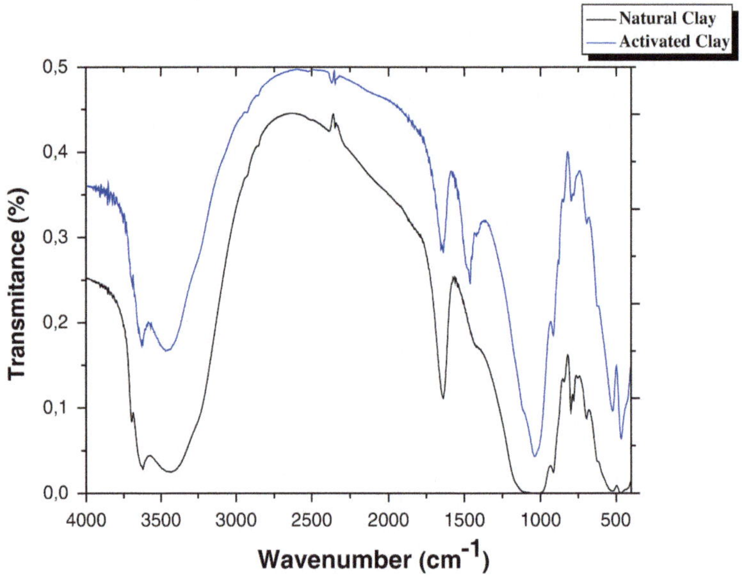

Fig. 2 FTIR spectra of natural and activated clay

The thermogravimetric results for *activated* clay (Fig. 3) revealed a weight loss corresponding to free and absorbed water on the outer surface of montmorillonite and organic materials [9] due to acid treatment in the range of 34–133 °C.

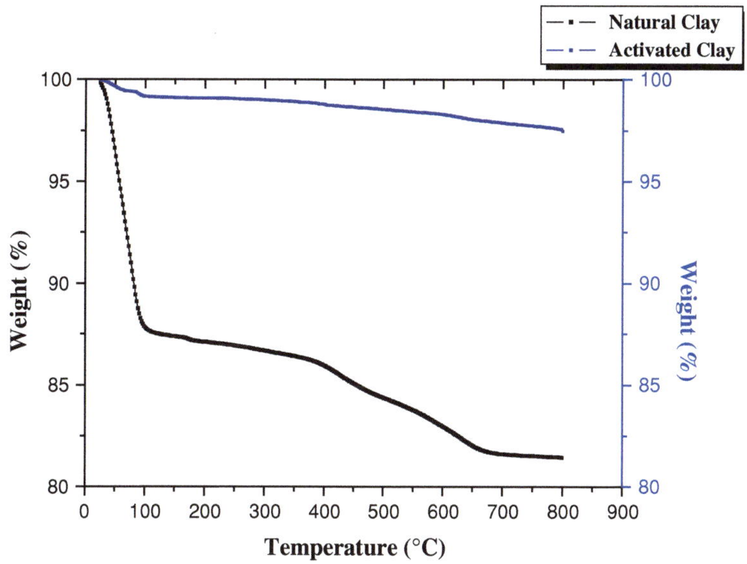

Fig. 3 TGA curve of: natural clay, treated clay

The second peak occurring between 397 and 536 °C refers to the loss of hydroxyl in the structure [9]. The corresponding weight loss for *natural* and *activated clay* were 8.69 and 2.04 %, respectively.

Results showed an endothermic peak in the DTA curve of *natural clay* in the range of 25 to 130 °C due to the dehydration of clay minerals are shown in Fig. 4. The second endothermic phenomenon took place between 350 and 600 °C, this peak is referring to the loss of hydroxyl in the structure [10]. The corresponding loss of bicarbonates and carbonates are related to the endothermic peak around 283 and 702 °C, respectively.

4 Adsorption of Copper Ions

4.1 Experimental Procedure

Solutions of Cu(II) concentrations in the range of 10–100 mgl^{-1} were prepared from a stock solution of $CuSO_4$, $5H_2O$. The pH was adjusted with 0.1 M NaOH or HNO_3.

Amounts of 0.2 g clay were dispersed in the different copper salt solutions and shaken during 2 h [11]. The dispersions were filtered, and the copper concentration was determined by spectrophotometry at a wave-length of 805 nm (Visible spectrophotometer using KBr disc method). The amounts of Cu(II) adsorbed were calculated from the concentration differences.

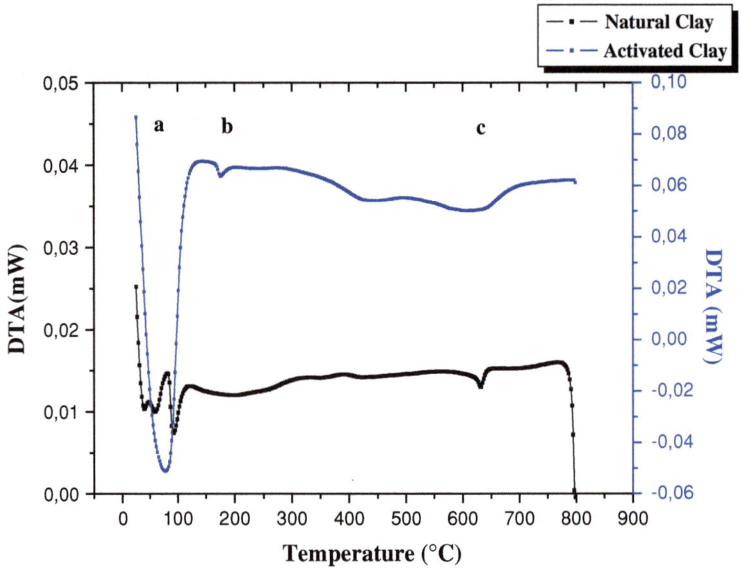

Fig. 4 DTA curve of natural clay and modified clay **a** dehydration, **b** dehydroxylation, **c** calcination

Adsorption of heavy metals ions is often pH-dependant [12, 13]. The influence of pH is shown in Fig. 5. Adsorption of copper ions was maximal at PH 6.5. Similar results were reported by [14]. After activation, the same effect was observed (Fig. 6). At lower pH, the adsorption of protons of water competed with the adsorption of copper ions, and at higher pH copper hydroxide was precipitated (Table 1).

4.2 Adsorption Isotherms

The Langmuir model is the simplest and the most commonly-used model to represent the adsorption from a liquid phase by a solid phase [15]. This model assumes a monolayer adsorption.

The obtained adsorption data were fitted by the linearized Langmuir equation:

$$\frac{C_{eq}}{(x/m)} = \frac{1}{(X_m \times b)} + \frac{C_{eq}}{X_m} \tag{1}$$

where C_{eq} is the equilibrium adsorptive concentration in solution, X_m the monolayer capacity *(x/m)* is the specific amount and b is the related to the adsorption energy. The Langmuir isotherm showed a fit with the experiment data (Figs. 7 and 8).

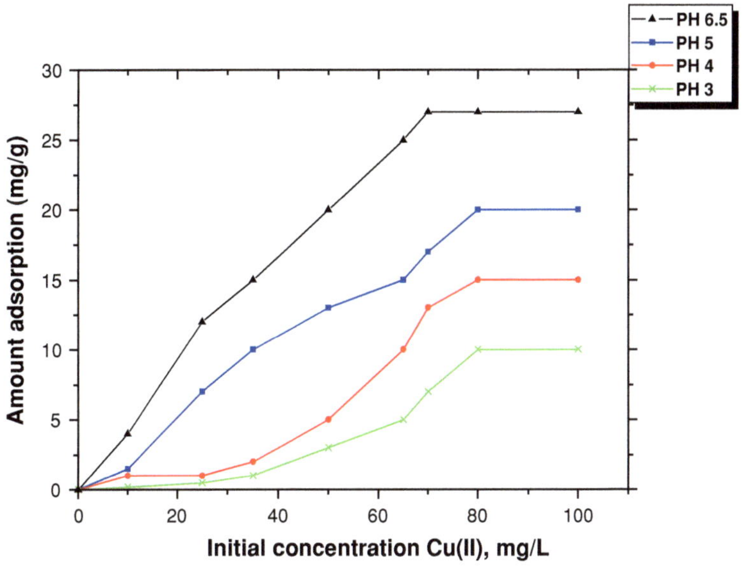

Fig. 5 Adsorption of Cu(II) on natural clay at different pH values

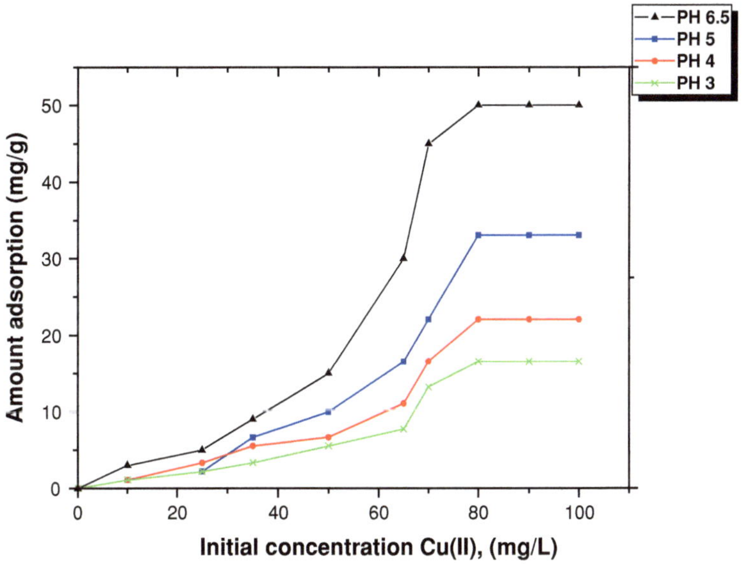

Fig. 6 Adsorption of Cu(II) on activated clay at different pH values

With increased temperature, the adsorption of Cu(II) increased (Fig. 7) [16] confirming that the process was exothermic. After activation of the clay, the same effect was observed (Fig. 8).

Table 1 Chemical composition of the natural clay and activated clay (in mass%)

Sample	Activated clay	Natural clay
SiO_2	67.27	54.782
Al_2O_3	21.25	30.794
SO_3	0.314	0.1
K_2O	2.10	1.199
MgO	2.204	1.761
CaO	0.037	0.015
Fe_2O_3	3.069	2.636
Na_2O	2.57	4.33
TiO_2	0.265	0.1
P_2O_5	0.015	0.03

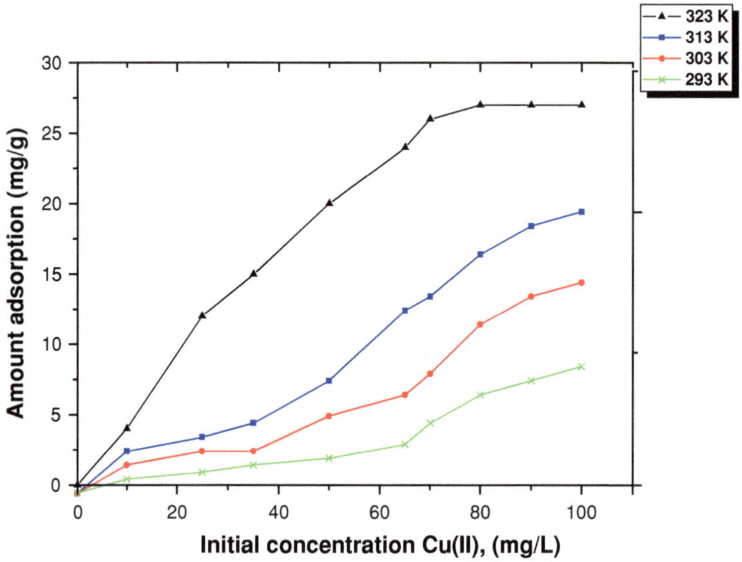

Fig. 7 Langmuir isotherms for natural clay at PH 6.5

The parameters derived from the least-squares fitting of the isotherms by the linearized Langmuir equation (Fig. 9) are given in Table 2.

Adsorption isotherms of Cu(II) on the natural clays and activated clay are shown in Fig. 10.

The monolayer capacity (X_m) for activated and natural clay was 15.40 and 12.22 mg.g^{-1} respectively. The higher b value of natural clay compared with that of activated clay showed that the adsorption of copper ions on the raw clay required more energy.

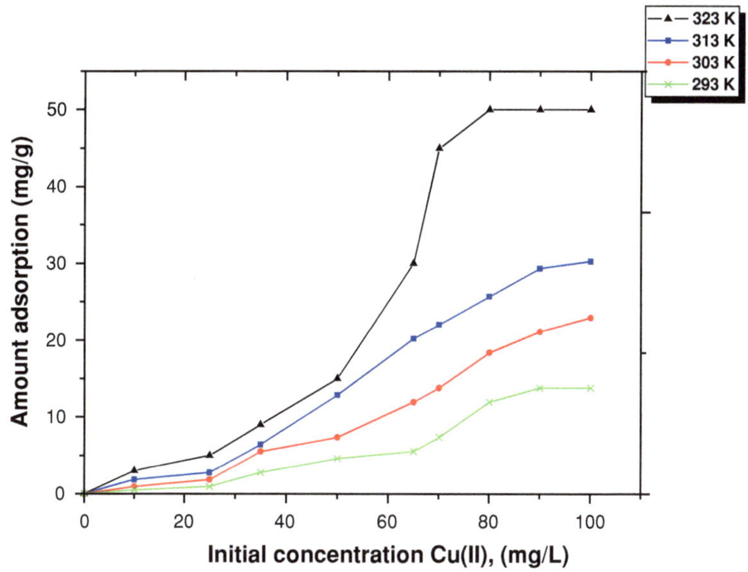

Fig. 8 Langmuir isotherms for activated clay at pH 6.5

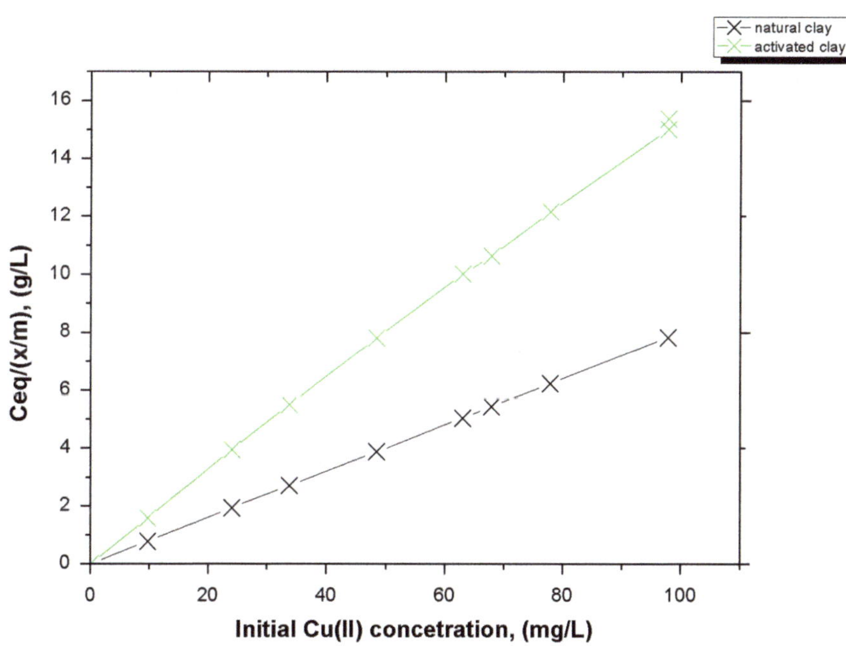

Fig. 9 Langmuir isotherms for Cu(II) adsorption at 20 °C (clay 0.2 g/l, pH 6.5, initial Cu(II) mg/L)

Table 2 Langmuir parameters of adsorption isotherms at 20 °C

Adsorbents	Langmuir equation		
	X_m(mg.g^{-1})	b(L.mg^{-1})	R^2
Natural clay	12.22	0.62	0.992
Activated clay	15.40	0.79	0.992

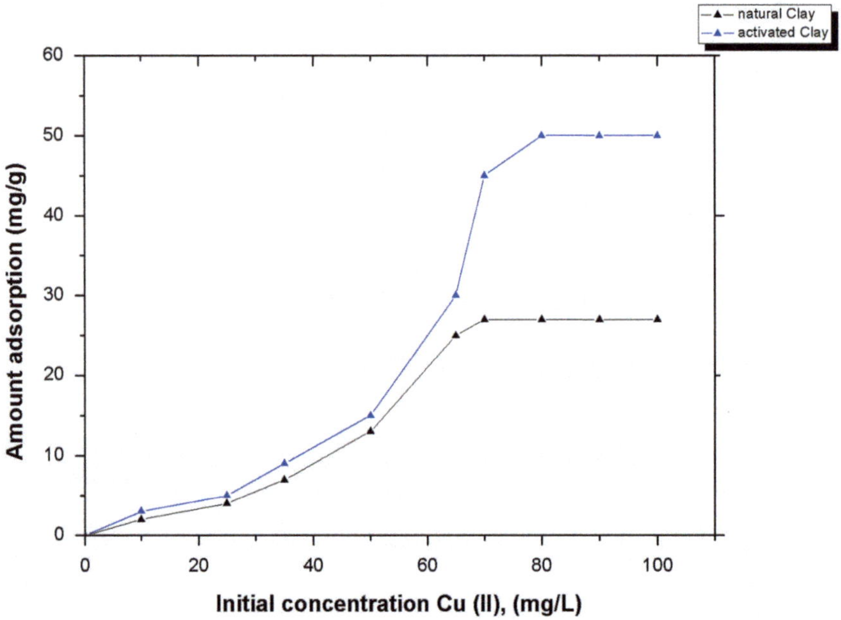

Fig. 10 Adsorption isotherms of Cu(II) on the natural clays and activated clay adsorbents at 20 °C, PH 6.5

5 Thermodynamic Parameters

The thermodynamic parameters for the adsorption of Cu(II) by natural and activated clay such as the enthalpy change $\Delta H°$, the Gibbs free energy change $\Delta G°$ and the entropy change $\Delta S°$ can be calculated from the variation of maximum adsorption with temperature T using the following basic thermodynamic relations [17].

$$K_{\text{ads}} = Qe.\frac{\left(\frac{m}{v}\right)}{\left[C0 - Qe\left(\frac{m}{v}\right)\right]} \tag{2}$$

$$\Delta G° = RT\ln K_{\text{ads}} \tag{3}$$

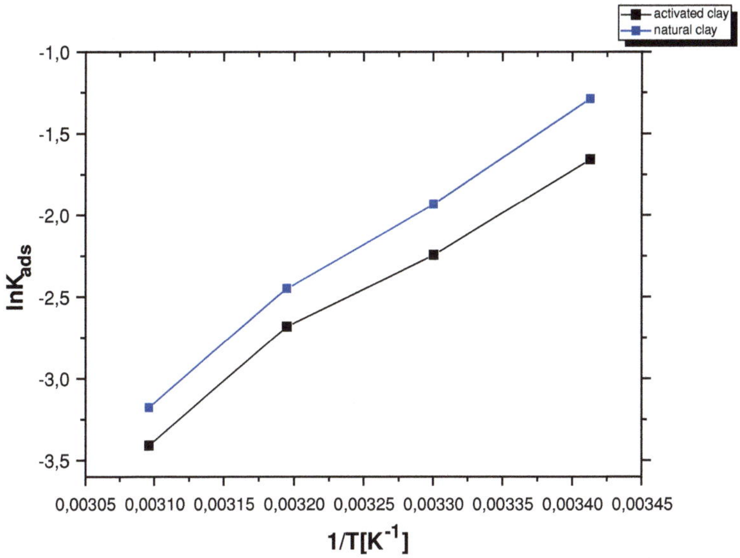

Fig. 11 ln K_{ads} as a function of $1/T$ at PH 6.5

$$\Delta S^\circ = \frac{(\Delta H^\circ - \Delta G^\circ)}{T} \qquad (4)$$

where R is the gas constant, R = 8.314×10^{-3} kJ mol^{-1} K^{-1}; K_{ads} is the equilibrium constant, T is the absolute temperature; ΔG° is the change in free energy, kJ mol^{-1}; ΔG° is the change in enthalpy, kJ mol^{-1}; ΔS° is the change in entropy, kJ mol^{-1}. According to Eq. (2), the mean value of the enthalpy change due do the adsorption of Cu(II) by natural and activated clay over the temperature range studied can be determined graphically by linear plotting of ln K_{ads} against $1/T$ using the least squares analysis shown in Fig. 11.

The mean enthalpy change can be determined from the slope of the straight line. The variation of Gibbs free energy and entropy change with temperature can be calculated using Eqs. (3) and (4), respectively, the results are arranged in Table 3. An important result that can be obtained from Table 3 is that the Gibbs free energy (ΔG°) is small and negative with its value decreasing with increasing temperature. This indicates that the adsorption processes of Cu(II) by natural and activated clay can be enhanced by decreasing temperature. The values of entropy change (ΔS°) are positive and remain almost constant with temperature. This gives an evidence that structural changes in Cu(II) and natural and activated clay occur during the adsorption process (Fig. 11). The negative values of enthalpy change (ΔH°) for the adsorption is lower than 80 kJ mol^{-1}, suggesting the physical nature of the sorption, i.e., physisorption conducted with van der Waals forces.

Table 3 Thermodynamic parameters

	$-\Delta G°$ (kJ.mol^{-1})				$-\Delta H°$(Kj.mol^{-1})	$\Delta S°$(J.mol.K^{-1})
	293 K	303 K	313 K	323 K		
Natural clay	3.140	4.873	6.375	8.534	48.572	176.306
Activated clay	4.039	5.649	6.981	9.155	33.391	127.769

The enthalpy change $\Delta H°$ of the adsorption were negative −33.3915 to −48.5720 kJ/mol indicating physical adsorption of Cu(II) on to natural and activated clay [17]. The values of ΔG° are negative (Table 3), which means that the reaction of Cu(II) adsorption is spontaneous exothermic [18].

6 Conclusion

Adsorption isotherms were both well described by the Langmuir model. A maximum capacity of Cu(II) adsorbed on natural and activated clay at equilibrium was 15.40 and 12.22 mg.g^{-1} respectively. The negative value of $\Delta H°$ indicated that the adsorption process was exothermic in natural and Activated Clay, the negative values of $\Delta G°$ at different temperatures (293–323 K) indicated the spontaneous nature of Cu(II) adsorption. The adsorption reached a maximum at a PH 6.5 and increased with temperature. The Cu(II) adsorbed by natural clay was lower compared with activated clay.

Acknowledgements The authors would like to thank the University of Technology Faculty of Chemistry, Oran, Algeria and Mascara's Laboratory University. The authors thank also Prof. F. Djafri for her assistance in XRD and IR measurements and educational experience, without forgetting Prof. A. Djafri for her helpful discussions. Finally, the authors offer a special thanks to Prof. M. Bouchekara for his guidance and support throughout of this research.

References

1. Unlu, N., Ersoz, M.: Adsorption characteristics of heavy metal ions onto a low cost biopolymeric sobent from aqueous solutions. J. Hazard. Mater. B **136**, 272–280 (2006)
2. Srivastava, V.C., Mall, I.D., Misha, I.M.: Characterisation of mesoporous rice husk ash (RHA) and adsorption kinetics of metal ions from aqueous solution onto RHA. J. Hazard. Mater. B **134**, 257–267 (2006)
3. Das, N., Kumar, J.R.: Adsorption of some bivalent heavy metal ions from aqueous solutions by manganese nodule leached residues. J. Colloid Interface Sci. **293**, 253 262 (2006)
4. Pusino, A., Micera, G., Gessa, C., Petretto, S.: Interaction of diclofop and diclofop–methyl with Al $^{3+}$, Fe $^{3+}$, and Cu $^{2+}$ saturated montmorillonite. Clays Clay Miner. **37**, 558–562 (1989)
5. Robert, M., Tessier, D.: Méthode de préparation des argiles des sols pour des études minéralogiques. Ann. Agron **25**, 859–882 (1974)

6. Foletto, E.L., Volzone, C., Porto, L.: Performance of an Argentinian acid-activated bentonite in the bleaching of soybean oil. Braz. J. Chem. Eng. **20**, 139–145 (2003)
7. Hajjaji, M., Kacim, S., Alami, A., El-Bouadili, A., El Mountassir, M.: Chemical and mineralogical Characterization of a clay taken from the Moroccan Meseta and a study of the interaction between its fine fraction and methyleneblue. Appl. Clay Sci. **20**, 1–12 (2001)
8. Madejova, J.: FTIR techniques in clay mineral studies. Vib. Spectrosc. **31**, 1–10 (2003)
9. Heller-Kallai, L., Bergaya, F., Theng, B.K.G., Lagaly, G.: Handbook of Clay Science. Developments in Clay Science, pp. 289–308, Elsevier, Amsterdam (2006)
10. Ghosh, D., Bhattacharyya, K.G.: Adsorption of methylene blue on kaolinite Appl. Clay Sci. **20**, 295–300 (2002)
11. Yavuz, O., Altunkaynk, Y., Güzel, F.: Removal of copper, nickel, cobalt and manganese from aqueous solution by kaolinite. Water Res. **37**, 948–952 (2003)
12. Stumm, W.: Chemistry of the Solid–Water interface, pp. 21–34, 229–232. Wiley, New York (1992)
13. Farrah, H., Pickeringf, W.F.: The effect of pH on the retention of Cu, Pb, Zn and Cd by clay-humic acid mixtures. Water Air Soil Pollut. **8**, 189–179 (1977)
14. Bellir, K., Bencheikh, L., et al.: Study of the retention of heavy metals by natural material used as liners in landfills. Desalination **185**, 111–119 (2005)
15. Boddu, V.M., Abburi, K., et al.: Removal of hexavalent chromium from wastewater using a new composite chitosan biosorbent. Environ. Sci. Technol. **37**, 4449–4456 (2003)
16. Benzina, M.: Contribution à l'étude cinétique et thermodynamique de l'adsorption des vapeurs organiques sur des argiles locales, Modélisation d'un adsorbeur à lit fixe. Thèse de Doctorat es-science physique, Faculté des sciences de Tunis 25 (1990)
17. Seki, Y., Yurdakoc, K.: Adsorption of promethazine hydrochloride with KSF montmorillonite. Adsorption **12**(1), 89–100 (2006)
18. Suhas, C.P.J.M., Ribeiro, C.M.M.L.: Lignin from natural adsorbent to activated carbon. a review. Bioresour. Technol. **98**, 2301–2312 (2007)

Residual Stress Mapping in Alumina by Cr^{3+} Fluorescence Spectroscopy

Fathi A. ElFallagh

Abstract The stresses around the Vickers indents in alumina [surface orientation $(1\bar{1}\ 0\ 2)$] were investigated by Cr^{3+} fluorescence spectroscopy, measuring the shifts in the Cr^{3+} R_1 and R_2 luminescence lines. Changes in R_1 peak position can be related directly to changes in residual stress of the alumina in the sampled volume, and changes in R_1 peak width arise from local stress variations due to microstructural defects such as cracks and dislocations. The intensity of the R_1 and R_2 fluorescence peaks was observed to decrease with increasing proximity to the indent centre due to increased scattering from residual surface and microstructural damage such as dislocations, cracks and surface debris.

1 Introduction

Micro- and nano-indentations on the surface of materials provide a useful and quick way of determining mechanical properties of materials. Microindentations alter the microstructure and state of residual stress at the surface of a material.

Residual stress in alumina can be measured using Cr^{3+} fluorescence spectroscopy (e.g. [1–4]). The presence of chromium ions in α-Al_2O_3, substituting on the Al^{3+} sublattice, gives rise to two sharp luminescence lines (R_1 and R_2) in the visible region, which shift linearly with the applied stress [4].

$$\Delta\nu_{R1} = 7.590P - 1.5S \qquad \Delta\nu_{R2} = 7.615P - 0.6S \tag{1}$$

where $\Delta\nu_{R1}$ and $\Delta\nu_{R2}$ are the shifts in the R_1 and R_2 lines of ruby (in cm^{-1}) and P and S are the hydrostatic and non-hydrostatic component of the stress (in GPa),

F. A. ElFallagh (✉)
Azzaytuna University, Soog Alahad, Libya
e-mail: Fathi_Fallagh@yahoo.com

A. Öchsner and H. Altenbach (eds.), *Design and Computation of Modern Engineering Materials*, Advanced Structured Materials 54,
DOI: 10.1007/978-3-319-07383-5_25, © Springer International Publishing Switzerland 2014

respectively [4]. If the stresses are purely hydrostatic (i.e. $S = 0$), both lines shift by nearly equal amounts with applied hydrostatic stress P.

The general principles of relating an observed line shift in fluorescence to the state of stress have been described previously [5].

Single crystal sapphire samples have been indented under a range of loads, generating highly localised regions of surface microstructural damage and associated residual stress. The measured changes in residual stress have been related to the changes in alumina topography and microstructure (including the 2D distribution of cracks observed around the indentation sites), induced by the Vickers microindentations.

2 Experimental Procedures

Alumina samples (R-cut) have been indented using diamond Vickers indenters. Optical microscopy and SEM were used to study surface damage on the samples. Cr^{3+} fluorescence spectra were recorded, using an inVia/H26081 Renishaw spectrometer for as received and indented alumina samples.

2.1 Materials

Samples used for these studies were: An optically polished single crystal sapphire (α-Al_2O_3) wafer 0.55 mm thick with rhombohedral surface orientation $(1\bar{1}02)$ ("R-cut"). Single crystal alumina samples were chosen to avoid any influence of local grain crystallography and grain boundaries on the microstructural crack and stress analyses.

2.2 Indentation Procedure

A microhardness tester (Mitutoyo) was used to indent the alumina samples under ambient conditions (24 °C, ~43 % relative humidity) using a diamond Vickers indenter tip loaded to 50, 100, 200, 300 g and dwell time 15 s, keeping a constant tip and sample orientation for each sample.

Arrays of Vickers microindentations with different loads were done on the R-cut (see Fig. 1), leaving enough space between sequential indents to prevent interaction among surface cracks from different indentation sites.

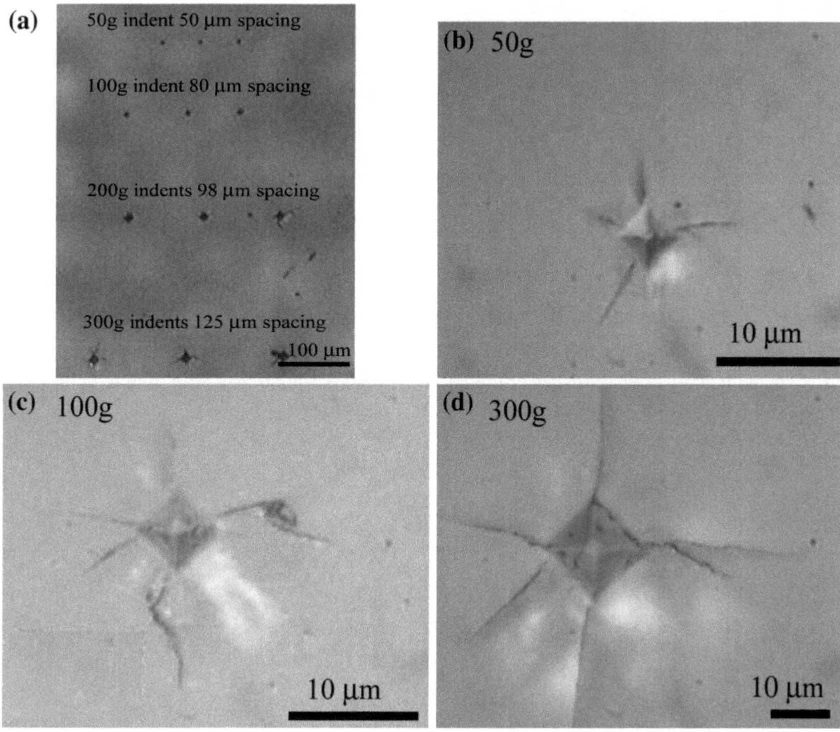

Fig. 1 **a** Optical image of a Vickers microindentation array on a single crystal sapphire sample (R-cut). **b**, **c** and **d** optical images of 50, 100 and 300 g indents in (**a**)

2.3 Cr^{3+} Fluorescence Spectroscopy

A Renishaw Cr^{3+} fluorescence spectrometer was used to study stress changes in alumina samples. Fluorescence spectra were recorded using an in Via/H26081 Renishaw spectrometer with a 20 mW Ar laser ($\lambda = 514.5$ nm) at 1 % power. A 50x objective lens was used which produces a spot size of 2 μm. Linear scans and 2D *x-y* map scans (Fig. 2) were done for (a) the optically polished single crystal alumina, (b) from 50, 100, 200 and 300 g Vickers microindentations. The line and map spectra were collected in steps of 2 μm, with the centres of the indents being the approximate centres of the *x-y* maps. The frequency scan range was from 14,300 to 14,500 cm^{-1}. Analysis of collected spectra for line scans and 2D square maps was done using both Renishaw software and Microcal Origin. The spectra (Fig. 3) were fitted with mixed Lorentzian and Gaussian curves, and the results for the R_1 and R_2 peak positions and widths (FWHM) were fitted using multi-peak functions and exponential decay functions respectively.

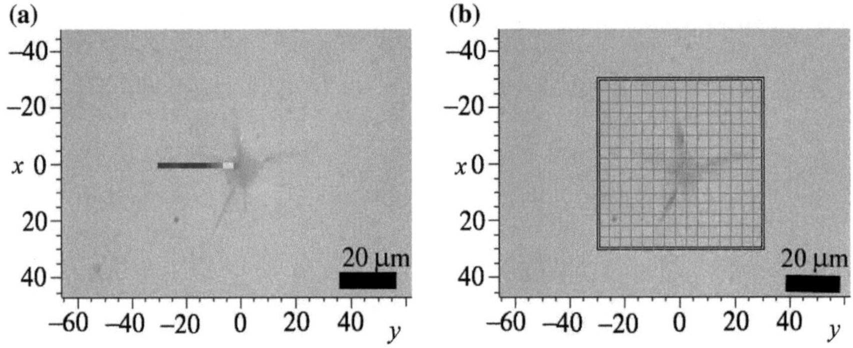

Fig. 2 **a** and **b** show optical images of the line scan and *x*-*y* square map scan used to collect data for the 100 g Vickers microindentation

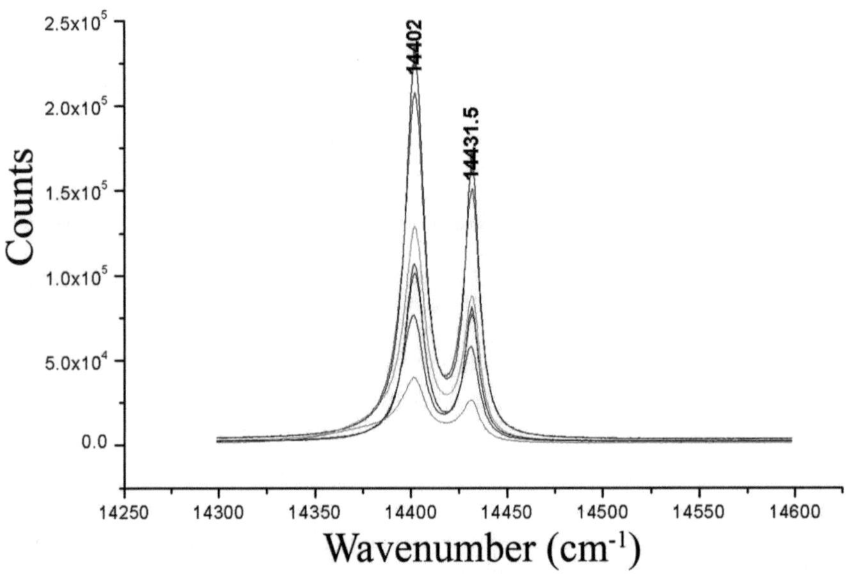

Fig. 3 Cr^{3+} fluorescence Peaks, R_1 and R_2 from line scan for a 300 g Vickers microindentation (starting point for the scan is the centre of microindentation site and moving out along one of the diagonals). The biggest R_1 and R_2 peaks correspond to the furthest scan point at 30 μm from the indent centre. Peak shifts are due to residual stress changes. The Intensity of the peaks is lowest the near centre of the microindentation due to scattering of the laser light

3 Results and Discussion

3.1 *Microindentation of Alumina*

The microindentation sites of the alumina samples (R-cut) were imaged using optical microscopy. Figure 4 shows the residual damage area with radiating radial cracks.

The microindentation tests of $(1\,\bar{1}\,0\,2)$ orientation alumina gave a Vickers Hardness of HV = 20.5 GPa ± 2 % averaged over four indents, which is within the expected range e.g. the reported value of 20.8 GPa for $(2\,\bar{1}\,\bar{1}\,0)$ sapphire [6]. The residual damage at the Vickers indentation sites, loads 50, 100, 200 and 300 g with identical diamond tip and crystal orientation, was imaged by SEM (no coating). Figure 5 shows the two types of cracks visible on the surface around these indentation sites (i) radial cracks, and (ii) circumferential cracks. The maximum length of surface cracks increased with indent load, e.g. for 100 g the radial cracks extended to a maximum of 12 μm from the corners of the residual indent impression, and for 300 g up to 20 μm. In brittle materials radial cracks typically initiate near the edges of the Vickers indenter tips where there is the highest stress concentration [6, 7]. The radial cracks in this chapter were located in very similar positions for loads 50, 100, 200 and 300 g and not exactly on the corners of the residual indent impression. This offset can be due to the local crystallography of the sample [7, 8]. Some linear surface traces, probably twins or slip bands, were also observed around the residual indent impression (Fig. 6) [6, 9]. The surface observations of the indentations are consistent with those previously reported in other studies of microindentation of alumina [6, 9].

3.2 *Crystallographic Analysis by Electron Backscattered Diffraction*

In order to define the orientations of microindentations sites with respect to the sample crystallographic directions, Crystallographic analysis by Electron Backscattered Diffraction (EBSD) scan were performed near these sites [10]. One of the diagonals of the microindentation sites was chosen to be parallel to the *y*-direction of the scan.

3.2.1 Pole Figure Calculations of the R-Cut Alumina Sample

A pole figure is a graphical representation of the orientation of an object in space. For example, pole figures in the form of stereographic projections are used to represent the orientation distribution of crystallographic lattice planes in crystallography [10, 11]. Figure 7 shows a Kikuchi pattern obtained near a 100 g

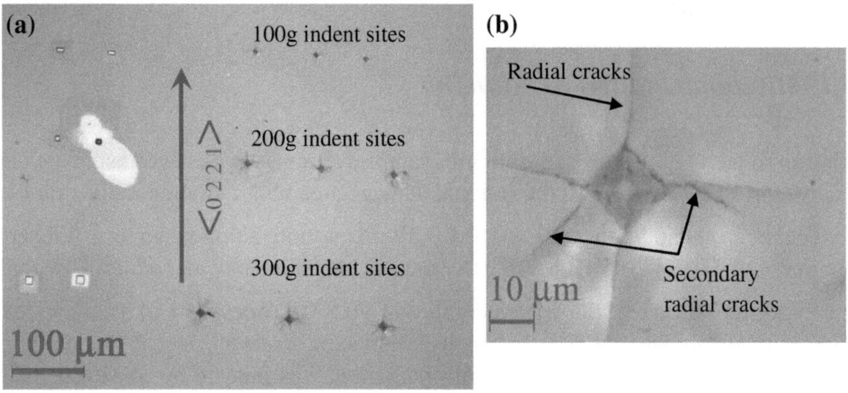

Fig. 4 Optical images of indentation sites in R-cut alumina sample **a** showing 100, 200 and 300 g indent sites. **b** 300 g indent site showing radial cracks

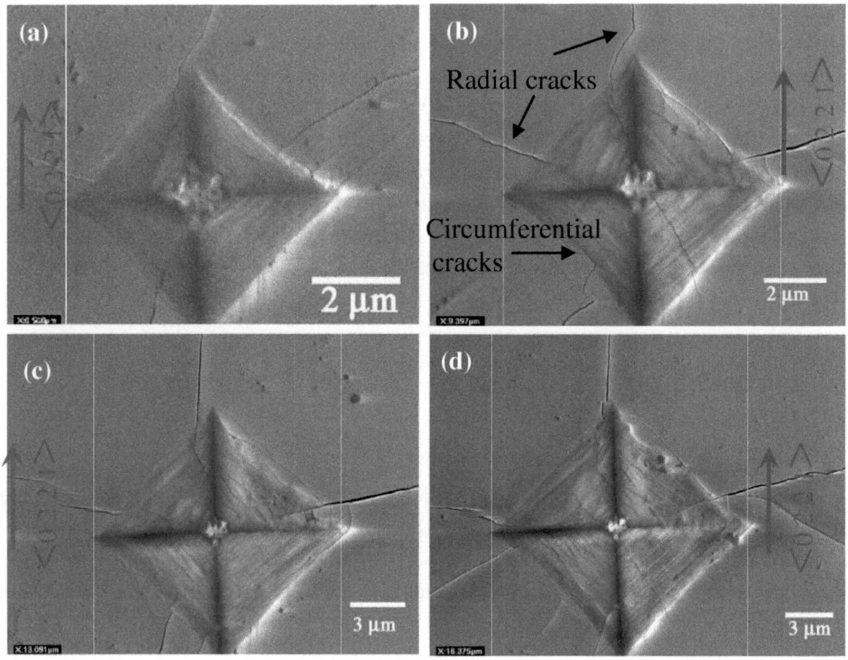

Fig. 5 **a**, **b**, **c** and **d** SE image of a 50, 100, 200 and 300 g Vickers indentation site at R-cut alumina sample, respectively, with radial and circumferential crack patterns

indentation site in R-cut alumina. The intersecting lines termed Kikuchi bands correspond to each of the lattice diffracting planes and the intersection points correspond to crystallographic directions.

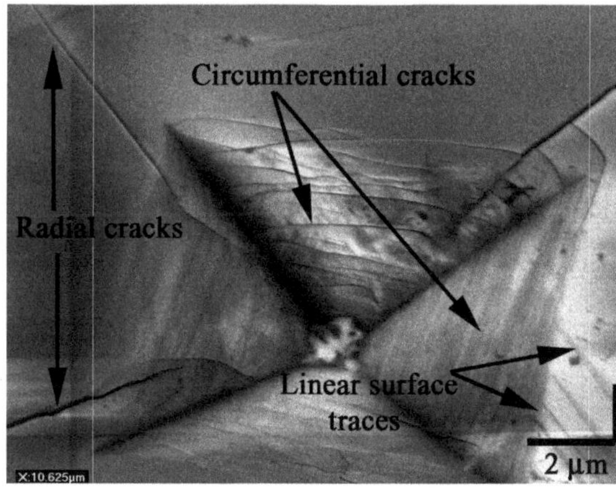

Fig. 6 SEM image of a 100 g Vickers indentation site with radial and circumferential crack patterns (depth of residual damage crater ~1.9 μm)

Using these Kikuchi patterns, pole figures were calculated to show the orientation distribution of crystallographic lattice planes, for example $\{1\bar{1}02\}$ planes which is the sample surface orientation (see Fig. 8). An inverse pole figure is the description of the sample orientation with respect to the crystal coordinate system [11].

From the pole figures shown in Figs. 8 and 9 and from the crystallography of alumina, it is determined that the sample y-direction is parallel to a $[02\bar{2}1]$ crystallographic direction (which is the only direction from this family lying on the $(1\bar{1}02)$ plane) for alumina. Therefore the vertical diagonal of the microindentation is parallel to this $[02\bar{2}1]$ direction (see Figs. 4 and 5).

3.3 Residual Stress Mapping by Cr^{3+} Fluorescence Spectroscopy

Microindentation on the surface of the alumina samples causes changes of the residual stress; in order to study these changes Cr^{3+} Fluorescence Spectroscopy was used.

3.3.1 As-Received and As-Indented Alumina

The stresses around the Vickers indents in alumina were investigated by Cr^{3+} fluorescence spectroscopy, measuring the shifts in the Cr^{3+} R_1 and R_2 luminescence lines away from the values obtained from the as-received material [1–4, 12, 13].

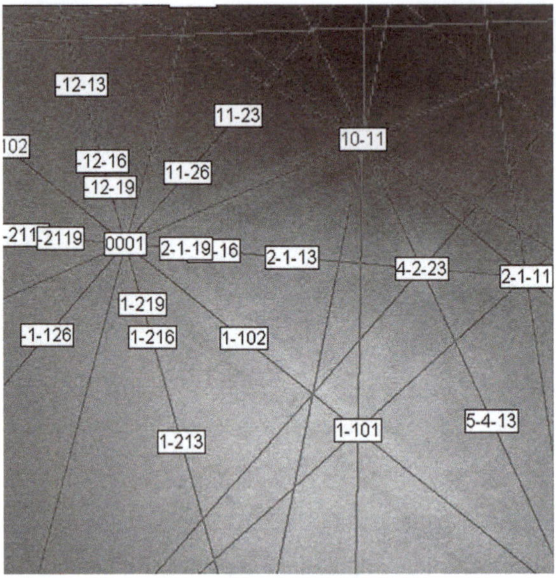

Fig. 7 Kikuchi Pattern for R-cut single crystal Sapphire

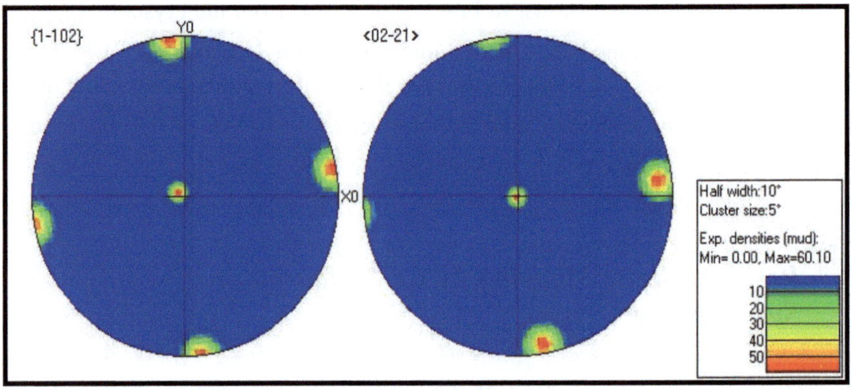

Fig. 8 **a** Calculated alumina $\{1\ \bar{1}\ 0\ 2\}<0\ 2\ \bar{2}\ 1>$ pole figure

The as-received optically polished single crystal alumina [surface orientation $(1\ \bar{1}\ 0\ 2)$], gave a R_1 peak position of 14401.6 cm^{-1} at room temperature. This is slightly lower than the values, 14,402 cm^{-1} [12] and 14402.5 ± 0.4 cm^{-1} [13] reported for unstressed alumina in the literature, which may be due to experimental calibration or a very small residual compressive stress existing in the polished alumina sample. The intensity of the R_1 and R_2 fluorescence peaks was observed to decrease with increasing proximity to the indent centre due to increased scattering

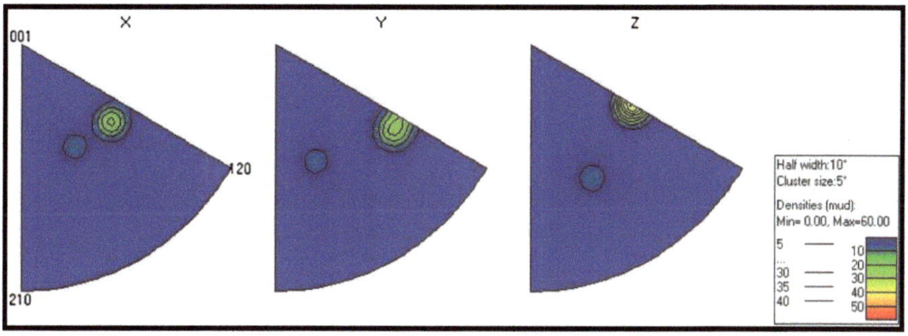

Fig. 9 Inverse pole figure for R-cut single crystal alumina

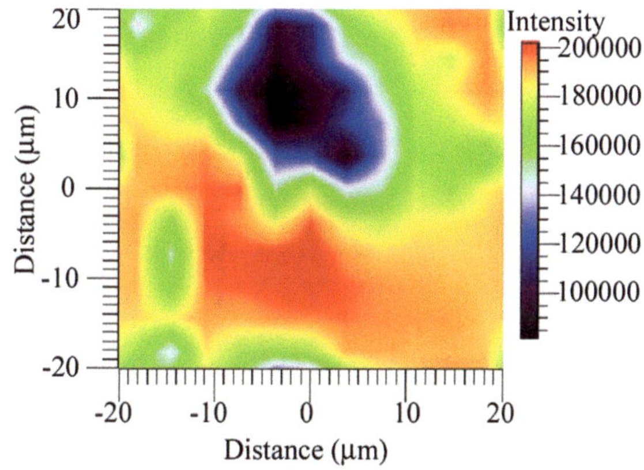

Fig. 10 Map of R_1 peak intensity around a 200 g Vickers microindentation site, indentation site marked by the *white diamond*

from residual surface and microstructural damage such as dislocations, cracks and surface debris (see Fig. 10). Maximum variations of the R_1 and R_2 peak intensity, position and width due to changes of focus, equivalent to sample height changes in the range of ±4 μm, were found to be ±11 %, ±0.05 cm^{-1} and ±0.08 cm^{-1} respectively.

Changes in R_1 and R_2 peak positions can be related directly to changes in residual stress of the alumina in the volume sampled, and changes in R_1 and R_2 peak widths arise from local stress variations due to microstructural defects such as cracks and dislocations [1–3]. Figure 11 shows the variation in measured R_1 peak position and width along a diagonal from the centre of a 200 g Vickers micro-indentation fitted using multi-peak curves. The R_1 luminescence line undergoes significant changes in its peak position and width in the vicinity of the indentation site. The line scan in Fig. 11a illustrates three zones where the R_1 peak position exhibits different displacement from the R_1 peak position in the as-received alumina; (i) a reduction in frequency, (ii) an increase in frequency, and (iii) the same as the as-received alumina. These zones correspond to a net compressive stress zone, a net tensile stress zone, and a neutral stress zone.

Fig. 11 Variations of **a**, **c** R_1 and R_2 peak positions and **b** R_1 peak width with distance along an indenter tip edge starting from the centre of a 200 g Vickers microindentation site

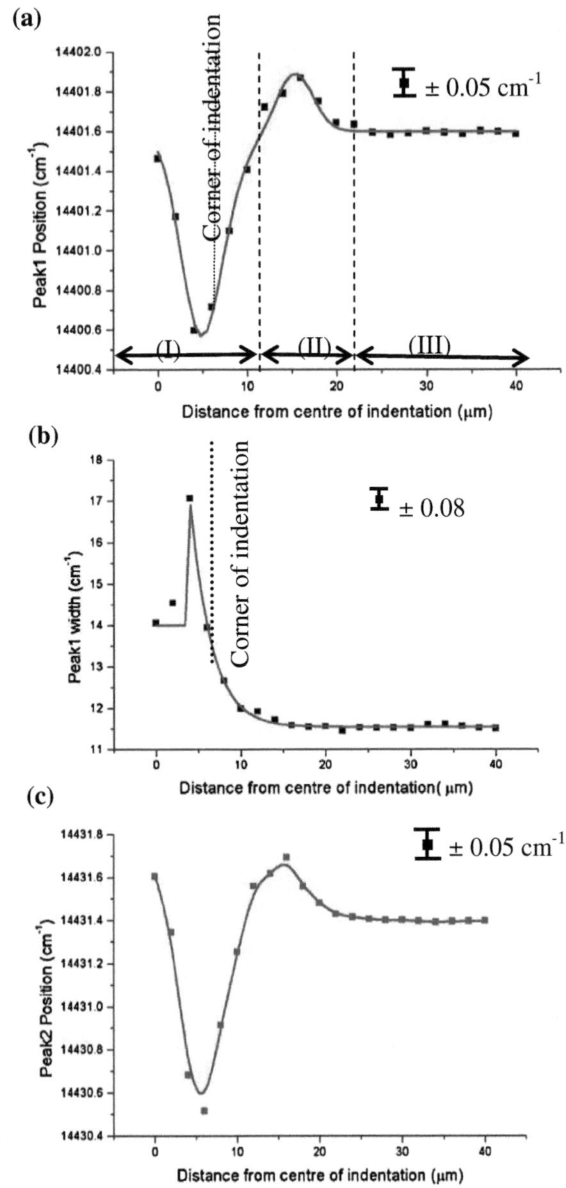

The residual stress around the indents was calculated from the shift in the frequency of the R_1 and R_2 peaks using Eq. (1) [1, 4]. Figure 12 shows the variation of the residual stress along the diagonal of the 200 g Vickers microindentation in Fig. 11a, b. The calculated stress is averaged over the volume excited by the laser beam. Given the laser power (0.2 mW), the laser spot size ~2 μm,

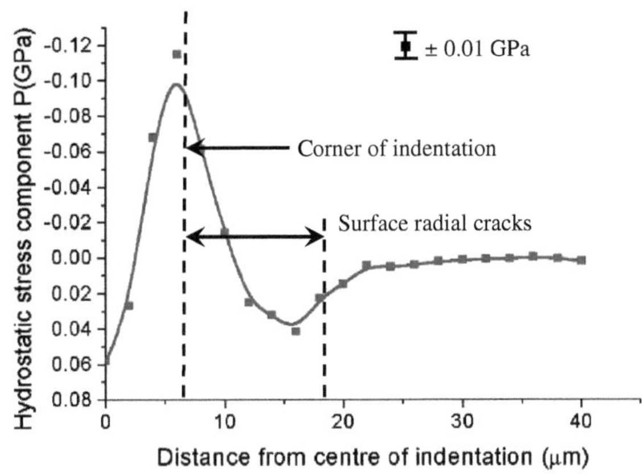

Fig. 12 Variation of the hydrostatic component of the residual stress with distance along an indenter tip edge starting from the centre of a 200 g Vickers microindentation site (from Fig. 11a)

and an approximate axial penetration depth of ~10 μm [1], the volume excited by the laser beam is approximately 31 μm^3.

In all the indents examined there is a significant compressive stress zone (I) underneath the residual indent impression. The compressive stress zone decreases from a peak within the indent impression, 120 MPa for the 200 g indent in Fig. 12, until the net stress changes to tensile (zone II). The commencement and maximum of this tensile zone are outside the residual indent impression, but within the range of the surface radial cracks (18 μm for the 200 g) observed propagating from the edge of the residual indentation impression. The maximum tensile stress measured was significantly less than the maximum compressive stress, being 40 MPa for the 200 g indent in Fig. 12 at 15 μm from the indent centre. The detectable tensile zone extended around 12–20 and 20–30 μm from the centre of 200 and 300 g Vickers indentations, respectively. This tensile zone, surrounding the central compressive stress zone, is expected to be a key driving force for crack propagation outwards from the indent centre.

The microstructural damage generated around the Vickers indent, such as cracks and dislocations, cause changes in R_1 peak width due to local stress variations in the volume excited by the laser beam. Figure 11b shows the changes in R_1 peak width for the 200 g Vickers microindentation in Figs. 11a and 12. At the centre of the indent, in the severe microstructural damage zone, there is significant peak broadening. From a maximum within the indent impression, the peak broadening then decreases exponentially, and at the edge of the compressive stress zone (I) is reduced to <10 % of its maximum value. It was consistently observed that the broadening of the R_1 peak is concentrated in the compressive stress zone.

Figure 13 shows 2D maps of R_1 peak position and width for a second 200 g Vickers microindentation site, which show the same features as the line scans (Fig. 11). Under the residual indent impression there is a significant compressive stress and a corresponding R_1 peak broadening zone. Outside the residual indent impression, there is a clear ring of net tensile stress, where the R_1 peak broadening is measurable but small. The central compressive stress zone and the surrounding

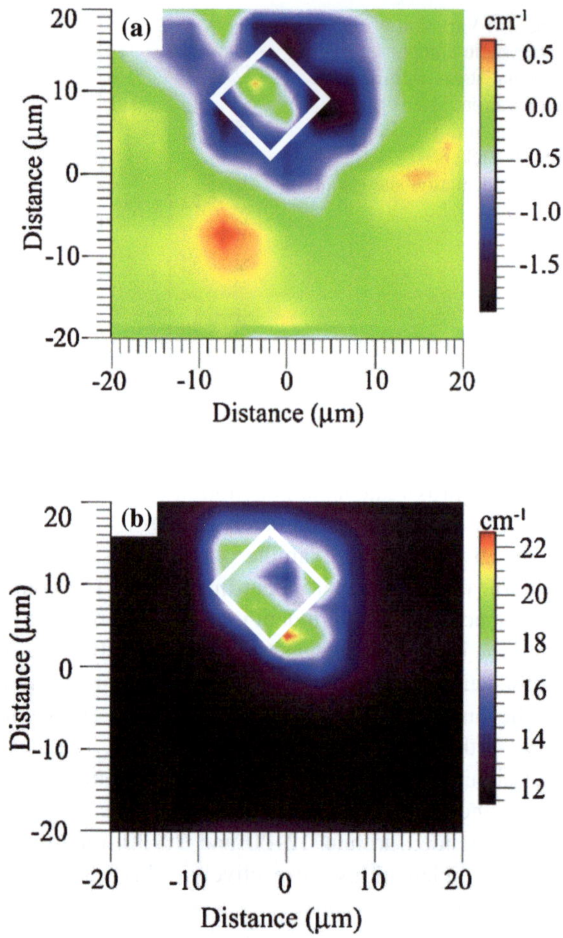

Fig. 13 2D Maps of **a** R_1 peak shift and **b** R_1 peak width, around a 200 g Vickers microindentation site, indentation site marked by the *white diamond*

tensile stress zone are consistent with previous studies of indentations and scratch tests [1, 4, 14, 15].

The Vickers indentation tip used to generate the surface damage in the alumina sample has four fold symmetry, but the microstructural damage and the measured residual stress around the indents do not exhibit the symmetry of the indenter tip (Fig. 13a). This is due to a number of factors including (i) the crystallography of the alumina, which determines which dislocations and crack systems can be activated, (ii) the mutually interactive process of dislocation and crack generation and propagation (which generates statistical variation), and (iii) the Cr^{3+} fluorescence measurement technique which averages over the 3D volume excited by the laser beam.

Interestingly the 200 g indents analysed in Figs. 11, 12 and 13 have a maximum value of measured compressive stress not at the centre of the indents, but about

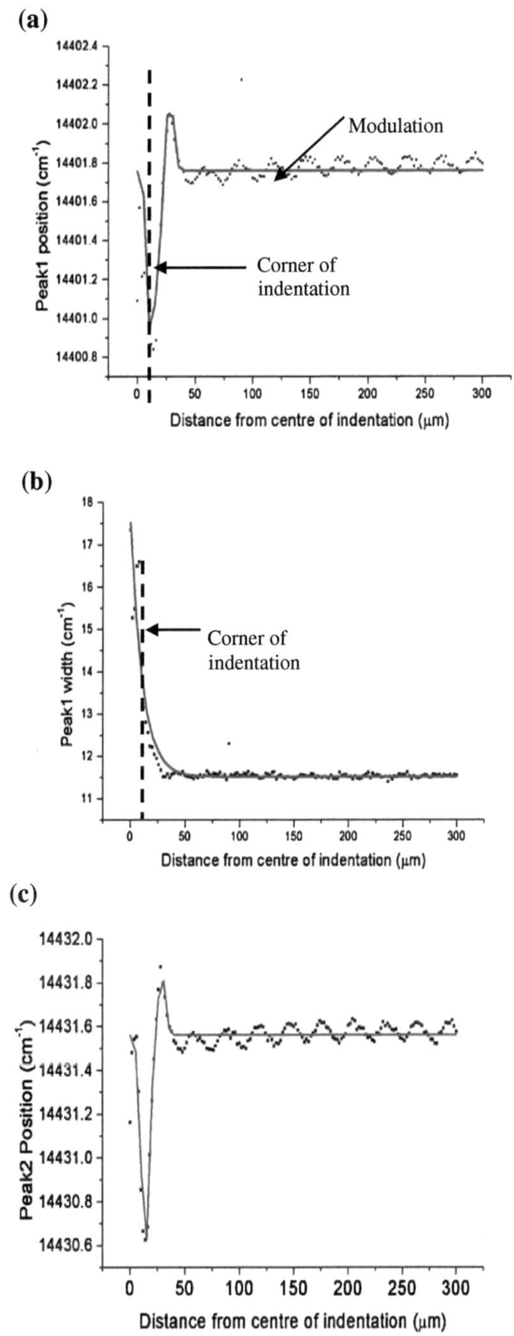

Fig. 14 **a**, **b** and **c** Show variations of peak1 position, peak1 width and peak2 position respectively, from the centre of 300 g Vickers microindentation site

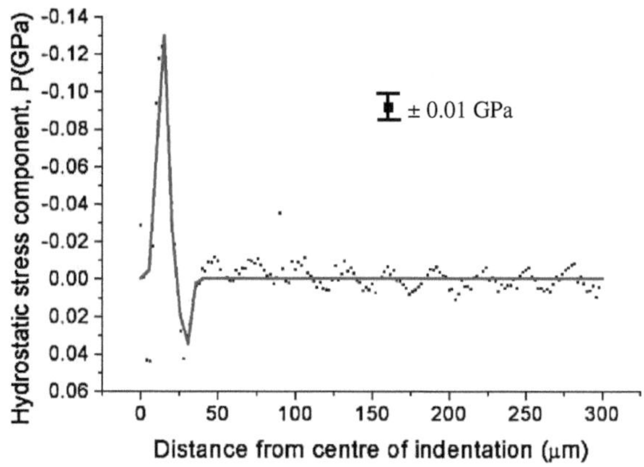

Fig. 15 Variation of the hydrostatic component of the residual stress with the distance from the centre of the 300 g Vickers microindentation

two thirds of the distance from the centre to the edge of the residual indent impression (120 MPa in Fig. 12). Similarly the maximum peak broadening is also observed towards the edge of the residual indent impression rather than at the centre, indicating increased local stress variations in the microstructure at that location. Further studies with better resolution (smaller spot size) and an increased number of data points, might better resolve the fluctuations in the residual stress near the centre of the indentation.

The results for the 300 g indentations sites (Figs. 14 and 15) were similar to 100 and 200 g except that the peak position base line had unexpected modulations due to equipment error.

4 Conclusions

- In this work, the surface damage caused by Vickers microindentation on R-cut alumina samples has been studied.
- Surface analyses reveal: Microindentation of single crystal alumina generates a complex central plastic deformation zone consisting of interconnecting cracks (radial, median, circumferential, lateral), planar faults (including basal and prism plane) and dislocations.
- The maximum length of surface cracks increased with indent load, e.g. for 100 g the radial cracks extended to a maximum of 12 μm from the corners of the residual indent impression, and for 300 g up to 20 μm
- The radial cracks in this chapter were located in very similar positions for loads 50, 100, 200 and 300 g and not exactly on the corners of the residual indent impression

Residual stress analysis by Cr^{3+} Fluorescence Spectroscopy reveals:

- The central zone of residual compressive stress extends laterally beyond the visible residual indent impression, and is surrounded by a zone of residual tensile stress which extends to beyond the range of the radial cracks. The maximum residual compressive stress measured did not occur at the geometrical centre of the indent.

- The Vickers indentation tip used to generate the surface damage in the alumina sample has four fold symmetry, but the microstructural damage and the measured residual stress around the indents do not exhibit the symmetry of the indenter tip.

References

1. Molis, S.E., Clarke, D.R.: Measurement of stresses using fluorescence in an optical microprobe: stresses around indentations in a chromium-doped sapphire. J. Am. Ceram. Soc. **73**(11), 3189–3194 (1990)
2. Ostertag, C.P., Robins, L.H., Cook, L.P.: Cathodoluminescence measurement of strained alumina single crystals. J. Eur. Ceram. Soc. **7**(2), 109–116 (1991)
3. Ma, Q., Clarke, D.R.: Piezospectroscopic determination of residual stresses in polycrystalline alumina. J. Am. Ceram. Soc. **77**(2), 298–302 (1994)
4. Banini, G.K., Chaudhri, M.M., Smith, T., et al.: Measurement of residual stresses around Vickers indentations in ruby crystal using a Raman luminescence. J. Phys. D Appl. Phys. **34**, L122–L124 (2001)
5. Grabner, L.: Spectroscopic technique for the measurement of residual stress in sintered Al_2O_3. J. Appl. Phys. **49**(2), 580–583 (1978)
6. Guillou, M.O., Henshall, J.L., Hooper, R.M.: Indentation fracture and soft impresser fatigue in sapphire and polycrystalline alumina. Int. J. Refract. Met. Hard Mater. **16**(4–6), 323–329 (1998)
7. Cook, R.F., Pharr, G.M.: Direct observation and analysis of indentation cracking in glasses and ceramics. J. Am. Ceram. Soc. **73**(4), 787–817 (1990)
8. Chan, H.M., Lawn, B.R.: Indentation deformation and fracture of sapphire. J. Am. Ceram. Soc. **71**(1), 29–35 (1988)
9. Farber, B.Y., Yoon, S.Y., Peter, K., et al.: Dislocation sources in sapphire (α-Al_2O_3) near microhardness indents. Mater. Res. Adv. Tech. **84**(6), 426–430 (1993)
10. Schwarzer, R.A.: Automated crystal lattice orientation mapping using a computer-controlled SEM. Micron **28**(3), 249–265 (1997)
11. Wang, Y.N., Huang, J.C.: Texture analysis in hexagonal materials. Mater. Chem. Phys. **81**(1), 11–26 (2003)
12. He, J., Clarke, D.R.: Determination of the piezospectroscopic coefficients for chromium-doped sapphire. J. Am. Ceram. Soc. **78**(5), 1347–1353 (1995)
13. Munro, R.G., Piermarini, G.J., Block, S., et al.: Model line-shape analysis for the ruby R lines used for pressure measurement. J. Appl. Phys. **57**(2), 165–169 (1985)
14. Inkson, B.J., Leclere, D., Elfallagh, F., et al.: The effect of focused ion beam machining on residual stress and crack morphologies in alumina. J. Phys: Conf. Ser. **26**, 219–222 (2006)
15. Wu, H.Z., Roberts, S.G., Mobus, G., et al.: Subsurface damage analysis by TEM and 3D FIB crack mapping in alumina and alumina/5vol.%SiC nanocomposites. Acta Mater. **51**(1), 149–163 (2003)

Superficial Parameters Determination of the Ti-6Al-4V Alloy Submitted to PIII Treatment in Different Times of Implantation

Susana Zepka, Danieli Aparecida Pereira Reis, Maria Margareth da Silva, Mario Ueda, Antonio Augusto Couto, Felipe Rocha Caliari and Adriano Gonçalves dos Reis

Abstract The advancement of technology leads to the development of new materials improving their tribology properties. It can be noticed in different areas like aerospace industry, chemical and oil that need resistant material in high temperatures and aggressive environments. In this case, it is important to think in its tribological properties, like wear, oxidation, toughness, and hardness. An effective mean, economic and with easy application is the plasma immersion ion implantation (PIII) technique. In the case of difficult shapes, the material can be equally treated. In this work, the Ti-Al-4V alloy was submitted to PIII during 2 and 3 h. The comparative analysis to determine which of which time was more efficient related to the tribological improvement measured by Auger, X-ray diffraction and wear. It will be observed images by MEV of the alloy submitted to PIII treatment.

S. Zepka · D. A. P. Reis (✉) · M. M. da Silva · F. R. Caliari · A. G. dos Reis
Instituto Tecnológico de Aeronáutica, Praça Mal.Eduardo Gomes, 50,
São José dos Campos, SP 12228-900, Brazil
e-mail: danielireis@gmail.com; danieli.reis@unifesp.br

D. A. P. Reis · F. R. Caliari
Universidade Federal de São Paulo—ICT, Rua Talim 330, São José dos Campos
SP 12231-280, Brazil

M. Ueda
Instituto Nacional de Pesquisas Espaciais, Avenida dos Astronautas, 1758,
São José dos Campos, SP 12227-010, Brazil

A. A. Couto
IPEN-CNEN/SP and Mackenzie Presbyterian University, Rua da Consolação,
930, São Paulo, SP 01302-907, Brazil

A. Öchsner and H. Altenbach (eds.), *Design and Computation of Modern Engineering Materials*, Advanced Structured Materials 54,
DOI: 10.1007/978-3-319-07383-5_26,

1 Introduction

The titanium alloy Ti-6Al-4V is the most widely used in engineering because it has very attractive properties as well as well as mechanical and microstructural aspects. However, during the deformation at high temperatures the alloy behavior changes with the thermodynamic parameters and the processing conditions. Although extraordinary advances in the development of titanium alloys with high tensile strength, ductility and creep resistance at high temperatures, the oxidation problems limit the use of these alloys at temperatures above 600 °C. Protective coatings that serve as barriers to the action of oxygen are necessary. The emerging technology of plasma immersion ion implantation (PIII) provides a simple and economical technique and can be used in structures of complex shapes in 3D, without the need of adherence of layers, training films and even chemical reactions. The study of the surfaces mechanical properties treated with the PIII technique is needed for the quantitative determination of improving their surface conditions. The Auger spectroscopy, X-ray diffraction and wear techniques were used in this work [1–9].

2 Materials and Methods

The plasma immersion ion implantation treatment consisted in mounting the sample of Ti-6Al-4V in the sample holder device, inserted and fixed in the reactor. It is sealed to form the vacuum, targeting the surface treatment, with a mechanical pump and after using a diffusing pump until the pressure up to 5.3×10^{-3} m bar, when the system is ready to begin the deployment implantation process. Nitrogen was injected (implantation of nitrogen ions in the sample) and to obtain the plasma, the pulser RUP-4 was turned on. The implantation of nitrogen treatments were performed at 250 °C, high voltage pulse of 8.5 kV, time average amplitude of 40 μs, frequency of 400 Hz, working pressure of 0.21 Pa during 120 and 180 min [2–12].

The determination of the atomic concentration as a function of depth for the samples with 2 and 3 h immersion was studied by the technique of Auger electron spectroscopy using the spectrometer FISIONS surface science instrument, Model F-310 MICROLAB, which was used to sputter the sample surfaces with argon. For the identification of the crystalline phases present in the samples was using a X-ray diffractometer, in a configuration θ/θ, with the equipment Panalytical hallmark and X'Pert Powder model. The analysis parameters were a pitch of 0.002, 2θ ranging between 10° and 90°, and a step time of 10 s.

The wear analysis was done using the pin-on-disk method. The equipment used was the CSM-Instruments Pin-on-disk Tribometer SN 18-313. The wear surface of the samples and the coefficient of friction was measured according to ASTM G99-95A [13]. The samples were prepared as single discs of 10 mm diameter and

3 mm in thickness, polished to a mirror finish, ultrasonically cleaned in acetone. The sample as received (without treatment PIII) was used as a reference for the characterization analysis. The width of the wear track was measured via optical microscopy, obtaining an average stretch of track. The cleaning of the samples and the pin was made with acetone; the analysis was performed at room temperature (18–22 °C) and relative humidity between 50 and 70 %. The pin is spherical of polycrystalline alumina with 99.85 % purity, with a hardness of 19,000 N/mm^2 and a diameter of 3 mm. The normal load used was 1.0 N, wear track radius of 3 mm distance traveled 94.25 m (5,000 cycles), the data acquisition rate of 5 Hz and a rate of 5 cm/s (150 rpm).

The fractographic analysis of implanted samples was done using a scanning electron microscope JEOL model JSM brand-5310.

3 Results and Discussions

The wear was determined by measuring the wear track width by optical microscope as seen in Fig. 1.

The wear is a function of the wear volume in mm^3, also called lost volume (V), is calculated by:

$$V = 2\pi R\left[r^2 \mathrm{sen}^{-1}(d/2r) - (d/4)\left(4r^2 - d^2\right)^{1/2}\right] \tag{1}$$

where R is the radius of the wear track, d is the width of track wear and r is radius of the pin. And the wear coefficient K is calculated by the expression:

$$K = V/(N \cdot L) \tag{2}$$

where V is the lost volume, N is the distance covered in meters and L is the applied load in Newtons [14–17].We then have the values of V and K for each immersion time shown in Table 1.

The results of the wear test is shown in Fig. 2, where it can be seen that the coefficient of friction of the treated sample is a function of number of cycles, and from the determinate number of cycles the average of the coefficient of friction, becomes of the untreated sample.

Comparing the untreated sample with the immersed for 120 min sample, it can be noted that the wear coefficient was 27.11 % less, the sample immersed for 180 min had a wear coefficient with respect to the untreated sample of the 34.22 % less. This result is consistent, if the wear coefficient decreases, the material hardness increases and then less material is lost. Comparing the results obtained in several other chapters, it can be said that they were consistent with the expected [12–16].

The X-ray diffraction shows through the Fig. 3, the surface modification with the nitrides formation. In Fig. 3a, it have the untreated sample which is observed

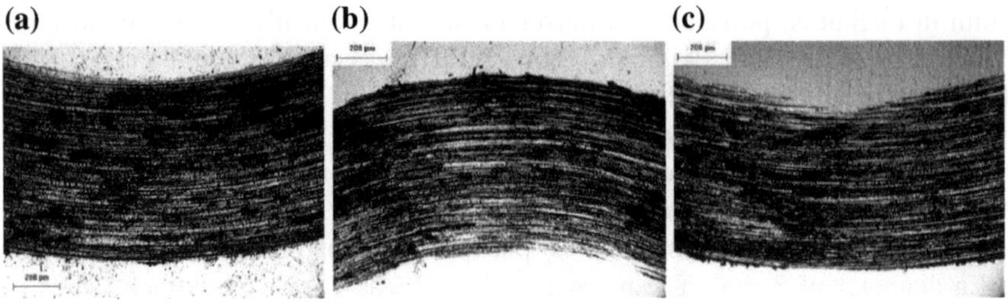

Fig. 1 Image of track wear obtained in the wear test: **a** Sample without treatment, **b** Sample with 2 h of implantation and **c** Sample with 3 h of implantation

Table 1 Values of *V* and *K* for each immersion time

Immersion time (min)	Track width *d* (mm)	Lost volume *V* (mm^3)	Wear coefficient $K \times 10^{-3}$ (mm^3/Nm)
0	0.73726	0.2124	20.25
120	0.66365	0.1547	1.64 (−27.11 %)
180	0.64341	0.1399	1.48 (−9.76 %)

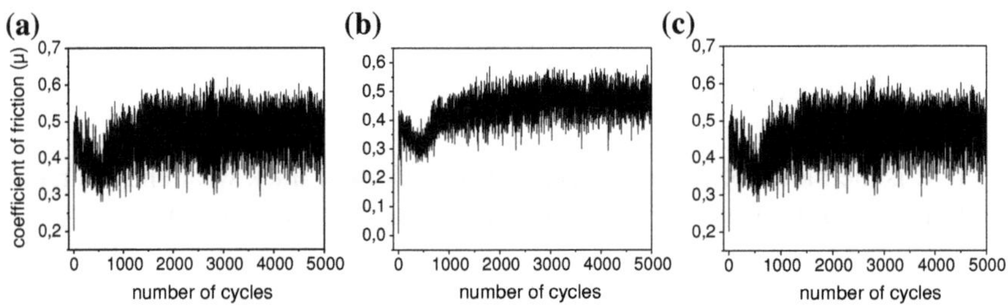

Fig. 2 Images of the results of wear tests: **a** untreated sample, **b** immersed for120 min sample and **c** immersed for180 min sample

only the presence of Tiα and Tiβ, in the Fig. 3b it have the result of sample with 120 min of implantation, where nitride formation appears, as well as in Fig. 3c, with more nitride formation and decreased Tiβ, since N is an alphagenic component [18, 19].

The Auger spectroscopy shows us through Fig. 4a and b, that with the PIII treatment, the layer obtained has the thickness of 18 nm for the sample at 2 h immersion and a layer of 36 nm for sample 3 h immersion. In addition we can observed in the sample which is for 2 h immersed that nitrogen atoms are measured up to a maximum concentration of 20 % to 6 nm and the sample 3 h immersion, 32 % of the maximum concentration to 18 nm.

It observe also the presence of the elements Ti, Al, and V in the alloy. The oxygen is found in the atmosphere and due to the strong affinity of titanium with

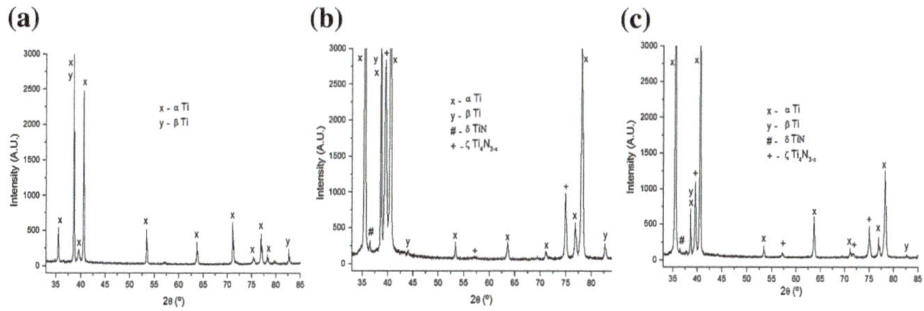

Fig. 3 **a** X-ray diffraction of the untreated sample, **b** X-ray diffraction of the sample with 2 h immersion and **c** X-ray diffraction of the sample with 3 h immersion

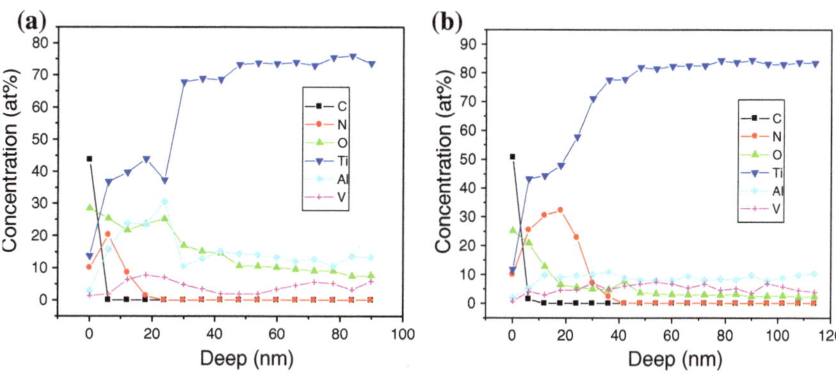

Fig. 4 Sample Auger spectroscopy with: **a** 2 h implantation, **b** 3 h implantation

oxygen, forming a passive oxide layer on its surface, which serves as a protective layer against corrosion. The thickness of this oxide layer may undergo some changes according to the ambient conditions and the treatment. In the sample treated for 3 h immersion, oxygen is detected within 10 nm with an atomic concentration maximum of approximately 25 % near the surface. It is observed that the concentrations of the atomic elements aluminum and vanadium are smaller near the surface due to the presence of oxygen and nitrogen. The presence of carbon at the surface is due possibly to the oil contamination of the PIII system vacuum pump [7, 10].

The scanning electron microscopy showed that the surface roughness increased, as observed in Fig. 5a–c. This roughness increases as the immersion time increases, for the formation of nitrides in the implantation surface [19].

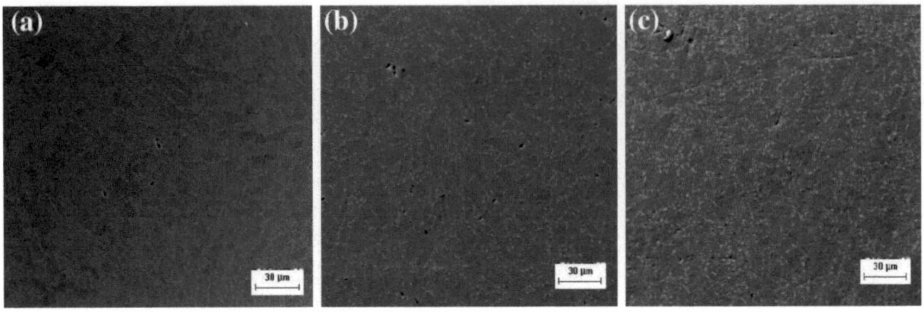

Fig. 5 **a** SEM image of the sample surface without treatment, **b** SEM image of the sample surface with 2 h of immersion and **c** SEM image of the sample surface with 3 h of immersion

4 Conclusions

The objective of this work was the evaluation of the wear resistance of Ti-6Al-4V alloy after plasma immersion ion implantation (PIII). The finality of this process was the modification of surface properties of the alloy to obtain better tribology properties. The samples were submitted to 120 and 180 min of implantation and analyzed by pin-on-disk process, where the lost volumes and wear coefficients were compared in the samples. Observed were the decreasing of attrite coefficient and the lost volume of the material during wear test. The implanted sample by 180 min was the wear coefficient in relation of the sample without treatment of 34.22 % lower, and 9.76 % lower in relation to the implanted sample by 120 min. It can be observed that the sample implanted by 180 min showed lower wear coefficient. The results confirmed the improvement of PIII treatment on mechanical resistance of the alloy.

Acknowledgments The authors thank FAPESP, CAPES and CNPq for supporting research.

References

1. Lee, W.S., Lin, C.F.: High–temperature deformation behavior of Ti-6Al-4V alloy evaluated by strain-rate compression tests. J. Mat. Proc. Tech. **75**, 127–136 (1998)
2. Welsch, G., Kahveci, A.I.: Oxidation of High Temperature. Intermetallics. TMS, Warrendale, PA (1988)
3. Silva, M.M., Ueda, M., Pichon, L., et al.: Surface modification of Ti-6Al-4V alloy by PIII at high temperatures: Effects of plasma potential. Nucl. Instrum. Methods Phys. Res. B **257**, 722–726 (2007)
4. Ueda, M., Silva, M.M., Lepienski, C.M., et al.: Temperature plasma immersion ion implantation of Ti6Al4V. Surf. Coat. Technol. **201**, 4953–4956 (2007)
5. Mello, C.B., Ueda, M., Silva, M.M., et al.: Tribological effects of plasma immersion ion implantation heating treatments on Ti-6Al-4V alloy. Wear **267**, 867–873 (2009)
6. Reis, D.A.P.: Effect of ceramic coatings and test atmosphere in the creep of refractory alloy (doctoral thesis). INPE, São José dos Campos, Brazil (2005)

7. Silva, M.M.: Modification of surface properties of Ti-6Al-4V by plasma-assisted processes at low and high temperatures (doctoral thesis) ITA, São José dos Campos, Brazil (2007)
8. Fouquet, V., Pichon, L., Straboni, A., et al.: Nitridation of Ti6Al4V by PBII: study of the nitrogen diffusion and of the nitride growth mechanism. Surf. Coat. Tech. **186**, 34–39 (2002)
9. Fouquet, V., Pichon, L., Drouet, M., et al.: Plasma assisted nitridation of Ti-6Al-4V. Appli. Surf. Sci. **221**, 248–258 (2004)
10. Silva, M.M., Ueda, M., Otani, C., et al.: Hybrid processing of Ti-6Al-4V using plasma immersion ion implantation combined with plasma nitriding mater. Research **9**, 97–100 (2005)
11. Raveh, A., Hansen, P.I., Avni, R. et al.: Mechanisms of r.f. plasma nitriding of Ti-6Al-4V alloy. Surf. Coat. Tech. **38**, 339 (1990)
12. Silva, L.I.G., Ueda, M., Silva, M.M., et al.: Results from experiments on hybrid plasma immersion ion implantation/nitriding processing of materials. IEEE Trans. Plasma Sci. **221**, 34–39 (2004)
13. ASTM G99–95a (2000) Standard Test Method for Wear Testing with a Pin-on-Disk Apparatus, Philadelphia
14. Rabinowicz, E.: Friccion and Wear of Materials. Wiley, Massachusetts (1964)
15. Silva, M.M., Pichon, L., Otubo, J.: Wear properties studies of NiTi shape memory alloy after nitrogen plasma based treatments. In: 21st International Congress of Mechanical Engineering, Brazil (2011)
16. Zepka, S., Reis, D.A.P., Silva, M.M. et al.: Study of Creep and Nanoindentation of Ti-6Al-4V Alloy after Plasma Immersion Ion Implantation (PIII). In: Proceedings of Materiais 2013 Congress, Coimbra (2013)
17. Ueda, M., Silva, M.M., Otani, C., et al.: Improvement of tribological properties of Ti6Al4V by nitrogen plasma immersion ion implantation. Surf. Coat. Tech. **169**, 408–410 (2003)
18. Mello, C.B., Ueda, M., Silva, M.M., et al.: Tribological effects of plasma immersion íon implantation heating treatments on Ti-6Al-4V alloy. Wear **267**, 867–873 (2009)
19. Nolan, D., Huang, S.W., Leskovsek, V., et al.: Slinding wear of titanium nitride thin films deposited on Ti-6Al-4V alloy by PVD and plasma nitriding processes. Surf. Coat. Tech. **200**, 5698–5705 (2006)

Investigation of Carbon Nanotube Defects on Its Strength Using Nonlinear Finite Element Modeling

Ehsan Mohammadpour and Mokhtar Awang

Abstract Due to their remarkable properties, carbon nanotubes have been used in many engineering applications in the form of composite. In some cases, however defects may occur and lead to poor strength of the composite. In this chapter, a finite element method has been developed to evaluate mechanical behavior of defective CNTs and graphene sheets. Stone-Wales defect (5-7-7-5) and vacancies in SWCNTs and graphene sheets with different hilarities were studied under axial load. The results show that CNTs and grapheme structures are sensitive to vacancies. However, the armchair grapheme structures have significant resistance to defect under tensile load. Present results are in good agreement with available literature.

Keywords Carbon nanotube · Vacancy · Defects · Nonlinear finite element method · Mechanical properties

1 Introduction

It was shown that carbon nanotube (CNT) defects can be found at different steps of its fabrication and purification process [1, 2], during device or composite production [3, 4], or under mechanical strains [5]. It was also reported that even few number of defects in the atomic structure of the CNT will result in deterioration of mechanical or electrical properties of the CNT [6, 7]. At present, the

E. Mohammadpour · M. Awang (✉)
Mechanical Engineering Department, Universiti Teknologi PETRONAS,
Seri Iskandar, Perak, Malaysia
e-mail: mokhtar_awang@petronas.com.my

E. Mohammadpour
e-mail: eh.mohammadpour@gmail.com

A. Öchsner and H. Altenbach (eds.), *Design and Computation of Modern Engineering Materials*, Advanced Structured Materials 54,
DOI: 10.1007/978-3-319-07383-5_27,

studies of the defective CNTs and atomistic effects of these imperfections has been a great challenge from both experimental and numerical aspects.

In order to simulate the mechanical behavior of carbon–carbon bonds beyond bond breaking, a complex interatomic potential function is needed. For instance, the Brenner potential function is mentioned as an accurate model [8]. A continuum mechanics approach directly incorporating the Brenner potential function has been developed by Zhang et al. [9] and Jiang et al. [10]. They modeled elastic properties and stress–strain relationships of carbon nanotubes based on a modified Cauchy–Born rule. An analytical molecular structural mechanics model [11] incorporating the modified Morse potential function [12] has been developed by model defect-free CNTs under tensile loadings.

From the available literature, continuum mechanics based models have not been fully developed for predicting the effects of defects on mechanical properties of CNTs and nanotube composites. The quantum mechanics and molecular dynamics (MD) simulations are the main approaches for modeling the effects of defects on mechanical [6, 7, 12–14] and thermal properties [15, 16] of CNTs. One attempt using a continuum based atomistic model to study defect nucleation in carbon nanotubes under mechanical loadings can be seen in the study by Jiang et al. [17]. The effect of defects on fracture of nanotubes has been studied by Tserpes et al. [18, 19] using the finite element (FE) based model where they assumed that the dimensions of the nanotube structures remain unchanged after the formation of defects, which may not be true as atoms redistribute to minimize energy.

In this chapter, a finite element method has been developed to evaluate mechanical behavior of defective CNTs and graphene sheets. Stone-Wales (SW) defect (5-7-7-5) [20] and vacancies in SWCNTs and graphene sheets with different chilarities were studied under axial load. Then, the stress–strain relationship of defective structures containing different number of defects is predicted by using the present finite element model based on the modified Morse potential function. Effect of defects on the tensile strength of carbon nanostructures are discussed. The predictions compare favorably to the corresponding published numerical results such as molecular dynamics for armchair and zigzag carbon nanotubes and graphene.

2 Finite Element Model

There are several different potential functions available [8, 21, 22] for describing C–C bond interaction other than simple harmonic functions. Among them, the modified Morse potential function [12] is accurate and practical for the present study. The modified Morse potential function was correlated to the Brenner potential function for strains below 10 %. The torsional energy term was neglected from the bond total energy due to its low contribution in tensile behavior of the nanotube and graphene sheets [23, 24]. The energy potential function is given as follow

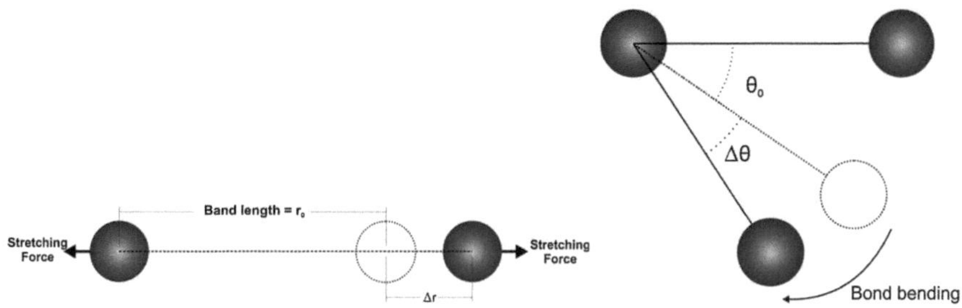

Fig. 1 C–C bond interaction according to modified morse functions

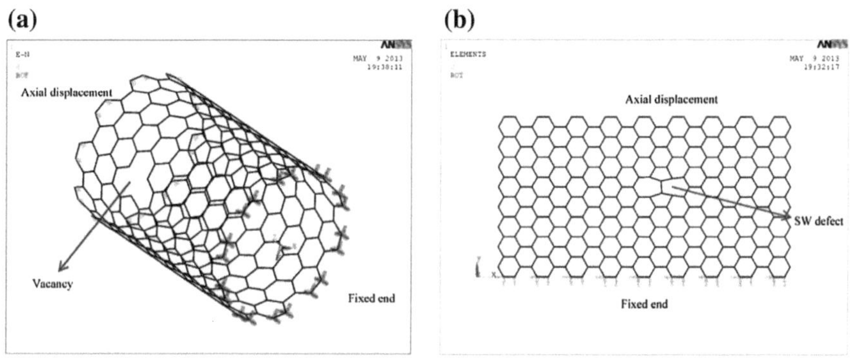

Fig. 2 **a** Vacancy in a SWCNT and **b** SW defect in a graphene sheet

$$E = E_{stretch} + E_{angle} \tag{1}$$

where, $E_{stretch}$ is the bond energy due to bond stretch Δr and E_{angle} is the bond energy due to bond angle variation $\Delta\theta$. The stretch force, the angle-variation moment can be obtained from the derivative of Eq. (1) as functions of bond stretch and bond angle variation, respectively:

$$F(\Delta r) = 2\beta D_e(1 - e^{-\beta\Delta r})e^{-\beta\Delta r} \tag{2}$$

$$M(\Delta\theta) = k_\theta\Delta\theta[1 + 3k_{sextic}\Delta\theta^4] \tag{3}$$

Finite element solutions for predicting nonlinear mechanical behavior of SWCNTs and graphene have been investigated using above equations for C–C bond variations. In this chapter, beam elements as suggested in [23, 25] were employed to model interatomic interactions. The bond interaction schematic is as shown in Fig. 1.

The boundary conditions for the models are shown in Fig. 2. The atoms on the bottom edge of the tube were fixed in all directions. The other end of the CNT was axially displaced incrementally to introduce load into the tube.

The interatomic force is calculated for all atoms along the end of the nanotube where the displacement is prescribed. The axial strain of the CNT is computed as $\varepsilon = \frac{\Delta l}{l_0}$, where l_0 is the initial length of the CNT or graphene. The force–strain relationship of nanotubes is predicted using the modified Morse potential as outlined above.

It should be noted that the present method does not require a tube thickness to be defined. However, in order to compare the results with published data, a conventional modulus and strengths can be calculated using wall thickness of 0.34 nm for CNTs and graphene sheets. The total force carried by the CNTs can be extracted from the simulation results and then the stress can be computed as $\sigma = \frac{F}{A}$ and the Young's modulus as $= \frac{F.l}{\Delta l.A}$, respectively.

The predicted nonlinear behaviors of nanotubes are very similar to results calculated from MD approach [12]. From both studies [11, 12], the predicted strengths and failure strains are significantly higher than the experimental values (11–63 GPa for strength and 10–13 % for failure strain) of Yu et al. [26]. This difference can be partially explained by the presence of defects that reduce stiffness and strength. The mechanical behavior of defective nanotubes is studied in the following section.

3 Results and Discussion

Various types of defects exist in CNTs and graphene such as vacancies and SW (5-7-7-5). Graphene sheets and SWCNTs of armchair (12, 12) and zigzag (20, 0) were modeled in length of 10 nm. It was shown before that the Young's moduli for both armchair and zigzag nanotubes decrease with decreasing tube diameter and approach the predicted graphite value when the tube diameter is increased. Effects of tube curvature were neglected because (12, 12) and (20, 0) nanotubes have similar tube diameter. In this work, the displacement was increased gradually until carbon–carbon (C–C) bonds reach inflation point and bonds breaks. Bond strain and stress can be extracted from the simulation results. Figure 3 shows effect of vacancies on the graphene (12, 12) and SWCNT (12, 12).

Different numbers of vacancies were analyzed to these structures in order to study the effect of vacancy density on the mechanical behavior of graphene sheets and nanotubes. Figure 4 shows stress-strain relationships for graphene (12, 12) and SWCNT (12, 12) containing different amount of vacancies. It can be seen when the number of vacancies increases from 0 to 50, both structures become less resistance to stress. The same behavior was observed from simulation results of graphene (20, 0) and SWCNT (20, 0). It has been reported that vacancies can badly deteriorate mechanical properties of carbon nanostructures.

Figure 5 shows the calculated Young's modulus of defective nanotubes and graphene sheets with different chilarities. Young's modulus of nanotubes and

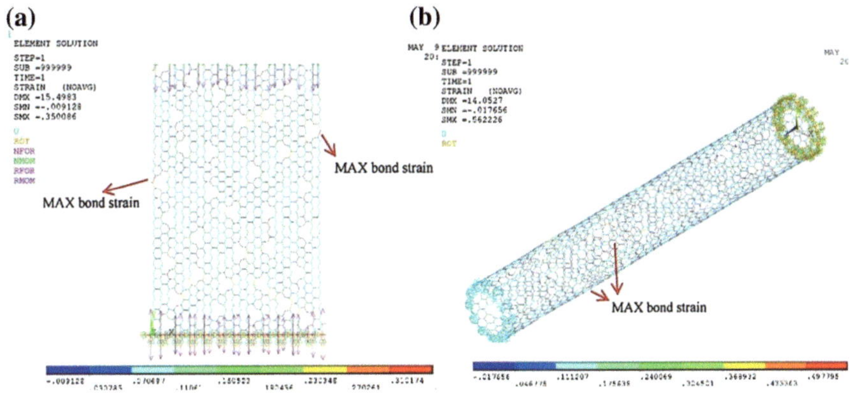

Fig. 3 Effect of 26 vacancies on the graphene and SWCNTs: **a** graphene (12, 12) and **b** SWCNT (12, 12)

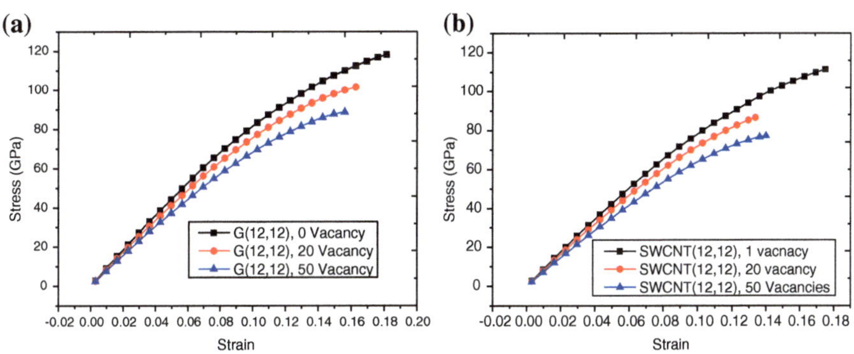

Fig. 4 Stress-strain curves for **a** Graphene (12, 12), **b** SWCNT (12, 12)

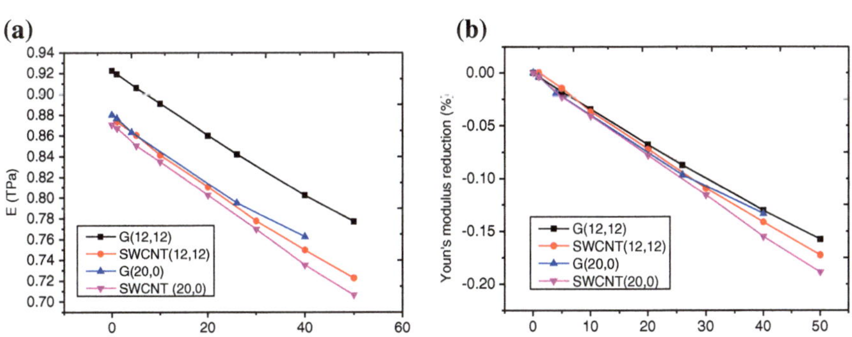

Fig. 5 **a** Effect of number of vacancies on the young's modulus of SWCNTs and graphene sheets, **b** reduction of young's modulus caused by vacancies

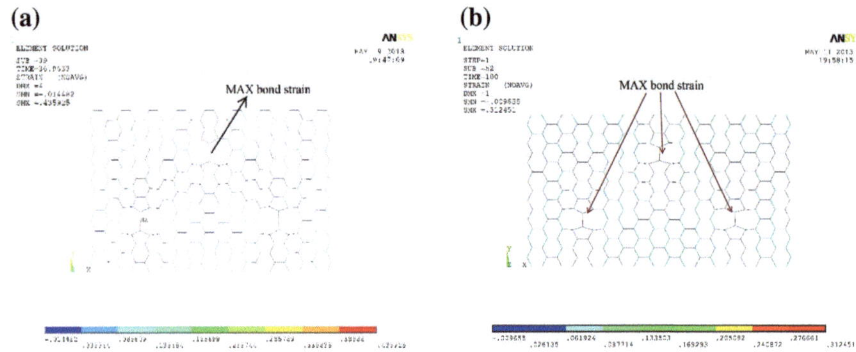

Fig. 6 Effect of SW defects on graphene (12, 12): **a** 1 SW and **b** 3 SW

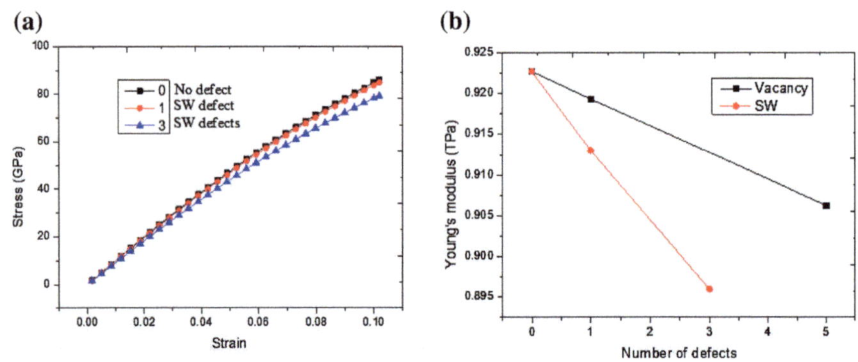

Fig. 7 **a** Stress-strain curves of defect-free and defective graphene sheet (12, 12), **b** Effect of different point defects on the Young's modulus of graphene sheet (12, 12)

graphene sheets are reduced from 0.88 to 0.72 TPa when the number of vacancies increased from 1 to 50, respectively.

Results shows that such reduction is highly dependent on the chirality. Armchair structures have greater mechanical strength comparing to zigzag structures. Obviously the graph of graphene sheet (12, 12) shows remarkable resistance to vacancies in comparison with other nanostructures. Zigzag configurations are weaker and more sensitive to defects as deduced from simulation results.

The Stone–Wales 5-7-7-5 defect involves the 90° rotation of a carbon bond with a new configuration. Figure 6 shows two graphene sheet (12, 12) with one and three SW defect. Bond strains under tensile loading are depicted in Fig. 7. It seems that the SW defect is the weakest part in a graphene sheet. In addition, the effect of the SW defect on the configuration is found out to be local and limited to atoms in the neighborhood of the defect.

Atoms far away from the defect undergo uniform deformation and their geometry configurations are the same as the defect-free tubes. Stress-strain curves of prefect graphene sheets and defected graphene sheets are shown in Fig. 7. Results show that SW defect has significant effect on the tensile behavior of graphene.

Figure 7a shows the calculated stress–strain relationships for armchair and zigzag nanotubes with and without SW defects. As can be seen from Fig. 7b, when 3 SW defect were introduced to a (12, 12) graphene sheet, its Young's modulus decreases from 0.9227 GPa to 0.896 TPa which is more effective than vacancies. The predicted tensile strength as 85.9 GPa of defect free graphene is much greater than that 75 GPa of defective graphene.

4 Conclusions

Several conclusions can be drawn from this research:

1. The present approach is capable of predicting Young's modulus, and stress–strain relationship of graphene sheets and nanotubes with or without vacancy/SW defects.
2. Results show that CNTs and grapheme structures are sensitive to vacancies. However armchair grapheme structures have significant resistance to defect under tensile load. Moreover, armchair structures for both CNTs and grapheme sheets are stronger than zigzag ones.
3. Effects of the SW defect on the Young's modulus, fracture and failure of defective CNTs have been investigated. Using the present approach, it is feasible to model multiple defects and their interaction in both SWCNT and MWCNT since the present approach is much simpler and computationally efficient than the classical molecular dynamics model.

Acknowledgements The authors wish to acknowledge the financial supports given by the Ministry of Science Technology and Innovation (MOSTI) under an E-science grant No. 158-200-147 and Universiti Teknologi PETRONAS.

References

1. Andrews, R., Jacques, D., Qian, D., et al.: Purification and structural annealing carbon nanotubes at graphitization temperatures. Carbon **39**, 1681 (2001)
2. Mawhinney, D.B., Naumenko, V., Kuznetsova, A., et al.: Surface defect site density on single walled carbon nanotubes by titration. Chem. Phys. Lett. **6**, 213 (2000)
3. Gojny, F.H., Nastalczyk, J., Roslaniec, Z., et al.: Influence of different carbon nanotubes on the mechanical properties of epoxy matrix composites–a comparative study. Chem. Phys. Lett. **370**, 820 (2003)
4. Gojny, F.H., Schulte, K.: Functionalisation effect on the thermo-mechanical behaviour of multi-wall carbon nanotube/epoxy-composites. Comp. Sci. Tech. **64**, 2303 (2004)
5. Nardelli, M.B., Fattebert, J.L., Orlikowski, D., et al.: Mechanical properties, defects and electronic behavior of carbon nanotubes. Carbon **38**, 1703 (2000)

6. Sammalkorpi, M., Krasheninnikov, A., Kuronen, A., et al.: Mechanical properties of carbon nanotubes with vacancies and related defects. Phys. Rev. B **70**, 245 (2004)
7. Lu, Q., Bhattacharya, B.: Effect of randomly occurring stone–wales defects on mechanical properties of carbon nanotubes using atomistic simulation. Nanotechnology **16**, 555 (2005)
8. Brenner, D.W.: Empirical potential for hydrocarbons for use in simulating the chemical vapor deposition of diamond films. Phys. Rev. B **42**, 9458 (1990)
9. Zhang, P., Huang, Y., Geubelle, P.H., et al.: The elastic modulus of single-wall carbon nanotubes: a continuum analysis incorporating interatomic potentials. Int. J. Solids Struct. **39**, 3893 (2002)
10. Jiang, H., Zhang, P., Liu, B., et al.: The effect of nanotube radius on the constitutive model for carbon nanotubes. Comput. Mater. Sci. **28**, 429 (2003)
11. Xiao, X.R., Gama, B.A., Gillespie, J.W.: An analytical molecular structural mechanics model for the mechanical properties of carbon nanotubes. Int. J. Solids Struct. **42**, 3075 (2005)
12. Belytschko, T., Xiao, S.P., Schatz, G.C., et al.: Atomistic simulations of nanotube fracture. Phys. Rev. B **65**, 235 (2002)
13. Rahmandoust, M., Öchsner, A.: Influence of structural imperfections and doping on the mechanical properties of single-walled carbon nanotubes. J. Nano Res. **6**, 185–196 (2009)
14. Ghavamian, A., Rahmandoust, M., Öchsner, A.: A numerical evaluation of the influence of defects on the elastic modulus of single and multi-walled carbon nanotubes. Comput. Mater. Sci. **62**, 110–116 (2012)
15. Huxtable, S.T., Cahill, D.G., Shenogin, S.: Interfacial heat flow in carbon nanotube suspensions. Nat. Mater. **2**, 731 (2003)
16. Shenogin, S., Bodapati, A., Xue, L.: Deformation of glassy polycarbonate and polystyrene: the influence of chemical structure and local environment. Appl. Phys. Lett. **85**, 2229 (2004)
17. Jiang, H., Feng, X.Q., Huang, Y.: Defect nucleation in carbon nanotubes under tension and torsion: stone–wales transformation. Comput. Methods Appl. Mech. Eng. **193**, 3419 (2004)
18. Tserpes, K.I., Papanikos, P., Tsirkas, S.A.: A progressive fracture model for carbon nanotubes. Compos. B: Eng. **37**, 662 (2006)
19. Tserpes, K.I., Papanikos, P.: The effect of stone–wales defect on the tensile behavior and fracture of single-walled carbon nanotubes. Comp. Struct. **79**, 581 (2007)
20. Stone, A.J., Wales, D.J.: Theoretical studies of icosahedral C60 and some related structures. Chem. Phys. Lett. **128**, 501 (1986)
21. Abell, G.C.: Empirical chemical pseudopotential theory of molecular and metallic bonding. Phys. Rev. B **31**, 6184 (1985)
22. Tersoff, J.: Empirical interatomic potential for carbon with applications to amorphous-carbon. Phys. Rev. Let. **61**, 2872 (1988)
23. Li, C.Y., Chou, T.W.: A structural mechanics approach for the analysis of carbon nanotubes. Int. J. Solids Struct. **40**, 2487 (2003)
24. Chang, T., Gao, H.: Size-dependent elastic properties of a single-walled carbon nanotube via a molecular mechanics model. J. Mech. Phys. Solids **51**, 1059 (2003)
25. Kalamkarov, A.L., Georgiades, A.V., Rokkam, S.K., et al.: Analytical and numerical techniques to predict carbon nanotubes properties. Int. J. Solids Struct. **43**, 6832 (2006)
26. Yu, M.F., Files, B.S., Arepalli, S., et al.: Tensile loading of ropes of single wall carbon nanotubes and their mechanical properties. Phys. Rev. Lett. **84**, 5552 (2000)

Multimodal Pushover Target Acceleration Method Versus Dynamic Response of R/C Frames

Ivan Balić, Ante Mihanović and Boris Trogrlić

Abstract A new *multimodal pushover target acceleration method* for the nonlinear analysis of reinforced concrete (R/C) frames subjected to seismic action is presented in chapter. The aim of research shown in this chapter was to find the influence of multimodal combinations in assessing the bearing capacity of R/C frames based on linear (L) combination of modes, and to compare the target ground acceleration $a_{gr,t}$ ($a_{gr,u}$) of the multimodal pushover target acceleration method with the failure peak ground acceleration $a_{gr,d}$ obtained by a dynamic transient response of R/C frames. The target acceleration presents the lowest seismic resistance and it is the minimum acceleration of the base that leads to the ultimate limit state of structure, and it is reached by an iterative procedure. In accordance with the Eurocode 8 rules, application of the pushover method favors access utilizing the first mode. Examples of presented 5-storey spatial R/C frame show the significant influence of higher modes. A formulation for determining the equivalent structural system damping by equalizing the dissipated energy during one cycle of vibration of the nonlinear system and the equivalent linear system is presented in this chapter. Results of the dynamic response of R/C frames are also presented in this chapter. As seismic excitation eight real earthquake accelerograms are taken. On the basis of the results obtained by nonlinear dynamic time-history analysis validation of the procedure of searching the target ground acceleration was made. Taking into account the elastic spectrum with calculated equivalent structural damping and usability of the capacity curve up to 3/3 of a displacement, the comparison of the target acceleration $a_{gr,u}$ of the multimodal

I. Balić · A. Mihanović · B. Trogrlić (✉)
Faculty of Civil Engineering, Architecture and Geodesy, University of Split,
Matice hrvatske 15, 21000 Split, Croatia
e-mail: boris.trogrlic@gradst.hr

I. Balić
e-mail: ivan.balic@gradst.hr

A. Mihanović
e-mail: ante.mihanovic@gradst.hr

A. Öchsner and H. Altenbach (eds.), *Design and Computation of Modern Engineering Materials*, Advanced Structured Materials 54,
DOI: 10.1007/978-3-319-07383-5_28,

pushover method with the failure peak ground acceleration $a_{gr,d}$, obtained by a dynamic response of the structure, shows a very good agreement between the target acceleration and the failure peak ground acceleration.

Keywords Target acceleration · Multimodal pushover method · Dynamic response · Seismic load · R/C frame · Linear combination of modes · Structural damping

1 Introduction

One of the methods for determining the seismic resistance of a structure is the nonlinear static pushover method which is a part of the European Norm EN 1998-1 [1]. In practical application, the pushover method based on form of the first load vector is dominant. The European Norm EN 1998-1 introduced nonlinear analysis of structures, which implementation and determination of target displacement are shown in Annex B. In numerous research papers, such as [2–10], the influence of higher modes on the results of nonlinear static analysis applied through the pushover method were analysed with the conclusion of significant impact of higher modes. In certain papers [11–13] comparison of results between nonlinear dynamic and static seismic analysis of structures were carried out.

A new multimodal pushover target acceleration method and the term target acceleration as a measure of the lowest seismic resistance is presented in this chapter. The method can be applied on a specific elastic spectrum. In examples presented in this chapter, spectrum type 1 and ground type *A* are chosen in accordance with the EC8 [1].

In this chapter, 5-storey spatial R/C frames are analyzed by nonlinear numerical model of stability and load-bearing capacity of R/C space frames with composite cross-sections [14, 15]. The global iterative-incremental numerical procedure is used with a monotonic increase of loading until the system collapse which is caused by occurring of mechanism and/or losing of stability. The space equilibrium path, i.e. the capacity curve and plastic hinge appearance of the cross-section are obtained as a result of the analysis. The possibility of incremental loading the first one (vertical-self weight, permanent load) and then the other load (horizontal earthquake action) enables the application of the nonlinear static pushover method in accordance with the European Norm EN 1998-1 [1].

On the example of a 5-storey spatial R/C frame, a new multimodal pushover target acceleration method is presented and a validation of the procedure of searching the target ground acceleration on the basis of the results obtained by nonlinear dynamic time-history analysis was made.

Figure 1 shows a schematic presentation of the aim of this chapter, which will be described below.

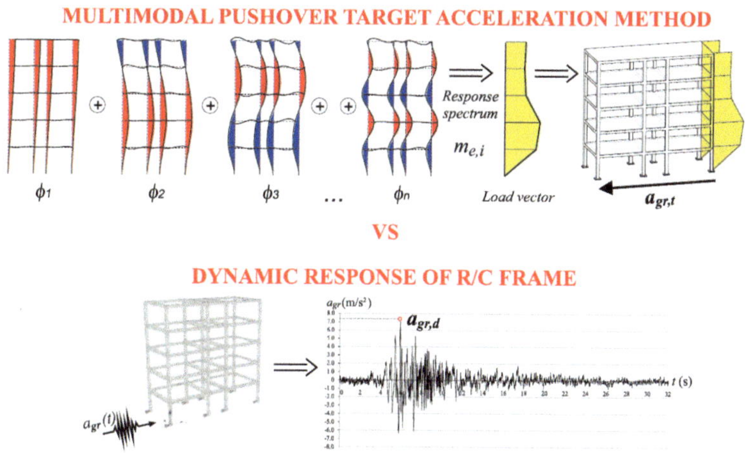

Fig. 1 Schematic presentation—multimodal pushover target acceleration method versus dynamic response of R/C frame

2 Multimodal Pushover Target Acceleration Method

The 5-storey spatial R/C frame with dimensions as shown in Fig. 2a is analysed. Dimensions and discretization of cross-sections of beams are shown in Fig. 2d, e and dimensions and discretization of cross-sections of columns are shown in Fig. 2f. Properties of cross-sections are assigned to the frame according to Fig. 2b.

For the computation of the eigenvectors, a self weight of model is assigned, additional dead load on plates (2.0 kN/m^2), variable load on plates (2.0 kN/m^2) and material module $E_{c0} = 30.5$ GPa, as initial elastic module of concrete.

To calculate the capacity curve, numerical models of concrete and reinforcing steel are used as shown in Fig. 2g, h. Distributed loading is assigned on beams according to Fig. 2c, and concentrated vertical forces (25.0 kN) are set in nodes on intersections of beams and columns. The total weight of the model is $W = 8{,}290.0$ kN, and the total mass of the model is $M = 845.0$ t.

2.1 Eigenvectors

By solving the linear dynamic analysis of the 5-storey R/C frame in x and y direction, eigenvectors ϕ_i with a data of corresponding periods T_i and participating modal masses $m_{e,i}$, are obtained as shown in Tables 1 and 2. The failure shear forces F_{fi} and the limit ground acceleration $a_{gr,i}$ for the each eigenvector are determined by the pushover procedure.

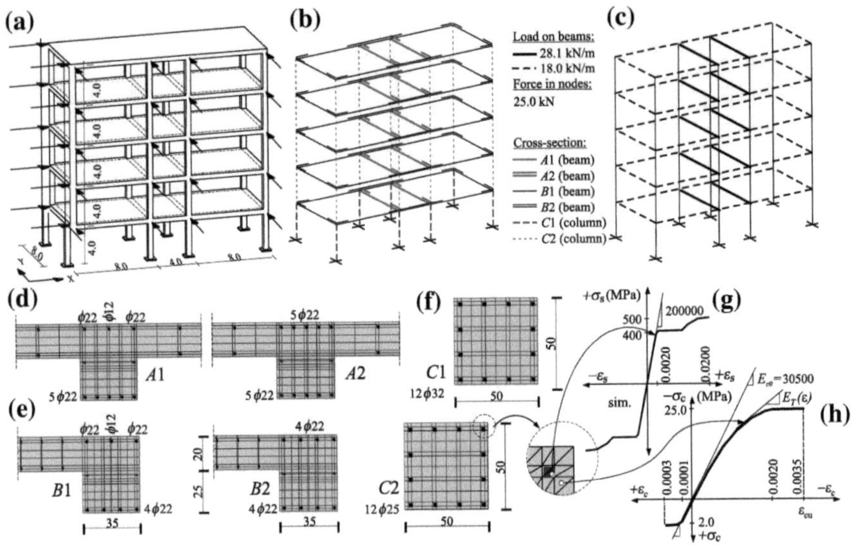

Fig. 2 *Example* **a** 5-storey spatial R/C frame, **b** cross-section marks, **c** load on beams, **d–f** cross-section discretisation, **g** reinforcing-steel model, **h** concrete model

Table 1 Horizontal load vector properties for 5-storey R/C frame in x direction

Vector	T_i (s)	$m_{e,i}$ (%)	$m_{e,i}$ (t)	F_{fi} (kN)	$a_{gr,I}$ (g)
$\phi 1$	0.839	81.47	688.42	1,735	0.405
$\phi 2$	0.265	10.51	88.81	1,823	1.75
$\phi 3$	0.146	4.12	34.81	2,786	7.90
$\phi 4$	0.097	1.98	16.73	1,954	6.32
$\phi 5$	0.076	0.67	5.66	1,838	41.70

Table 2 Horizontal load vector properties for a 5-storey R/C frame in y direction

Vector	T_i (s)	$m_{e,i}$ (%)	$m_{e,i}$ (t)	F_{fi} (kN)	$a_{gr,I}$ (g)
ϕ_1	0.953	80.34	678.87	1,427	0.39
ϕ_2	0.293	10.94	92.44	1,572	1.65
ϕ_3	0.156	4.51	38.11	2,593	4.62
ϕ_4	0.101	2.27	19.18	2,146	8.08
ϕ_5	0.077	0.78	6.59	1,576	17.70

2.2 Determination of the Target Acceleration

A defined shape of the lateral loading vector is monotonically increased in the nonlinear analysis, up to failure, i.e. achieving the critical load factor. The shape of the first load vector in limit state, corresponding capacity curve and the order of

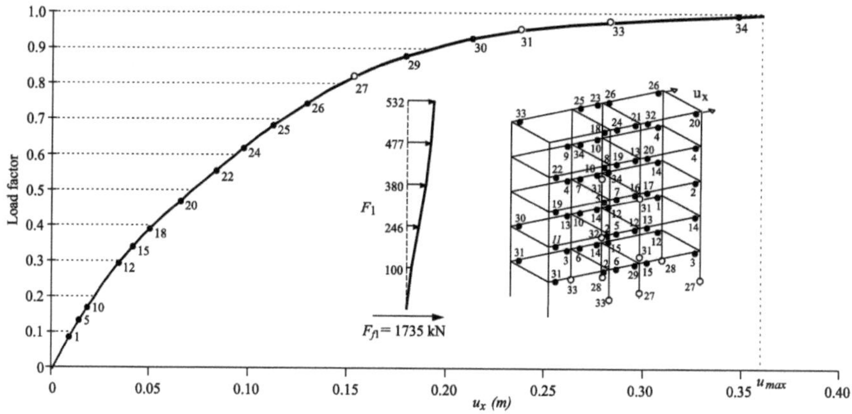

Fig. 3 Capacity curve for the first vector ϕ_1 (direction x)

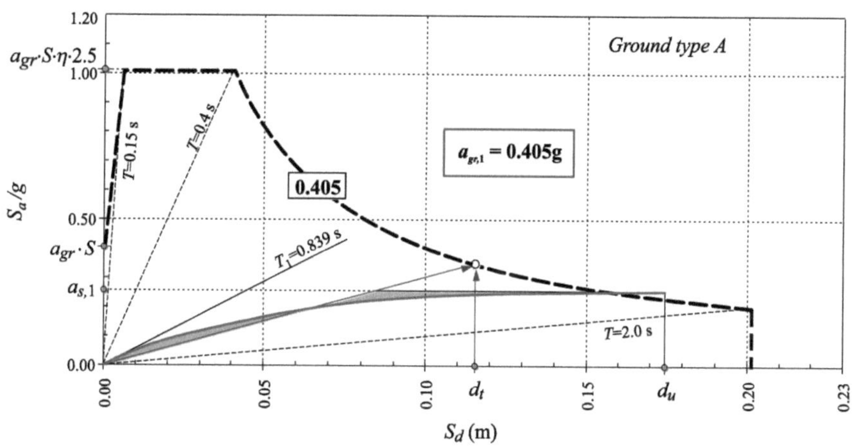

Fig. 4 Determination of seismic resistance for the first vector ϕ_1

beginning of plastification and plastic hinge positions for x direction are shown in Fig. 3. The determination of capacity curves of all higher significant modes is analogous.

Conversion of the capacity curve into the Acceleration-Displacement Response Spectrum (ADRS) format according to the previously selected spectrum, determination of the elasto-plastic equivalent and limit ground acceleration of the first vector $a_{gr,1}$, according to EC8 [1], are all shown in Fig. 4. Spectrum type 1 with 5 % damping for the ground type A according to EN 1998 is used. The obtained target acceleration has a value of $a_{gr,1}$ = 0.405 g, which is the seismic resistance of the first vector for a 5-storey R/C frame in x direction.

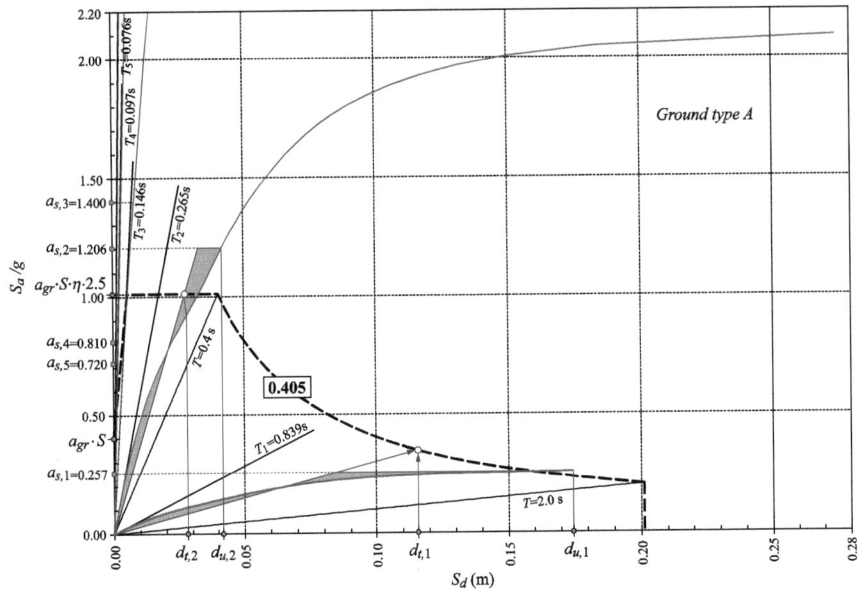

Fig. 5 Determination of spectral acceleration of modes for 5-storey R/C frame in *x* direction, on elastic spectrum with 5 % damping for ground acceleration 0.405 g

The limit ground acceleration obtained for the first vector ϕ_1, i.e. $a_{gr,1} = 0.405$ g, is assumed in the first step as the initial ground acceleration of an iterative procedure in determining the target ground acceleration $a_{gr,t}$.

Determination of the spectral acceleration $a_{s,i}$ of the analyzed five modes for the 5-storey R/C frame in *x* direction, for the initial acceleration $a_{gr,1} = 0.405$ g, i.e. the first step in the iterative procedure of determining the target acceleration for a linear (L) combination of modes is shown in Fig. 5. The figure shows periods T_i and capacity curves of analyzed modes. Capacity curves of the first and the second mode are completely visible in the figure, while the curves of higher modes are cut off above the spectral acceleration of 2.20 g and above the spectral displacement of 0.28 m.

Spectral acceleration $a_{s,i}$ of the analyzed five modes are obtained from the capacity curves of vectors ϕ_i, respecting a 2/3 usability of displacement d_u on a capacity curve, i.e. taking into account the target displacement d_t according to the EN 1998-1 (Annex B) [1].

A linear (L) combination of modes is used for analysis in this chapter, because the test of linear (L) and Square Root of the Sum of Square (SRSS) combination of modes shows that the linear combination is more probable [16]. The initial limit load $F_L = \sum F_i$ for the linear (L) combination of modes can be described with the following expression.

Table 3 Target acceleration determination for linear (L) combination of modes in x direction

a_{gr}	$a_{s,1}$	F_1	$a_{s,2}$	F_2	$a_{s,3}$	F_3	$a_{s,4}$	F_4	$a_{s,5}$	F_5	$F_L = \sum F_i$	F_f	F_f/F_L
0.405	0.257	1,735	1.206	1,051	1.400	478	0.810	133	0.720	40	3,437	2,303	0.67
0.27	0.244	1,648	0.815	710	0.930	318	0.534	88	0.479	27	2,790	2,232	0.80
0.22	0.228	1,540	0.700	610	0.750	256	0.443	73	0.390	22	2,500	2,175	0.87
0.19	0.212	1,432	0.620	540	0.667	228	0.383	63	0.330	18	2,281	2,190	0.96
0.18	0.206	1,391	0.592	516	0.628	214	0.352	58	0.320	18	2,197	2,197	1.00
0.17	0.198	1,337	0.585	510	0.608	208	0.338	56	0.308	17	2,127	2,191	1.03

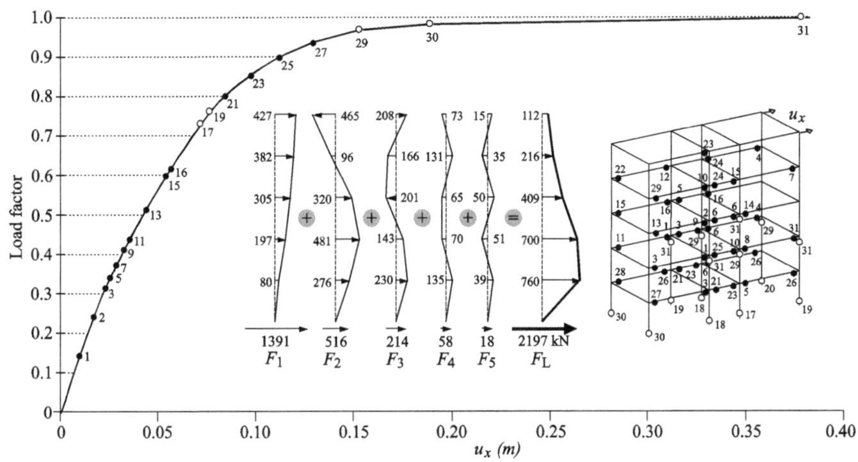

Fig. 6 Capacity curve for linear (L) combination of modes (x direction)

$$F_L = \sum F_i = \sum_{i=1}^{n} \pm\, m_{e,i}\, a_{s,i}(a_{gr})\, F_{fi}/\left|F_{fi}\right| \tag{2.1}$$

Table 3 shows an iterative procedure for determining the target ground acceleration for linear (L) combination of modes of a 5-storey R/C frame in x direction. The obtained target acceleration amounts to $a_{gr,t} = 0.18$ g, and the corresponding shear force in the cross-section at the base of the calculation model is $F_L = 2{,}197$ kN.

The iterative procedure is considered completed when the load equality is obtained from the general failure vector F_L and the design limit load F_f. The obtained result is the target ground acceleration $a_{gr,t}$ which defines the lowest seismic resistance of the structure.

Load vectors for all participating modes, capacity curve, and the order of occurrence of plastification hinges for linear (L) combination of modes for a 5-storey R/C frame in x direction, are presented in Fig. 6.

Table 4 Target acceleration determination for linear (L) combination of modes in y direction

a_{gr}	$a_{s,1}$	F_1	$a_{s,2}$	F_2	$a_{s,3}$	F_3	$a_{s,4}$	F_4	$a_{s,5}$	F_5	$F_L = \sum F_i$	F_f	F_f/F_L
0.39	0.214	1,427	1.142	1,036	1.388	519	0.803	151	0.698	45	3,178	2,097	0.66
0.25	0.202	1,345	0.722	655	0.843	315	0.510	96	0.443	29	2,440	1,927	0.79
0.19	0.184	1,225	0.560	508	0.713	267	0.390	73	0.330	21	2,094	1,927	0.92
0.17	0.173	1,152	0.505	456	0.638	239	0.345	65	0.308	20	1,934	1,934	1.00

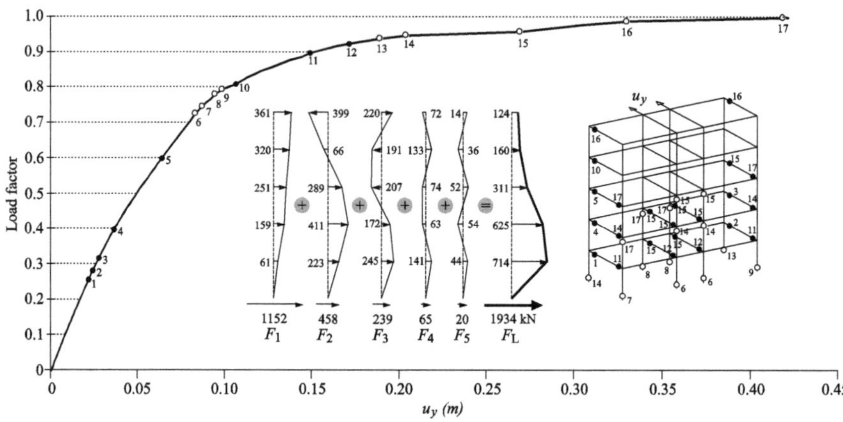

Fig. 7 Capacity curve for linear (L) combination of modes (y direction)

Determination of the target ground acceleration for linear (L) combination of modes of a 5-storey R/C frame in y direction is shown in Table 4. Limit ground acceleration obtained for the first vector ϕ_1, which amounts $a_{gr,1} = 0.39$ g, is assumed as the initial ground acceleration in the first step of an iterative procedure. The obtained target ground acceleration is $a_{gr,t} = 0.17$ g.

Figure 7 shows load vectors for all participating modes, capacity curve, and the order of occurrence of plastification hinges for linear (L) combination of modes for a 5-storey R/C frame in y direction.

Comparison of the target ground acceleration and corresponding shear force, for the first vector analysed individually and for linear (L) combination of modes for a 5-storey R/C frame in x and y directions are shown in Table 5.

The obtained results show that the target acceleration $a_{gr,t}$ obtained by the multimodal pushover method for the linear (L) combination of modes with the target displacement d_t, i.e. 2/3 usability of the displacement d_u on the capacity curve, and the elastic spectrum with 5 % damping is several times lower than a single mode acceleration $a_{gr,1}$ obtained by the criteria of the target displacement for the first vector according to EN 1998-1 (Annex B) [1].

Table 5 Comparison of the target acceleration of the first mode and linear (L) combination of modes with 2/3 usability of displacement on the capacity curve and the elastic spectrum with 5 % damping

R/C frame	Direction	Displacement = d_t = **2/3** d_u				
		Damping (%)	$a_{gr,1}$ (g)	F_1 (kN)	$a_{gr,t}$ (g)	F_L (kN)
5-storey	x	5	0.405	1,735	0.18	2,197
5-storey	y	5	0.39	1,427	0.17	1,934

2.3 Determination of Equivalent Structural Damping

A new formulation of the equivalent structural damping of a system by equalization of the dissipated energy during one cycle of vibration of a nonlinear system and the equivalent linear system, provided that the executed work of a nonlinear system is equal to the executed work of the equivalent linear system with secant stiffness is presented in this chapter.

The capacity curve of the first vector ϕ_1, which is defined by a failure shear force of the first vector F_{f1} and the maximum horizontal displacement u_{max}, is adopted as the relevant curve to determine the hysteresis capacity curve of a structure. A return part of the hysteretic capacity curve to the zero force is a linear function, which slope is defined from the equality of areas A_1 and A_2.

The equivalent structural damping ratio is defined by Eq. (2.2), according to [17]

$$\xi = \frac{E_D}{4\pi E_S} \tag{2.2}$$

where ξ is the equivalent structural damping, E_D is the dissipated energy during one cycle of vibration of a nonlinear system, which is equal to the surface within the hysteresis loop and E_S is the dissipated energy during one cycle of vibration of the linear system, which is equal to the surface of a triangle, see Fig. 8.

Determination of the equivalent structural damping by equalization of the dissipated energy during one cycle of vibration of a nonlinear system and the equivalent linear system with secant stiffness k_{sec}, on the example of a 5-storey R/C frame in x direction, is shown in Fig. 8.

The equivalent structural damping of the observed 5-storey R/C frame in x direction is determined according to the expression (2.3), and 28 % equivalent structural damping was obtained.

$$\xi = \frac{E_D}{4\pi E_S} = \frac{0.644}{4 \cdot \pi \cdot 0.180} = 0.28 \tag{2.3}$$

Determination of the equivalent structural damping for a 5-storey R/C frame in y direction is analogous. The equivalent structural damping of the analyzed

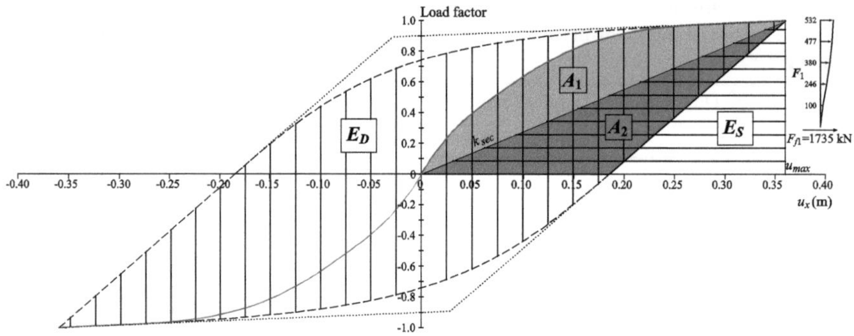

Fig. 8 Determination of the equivalent structural damping based on hysteretic energy dissipation of linear and nonlinear system of 5-storey R/C frame (x direction)

5-storey R/C frame in y direction is determined according to the expression (2.4), and 29 % equivalent structural damping was obtained.

$$\xi = \frac{E_D}{4\,\pi\,E_S} = \frac{0.796}{4 \cdot \pi \cdot 0.215} = 0.29 \tag{2.4}$$

2.4 Target Acceleration for Linear (L) Combination of Modes with Elastic Spectrum with Real Structural Damping and 3/3 Usability of Displacement on Capacity Curve

The results obtained by the target acceleration method for linear (L) combination of modes at full usability, i.e. 3/3 usability, of the displacement d_u on the capacity curve and with elastic spectrum for calculated equivalent structural damping will be used for comparison with the results of the dynamic response of the same analyzed structure.

Determination of the spectral acceleration of modes $a_{s,i}$ was conducted for 3/3 usability of the displacement d_u on a capacity curve, as shown in Fig. 9.

In the elastic spectrum, a damping correction factor η is defined by expression (2.5) where ξ is the calculated equivalent structural damping given as a percentage.

$$\eta = \sqrt{\frac{10}{5 + \xi}} \tag{2.5}$$

Comparison of the elastic spectrum with 5 and 28 % damping for ground acceleration $a_{gr} = 0.30$ g is shown in Fig. 10.

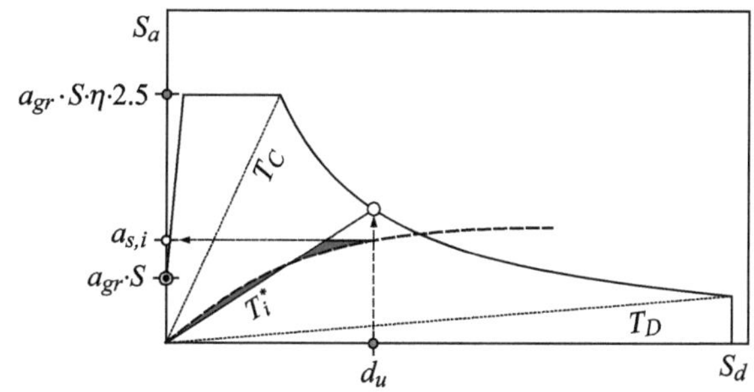

Fig. 9 Determination of spectral acceleration $a_{s,i}$ with 3/3 usability of displacement

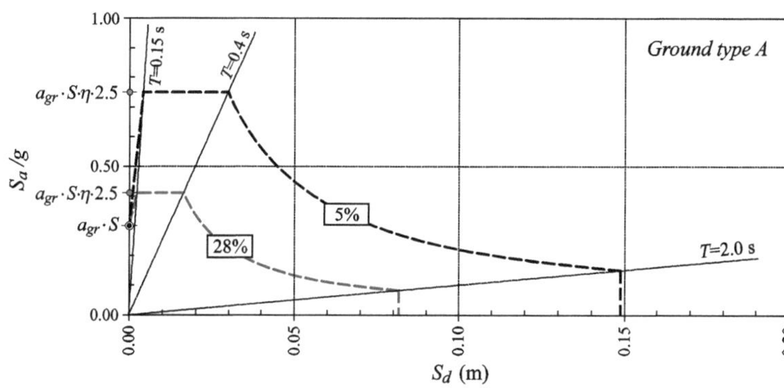

Fig. 10 Comparison of the elastic spectrum with 5 and 28 % damping for ground acceleration $a_{gr} = 0.30$ g

Determination of the spectral acceleration $a_{s,i}$ of analyzed five modes for a 5-storey R/C frame in x direction, for the acceleration $a_{gr} = 0.46$ g, with 3/3 usability of displacement d_u on capacity curve and the elastic spectrum for 28 % damping is shown in Fig. 11.

Table 6 shows the determination of the target ground acceleration for a linear (L) combination of modes of a 5-storey R/C frame in x direction with 3/3 usability of the displacement d_u on a capacity curve and the elastic spectrum for 28 % damping. A ground acceleration $a_{gr} = 0.45$ g is assumed as the initial ground acceleration in the first step of the iterative procedure. The obtained target ground acceleration is $a_{gr,u} = 0.46$ g and the corresponding shear force in the cross-section at the base of the calculation model is $F_L = 2{,}145$ kN.

Load vectors for all participating modes and capacity curve for linear (L) combination of modes for a 5-storey R/C frame in x direction with 3/3 usability of displacement d_u on a capacity curve and the elastic spectrum for 28 % damping, are presented in Fig. 12.

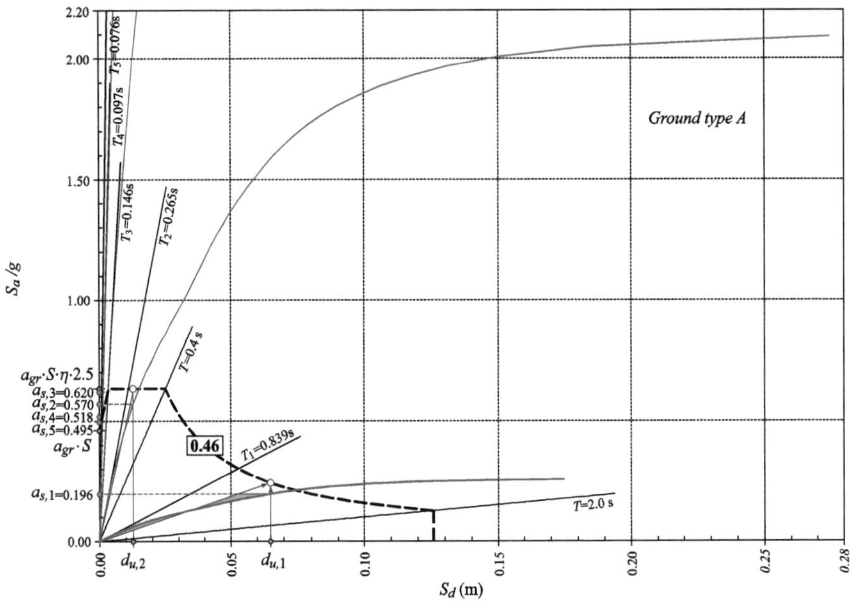

Fig. 11 Determination of spectral acceleration $a_{s,i}$ of modes for 5-storey R/C frame in x direction, on elastic spectrum with 28 % damping for ground acceleration 0.46 g and with 3/3 usability of displacement d_u on capacity curve

Table 6 Target acceleration for linear (L) combination of modes for 5-storey R/C frame in x direction, with 3/3 usability of displacement on a capacity curve and the elastic spectrum for 28 % damping

a_{gr}	$a_{s,1}$	F_1	$a_{s,2}$	F_2	$a_{s,3}$	F_3	$a_{s,4}$	F_4	$a_{s,5}$	F_5	$F_L = \sum F_i$	F_f	F_f/F_L
0.45	0.193	1,303	0.568	495	0.608	208	0.509	84	0.482	27	2,116	2,180	1.03
0.46	0.196	1,324	0.570	497	0.620	212	0.518	85	0.495	27	2,145	2,145	1.00

Table 7 shows the determination of the target ground acceleration for a linear (L) combination of modes of a 5-storey R/C frame in y direction with 3/3 usability of displacement d_u on a capacity curve and the elastic spectrum for 29 % damping. The obtained target acceleration is $a_{gr,u} = 0.455$ g, and the corresponding shear force in the cross-section at the base of the model is $F_L = 1{,}914$ kN.

Load vectors for all participating modes and capacity curve for a linear (L) combination of modes for 5-storey R/C frame in y direction with 3/3 usability of the displacement d_u on a capacity curve and the elastic spectrum for 29 % damping, are presented in Fig. 13.

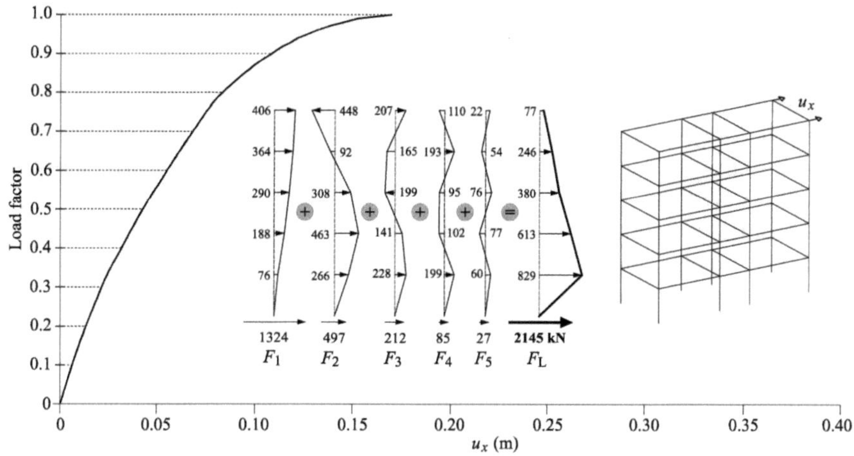

Fig. 12 Capacity curve for linear (L) combination of modes for 5-storey R/C frame in x direction, with 3/3 usability of displacement on a capacity curve and the elastic spectrum for 28 % damping

Table 7 Target acceleration for linear (L) combination of modes for 5-storey R/C frame in y direction, with 3/3 usability of displacement on a capacity curve and the elastic spectrum for 29 % damping

a_{gr}	$a_{s,1}$	F_1	$a_{s,2}$	F_2	$a_{s,3}$	F_3	$a_{s,4}$	F_4	$a_{s,5}$	F_5	$F_L = \sum F_i$	F_f	F_f/F_L
0.45	0.168	1,119	0.496	450	0.585	219	0.511	96	0.480	31	1,915	1,953	1.02
0.455	0.169	1,125	0.504	457	0.545	204	0.515	97	0.486	31	1,914	1,914	1.00

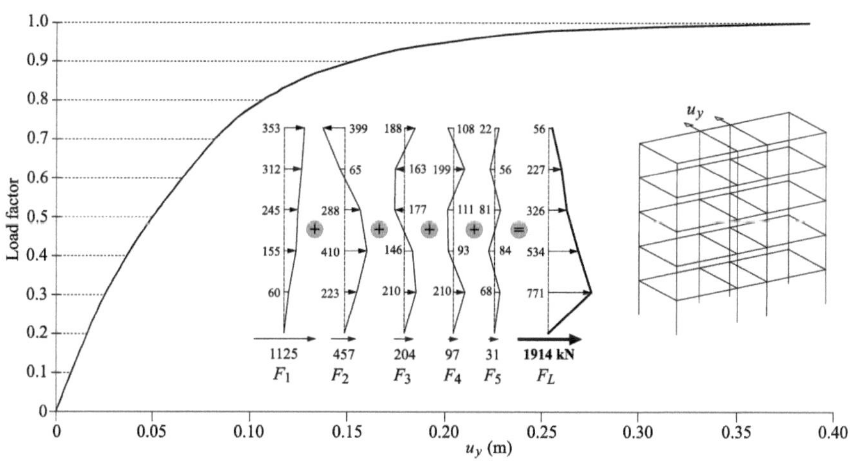

Fig. 13 Capacity curve for linear (L) combination of modes for 5-storey R/C frame in y direction, with 3/3 usability of displacement on capacity curve and with elastic spectrum for 29 % damping

3 Dynamic Response of 5-Storey Spatial R/C Frame

The calculation of the seismic resistance by an incremental dynamic analysis was carried out for a 5-storey spatial R/C frame which is shown in Fig. 2. The dynamic response of the 5-storey R/C frame was carried out by *SeismoStruct* [18] software package for nonlinear analysis.

As seismic excitation, eight real earthquake accelerograms are adopted from the *European Strong-Motion Database* [19]. The analyzed spatial frame was loaded by accelerogram records of eight earthquakes. Earthquake amplitudes were increased incrementally until a failure peak ground acceleration $a_{gr,d}$ occurred, i.e. until the collapse of the structure. All selected earthquake accelerograms were recorded at the sites which soil characteristics corresponds to the ground type *A* according to EN 1998-1 [1]. The selected real earthquakes are the following: Montenegro (Montenegro)-1979, Campano Lucano (Italy)-1980, Aigion (Greece)-1995, Olfus (Island)-2008, Calabria (Italy)-1978, Valnerina (Italy)-1979, Banja Luka (B&H)-1981 and Sicilia Orientale (Italy)-1990, as shown in Fig. 14.

The model of the 5-storey spatial R/C frame from the program *SeismoStruct* [18] with marked sites of earthquake action is shown in Fig. 15.

As a result of the dynamic response, failure peak ground acceleration $a_{gr,d}$ and total failure shear force on the base F_d are shown.

Results of the analysis of dynamic response of a 5-storey R/C frame in x direction are presented in Table 8, and results of dynamic response of the same frame in y direction are presented in Table 9.

A maximum value of the failure peak ground acceleration $a_{gr,d}$ for the earthquake Valnerina (*) is discarded, since it deviates the most from the average. A relevant result was calculated as the average value of dynamic response results for a series of the remaining seven earthquake records.

The obtained seismic resistance of a 5-storey R/C frame in x direction is 0.46 g and the average total failure shear force on the base of the model is 1,821 kN.

The obtained seismic resistance of a 5-storey R/C frame in y direction is 0.43 g and the average total failure shear force on the base of the model is 1,695 kN.

4 Comparison of Results

Comparison between the target ground acceleration $a_{gr,t}$ ($a_{gr,u}$) of the multimodal pushover target acceleration method with the failure peak ground acceleration $a_{gr,d}$ obtained by a dynamic transient response of R/C frames is shown in Table 10.

Table 10 shows a comparison of results of seismic resistance for a 5 storey spatial R/C frame.

The results of seismic resistance show that the target acceleration $a_{gr,u}$, obtained by the multimodal pushover target acceleration method for the linear (L) combination of modes of a 5-storey R/C frame with 3/3 usability of the displacement d_u

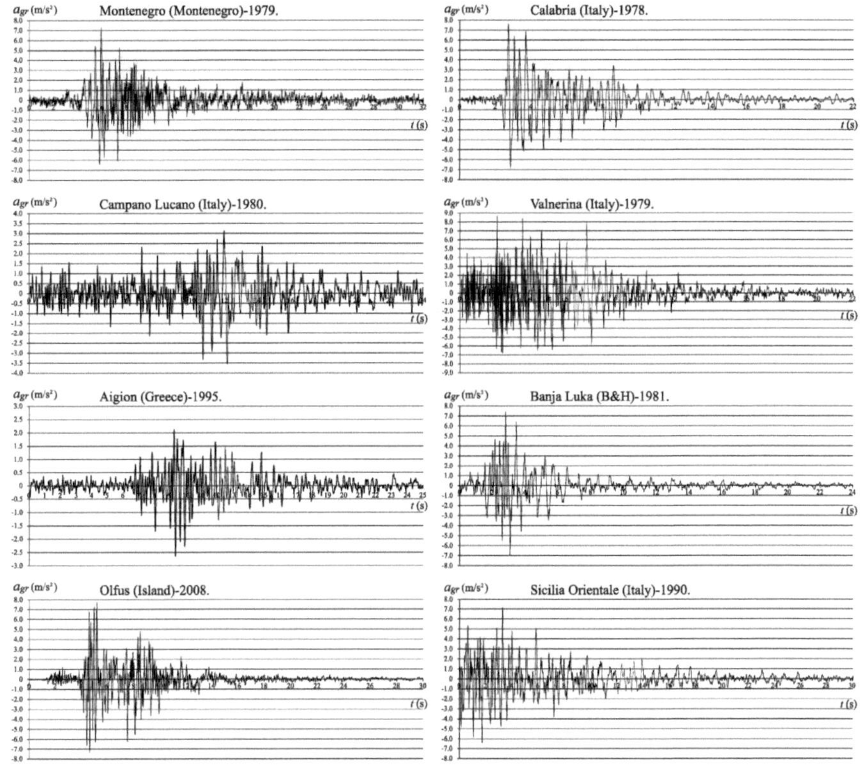

Fig. 14 Real earthquake accelerograms [19]

Fig. 15 The model of 5-storey R/C frame from the program *SeismoStruct* [18]

Table 8 Dynamic response of 5-storey spatial R/C frame in x direction

	Earthquake	$a_{gr,d}$ (g)	$F_{d,x}$ (kN)
1.	Montenegro	0.41	1,808
2.	Campano Lucano	0.29	1,524
3.	Aigion	0.41	1,527
4.	Olfus	0.43	1,935
5.	Calabria	0.62	2,084
*	Valnerina	0.67	2,302
6.	Banja Luka	0.48	1,854
7.	Sicilia Orientale	0.57	2,015
Average of 7 earthquakes		0.46	1,821

Table 9 Dynamic response of 5-storey spatial R/C frame in y direction

	Earthquake	$a_{gr,d}$ (g)	$F_{d,y}$ (kN)
1.	Montenegro	0.37	1,660
2.	Campano Lucano	0.33	1,892
3.	Aigion	0.48	1,750
4.	Olfus	0.27	1,308
5.	Calabria	0.55	1,703
*	Valnerina	0.69	1,534
6.	Banja Luka	0.43	1,820
7.	Sicilia Orientale	0.55	1,733
Average of 7 earthquakes		0.43	1,695

on a capacity curve and with elastic spectrum for calculated equivalent structural damping is significantly higher than the target acceleration $a_{gr,t}$ obtained by the multimodal pushover target acceleration method with elastic spectrum for 5 % damping and 2/3 usability of the displacement d_u on a capacity curve. The failure loads F_L are approximately the same in both cases.

The results in Table 10 also show that the ground acceleration $a_{gr,1}$ obtained by the criteria of the target displacement for the first vector observed individually according to EN 1998-1 (Annex B) [1] is lower than the target acceleration $a_{gr,u}$ of a 5-storey R/C frame for x and y directions obtained by the multimodal pushover method.

Finally, there is a very good agreement between the failure peak ground acceleration $a_{gr,d}$ obtained by the dynamic response of the structure and the target ground acceleration $a_{gr,u}$ of the multimodal pushover target acceleration method for linear (L) combination of modes with 3/3 usability of displacement d_u on capacity curve and with elastic spectrum for calculated equivalent structural damping, what also proves that the calculated equivalent structural damping is real.

Table 10 Comparison of obtained seismic resistance for a 5-storey spatial R/C frame

R/C frame	Direction	Displacement = $d_t = 2/3\ d_u$					Displacement = $d_t = 3/3\ d_u$			Dynamic response	
		Damping (%)	$a_{gr,1}$ (g)	F_1 (kN)	$a_{gr,t}$ (g)	F_L (kN)	Damping (%)	$a_{gr,u}$ (g)	F_L (kN)	$a_{gr,d}$ (g)	F_d (kN)
5-storey	x	5	0.405	1,735	0.18	2,197	28	0.46	2,145	0.46	1,821
5-storey	y	5	0.39	1,427	0.17	1,934	29	0.455	1,914	0.43	1,695

5 Conclusions

This chapter presents a new multimodal pushover target acceleration method for the nonlinear static analysis of resistance of R/C structures exposed to seismic action. The method determines the target acceleration, which presents the lowest seismic resistance of a structure. A comparison between the target ground acceleration $a_{gr,t}$ ($a_{gr,u}$) of the multimodal pushover target acceleration method with the failure peak ground acceleration $a_{gr,d}$ obtained by a dynamic transient response of the structure is shown on the example of a 5-storey R/C frame.

This chapter also presents a new formulation for determining the equivalent structural damping, which is obtained by equalization of the dissipated energy during one cycle of vibration of the nonlinear system and the equivalent linear system, provided that the executed work of nonlinear system is equal to the executed work of the equivalent linear system with the secant stiffness.

Based on the results of the analyzed 5-storey spatial R/C frame, the following conclusion can be drawn:

- The influence of higher modes, especially the second and third one, is very significant.
- The general failure vector which belongs to the target acceleration differs significantly by shape from the shape of the load of the first vector, and also from the shape of the load with uniform acceleration by the height of the structure, as given in EN 1998-1 [1].
- There is a very good agreement between the failure peak ground acceleration $a_{gr,d}$ obtained by dynamic response of structure and the target acceleration $a_{gr,u}$ of the multimodal pushover method for linear (L) combination of modes, with taking into account the elastic spectrum for calculated equivalent structural damping ξ and with 3/3 usability of displacement d_u on capacity curve.

Acknowledgments The partial financial support provided by the Ministry of Science, Education and Sports of the Republic of Croatia for the *Modelling failure and stability of structures under large displacement* project, Grant. No. 083-0831541-1545, is gratefully acknowledged.

References

1. CEN: Eurocode 8: Design of Structures for Earthquake Resistance. Part 1: General Rules, Seismic Actions and Rules for Buildings. European Committee for Standardization, Brussels (2004)
2. Chopra, A.K., Goel, R.K.: A modal pushover analysis procedures for estimating seismic demands for buildings. Earthq. Eng. Struct. D **31**(3), 561–582 (2002)
3. Chopra, A.K., Goel, R.K., Chintanapakdee, C.: Evaluation of a modified MPA procedure assuming higher modes as elastic to estimate seismic demands. Earthq. Spectra **20**(3), 757–778 (2004)

4. Diotallevi, P.P., Landi, L., Pollio, B.: Evaluation of conventional and advanced pushover procedures for regular and irregular RC frames. In: Proceedings of the 14th World Conference on Earthquake Engineering. Beijing, China (2008)
5. Fajfar, P.: A nonlinear analysis method for performance based seismic design. Earthq. Spectra **16**(3), 573–592 (2000)
6. Goel, R.K., Chopra, A.K.: Evaluation of modal and FEMA pushover analyses—SAC buildings. Earthq. Spectra **20**(1), 225–254 (2004)
7. Goel, R.K., Chopra, A.K.: Role of higher-"mode" pushover analyses in seismic analysis of buildings. Earthq. Spectra **21**(4), 1027–1041 (2005)
8. Kalkan, E., Kunnath, S.K.: Adaptive modal combination procedure for nonlinear static analysis of building structures. J. Struct. Eng.-ASCE **132**(11), 1721–1731 (2006)
9. Kalkan, E., Kunnath, S.K.: Assessment of current nonlinear static procedures for seismic evaluation of buildings. Eng. Struct. **29**, 305–316 (2007)
10. Sasaki, K.K., Freeman, S.A., Paret, T.F.: Multi-mode pushover procedure (MMP)—a method to identify the effects of higher modes in a pushover analysis. In: Proceedings of the 6th U.S. National Conference on Earthquake Engineering. Seattle, USA, pp. 1–12 (1998)
11. Athanassiadou, C.J.: Seismic performance of R/C plane frames irregular in elevation. Eng. Struct. **30**(5), 1250–1261 (2008)
12. Čaušević, M., Mitrović, S.: Comparison between non-linear dynamic and static seismic analysis of structures according to European and US provisions. Bull. Earthq. Eng. **9**(2), 467–489 (2011)
13. Mwafy, A.M., Elnashai, A.S.: Static pushover versus dynamic collapse analysis of RC buildings. Eng. Struct. **23**(5), 407–424 (2001)
14. Trogrlić, B.: Nonlinear numerical model of stability and load-bearing capacity of reinforced concrete space frames. Ph.D. dissertation. University of Split, Croatia (2003)
15. Trogrlić, B., Mihanović, A.: The comparative body model in material and geometric nonlinear analysis of space R/C frames. Eng. Comput. **25**(2), 155–171 (2008)
16. Balić, I., Mihanović, A., Trogrlić, B.: Target acceleration in multimodal pushover method for R/C frames. Gradev **65**(4), 305–318 (2013)
17. Chopra, A.K.: Dynamics of Structures: Theory and Applications to Earthquake Engineering, 3rd edn. University of California at Berkeley, Prentice Hall, New Jersey (2007)
18. Pinho, R.: SeismoStruct—a computer program for static and dynamic nonlinear analysis of framed structures. http://www.seismosoft.com. Accessed 02 April 2013
19. European Strong-Motion Database. http://www.isesd.hi.is/ESD_Local/frameset.htm. Accessed 02 April 2013

Hardening and Roughness Reduction of Carbon Steel by Laser Polishing

Stefan Stein, Rainer Börret, Andreas Kelm, Elvira Reiter, Gerhard Schneider and Harald Riegel

Abstract Laser polishing (LP) is a non-contact method to smooth surfaces. Its benefits are a fast and automated surface modification. The high-quality tempered steel 1.7225 refined by electro-slag remold is investigated. Laser polishing studies were carried out with a shielding gas enclosure to ensure an oxidation free condition. The roughness of the initial and laser polished surface have been determined by microscopy, roughness spectroscopy and white light interferometry (WLI). Finite element method (FEM) simulation is carried out to determine the heating and cooling rates of the steel surface. Scanning electron microcopy (SEM) and energy-dispersive X-ray spectroscopy (EDX) has been used to analyze irregularities on the treated surface. The results show an overall roughness reduction by laser remelting by 47 % from $R_{z,i} = 5.42$ to $R_{z,f} = 2.87$ μm. Roughness spectroscopy results show smooth, polished surface with values of $R_z = 0.5$ μm for wavelengths smaller than 100 μm. At wavelengths above 200 μm the surface structure is dominated by a few small craters. The composition inside the craters has been analyzed. SEM results show small droplets inside the irregularities. EDX analysis of the impurities indicates a composition comparable to slag. The slag might origin from the electro slag remold process during steel production.

S. Stein (✉) · R. Börret · A. Kelm · E. Reiter · G. Schneider · H. Riegel
Aalen University, Beethovenstraße 1, 73430 Aalen, Germany
e-mail: stefan.k.stein@gmail.com

R. Börret
e-mail: rainer.boerret@htw-aalen.de

A. Kelm
e-mail: andreas.kelm@htw-aalen.de

E. Reiter
e-mail: elvira.reiter@htw-aalen.de

G. Schneider
e-mail: gerhard.schneider@htw-aalen.de

H. Riegel
e-mail: harald.riegel@htw-aalen.de

A. Öchsner and H. Altenbach (eds.), *Design and Computation of Modern Engineering Materials*, Advanced Structured Materials 54,
DOI: 10.1007/978-3-319-07383-5_29,

Keywords Laser polishing · Laser remold · Electro slag remold steel · Surface hardening · Laser grading

1 Introduction

Laser polishing (LP) is based on remelting a micro layer on the ground surface. A thin layer just big enough to extend from the maximum to the minimum topographic extensions gets liquefied. Subsequently the surface tension evens the surface of the liquid. Finally the molten material solidifies in a smooth surface. LP is non-abrasive, which is an advantage compared to conventional polishing techniques [1–3]. The investigated steel is 1.7225 a carbon steel. The steel was cleaned during a so-called electro slag remolding process. This process reduces impurities, especially sulfide inclusions. Sulfide impurities have a low evaporation temperature and according to [4–7] are the cause of formation of craters on the laser polished surface. A shielding gas atmosphere can prevent the formation of waves on the surface by avoiding a turbulent gas flow on the surface [1]. LP influences the microstructure of the material by remelting and hardening. The microstructure formed after a treatment with a Nd:YAG laser shows a high homogeneity level and refinement [8]. High thermal energy is exposed to the surface resulting in a large variety of effects, e.g. evaporating inclusions, decomposing carburizations or annealing of carbon steels can occur [9]. The interaction between laser and impurities can lead to the formation of craters. The cause of this local explosive reaction is the low evaporation temperature of theses in homogeneities such as nonmetallic inclusions like manganese sulfur or segregations [4, 10]. In [10] manganese sulphur inclusions are found as the root cause for craters.

Surface topography using a Fourier transform of the surface topography was carried out from Hafiz [1] and Pfefferkorn [2]. Applying a filter in the Fourier domain followed by a back transformation enables an examination of the periodical structures of the roughness to determine the roughness Rz of different spectra. In particular, short-wave structures are being smoothed to a high extent by the machining with laser radiation [1].

2 Experimental Setup

2.1 Used Laser and Measurement Devices

The experimental set-up consists of a pulsed Nd:YAG Laser with a Rayleigh length of 1.88 mm and a minimal spot diameter of 292 μm. The Laser was operated with a defocus of 2 mm above the probe surface which results in a diameter of 444 μm on the probe surface. The samples are protected from ambient air by Argon in a gas shielding chamber.

The surface of the samples is characterized tactile by a Mitutoyo Surftest SJ-301. Further measurement devices are a Zygo New View 500 WLI and a scanning electron beam microscope Leo Gemini 1525.

2.2 Used Steel

The steel used in the experiments is a carbon steel 1.7225, which is electro slag remold. Sulphur inclusions get reduced by this production process. The slag binds nonmetallic elements and improves the steel quality of the solidified bulk material.

The initial surface of the bulk material was ground. Ruby corundum with a grain size between 370 and 310 μm was used. The specimen was quenched and tempered. Figure 1 (left) shows a scanning electron microscope (SEM) image of the initial surface. Grinding grooves are the dominant surface structures. The ground surface has a roughness of $R_z = 5.14$ μm.

3 Numerical Simulation

During laser polishing the molten steel surface solidifies by self quenching. When processing carbon steel the structure of the surface layer transforms into the bridal martensitic structure. The formation of martensitic depends on the time-temperature-characteristic of the process which is very fast and short when processed with laser radiation [8]. The cooling rate needs to be sufficiently high to prevent the carbon atoms from diffusing. In order to estimate the heating and cooling rates a FEM-simulation was carried out. Results are shown in Fig. 2. Due to the small volume of the used model (2 cm^3) a continuous increase of temperature of the sample is observed starting at 300 K and ending after 70 s at about 850 K.

In the simulation model the laser spot moves in meandering patterns over the surface. Figure 2 shows several temperature peaks indicating the different processing traces. Each maximum corresponds to the closest distance of the laser spot to the measuring point in its corresponding trace. The three biggest maxima correspond to traces where the laser spot touches the measuring point on the surface at z = 0 mm. The trace in Fig. 2 with the biggest maximum is surrounded by a rectangular box and displayed in Fig. 2 (inlet) with enlarged time scale. The simulation results are marked as crosses and the solid line represents an interpolation. Two calculations are run during one laser period, one with laser pulse on and the second one with laser pulse off. The modulation of the calculated temperature shows a big temperature variation between the laser pulses on one trace.

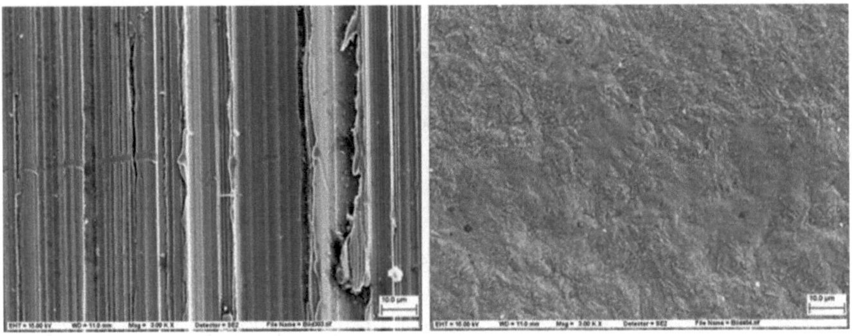

Fig. 1 SEM images of the initial surface (*left*) and the surface after remelting (*right*)

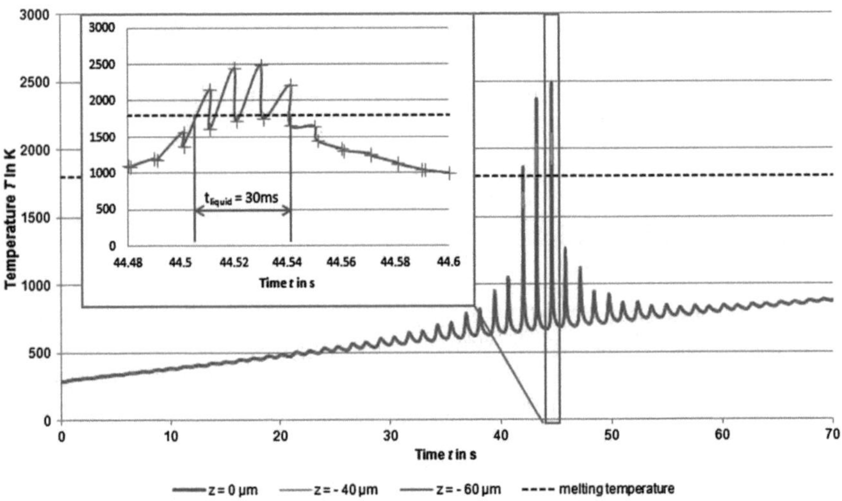

Fig. 2 Temperature-time characteristic of the laser polishing process at three different depths

Temperatures above the melting temperature indicate the formation of liquid metal. Melting enthalpy however has not been taken into consideration. According to the simulation results a melt pool at the surface at z = 0 is to be expected for three traces. The melt pool exists for a time period of about $\Delta t_{\text{liquid}} = 30$ ms during four laser pulses. Already after about 100 ms the material is cooled down by $\Delta T = 1{,}000$ K, which corresponds to a cooling rate of 10,000 K/s. The simulation results indicate very high heating and cooling rates.

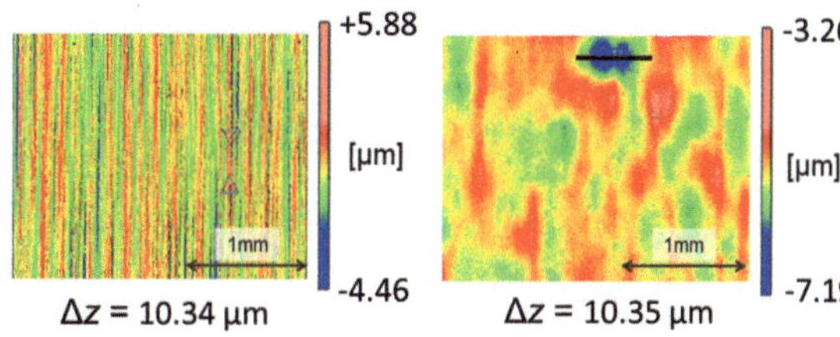

Fig. 3 White light interferometry (WLI) images of the initial (*left*) and remelted surface (*right*)

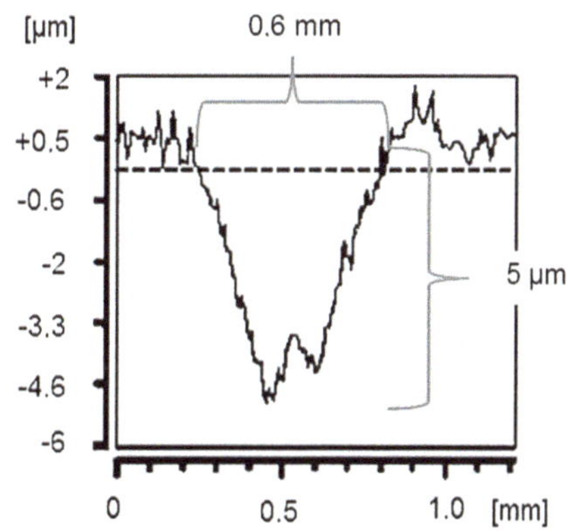

Fig. 4 Cross section of a crater

4 Experimental Results

4.1 Roughness Reduction

SEM images of the steel surface are displayed in Fig. 1. On the right side of Fig. 1 the laser remelted surface is illustrated. Martensitic needles are the prevailing structures. Figure 3 shows results done by white light interferometry of the ground (left) and the laser remelted surfaces (right).

The remelted surface exhibits a crater in the upper center of the image, marked by a horizontal bar. The crater on the treated surface exhibits a diameter of 600 µm and a depth of 5 µm, Fig. 4. The depth contributes to half of the peak to valley value in Fig. 3. The crater contributes considerably to the surface roughness R_z and dominates the surface quality.

The surface roughness according to EN ISO 3274 has been determined. All measurements were carried out perpendicularly to the grinding grooves or the laser processing marks. In addition a roughness-spectrum analysis has been carried out. The analyzed wavelengths start according to EN ISO 3274 at a minimum wavelength of 2.5 µm and ending at 800 µm. According to EN ISO 3274 the roughness is determined by using a Gaussian filter, whereas the roughness spectrum is done

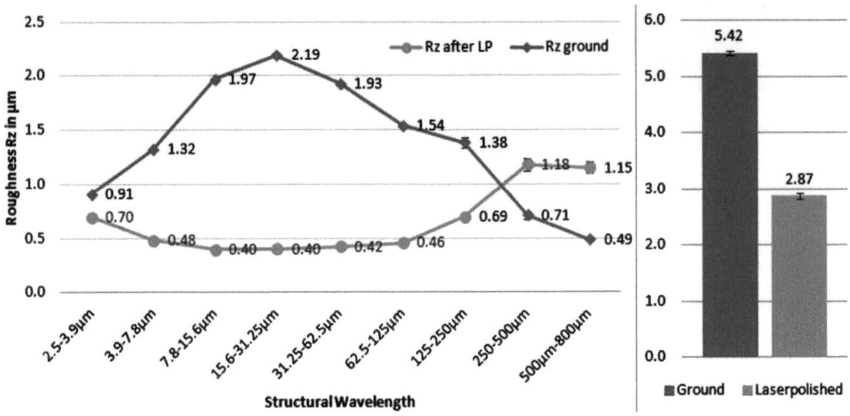

Fig. 5 Roughness spectrum Rz (*left*) and according to EN ISO 3274 (*right*)

with a rectangular filter in the spatial frequency domain, Fig. 5. Two roughness spectra are displayed, the initial ground surface (diamonds) and the surface after laser polishing (circles). Each displayed value is a mean value out of five measurements at different sample locations. The standard deviation is displayed as well and is mostly negligible, see Fig. 5. Figure 5 (right) shows the roughness R_z according to EN ISO 3274. The roughness is reduced by laser polishing from $R_z = 5.42$ µm to $R_z = 2.87$ µm.

The roughness spectrum of the ground surface exhibits a maximum value of 2.19 µm at wavelengths between 15.6 and 31.25 µm. The roughness at the wavelength limits is much lower, that is 0.91 µm at small wavelength and 0.49 µm at the upper end. The spectrum of the laser polished surface has almost an inverse structure, leading to very low roughness in an extended area from 3.9 to 125 µm of wavelength. The roughness at structures above a wavelength of 250 µm is significantly higher than the ground surface. In this domain the measurement results show also the highest standard deviations. Irregular distributed craters are the cause for the significant attitudes at long structural wavelengths.

4.2 Formation of Craters

The formation of craters is caused by an eruption inside the melt pool. The liquid metal gets ejected out of the surface. The eruption is caused by a low vaporizing impurity. Figure 6 (left) shows a SEM-picture of the inside of a crater. In the center the track of the overlapping laser beam pulses can be observed. The areas marked by two ellipses are covered by droplets. The droplet marked with a square is displayed enlarged on the right side of Fig. 6.

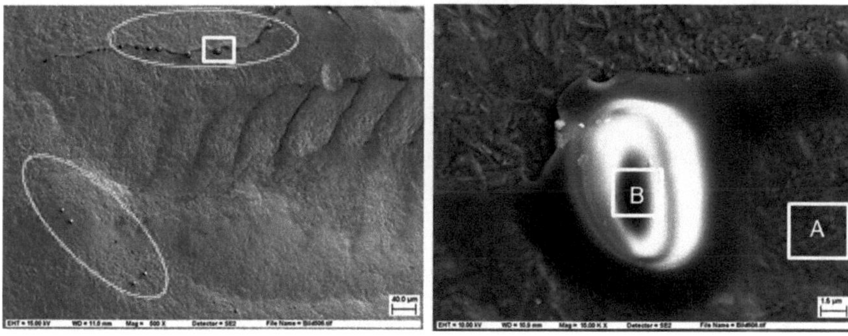

Fig. 6 Analysis of a crater by SEM. Drop shaped structures can be observed at the edge

Table 1 Results of EDX-analysis at positions A and B according to Fig. 6 (right)

Element	O	Al	Si	Ca	Cr	Mn	Fe
A, steel	–	–	0.88	–	1.2	1.2	96.8
B, drop	40.2	5.3	16.7	3.5	1.1	20.8	12.4

The dimensions of this droplet are about 7 times 10 μm. Two EDX spectra were taken at the marked positions A and B. Position B is directly on the droplet and A is in the bulk inside the martensitic needle structure. The results of the EDX-analysis with detected elements and their relative occurrence are displayed in Table 1.

The EDX spectrum on the steel surface at position A shows a content of iron with 96.8 % followed by manganese and chrome as the next prevalent elements. The EDX-spectra of the droplet at position B shows a high concentration of the elements oxygen, aluminum, silicon and calcium. Slag consists largely of the elements silicon dioxide, calcium oxide, iron (II) oxide and aluminum oxide [11]. Slag is a glassy or crystalline residue of solidified nonmetallic elements. Considering the results of the EDX analysis, it can be assumed, that the inclusions found are slag. If parts of the slag get dragged during the production process into the clean metal it will also solidify and stay as an impurity in the ESR steel. That implies that the impurities found in the crater are caused by the residues of slag.

4.3 Structural Changes and Hardness Modification

The surface is characterized by its shape and its micro-structure. The shape is dominated by martensitic needle structure on a microscopic scale and craters on a macroscopic scale. The microstructure has been analyzed by the use of a polished cross section, Fig. 7 (left). Two zones are marked. Zone 1 with a thickness of 53.4 μm is affected by the laser beam with a refined structure. Zone 2 represents the bulk.

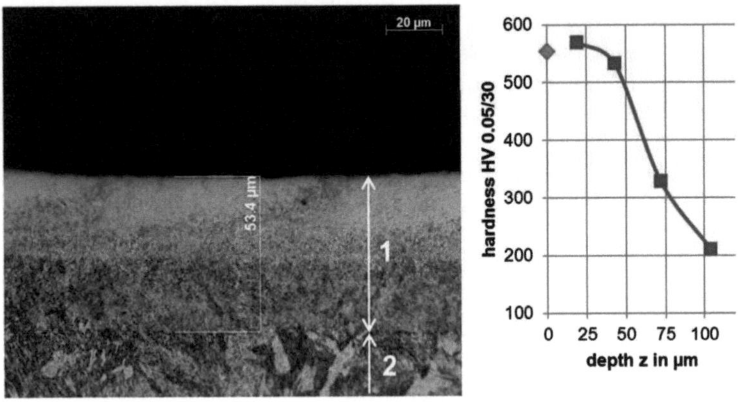

Fig. 7 Cross section of the laser polished surface with quenched and tempered bulk

The Vickers hardness is plotted in Fig. 7 (right). The diamond at a depth of z = 0 µm represents hardness on the surface HV 1/30. The squares show HV as a function of the depth z with HV 0.05/30. The laser remelted zone exhibits a largely increased hardness. After a depth of 100 µm the hardness approaches the bulk hardness of about 200 HV. The hardness of the surface is increased by a factor of 2.7.

The results of numerical simulations in Fig. 2 showed already large temperature gradients and a melting temperature up to a depth of 60 µm. As a result a martensitic structure with high hardness up to the same depth is expected. This can be confirmed by the results in Fig. 7.

5 Conclusion

Electro slag remold 1.7225 steel has been used for surface remelting by a pulsed laser beam. An FEM-simulation was carried out to estimate the temperature-time characteristics during the process. The results show a large temperature gradient in the order of 10,000 K/s and melting temperature down to a depth of about 60 µm. Polished cross sections were investigated with microscopy and hardness measurements. The remelted area extends 50 µm deep into the material. The increased hardness, the extent of the melted area and the large temperature gradient correlate and indicate a martensitic surface layer.

The experiments had been carried out with a shielding argon gas atmosphere. The used laser was an Nd:YAG Laser. The initial roughness of R_z = 5.86 µm was decreased by laser polishing to R_z = 2.87 µm, according to EN ISO 3274. The roughness spectrum of the initial surface shows a maximum at structural wavelengths between 15.6 and 31.25 µm. The laser polished surface exhibits smooth topographies over a wide range of measurement. The roughness R_z as a function of structural wavelength shows Rz values smaller than 0.50 μm from 3.9 to 125 μm. At structural wavelengths above 200 μm the topography is dominated by craters resulting in a higher roughness.

Craters develop due to impurities. Evaporation temperature of the contamination is close to melting temperature of steel. The evaporation results in a drastic volume change. The liquid metal gets ejected, leaving a crater behind. The craters have been analyzed with scanning electron beam microscopy. Unexpected droplets have been found at the edges of the craters. The droplets have a diameter of about 10 μm. An energy-dispersive X-ray spectroscopy of the droplets showed a high concentration of oxygen, aluminum, silicon and calcium. It is concluded, that the origin of the craters are inclusions of slag. This might be remains of the ESR-manufacturing process of the steel.

References

1. Hafiz, A.M.K., Bordatchev, E.V., Tutunea-Fatan, R.O.: Influence of Overlap Between the Laser Beam Tracks on Surface Quality in Laser Polishing of AISI H13 Tool Steel. J. Manuf. Process. **14**, 425–434 (2012)
2. Pfefferkorn, F.E., Duffie, N.A., Li, X., Vadali, M., Ma, C.: Improving surface finish in pulsed laser micro polishing using thermocapillary flow. CIRP Ann.—Manuf. Tech. **62**(1), 203–205 (2013)
3. Ukar, E., Lamikiz, A., de López Lacalle, L.N., del Pozo, D., Arana, J.L.: Laser polishing of tool steel with CO2 laser and high-power diode laser. Int. J. Mach. Tools Manuf. **50**, 115–125 (2010)
4. Kiedrowski, T.: Oberflächenstrukturbildung beim Laserstrahlpolieren von Stahlwerkstoffen. Shaker, Aachen (2009)
5. Liebing, C.: Beeinflussung funktioneller Oberflächeneigenschaften von Stahlwerkstoffen durch Laserpolieren, Aachen Techn Hochsch (2010)
6. Willenborg, E.: Polieren von Werkzeugstählen mit Laserstrahlung—Berichte aus der Lasertechnik. Shaker, Aachen (2006)
7. Willenborg, E.: Polishing with laser radiation. In: Poprawe, R. (ed.) Tailored Light 2—Laser Application Technology. Springer, Berlin (2011)
8. Kac, S., Kusinski, J.: SEM structure and properties of ASP2060 steel after laser melting. In: Proceedings of Symposium G on Protective Coatings and Thin Films-03 of the E-MRS 2003 Spring Conference: 180–181 and 611–615 (2004)
9. Martinez, S., Lamikiz, A., Tabernero, I., Ukar, E.: Laser Hardening Process with 2D Scanning Optics. Laser Assist. Net Shape Eng. 6 Proc. LANE 2010 Part 2, **39**, 309–317 (2012)
10. Avilés, R., Albizuri, J., Lamikiz, A., Ukar, E., Avilés, A.: Influence of laser polishing on the high cycle fatigue strength of medium carbon AISI 1045 steel. Int. J. Fatigue **33**, 1477–1489 (2011)
11. Giesserei Lexikon, http://www.giessereilexikon.com/index.php?option=com_content&view=article&id=1157%3Aschlacke&catid=19%3As&lang=de Accessed 30 Oct 2013

Effect of Austenization Temperatures and Times on Hardness, Microstructure and Corrosion Rate of High Carbon Steel

Mohamed A. Gebril, Mohammad. S. Aldlemey and Abdessalam F. Kablan

Abstract High carbon steel is characterized by good toughness and high hardness, especially after an appropriate heat treatment. In this work heat treatment of 0.75 % carbon steel with 4.50 % Mo, including heating to a different austenitic phase (850, 900 and 950 °C) and holding for different times (30 min and 1 h), then cooling by different media is considered. The results show no noticeable increase in the hardness at 850 °C, but at 900 °C changing in hardness values can be observed clearly with oil and water quenched specimens, at 950 °C there is a significant changing in hardness values in all quenching media even at air and furnace cooling. As Austenization temperature increase, the hardness of the samples increases especially for samples cooled with oil and water at 950 °C. Tests showed that the microstructures contain carbides, that carbides did not completely melted even at the temperature of 1,050 °C due to percentage of molybdenum which produces stable carbides until melting.

Keywords Austenization temperature · Austenizing time · High carbon steel · Corrosion rate

M. A. Gebril (✉)
Mechanical Department, Engineering Faculty,
Benghazi University, Benghazi, Libya
e-mail: magebril@yahoo.com

Mohammad. S. Aldlemey
Technical Collage of Mechanical Engineering, Benghazi, Libya
e-mail: maldleme@yahoo.com

A. F. Kablan
University of Technical Research for Engineering Science, Benghazi, Libya
e-mail: Utres_edu@yahoo.com

A. Öchsner and H. Altenbach (eds.), *Design and Computation of Modern Engineering Materials*, Advanced Structured Materials 54,
DOI: 10.1007/978-3-319-07383-5_30,

1 Introduction

Steel is an alloy of iron and carbon, where other elements are present in quantities very small compared to the major constituents. The other alloying elements allowed in plain-carbon steel are manganese (1.65 % max) and silicon (0.60 % max) [1]. Steel with low carbon content has the same properties as iron, soft but easily formed. As carbon content rises, the metal becomes harder and strong but less ductility and more difficult to weld. Higher carbon content lowers the steel melting point and its temperature resistance in general [2]. Heat treatment involves the application of heat, to material to obtain desired material properties (e.g. mechanical, corrosion, electrical, magnetic, etc.). During the heat treatment process, the material usually undergoes phase microstructural and crystallographic changes [3]. The purpose of heat treating carbon steel is to change the mechanical properties of steel, usually ductility, hardness, yield strength, tensile strength and impact resistance. The electrical, corrosion and thermal conductivity are also slightly altered during the heat treatment process [4]. If a sample of plain carbon steel in the austenitic condition is rapidly cooled to room temperature by quenching in water, its structures will be changed from austenite to martensite. Martensite in plain carbon steels is a metastable phase consisting of a super saturated interstitial solid solution of carbon in body centered cubic iron or body centered tetragonal iron [5].

Molybdenum is a relatively expensive alloying element, has a limited solubility in austenite and ferrite iron, and is a strong carbide former. Molybdenum has a strong effect on hardenability and, like chromium, increases the high-temperature hardness and strength of steels. Steel containing molybdenum are less susceptible to temper brittleness than other alloy steel [6]. The transformation of austenite to martensite by diffusionless shear type transformation in quenching is also responsible for higher hardness obtained and this property is attributed to the effectiveness of the interstitial carbon in hindering the dislocation motion [7]. When the steel has reached the hardening temperature it is austenitic provided that the temperatures has been correctly chosen. The time of holding at the hardening temperature depends on the desired degree of carbide dissolution and acceptable grain size. Since the amount of carbide is different for different types of steel the time of holding is also dependant on the grade of steel. However, the holding time is not only dependent on the hardening temperature but also on the rate of heating. With very slow heating, hypoeutectoid steels completely austenitic immediately above Ac3, and the holding time should not be necessary. With more rapid heating some holding time should be required to ensure temperature equalization carbide dissolution. Alternatively, a higher temperature might be used. Plain carbon and low alloy structural steels which contain easily dissolved carbides require only a few minutes holding time after they have reached the hardening temperature. In order to make certain that there has been sufficient carbide dissolution, holding time 5–15 min is quite sufficient [8].

2 Experimental Work

The material used in this study is high carbon, with the chemical composition of the material as shown in Table 1.

2.1 Treatment Processes

The steel samples were divided to six groups, first three groups, the samples were heated to the austenizing temperature of 850, 900 and 950 °C, soaked for 30 min for each austenizing temperature, then all samples subjected to annealing, normalizing processes, quenched in oil and quenched in water. Second three groups, the samples were heated at same austenizing temperatures but soaking time was 60 min and subjected to annealing, normalizing processes, quenched in oil and quenched in water.

2.2 Strength Test

The hardness values of the samples were determined using a digital hardness testing machine. The surfaces were thoroughly polished before samples were tested. The various hardness values were recorded in HRC. Tensile test was measured using a universal testing machine.

2.3 Corrosion Rate

The corrosion rates of the samples were determined using the weight loss method, samples were immersed in HCL solution (concentration 8 %).

2.4 Metallographic Examination

The samples were ground on a water lubricated hand grinding set-up of abrasive chapters, progressing through from the coarsest to the finest grit size. The 240, 320, 400, 500 and 800 grades were used in the order. Polishing was carried out on a rotating disc of synthetic velvet polishing cloth impregnated with 0.5 micron

Table 1 Chemical composition of material high carbon steel

Element	%C	%Si	%Mn	%P	%S	%Cr	%Ni	%Mo
	0.75	0.36	0.26	0.018	0.010	3.30	0.12	4.50

alumna paste. The specimens were then etched with the standard 2 % nital. The optical microscopic examination were carried out on a metallurgical microscope at a magnification of 100 X.

3 Results and Discussion

Hardness: The results of the hardness of the steel specimens after various heat treatment temperatures and different heat treatment processes are shown in Figs. 1, 2 and 3.

Corrosion Rate: The result of the corrosion rate investigations is shown in Fig. 4.

Microstructures analysis: The results of the microstructure analysis is shown in Figs. 5, 6 and 7.

4 Discussion

Figures 1, 2 and 3 shows the hardness value for steel specimens after various heat treatment temperatures and different heat treatment processes. Specimens heated to 850 °C with both different soaking time at austenite phase and cooled in different medium shows no noticeable increase in the hardness. But at 900 °C changing in hardness values can be observed clearly with oil and water quenched specimens, from 20 to 24 HRC in oil coolant and from 23 to 32 HRC at soaking time 30 and 60 min respectively. At Fig. 3 there is a significant changing in hardness values in all quenching media even at air and furnace cooling, the hardness values were changing from 35 to 42 HRC at slow cooling rate at furnace cooling, from 38 to 45 HRC at air cooling, from 40 to 49 HRC at oil quenching and from 46 to 56 HRC at water quenching due to clear stable carbides formation and with increasing temperature and soaking time the carbides particles become smaller in size and increasing hardness values which is also manifested in the microstructure. Figure 4 shows the corrosion rate of the annealing and normalizing process is a lower than the corrosion rate at quenching due to less of stressed inside the samples due to martensite structure formation, but the corrosion rate will be increases with quenching process after heat treatment do to formation of spheroidal graphite particles that contain a strong two phase with martensite. Figures 5,

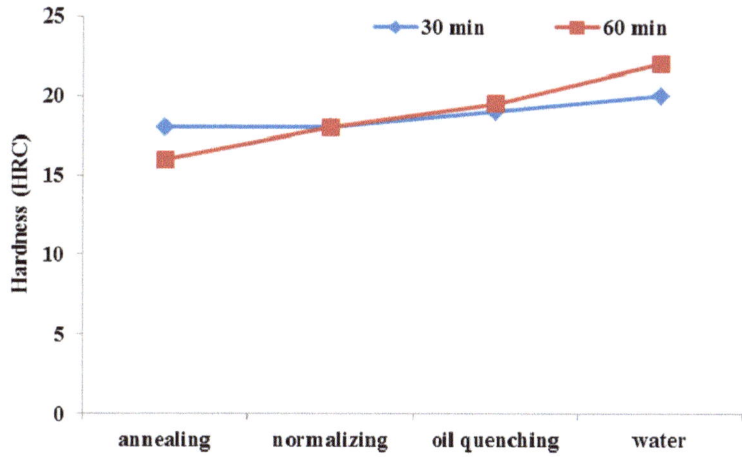

Fig. 1 Hardness values at 850 °C for different types of heat treatment at different soaking time

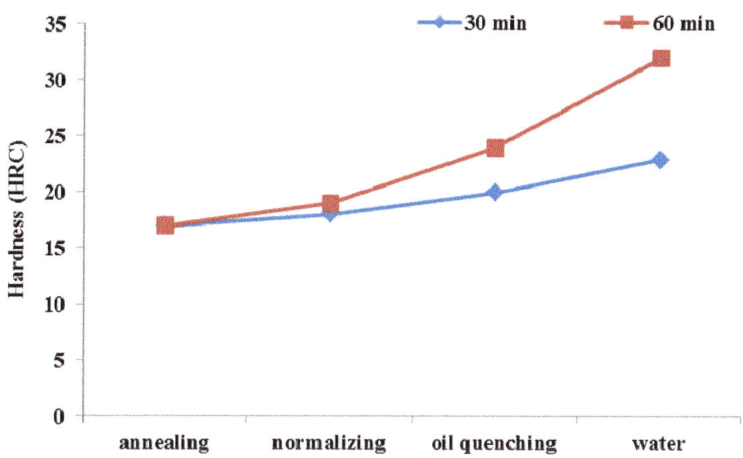

Fig. 2 Hardness values at 900 °C for different types of heat treatment at different soaking time

6 and 7 show the microstructures of specimen for three different heating at 60 min soaking time. The structure consists of spheroidal carbides particles in a martensite matrix, also it is clear with increasing heating temperature the spheroidal carbides particles become smaller and this leads to an increase in hardness values.

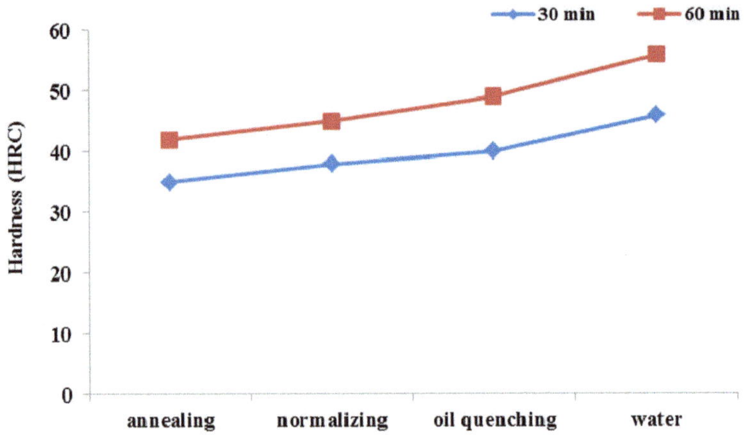

Fig. 3 Hardness values at 950 °C for different types of heat treatment at different soaking time

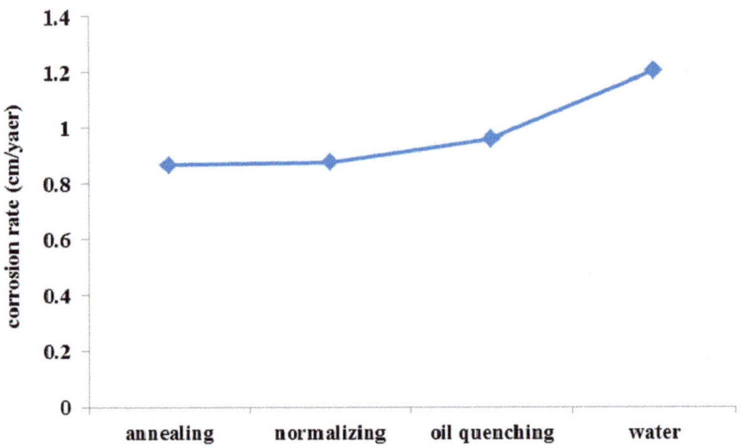

Fig. 4 Corrosion rate for samples after heat treatment at 900 °C

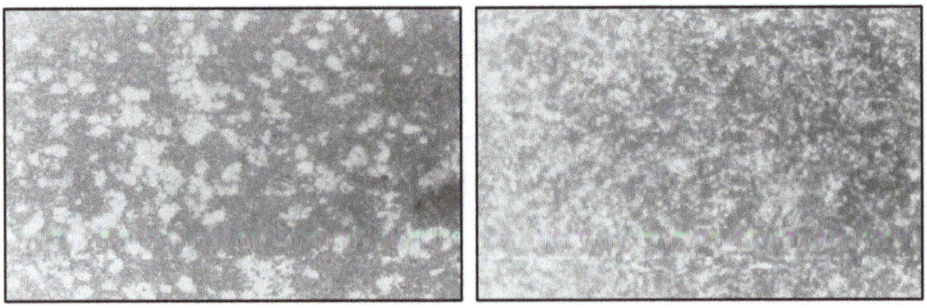

Fig. 5 At 900 °C water and oil quenching, consisting spheroidal carbides particles in martensite matrix

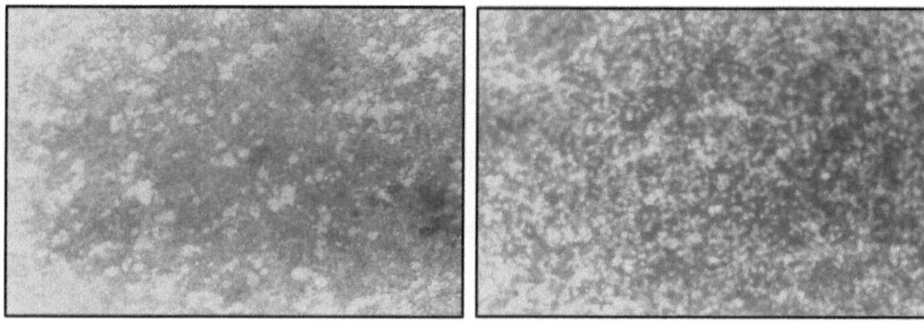

Fig. 6 At 950 °C water and oil quenching, structure is martensite with some of spheroidal–carbide particles (*white dots*)

Fig. 7 At 1,050 °C water quenching, 1 h soaking time

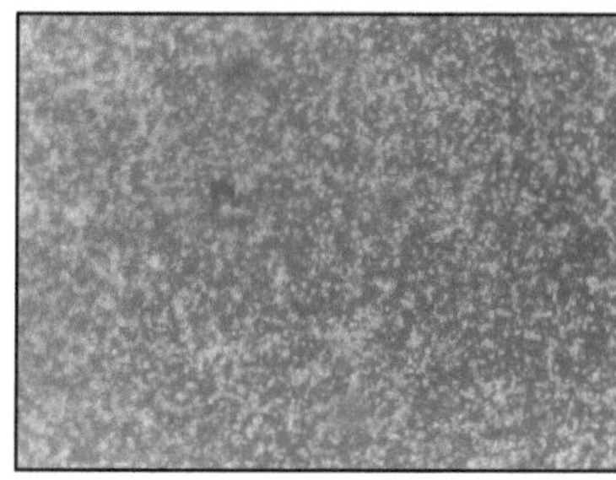

5 Conclusion

The austenizing temperature is an important factor during heat treatment specially with high carbon steel to reach homogeneous austenite. The clear effect of the austenizing temperature and soaking time are effective with increases of the heat treatment temperature in coincidence of the soaking time specially with high carbon content in the steel and presence of the carbide former metal alloy, that carbides are stable even at high temperature and leads to significant increasing in hardness. When stable carbides are formed it will be decreasing in size with increasing the austenizing temperature and soaking time. With advantage of increasing the hardness and strength of samples, disadvantage is will be more susceptible to corrosion due to carbide formation.

References

1. Oberg, E. et al.: Machinery's Handbook (25th ed), Industrial press Inc, New York (1996)
2. Smith, W., Hashemi, J.: Foundations of Material science and Engineering, (4th edn) p. 28–36. McGraw's—Hill Book, Boston (2006)
3. Rajan, T.V., Sharma, C.P., Sharma, A.: Heat Treatment Principles and Techniques, pp. 36–58. PHI, Delhi (1989)

4. Mamoru, O., Yukito, T. Hitoshi, K, and Yuji, F.: Development of New Steel Plates for Building Structural use, Nippon Steel Technical Report, No.44, p. 5–18 (1990)
5. Smith, W.F.: Foundations of Materials Science and Engineering, (3rd edn), p. 442. McGRAW-HILL. New York (2004)
6. Avner S.H.: Introduction to Physical Metallurgy, (2nd edn) pp. 60–361. McGRAW-HILL. New York (1974)
7. Callister Jr.W.O.: Material Science and Engineering An introduction, pp.322–328 Willey, USA (2003)
8. Smith J.L. Heat Treatment of Metals, (1st edn) pp. 211–212. CBS. New Delhi. Bangalore, Pune (2008)

Statistical Analysis of Automatic Scanning of a Car Roof

Islam Sharaf, Said M. Darwish and Abdulrahman Al-Ahmari

Abstract Reverse engineering (RE) is a design technique where a computer aided design (CAD) model can be obtained from an existing part. It consists of capturing 3D point clouds of an existing part and generating its CAD model. There might be a degradation in the resulting accuracy of the digitized shape due to non-uniform scanning speed coupled with non-stable movement around the scanning object, when the 3D laser scanners operated manually. In the present work, a CAD model of a car roof is obtained using automatic scanning. A five-axis CNC machine is used to guide a non-contact 3D laser scanner in capturing the digitized shape of the car roof. This chapter demonstrates the differences in accuracy of the digitized shape considering scanning speed, laser power, resolution and shutter time.

1 Introduction

Nowadays, reverse engineering (RE) is used in various applications varying from manufacturing to medical applications. The RE technique is performed in order to obtain a geometric CAD model from 3-D points acquired by scanning/digitizing existing parts/products as shown in Fig. 1.

I. Sharaf (✉) · S. M. Darwish · A. Al-Ahmari
Advanced Manufacturing Institute, King Saud University, Riyadh, Saudi Arabia
e-mail: eng.isharaf@gmail.com

S. M. Darwish
e-mail: darwish@ksu.edu.sa

A. Al-Ahmari
e-mail: alahmari@ksu.edu.sa

I. Sharaf · S. M. Darwish · A. Al-Ahmari
Industrial Engineering Department, King Saud University, PO Box 800, Riyadh 11421, Saudi Arabia

A. Öchsner and H. Altenbach (eds.), *Design and Computation of Modern Engineering Materials*, Advanced Structured Materials 54,
DOI: 10.1007/978-3-319-07383-5_31,

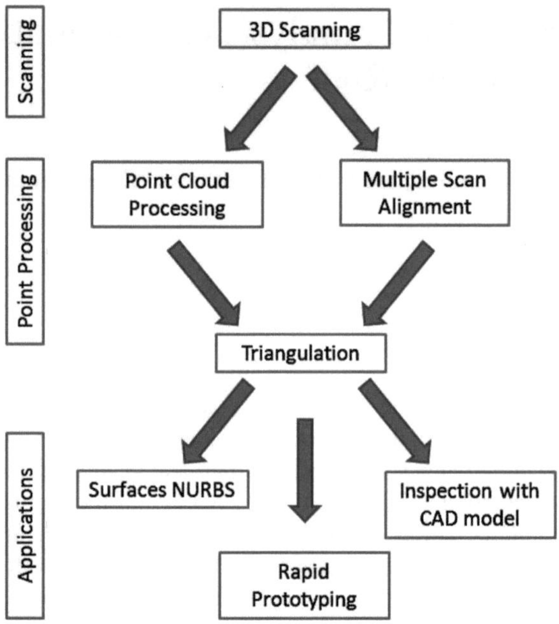

Fig. 1 Concept of RE

A 3D laser scanner is a non-contact RE tool, where the laser beam from the scanner is directed on to the object to collect the data points [1, 2]. Nowadays, a variety of 3D scanners is available to digitize the objects that vary considerably in size from microscopic to large objects.

According to our literature survey, there might be a degradation in the resulting accuracy of the digitized shapes due to non-uniform scanning speeds and non-stable movement around the scanning object when the 3D scanners are operated manually [3]. It has also been found in the previous researches that people using robots to automate portable scanning of small objects achieved better accuracy in less time in comparison to manual scanning [4–6]. However, to improve its scanning accuracy and reduce its time, there is a need to automate the process [5–7].

Therefore, there have been many applications that successfully utilized robotically controlled motion to computerize the scanning process to get better results. For example, a robotically-controlled 3D laser scanning system was applied to scan the turbine blades of an aero engine [8]. A robotically-controlled turntable was used to automatically plan and control the motion of a laser profile scanner to measure the dimensions of an object of an unknown shape [9]. The works related to automatic data capturing using robots can also be found for large parts but the main purpose was for inspection [10]. The current approaches for scanning large objects using 3D portable laser scanners take longer time and result in lower accuracy of digitized shapes which requires extensive post-processing [5, 11].

In the present work, the automated scanning of a car roof will be evaluated by experimental design technique using the ANOVA test.

2 Experimental Setup

2.1 Saudi Gazal-2 Sedan Car

The Advanced Manufacturing Institute (AMI) of King Saud University (KSU) introduced the concept model of its second automotive product, i.e. the Gazal 2 Sedan.

The following Fig. 2 shows the proposed model of the Gazal 2. The concept model was designed and styled on the platform of the Proton Saga, i.e. a Malaysian car.

Figure 3 shows the 2D views of the Gazal 2 model. It is worth noting that before the styling phase the chassis and the mechanics of the Saga were all scanned manually using an Optotrak scanner from NDI company.

Throughout the manual scanning stage, the chassis was divided into parts where each part has its local coordinates. After this, different chassis parts were scanned and all elements were assembled using global coordinates as shown in Fig. 4.

2.2 Equipment Involved for Automatic Scanning

A 3D laser non-contact scanner (ExaScan®, Creaform), a 5-axis CNC machine (CMS, Poseidon), a workstation Laptop (HP EliteBook 8,560 w) and the statistical software Minitab 16 were used to conduct the task.

2.3 Automatic Scanning Procedures

The 3D laser non-contact scanner (ExaScan®, Creaform) is attached to the head of a 5-axis CNC machine (CMS, Poseidon) to get benefit from the movement of the head in scanning the car roof, see Fig. 5. The scanner is attached to a workstation laptop to get the digitized shape.

As the scanner comes with a self-positioning system and as it uses self-technology positioning, the scanner needs positioning targets to be put on the surface which is to be scanned in a triangular order as shown in Fig. 6.

Then a G and M code is generated to make the path of the head of the CNC with suitable distances.

Fig. 2 Gazal 2

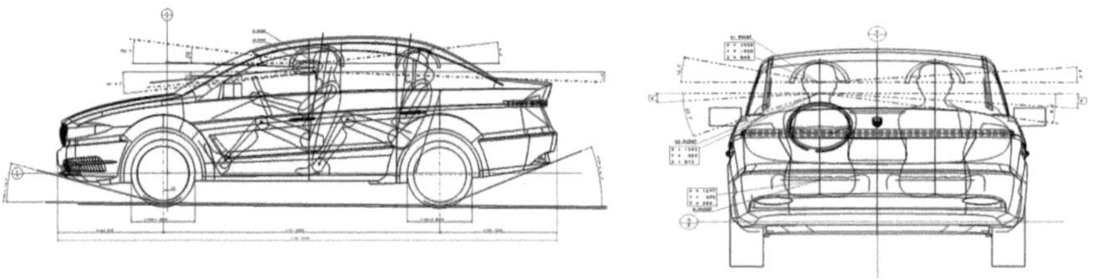

Fig. 3 2D views

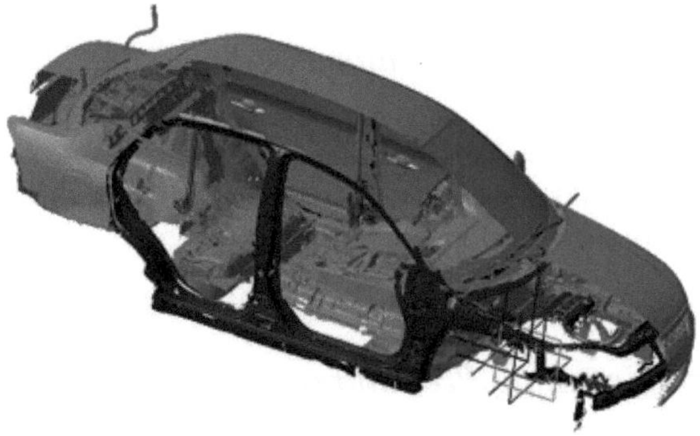

Fig. 4 Global coordinates

2.4 Design of Experiments

The experiments consist of four factors where each factor contains three levels (see Table 1). The ANOVA test was used to analyze the variances of the parameters and to decide which factors interact with the response that is the point clouds of the digitized shape of the car roof.

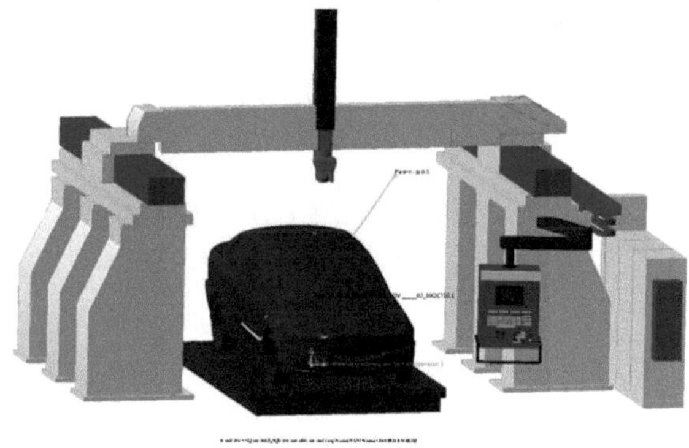

Fig. 5 The experimental setup

Fig. 6 Targets on the car roof

Table 1 Main factors and Levels

Levels	Parameters			
	Speed of scanning (mm/min)	Laser power (%)	Shutter speed (ms)	Resolution (mm)
1	2000	46	2	1
2	4000	65	8.7	2
3	5000	95	10.7	3

2.5 Response Measurement

The response for the experiments determined by the triangle counts (mesh) of the digitized shape of the car roof is shown in Fig. 7 as an example of the output file.

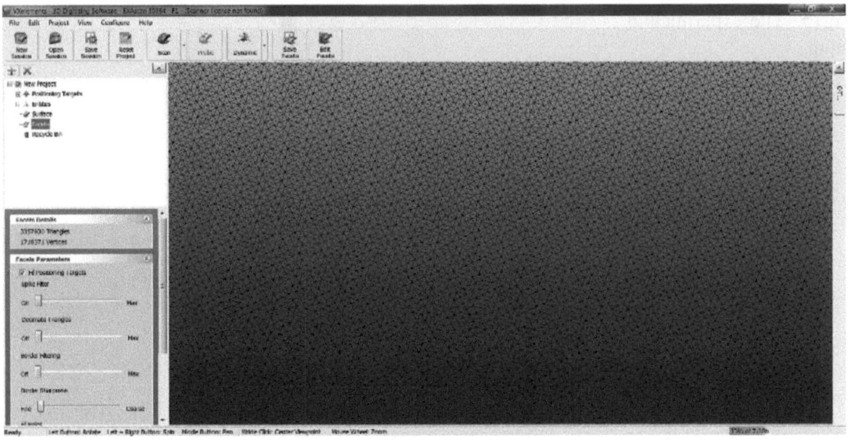

Fig. 7 The triangles shape (Mesh)

3 Results and Discussions

The experiments were carried 81 times in order to see the effect of the parameters according to the design of experiments as shown in Table 1 on the output digitized shape. The best parameters will be the most accurate digitized shape according to the largest number of triangle counts, sample of the data are shown in Table 2.

As shown in Table 2, the largest number of triangle counts is marked in grey; it took nearly 20 min to scan the whole car roof at the best parameters that saves time if scanned manually. In addition to, it gave a high detailed digitized shape for the roof, these parameters gives the best output digitized shape as shown in Fig. 8.

3.1 ANOVA Test

The ANOVA test is used to check the significance of the data to the response, which is the triangle counts as shown in Table 3.

From the ANOVA table it can be concluded that:

(1) The significant factors ($p < 0.05$) is the resolution.
(2) Insignificant factors ($p > 0.05$) are all the other factors and their interactions.
(3) There is no interactions between the parameters and the response.

Table 2 Data collected

Input parameters				Response
Speed of CNC	Laser power	Shutter time	Resolution	Triangle counts(Mesh)
2,000	46	2	1	3189,513
2,000	46	2	2	815,737
2,000	46	2	3	362,631
2,000	46	8.7	1	3264,595
2,000	46	8.7	2	829,592
……	……	……	……	……
4,000	65	2	1	3229,565
4,000	65	2	2	832,048
4,000	65	2	3	353,049
4,000	95	2	2	833,178
4,000	95	2	3	352,589
4,000	95	8.7	1	3261,939
……	……	……	……	……
5,000	46	2	2	877,859
5,000	46	2	3	371,280
5,000	46	8.7	1	3513,583
5,000	46	8.7	2	897,734
5,000	95	2	3	345,940
5,000	95	8.7	1	3288,819
5,000	95	8.7	2	832,117

Fig. 8 Digitized output of the car roof

3.2 Residual Plots

Residual plot graphs are presented in Fig. 9. From the residual plots, the data considered is to be normally fitted.

Table 3 Analysis of the data

Source	F	P
CNC speed	0.65	0.533
Laser power	0.53	0.597
Shutter time	1.17	0.336
Resolution	2971.63	0.000
CNC speed*laser power	0.23	0.915
CNC speed*shutter time	1.42	0.271
CNC speed*resolution	0.42	0.791
Laser power*shutter time	1.40	0.280
Laser power*resolution	0.44	0.781
Shutter time*resolution	0.72	0.593
CNC speed*laser power*shutter time	1.76	0.160
CNC speed*laser power*resolution	0.24	0.976
CNC speed*shutter time*resolution	0.89	0.548
Laser power*shutter time*resolution	0.80	0.611

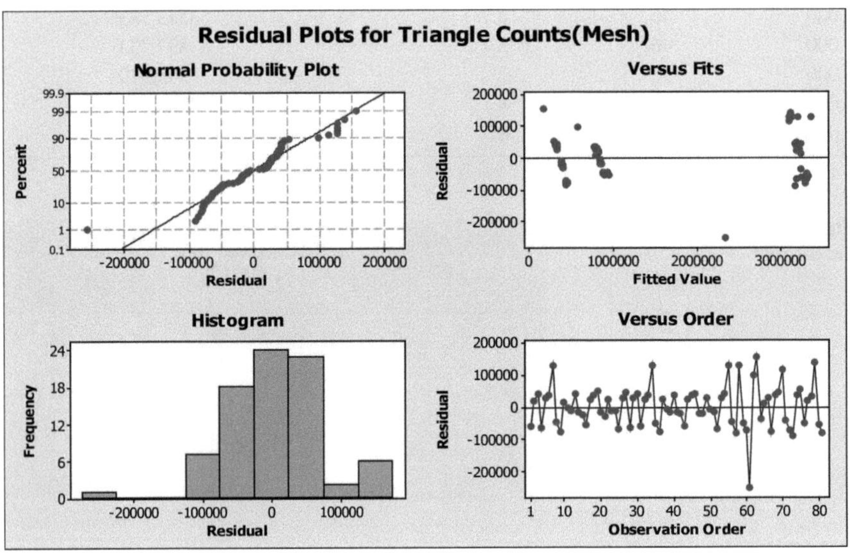

Fig. 9 Residual plots

3.3 Main Effect Plots

Based on the main effect plots in Fig. 10, it can be concluded that only the resolution has a significant effect on the response, i.e. triangle counts, as all the other factors have no effect on the triangle counts. These results as shown are biased.

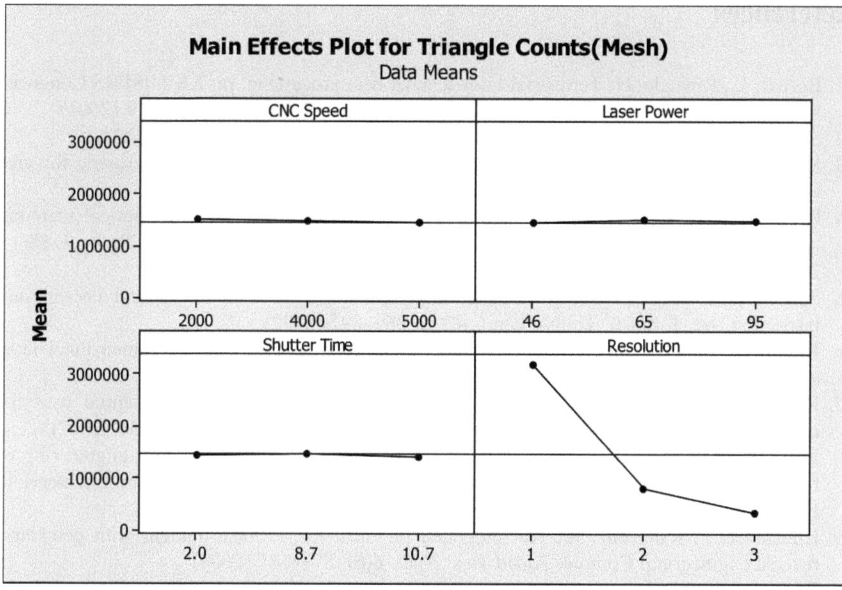

Fig. 10 Main effect plots

4 Conclusions

We could conclude from the above results the following:

(1) The integration of the CNC with the 3D laser scanner is an effective way in scanning large parts with constant speed and stable movement.
(2) The results appeared to be biased so the levels of the parameters needs to be adjusted.
(3) This integration can be used in inspection purposes in aviation or automotive applications.
(4) The analysis of the data showed that the only factor that affects the triangle counts is the resolution of the shape.
(5) The best parameter is a speed of 5,000 mm/sec that took only 20 min to scan the whole roof of the car with an excellent digitized output.

5 Future Work

Some further work will be done as the results were biased and will be adjusted. In addition, another response will be added to get better results, which is the deviation of the output-digitized shape with the original CAD model of the car roof.

References

1. Bornaz, L., Rinaudo, F.: Terrestrial laser scanner data processing. In: XXth ISPRS Congress: Proceedings of Commission V 2004, Citeseer: Istanbul, Turkey. pp. 514–519 (2004)
2. Schofield, Engineering Surveying, vol. 5, p. 544. Butterworth-Heinemann. (2001)
3. Rahayem, M.: An Industrial Robot as Part of an Automatic System for Geometric Reverse Engineering. Robot manipulators (2008)
4. Reinhart, G., Tekouo, W.: Automatic programming of robot-mounted 3D optical scanning devices to easily measure parts in high-variant assembly. CIRP Ann. Manuf. Technol. **58**(1), 25–28 (2009)
5. Son, S., Park, H., Lee, K.H.: Automated laser scanning system for reverse engineering and inspection. Int. J. Mach. Tools Manuf. **42**(8), 889–897 (2002)
6. Kau, C.H., et al.: Reliability of measuring facial morphology with a 3-dimensional laser scanning system. Am. J. Orthod. Dentofac. Orthop. **128**(4), 424–430 (2005)
7. Kuzminsky, S.C., Gardiner, M.S.: Three dimensional laser scanning: potential uses for museum conservation and scientific research. J. Archaeol. Sci. **38**, 2744–2751 (2012)
8. Zhao, Y., et al.: Development of a robotic 3D scanning system for reverse engineering of freeform part. In: International Conference on Advanced Computer Theory and Engineering. Phuket, Thailand: IEEE (2008)
9. Kjellander, J., Rahayem, M.: An integrated platform for 3d measurement with geometric reverse engineering. Comput. Aided Des. Appl. **6**(6), 877–887 (2009)
10. Paoli, A., Razionale, A.V.: Large yacht hull measurement by integrating optical scanning with mechanical tracking-based methodologies. Rob. Comput. Integr. Manuf. **28**(5), 592–601 (2012)
11. Racasan, R., et al.: Integrating the concept of reverse engineering in medical applications. In: Automation, Quality and Testing, Robotics. vol. 2, pp. 28–30 May IEEE (2010)

MIX
Papier aus verantwortungsvollen Quellen
Paper from responsible sources
FSC® C105338

If you have any concerns about our products,
you can contact us on
ProductSafety@springernature.com

In case Publisher is established outside the EU,
the EU authorized representative is:
Springer Nature Customer Service Center GmbH
Europaplatz 3, 69115 Heidelberg, Germany

Printed by Libri Plureos GmbH
in Hamburg, Germany